Nonequilibrium Statistical Physics

Statistical mechanics has been proven to be successful at describing physical systems at thermodynamic equilibrium. Since most natural phenomena occur in nonequilibrium conditions, the present challenge is to find suitable physical approaches for such conditions. This book provides a pedagogical pathway that explores various perspectives. The use of clear language and explanatory figures and diagrams to describe models, simulations, and experimental findings makes it a valuable resource for undergraduate and graduate students and for lecturers teaching at varying levels of experience in the field. Written in three parts, it covers basic and traditional concepts of nonequilibrium physics, modern aspects concerning nonequilibrium phase transitions, and application-orientated topics from a modern perspective. A broad range of topics is covered, including Langevin equations, Lévy processes, directed percolation, kinetic roughening, and pattern formation.

Roberto Livi is Professor of Theoretical Physics at the University of Florence where he teaches courses on statistical physics and thermodynamics. He is also the Director of the Interdepartmental Center for the Study of Complex Dynamics and an associate member of the National Institute of Nuclear Physics (INFN) and of the Institute for Complex Systems of the National Research Council (CNR).

Paolo Politi is Head of the Florence Unit of the Institute for Complex Systems, National Research Council (CNR), and teaches Statistical Physics at the University of Florence. He is Fellow of the Marie Curie Association, of the Alexander von Humboldt Foundation and of the Japan Society for the Promotion of Science. He was awarded the Outreach Prize of the Italian Physical Society for promoting the public discussion of science through Science Cafés.

Nonequilibrium Statistical Physics

A Modern Perspective

Roberto Livi
University of Florence

Paolo Politi
Institute for Complex Systems (CNR), Florence

CAMBRIDGE
UNIVERSITY PRESS

University Printing House, Cambridge CB2 8BS, United Kingdom

One Liberty Plaza, 20th Floor, New York, NY 10006, USA

477 Williamstown Road, Port Melbourne, VIC 3207, Australia

314-321, 3rd Floor, Plot 3, Splendor Forum, Jasola District Centre, New Delhi - 110025, India

79 Anson Road, #06-04/06, Singapore 079906

Cambridge University Press is part of the University of Cambridge.

It furthers the University's mission by disseminating knowledge in the pursuit of education, learning and research at the highest international levels of excellence.

www.cambridge.org
Information on this title: www.cambridge.org/9781107049543
DOI: 10.1017/9781107278974

First published 2017

A catalogue record for this publication is available from the British Library

Library of Congress Cataloging in Publication data
Names: Livi, Roberto, author. | Politi, Paolo, 1965- author.
Title: Nonequilibrium statistical physics : a modern perspective / Roberto Livi (Università di Firenze), Paolo Politi (Istituto dei Sistemi Complessi, Firenze).
Description: Cambridge, United Kingdom ; New York, NY : Cambridge University Press, 2017. | Includes bibliographical references and index.
Identifiers: LCCN 2017023470| ISBN 9781107049543 (hardback ; alk. paper) | ISBN 1107049547 (hardback ; alk. paper)
Subjects: LCSH: Nonequilibrium statistical mechanics.
Classification: LCC QC174.86.N65 L58 2017 | DDC 530.13–dc23
LC record available at https://lccn.loc.gov/2017023470

ISBN 978-1-107-04954-3 Hardback

To Claudia and Daniela

Contents

Preface page xiii
Acknowledgements xv
Notations and Acronyms xvi

1 Brownian Motion, Langevin and Fokker–Planck Equations 1
1.1 Introduction 2
1.2 Kinetic Theory 4
1.2.1 The Ideal Gas 4
1.2.2 Random Walk: A Basic Model of Diffusion 7
1.3 Transport Phenomena 10
1.4 Brownian Motion 13
1.4.1 The Langevin Equation for the Brownian Particle 15
1.4.2 The Fokker–Planck Equation for the Brownian Particle 19
1.5 Discrete Time Stochastic Processes 21
1.5.1 Markov Chains 22
1.5.2 Useful Examples of Markov Chains 25
1.5.3 Ergodic Markov Chains 30
1.5.4 Master Equation and Detailed Balance 32
1.5.5 Monte Carlo Method 33
1.6 Continuous Time Stochastic Processes 36
1.6.1 Stochastic Differential Equations 37
1.6.2 General Fokker–Planck Equation 40
1.6.3 Physical Applications of the Fokker–Planck Equation 42
1.6.4 A Different Pathway to the Fokker–Planck Equation 47
1.6.5 The Langevin Equation and Detailed Balance 54
1.7 Generalized Random Walks 55
1.7.1 Continuous Time Random Walk 56
1.7.2 Lévy Walks 64
1.8 Bibliographic Notes 68

2 Linear Response Theory and Transport Phenomena 70
2.1 Introduction 71
2.2 The Kubo Formula for the Brownian Particle 73
2.3 Generalized Brownian Motion 75
2.4 Linear Response to a Constant Force 77
2.4.1 Linear Response and Fluctuation–Dissipation Relations 79

2.4.2 Work Done by a Time-Dependent Field 83
2.4.3 Simple Applications of Linear Response Theory 84
2.5 Hydrodynamics and the Green–Kubo Relation 89
2.6 Generalized Linear Response Function 95
2.6.1 Onsager Regression Relation and Time Reversal 96
2.7 Entropy Production, Fluxes, and Thermodynamic Forces 98
2.7.1 Nonequilibrium Conditions between Macroscopic Systems 99
2.7.2 Phenomenological Equations 101
2.7.3 Variational Principle 105
2.7.4 Nonequilibrium Conditions in a Continous System 107
2.8 Physical Applications of Onsager Reciprocity Relations 109
2.8.1 Coupled Transport of Neutral Particles 109
2.8.2 Onsager Theorem and Transport of Charged Particles 113
2.9 Linear Response in Quantum Systems 125
2.10 Examples of Linear Response in Quantum Systems 129
2.10.1 Power Dissipated by a Perturbation Field 129
2.10.2 Linear Response in Quantum Field Theory 131
2.11 Bibliographic Notes 134

3 From Equilibrium to Out-of-Equilibrium Phase Transitions 136
3.1 Introduction 136
3.2 Basic Concepts and Tools of Equilibrium Phase Transitions 138
3.2.1 Phase Transitions and Thermodynamics 138
3.2.2 Phase Transitions and Statistical Mechanics 141
3.2.3 Landau Theory of Critical Phenomena 145
3.2.4 Critical Exponents and Scaling Hypothesis 149
3.2.5 Phenomenological Scaling Theory 152
3.2.6 Scale Invariance and Renormalization Group 154
3.3 Equilibrium States versus Stationary States 159
3.4 The Standard Model 161
3.5 Phase Transitions in Systems with Absorbing States 164
3.5.1 Directed Percolation 164
3.5.2 The Domany–Kinzel Model of Cellular Automata 168
3.5.3 Contact Processes 172
3.6 The Phase Transition in DP-like Systems 173
3.6.1 Control Parameters, Order Parameters, and Critical Exponents 173
3.6.2 Phenomenological Scaling Theory 176
3.6.3 Mean-Field Theory 177
3.7 Bibliographic Notes 181

4 Out-of-Equilibrium Critical Phenomena 183
4.1 Introduction 183
4.2 Beyond the DP Universality Class 184
4.2.1 More Absorbing States 185
4.2.2 Conservation Laws 188

4.3 Self-Organized Critical Models 189
4.3.1 The Bak–Tang–Wiesenfeld Model 191
4.3.2 The Bak–Sneppen Model 192
4.4 The TASEP Model 195
4.4.1 Periodic Boundary Conditions 195
4.4.2 Open Boundary Conditions 197
4.5 Symmetry Breaking: The Bridge Model 203
4.5.1 Mean-Field Solution 205
4.5.2 Exact Solution for $\beta \ll 1$ 207
4.6 Bibliographic Notes 212

5 Stochastic Dynamics of Surfaces and Interfaces 213
5.1 Introduction 214
5.2 Roughness: Definition, Scaling, and Exponents 216
5.3 Self-Similarity and Self-Affinity 221
5.4 Continuum Approach: Toward Langevin-Type Equations 223
5.4.1 Symmetries and Power Counting 223
5.4.2 Hydrodynamics 226
5.5 The Random Deposition Model 228
5.6 The Edwards–Wilkinson Equation 229
5.6.1 Dimensional Analysis 231
5.6.2 The Scaling Functions 233
5.7 The Kardar–Parisi–Zhang Equation 238
5.7.1 The Galilean (or Tilt) Transformation 242
5.7.2 Exact Exponents in $d = 1$ 244
5.7.3 Beyond the Exponents 247
5.7.4 Results for $d > 1$ 249
5.8 Experimental Results 250
5.8.1 KPZ $d = 1$ 250
5.8.2 KPZ $d = 2$ 251
5.9 Nonlocal Models 252
5.10 Bibliographic Notes 258

6 Phase-Ordering Kinetics 259
6.1 Introduction 260
6.2 The Coarsening Law in $d = 1$ Ising-like Systems 264
6.2.1 The Nonconserved Case: Spin-Flip (Glauber) Dynamics 265
6.2.2 The Conserved Case: Spin-Exchange (Kawasaki) Dynamics 266
6.3 The Coarsening Law in $d > 1$ Ising-like Systems 269
6.3.1 The Nonconserved Case 270
6.3.2 The Conserved Case 275
6.4 Beyond the Coarsening Laws 280
6.4.1 Quenching and Phase-Ordering 280
6.4.2 The Langevin Approach 286
6.4.3 Correlation Function and Structure Factor 288

6.4.4 Domain Size Distribution 291
6.4.5 Off-Critical Quenching 294
6.5 The Coarsening Law in Nonscalar Systems 296
6.6 The Classical Nucleation Theory 303
6.6.1 The Becker–Döring Theory 304
6.7 Bibliographic Notes 309

7 Highlights on Pattern Formation 311
7.1 Pattern Formation in the Laboratory and the Real World 311
7.2 Linear Stability Analysis and Bifurcation Scenarios 315
7.3 The Turing Instability 321
7.3.1 Linear Stability Analysis 322
7.3.2 The Brusselator Model 326
7.4 Periodic Steady States 328
7.5 Energetics 334
7.6 Nonlinear Dynamics for Pattern-Forming Systems: The Envelope Equation 336
7.7 The Eckhaus Instability 339
7.8 Phase Dynamics 344
7.9 Back to Experiments 347
7.10 Bibliographic Notes 349

Appendix A The Central Limit Theorem and Its Limitations 351

Appendix B Spectral Properties of Stochastic Matrices 353

Appendix C Reversibility and Ergodicity in a Markov Chain 355

Appendix D The Diffusion Equation and Random Walk 357

Appendix E The Kramers–Moyal Expansion 366

Appendix F Mathematical Properties of Response Functions 368

Appendix G The Van der Waals Equation 372

Appendix H The Ising Model 376

Appendix I Derivation of the Ginzburg–Landau Free Energy 384

Appendix J Kinetic Monte Carlo 387

Appendix K The Mean-field Phase Diagram of the Bridge Model 389

Appendix L The Deterministic KPZ Equation and the Burgers Equation 393

Appendix M The Perturbative Renormalization Group for KPZ: A Few Details 398

Appendix N The Gibbs–Thomson Relation 402

Appendix O The Allen–Cahn Equation 404

Appendix P The Rayleigh–Bénard Instability 406

Appendix Q General Conditions for the Turing Instability 412

Appendix R Steady States of the One-Dimensional TDGL Equation 414

Appendix S Multiscale Analysis 415

Index 418

Preface

Since 2006 the authors of this book have been sharing the teaching of an advanced course on nonequilibrium statistical physics at the University of Florence, Italy. This is an advanced course for students in the last year of the master's thesis curriculum in physics, which is attended also by PhD students. If the reader of this Preface is a colleague or a student who already attended a similar course, he or she should not be astonished by the following statement: this book was primarily conceived to organize the contents of our course, because the offer of textbooks on nonequilibrium statistical physics is typically much more limited and specialized than in the case of equilibrium statistical physics. From the very beginning it was clear in our minds that we had to aim at a textbook written for students, neither for experts nor for colleagues. In fact, we believe that a textbook on advanced topics written for the benefit of students can also be useful for colleagues who want to approach these topics, while the contrary does not hold. We dare, indeed, to say that if a book is written devoting special consideration for the understanding of students, it could be beneficial for the whole scientific community. After these preliminary remarks, which should be taken as an omen more than a statement, we want to illustrate and justify the contents of this textbook.

When we started to think about that, the first question that came to our mind was the following: what should it contain? As one could expect, the answer was not unique. It depends not only on our personal taste and interest, but it is also inspired by a sort of "tradition," which changes significantly from university to university as well as from country to country. This is why, after having produced a preliminary list of topics, we asked for the opinion of some Italian and foreign colleagues. Honestly, we did not change that much of the preliminary list after having received their advice, partly because they did not raise strong criticisms and also because we eventually obtained quite incoherent suggestions. As a consequence, if the reader does not agree with our choices, we must admit that we lack strong arguments to tackle their opinion, but we also expect that the reader should not be in the position of raising arguments that are too negative.

A further remark concerns the bibliography. We have decided, as in most textbooks, to limit the number of references. In fact, we have tried to always provide the derivation of the results step by step, and unproved statements are very rare. So, rather than spreading references throughout the book we have preferred to collect them at the end of each chapter, in a section called *Bibliographic Notes*. Here, we give first some general references (either a book or a review), then we give more specific references, always accompanied by a short comment about their relevance to the specific chapter.

Now we come to the very structure of this book. It can be thought as subdivided into three parts. The first two chapters deal with those basic topics that one naturally expects to find in such a text: Brownian motion, Markov chains, random walks, Langevin and Fokker–Planck equations, linear response theory, Onsager relations, Lévy processes, etc. The third and fourth chapters are devoted to topics that must be present in a *modern* textbook on nonequilibrium physics, namely, nonequilibrium phase transitions. Finally, the last three chapters stem from our personal interests and each one could be sufficient for occupying a volume by itself: kinetic roughening, phenomena of phase-ordering, and pattern formation. Rather than giving here further details about the content of the book, we invite the interested reader to browse the opening to each chapter.

The book is accompanied by a website, https://sites.google.com/site/nespbook/ that offers additional material. You are invited to report errata or send comments by writing to nespbook@gmail.com.

Acknowledgements

Several colleagues have read and commented on one or more chapters of the book. So it is a great pleasure to thank Daniel ben-Avraham, Federico Corberi, Joachim Krug, Stefano Lepri, Chaouqi Misbah, Stephen Morris, and Jacques Villain. Filippo Cherubini is acknowledged for having produced many figures of the book, Gian Piero Puccioni for the continuing technical help and Ruggero Vaia for deep discussions on analytic questions. RL thanks the Max Planck Institute for the Physics of Complex Systems in Dresden for financial support and for the kind hospitality over a few months, when part of this book was written. PP thanks the Alexander von Humboldt Foundation, the University of Grenoble Alpes, and the CNRS for financial support, allowing him to spend a few months in Cologne (at the Institute for Theoretical Physics, University of Cologne), in Dresden (at the Max Planck Institute for the Physics of Complex Systems) and in Grenoble (at the Laboratoire Interdisciplinaire de Physique) to work on this book. These institutes are also acknowledged for the pleasant environments they provided for such work. PP warmly thanks Carlo and Monica for their hospitality in Femminamorta, where this book was written during the hottest summer periods.

Notations and Acronyms

Notations

Throughout this book we assume the Boltzmann constant ($K_B = 1.380658\ 10^{-23}\ \mathrm{J K^{-1}}$) to be equal to one, which corresponds to measuring the temperature in joules or an energy in Kelvin.

As for the Fourier transform, we use the same symbol in the real and the dual space, using the following conventions:

$$h(\mathbf{k}) = \int d\mathbf{x} \exp(-i\mathbf{k} \cdot \mathbf{x}) h(\mathbf{x}),$$

$$h(\mathbf{x}) = \frac{1}{(2\pi)^d} \int d\mathbf{k} \exp(i\mathbf{k} \cdot \mathbf{x}) h(\mathbf{k}).$$

Similarly, for the Laplace transform,

$$\omega(s) = \int_0^\infty dt e^{-st} \omega(t),$$

$$\omega(t) = \frac{1}{2\pi i} \mathrm{PV} \int_{a-i\infty}^{a+i\infty} ds e^{st} \omega(s).$$

Here above, PV is the principal value of the integral and $a > a_c$, a_c being the abscissa of convergence.

The friction coefficient of a particle of mass m, i.e. the ratio between the force F_0 acting on it and its terminal drift velocity v_∞, is indicated by the symbol $\tilde{\gamma} = F_0/v_\infty$, but we frequently use the reduced friction coefficient, $\gamma = \tilde{\gamma}/m$, which has the dimension of the inverse of time. Similar reduced quantities are used in the context of Brownian motion.

The Helmholtz free energy is indicated by the symbol $F = U - TS$, and the free energy density is indicated by f. We often use a free energy functional, also called pseudo free energy functional or Lyapunov functional, and it is indicated by $\mathcal{F}$. It is the space integral of a function f, $\mathcal{F} = \int d\mathbf{x} f$. The susceptibility, indicated by χ, may be an extensive as well as an intensive quantity, depending on the context.

The space dimension is indicated by d.

Acronyms

ASEP	asymmetric exclusion process
BD	ballistic deposition
BS	Back–Sneppen
BTW	Back–Tang–Wiesenfeld
CA	cellular automaton
CDP	compact directed percolation
CH	Cahn–Hilliard
CIMA	chlorite-iodide-malonic-acid
CTRW	continuous time random walk
DK	Domany–Kinzel
dKPZ	deterministic Kardar–Parisi–Zhang
DLA	diffusion limited aggregation
DLG	driven lattice gas
DP	directed percolation
DyP	dynamical percolation
DW	domain wall
emf	electromotive force
EW	Edwards–Wilkinson
GL	Ginzburg–Landau
GOE	Gaussian orthogonal ensemble
GUE	Gaussian unitary ensemble
HD	high density
KMC	kinetic Monte Carlo
KPZ	Kardar–Parisi–Zhang
LD	low density
LG	lattice gas
LW	Levy walk
MC	maximal current
MF	mean-field
NESS	nonequilibrium steady state
PC	parity conserving
PV	principal value
QFT	quantum field theory
RD	random deposition
RDR	random deposition with relaxation
RG	renormalization group
RSOS	restricted solid on solid
SH	Swift–Hohenberg
SOC	self-organized criticality
TASEP	totally asymmetric exclusion process
TDGL	time-dependent Ginzburg Landau
TW	Tracy–Widom

1 Brownian Motion, Langevin and Fokker–Planck Equations

In this chapter the reader can find the basic ingredients of elementary kinetic theory and of the mathematical approach to discrete and continuous stochastic processes, all that is necessary to establish a solid ground for nonequilibrium processes concerning the time evolution of physical systems subject to a statistical description. In fact, from the first sections we discuss problems where we deal with the time evolution of average quantities, such as in the elementary random walk model of diffusion. We also illustrate the bases of transport phenomena that allow us to introduce the concept of transport coefficients, which will be reconsidered later in the framework of a more general theory (see Chapter 2). Then we focus on the theory of Brownian motion, as it was originally formulated by Albert Einstein, and how this was later described in terms of the Langevin and of the Fokker–Planck equations, specialized to a Brownian particle. The peculiar new ingredient that was first introduced in the Langevin formulation is noise, which epitomizes the effect of incoherent fluctuations of the Brownian particle due to the interaction with the solvent particles, which are subject to a thermal motion. Averaging over thermal noise allows one to obtain a statistical inference on the diffusive behavior of a Brownian particle. The Fokker–Planck formulation tells us that we can obtain an equivalent description by considering the evolution of the space-time probability function of a Brownian particle, rather than averaging over its noisy trajectories.

The mathematical formulation of stochastic processes in discrete space and time (Markov chains) is illustrated, together with many examples and applications, including random walk processes and the Monte Carlo procedure. This important mathematical theory provides us with the tools for a general formulation of stochastic processes in continuous space and time. This is not at all a straightforward step, since the presence of noise needs a suitable mathematical procedure, when passing to a continous time description. In particular, we have to establish a consistent relation between infinitesimal time and noise increments, which allows for two possible different formulations of continuous time Langevin–like equations. In this general framework we can derive also the Fokker–Planck equation for general stochastic processes, rather than for the mere description of the diffusive motion of the Brownian particle. We discuss some interesting applications of this equation to point out the physical importance of this general formulation of stochastic processes and, specifically, its relevance for nonequilibrium processes. In the last part of this chapter we introduce a description of those stochastic processes that do not exhibit a standard diffusive behavior. More precisely, we discuss the so-called continuous time random walk model and we focus our considerations on processes named Lévy flights and Lévy walks, which play an increasing importance in the modern applications of nonequilibrium processes.

1.1 Introduction

The idea that thermodynamics could be related to a mechanical theory of matter dealing with a large number of particles, i.e., atoms and molecules, was speculated on from the very beginning of kinetic theory on the middle of the nineteenth century. In a historical perspective, we could say that such an idea was a natural consequence of the formulation of the first principle of thermodynamics by the German natural philosopher Julius Robert von Mayer, establishing the equivalence between mechanical work and heat. This was checked experimentally in the famous experiment by James Prescott Joule and many contemporary physicists, among which Rudolf Clausius, August Karl Krönig, William Thomson (Lord Kelvin), James Clerk Maxwell, and Ludwig Eduard Boltzmann devoted a good deal of their efforts to develop the foundations of kinetic theory.

The reader should consider that these scientists were assuming the validity of the atomic hypothesis, despite no direct experimental evidence of the existence of atoms and molecules available at that time. Accordingly, the reader should not be surprised that such a nowadays "obvious" concept was strongly opposed by a large part of the scientific community in the last decades of the nineteenth century, as a reaction to a mechanistic foundation of science that, on the one hand, supported a materialistic and, apparently, deterministic basis of natural phenomena and, on the other hand, raised serious conceptual paradoxes, most of which related to the time reversibility of mechanical laws. In fact, the other cornerstone of thermodynamics is the second principle, which amounts to establishing the irreversibility of thermodynamic processes, due to the natural tendency of thermodynamic systems to evolve toward a well-defined equilibrium state in the absence of energy supplied by some external source.

The mechanistic approach to thermodynamics was pushed to its extreme consequences in the work by Ludwig Eduard Boltzmann. His celebrated transport equation represents a breakthrough in modern science and still today we cannot avoid expressing our astonishment about the originality and deep physical intuition of the Austrian physicist. Despite being inspired by a specific model, namely the ideal gas, the main novelty of Boltzmann's equation was that it represents the evolution of a distribution function, rather than the trajectories of individual particles in the gas. Boltzmann realized quite soon that the only way to describe the behaviour of a large number of particles (a mole of a gas contains an Avogadro number of particles, approximately equal to $N_A \simeq 6.022 \times 10^{23}$) was to rely on a statistical approach, where the laws of probability had to be merged into the description of physical laws. We want to point out that the success of Boltzmann's equation is not limited to establishing the foundations of equilibrium statistical mechanics. In fact, it also provides a description of the evolution toward equilibrium by the derivation of hydrodynamic equations associated with the conservation of mechanical quantities, i.e., number, momentum, and energy of particles. They are found to correspond to the continuity equation and to two more phenomenological equations, i.e., the Navier–Stokes and the heat ones. These equations provide a mathematical basis for the theory of transport phenomena and a physical definition of transport coefficients in terms of basic quantities of kinetic theory, such as the mean free path, the average speed of particles, the heat capacity, etc.

On top of that, since 1827 the experiment performed by English botanist Robert Brown describing the phenomenon known as Brownian motion challenged the scientific community. In fact, a pollen particle suspended in water (or any similar solvent) was found to exhibit an erratic motion that, apparently, could not be reconciled with any standard mechanical description. Even assuming the atomistic hypothesis and modeling the motion of the pollen particle as a result of collisions with the atoms of the solvent seemed to fail to provide a convincing explanation. In fact, at the microscopic level one might argue that elastic collisions with the atoms of the solvent could transmit a ballistic motion to the pollen particle. However, the conclusion would be that the combined effect of all of these collisions, occurring for symmetry reasons in any direction, vanishes to zero. On the contrary, the experimental observation of the erratic motion of the pollen particle indicated that the distance of the particle from its original position grew over sufficiently long time intervals as the square root of time, thus showing the diffusive nature of its motion. Repeating many times the same experiment, where the pollen particle, the solvent, and the temperature of the solvent are the same, the particle in each realization follows different paths, but one can perform a statistical average over these realizations that enforces the conclusion that the particle exhibits a diffusive motion.

The universal character of this phenomenon was confirmed by the experimental observations that a diffusive behavior was found also when the type of Brownian particle, the solvent, and the temperature were changed, yielding different values of the proportionality constant between time and the average squared distance of the particle from its initial position. A convincing explanation of Brownian motion had to wait for the fundamental contribution of Albert Einstein, which appeared in 1905, the same year as his contributions on the theories of special relativity and the photoelectric effect. Einstein's phenomenological theory of Brownian motion, relying on simple physical principles, inspired the French scientist Paul Langevin, who proposed a mechanistic approach. The basic idea was to write a Newton-like ordinary differential equation where, for the first time, a force was attributed a stochastic nature. In fact, the microscopic forces exerted by the solvent particles through elastic collisions with the Brownian particle are represented as uncorrelated fluctuations in space and time, whose square amplitude is assumed to be proportional to the thermal energy; according to kinetic theory, this amounts to the solvent temperature T, provided the Brownian particle is at thermodynamic equilibrium with the solvent. Some years later Adriaan Daniël Fokker and Max Planck proposed an alternative formulation of the Brownian particle problem, based on a partial differential equation, describing the evolution of the probability distribution of finding a Brownian particle at position $\mathbf{x}$ at time t, in the same spirit of Boltzmann's equation for an ideal gas. In fact, the Fokker–Planck equation was derived as a master equation, where the rate of change in time of the distribution function depends on favorable and unfavorable processes, described in terms of transition rates between different space-time configurations of the Brownian particle. Making use of some simplifying assumptions, this equation was cast into a form where the diffusive nature of the problem emerges naturally, while it allows one to obtain an explicit solution of the problem.

On the side of mathematics, at the end of the nineteenth century the Russian Andrej Andreevič Markov developed a new mathematical theory concerning stochastic processes, nowadays known as Markov chains. The original theory takes advantage of some simplifying assumptions, like the discreteness of space and time variables as well as of the numerability of the possible states visited by the stochastic process. It was the first time a dynamical theory was assumed to depend on random uncorrelated events, typically obeying the laws of probability. Despite the scientific motivations of Markov, which were quite different from those that moved the above-mentioned physicists to tackle the problem of Brownian motion, some decades later a more general theory of stochastic processes in continuous space and time emerged from the fruitful combination of these different scientific pathways. This allowed the scientific community to unveil the great potential contained in this theory, which could be applied to a wide spectrum of mathematical and physical problems concerning the evolution in time of statistical systems and thus providing the conceptual foundations of nonequilibrium statistical mechanics. Typical modern aspects of this field of physics are contained in the theory of continuous time random walk, discussed at the end of this chapter. It provides the mathematical tools for describing a wide range of stochastic processes, which overtake the limits of standard diffusive behavior, allowing for subdiffusive and superdiffusive regimes. These have been recently recognized as almost ubiquitous in nature, since they have been found to characterize a wide range of phenomena of interest not only for physics, but also for biology, chemistry, geology, finance, sociology, etc. All the following chapters will be devoted to an illustration of the many aspects concerning nonequilibrium statistical mechanics and their relevance for physical science. The introductory and pedagogical character of this book cannot allow us to account for the interdisciplinary potential of this approach, which overtakes, by far, any other domain of modern physics.

1.2 Kinetic Theory

1.2.1 The Ideal Gas

The basic model for understanding the mechanical foundations of thermodynamics is the ideal gas of Boltzmann. It is a collection of N identical particles of mass m that can be represented geometrically as tiny homogeneous spheres of radius r. One basic assumption of the ideal gas model is that we are dealing with a diluted system; i.e., the average distance δ between particles is much larger than their radius,

$$\delta = \left(\frac{1}{n}\right)^{\frac{1}{3}} \gg r, \tag{1.1}$$

where $n = N/V$ is the density of particles in the volume V occupied by the gas.[1] In the absence of external forces particles move with constant velocity[2] until they collide

[1] For a real gas of hydrogen molecules at room temperature (300 K) and atmospheric pressure (1 atm), $\delta \sim 10^{-6}$ m and $r \sim 10^{-10}$ m.

[2] One could argue that at least gravity should be taken into account, but its effects are generally negligible in standard conditions. An example where gravity has relevant, measurable effects will be studied in Section 1.4: it is the Brownian motion of colloidal particles; see Fig. 1.6.

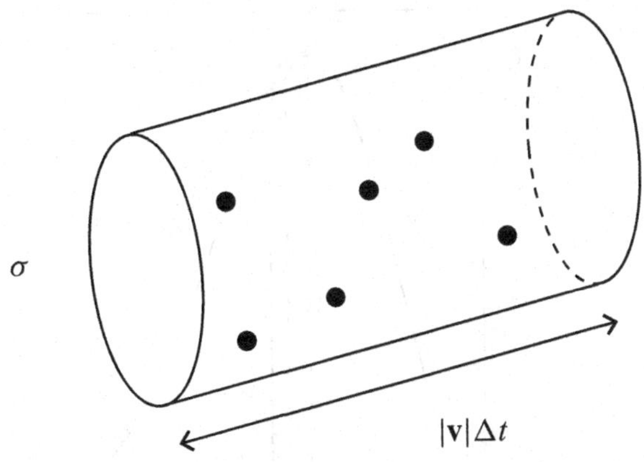

Fig. 1.1 Illustration of the concept of cross section. The black dots in the cylinder spanned by the cross section σ represent the centers of molecules hit in the time interval Δt by a molecule moving at speed $\mathbf{v}$.

pairwise, keeping their total momentum and energy constant (elastic collisions[3]). It can be easily realized that in such a diluted system multiple collisions are such rare events that they can be neglected for practical purposes.

Now we want to answer the following question: what is the rate of these collisions and the average distance run by a particle between subsequent collisions? We can estimate these quantities by considering that a particle moving with velocity $\mathbf{v}$ in a time interval Δt can collide with the particles that are contained in a cylinder of basis $\sigma = 4\pi r^2$ (called cross section) and height $|\mathbf{v}|\Delta t$; see Fig. 1.1. For the sake of simplicity we can assume that all the particles inside the cylinder are at rest with respect to the moving particle, so that we can estimate the number of collisions as

$$\mathcal{N}_{coll} = n\sigma|\mathbf{v}|\Delta t. \tag{1.2}$$

Accordingly, the number of collisions per unit time is given by the expression

$$\frac{\mathcal{N}_{coll}}{\Delta t} = n\sigma|\mathbf{v}| \tag{1.3}$$

and the average time between collisions reads

$$\tau \equiv \frac{\Delta t}{\mathcal{N}_{coll}} = \frac{1}{n\sigma|\mathbf{v}|}. \tag{1.4}$$

A quantitative estimate of τ can be obtained by attributing to $|\mathbf{v}|$ the value $\langle v\rangle$ of the equilibrium average of the modulus of the velocity of particles, v, in the ideal gas, according to Maxwell's distribution (see Fig. 1.2),

$$P(v) = \frac{4}{\sqrt{\pi}}\left(\frac{m}{2T}\right)^{3/2} v^2 \exp\left(-\frac{mv^2}{2T}\right), \tag{1.5}$$

[3] This hypothesis amounts to assuming that the particles of the gas are rigid spheres, that they do not suffer any deformation in the collision process. In fact, in a real gas the energy transferred to the internal degrees of freedom of the molecules can be practically neglected in standard conditions.

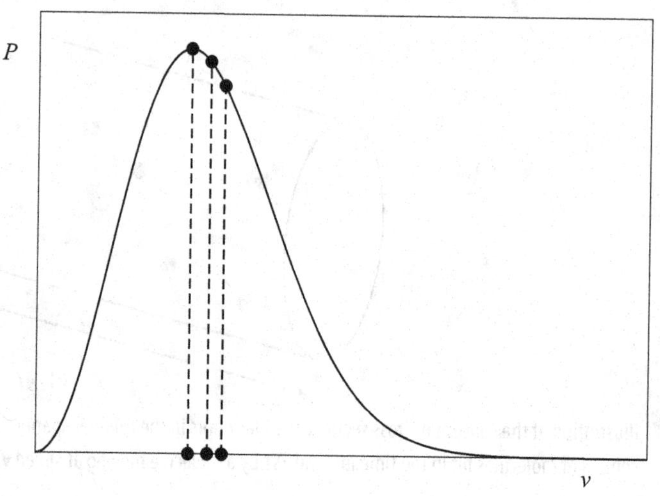

Fig. 1.2 The Maxwell distribution, Eq. (1.5). We indicate, from left to right, the most likely velocity v_{max}, the average velocity $\langle v \rangle$, and the square root of the average square velocity, $\langle v^2 \rangle^{1/2}$, whose expressions are given in Eq. (1.6).

where T is the temperature of the ideal gas at equilibrium. Using such distribution, we obtain the expressions

$$v_{\max} = \sqrt{\frac{2T}{m}}, \qquad \langle v \rangle = \sqrt{\frac{8T}{\pi m}} = \frac{2}{\sqrt{\pi}} v_{\max}, \qquad \langle v^2 \rangle^{1/2} = \sqrt{\frac{3T}{m}} = \sqrt{\frac{3}{2}} v_{\max}, \tag{1.6}$$

for the most likely velocity, the average velocity and the square root of the average square velocity, respectively.

We can now rewrite (1.4) as

$$\tau = \frac{1}{n\sigma \langle v \rangle} \tag{1.7}$$

and determine the average distance run by a particle between two collisions, i.e., its mean free path, by the expression

$$\lambda = \langle v \rangle \tau = \frac{1}{n\sigma}. \tag{1.8}$$

This formula corresponds to the case of a single moving particle colliding with target particles that are supposed to be immobile. But this is not the case, because in reality the target particles also move and a better estimate of τ and λ can be obtained using the formula

$$\tau = \frac{1}{n\sigma \langle v_r \rangle}, \tag{1.9}$$

where v_r is the modulus of the relative velocity $\mathbf{v}_r$, which follows the distribution

$$P_r(v_r) = \sqrt{\frac{2}{\pi}} \left(\frac{m}{2T} \right)^{3/2} v_r^2 \exp\left(-\frac{m v_r^2}{4T} \right). \tag{1.10}$$

This formula is a consequence of the general observation that the sum (or the difference) of two Gaussian variables is a Gaussian variable whose variance is the sum of their variances.

In this case, $\mathbf{v}_r = \mathbf{v}_1 - \mathbf{v}_2$, with $\mathbf{v}_{1,2}$ satisfying the Maxwell distribution (1.5) and the doubling of the variance explains why the exponent $(mv^2/2T)$ in Eq. (1.5) now becomes $(mv_r^2/4T)$. Then, the prefactor changes accordingly, in order to keep $P_r(v_r)$ normalized.

With Eq. (1.10) at hand, we can evaluate

$$\langle v_r \rangle = \sqrt{\frac{16T}{\pi m}} = \sqrt{2}\langle v \rangle \tag{1.11}$$

and obtain

$$\tau = \frac{1}{\sqrt{2} n\sigma \langle v \rangle}, \tag{1.12}$$

from which we can evaluate the mean free path,

$$\lambda = \langle v \rangle \tau = \frac{1}{\sqrt{2} n\sigma}. \tag{1.13}$$

It is worth noting that the ratio between λ and τ gives $\langle v \rangle$, not $\langle v_r \rangle$, because one particle travels an average distance λ in time τ.

We can finally use the formula (1.13) to evaluate the mean free path for a gas at room temperature and pressure. In this case λ is typically $O(10^{-7}\text{m})$, which is three orders of magnitude larger than the typical size r of a particle, $O(10^{-10}\text{m})$.

1.2.2 Random Walk: A Basic Model of Diffusion

We consider an ideal gas at thermal equilibrium with a heat bath at temperature T. If we fix our attention on one particle, we observe that collisions with the other particles produce a stepwise irregular trajectory, i.e., a sort of random walk. Beyond this qualitative observation we would like to obtain a quantitative description of this random walk. In principle the problem could be tackled by applying the laws of classical mechanics. In practice such a program is unrealistic, because one should know not only the initial velocity of the particle under examination, but also the velocities of all the particles that it will collide with. Such a computation is practically unfeasible, if we have to deal with a very large number of particles, like those contained in a mole of a gas.

In order to overcome such a difficulty we can introduce a suitable model, based on simplifying hypotheses. We assume that in between two collisions the observed particle keeps constant the modulus of its velocity, v. Moreover, the distance run by the particle between two collisions is also assumed to be constant and equal to ℓ. Finally, the direction along which the particle moves after a collision is completely uncorrelated with the one it was moving along before the collision. The latter hypothesis amounts to assuming that collisions can actually be considered as random events, thus contradicting the fully mechanical, i.e., deterministic, origin of the problem.[4] Without prejudice of generality, we

[4] We want to point out that in this way we introduce a statistical concept into the description of a purely mechanical process. This conceptual step has been at the origin of a long-standing debate in the scientific community over more than a century. Nowadays, it has been commonly accepted and it is a cornerstone of modern science. Anyway, this basic assumption still today relies more on its effectiveness in predicting observed phenomena, rather than on its logical foundations. On the other hand, the need of a stochastic approach

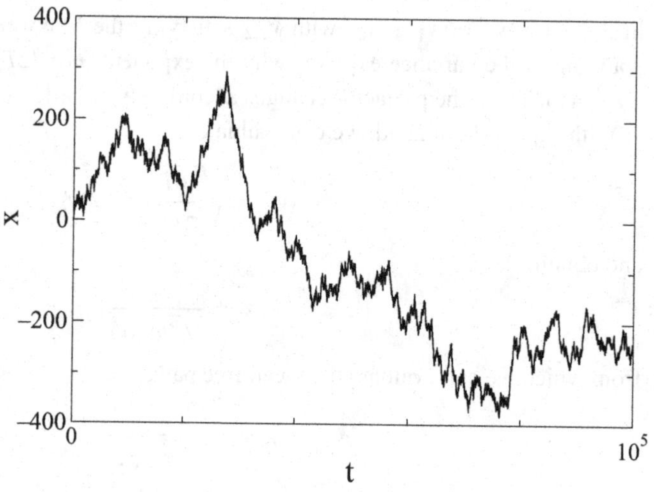

Fig. 1.3 Plot of a one-dimensional random walk, corresponding to $\ell = 1$.

assume that the selected particle at time $t = 0$ is at the origin of a Cartesian reference frame and we call $\mathbf{X}(t)$ the vector that identifies its position at time t. After having gone through N collisions, we can write

$$\mathbf{X} = \sum_{i=1}^{N} \mathbf{x}_i, \tag{1.14}$$

where $\mathbf{x}_i$ is the ith segment run by the particle after the ith collision ($|\mathbf{x}_i| = \ell, \forall i$), whose direction is random, i.e. uniformly distributed in the solid angle 4π. It is intuitive to conclude that as $N \to \infty$ the average value $\mathbf{X}/N \to 0$. For $d = 1$, in Fig. 1.3 we plot the resulting space-time trajectory, corresponding to $\ell = 1$.

As for the square displacement, we can write

$$X^2 \equiv \mathbf{X} \cdot \mathbf{X} = \sum_{i=1}^{N} \sum_{j=1}^{N} \mathbf{x}_i \cdot \mathbf{x}_j = \sum_{i=1}^{N} \sum_{j=1}^{N} \ell^2 \cos(\theta_{ij}) \tag{1.15}$$

where θ_{ij} is the angle in between the directions of segments $\mathbf{x}_i$ and $\mathbf{x}_j$. Since ℓ is a constant we can write the previous expression in a more convenient form,

$$X^2 = \ell^2 \sum_{i=1}^{N} \left(\sum_{j=1}^{N} \cos(\theta_{ij}) \right). \tag{1.16}$$

If $j = i$, then $\theta_{ij} = 0$, i.e. $\cos(\theta_{ij}) = 1$, and the previous equation can be written

$$X^2 = \ell^2 \sum_{i=1}^{N} \left(1 + \sum_{j \neq i} \cos(\theta_{ij}) \right). \tag{1.17}$$

could be also justified by invoking the contribution of dynamical details, such as the finite size of the particles or their internal rotational or vibrational degrees of freedom, that are usually neglected.

The values taken by $\cos(\theta_{ij})$ can be thought as random numbers, distributed in the interval $[-1,+1]$. If we compute the average of X^2 over a very large number of different realizations (replicas) of this random walk the sum $\sum_{j\neq i}\cos(\theta_{ij})$ is negligible and one can finally write

$$\langle X^2 \rangle = \ell^2 N, \tag{1.18}$$

where the symbol $\langle\ \rangle$ denotes the average over replicas. Notice that the larger the number of replicas, the better the statistical estimate $\langle X^2 \rangle$ for any N.

This result can be generalized by assuming the less strict hypothesis that the length of runs in between subsequent collisions is distributed according to some normalized distribution $g(\ell)$. A "real" example in two dimensions is discussed later on, in the context of Brownian motion (see Section 1.4, Fig. 1.5). If $\mathbf{x}_i = \ell_i \hat{\mathbf{x}}_i$, where $\hat{\mathbf{x}}_i$ is the unit vector in the direction of $\mathbf{x}_i$, we can write

$$\langle \mathbf{x}_i \cdot \mathbf{x}_j \rangle = \langle \ell_i \ell_j \rangle \langle \hat{\mathbf{x}}_i \cdot \hat{\mathbf{x}}_j \rangle = \langle \ell_i^2 \rangle \delta_{ij} \tag{1.19}$$

and

$$\langle X^2 \rangle = \langle \ell^2 \rangle N. \tag{1.20}$$

If $g(\ell)$ is a Poisson distribution, which corresponds to independent random events,

$$g(\ell) = \frac{1}{\lambda} \exp\left(-\frac{\ell}{\lambda}\right), \tag{1.21}$$

where λ is the mean free path defined in Eq. (1.13), we have to substitute ℓ^2 with $\langle \ell^2 \rangle = 2\lambda^2$ in Eq. (1.18), thus obtaining

$$\langle X^2 \rangle = 2\lambda^2 N. \tag{1.22}$$

Notice that $\lambda N = L = \langle v \rangle t$ is the total length run by the particle after N collisions, so we can write

$$\langle X^2 \rangle = 2\lambda \langle v \rangle t. \tag{1.23}$$

This relation indicates that in the random walk, the particle that was at the origin at time $t = 0$ is found at time t at an average distance from the origin, $\sqrt{\langle X^2 \rangle}$, that grows proportionally to $\sqrt{t}$. The proportionality constant between $\langle X^2 \rangle$ and t is usually written as $2\lambda\langle v \rangle = 2dD$, where D is the diffusion coefficient of the random walk in d space dimensions. Diffusion in real situations is quite a slow process. For instance, if we consider air molecules at $T = 20°$C we have $\langle v \rangle \sim 450$ m/s, while $\lambda \sim 0.06$ μm. Accordingly, a diffusing air molecule in these conditions runs a distance of 1 m in approximately 5 h and a distance of 10 m in approximately 20 days (a quasi-static situation, if convective or turbulent motions do not occur).

1.3 Transport Phenomena

The random walk model of a particle in an ideal gas is the appetizer of the general problem of transport processes. They concern a wide range of phenomena in hydrodynamics, thermodynamics, physical chemistry, electric conduction, magnetohydrodynamics, etc. They typically occur in physical systems (gases, liquids, or solids) made of many particles (atoms or molecules) in the presence of inhomogeneities. Such a situation can result from nonequilibrium conditions (e.g., the presence of a macroscopic gradient of density, velocity, or temperature), or simply from fluctuations around an equilibrium state.

The kinetic theory of transport phenomena provides a unified description of these apparently unlike situations. It is based on the assumption that even in nonequilibrium conditions gradients are small enough to guarantee that local equilibrium conditions still hold. In particular, the kinetic approach describes the natural tendency of the particles to transmit their properties from one region to another of the fluid by colliding with the other particles and eventually establishing global or local equilibrium conditions.

The main success of the kinetic theory is the identification of the basic mechanism underlying all the above-mentioned processes: the transport of a microscopic quantity (e.g., the mass, momentum or energy of a particle) over a distance equal to the mean free path λ of the particles, i.e. the average free displacement of a particle after a collision with another particle (see Eq. (1.13)). By this definition we are implicitly assuming that the system is a fluid, where each particle is supposed to interact with each other by collisions and propagate freely between successive collisions, the same conditions that we have discussed for the ideal gas model in Section 1.2.1.

Here we assume that we are dealing with a homogeneous isotropic system, where λ, the mean free path, is the same at any point and in any direction in space. Without prejudice of generality we consider a system where a uniform gradient of the quantity $A(\mathbf{x})$ is established along the z-axis, and $A(x,y,z) = A(x',y',z) = A(z)$ for any x, x', y, and y'. In particular, we assume that $A(z)$ is a microscopic quantity, which slowly varies at constant rate along the coordinate z of an arbitrary Cartesian reference frame. We consider also a unit surface S_1 located at height z and perpendicular to the z-axis; see Fig. 1.4(a). Any particle crossing the surface S_1 last collided at an average distance $\pm\lambda$ along the z-axis, depending on the direction it is moving. The net transport of the quantity $A(z)$ through S_1 amounts to the number of crossings of S_1 from each side in the unit time. Consistently with the assumption of local equilibrium we attribute the same average velocity $\langle v\rangle$ to all particles crossing S_1. Isotropy and homogeneity of the system imply also that one-third of the particles move on average along the z-axis, half of them upward and half downward. Accordingly, S_1 is crossed along z in the unit time interval by $\frac{1}{6}n\langle v\rangle$ particles in each direction.

The net flux of $A(z)$ through S_1 is given by

$$\Phi(A) = \frac{1}{6}\langle v\rangle\,[n(z-\lambda)A(z-\lambda) - n(z+\lambda)A(z+\lambda)]. \tag{1.24}$$

Since n and A vary weakly on the scale λ, one can use a first-order Taylor expansion and rewrite Eq. (1.24) as

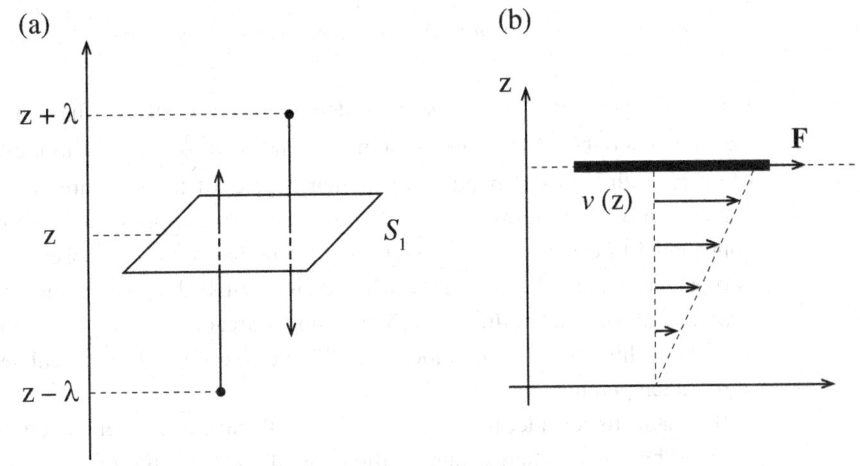

Fig. 1.4 (a) The surface S_1, normal to the $\hat{z}$ axis, is crossed by particles from both sides. Assuming particles move along the $\hat{z}$ axis, their most recent collision occurred at height $z - \lambda$ $(z + \lambda)$ if their speed is positive (negative). (b) A gradient of velocity along the z-axis in a liquid, produced by a plate at the liquid surface that is constrained to move at a finite speed by the application of a force **F**.

$$\Phi(A) = -\frac{1}{3}\langle v\rangle\,\lambda\,\frac{\partial(nA)}{\partial z}. \tag{1.25}$$

This calculation can be performed more carefully by introducing explicitly the Maxwell distribution function of particle velocities at equilibrium. Nonetheless, one recovers the same result.

The simplest case is the transport of mass, because of a density gradient along the z-axis, in which case $A(z)$ is a constant and

$$\Phi(n) = -\frac{1}{3}\langle v\rangle\,\lambda\,\frac{\partial n}{\partial z} = -D\,\frac{\partial n}{\partial z}, \tag{1.26}$$

where the quantity $D = \frac{1}{3}\langle v\rangle\lambda$ defines the diffusion coefficient of particles inside the fluid. This expression is equal to the definition of D through Eq. (1.23), $D = \lambda\langle v\rangle/d$, because in the calculation here above $d = 3$. In a real physical situation D depends both on the diffusing substance and the medium of diffusion. At room temperature a gas in air typically has $D \simeq 0.3\ \text{cm}^2\text{s}^{-1}$; the diffusion coefficient of a liquid in water is typically of the order $D \simeq 10^{-5}\ \text{cm}^2\text{s}^{-1}$; a gas in a solid has a much smaller diffusivity, of the order $D \simeq 10^{-9}\ \text{cm}^2\text{s}^{-1}$.

Other cases of physical interest correspond to the situations where a gradient of velocity or temperature is present and the density n is assumed to be constant, so $\Phi(A) = -\frac{1}{3}\,n\langle v\rangle\,\lambda\,\frac{\partial A}{\partial z}$. If there is a gradient of velocity, we assume that the fluid flows with constant macroscopic velocity $v(z)$ parallel to the (x, y)-plane. In such a situation there is a net transport of kinetic momentum $mv(z)$ (m is the mass of a particle), yielding a shear stress $\Phi(mv(z))$ between the fluid layers laying on the (x, y) plane (see Fig. 1.4(b)):

$$\Phi(mv(z)) = -\frac{1}{3}\, n\, m\langle v\rangle\, \lambda\, \frac{\partial v(z)}{\partial z} = -\eta\, \frac{\partial v(z)}{\partial z}, \tag{1.27}$$

where the quantity $\eta = \frac{1}{3} n\, m\, \langle v\rangle\, \lambda$ defines the viscosity of the fluid. By substituting Eq. (1.13) into the previous expression, one finds $\eta = \frac{m\langle v\rangle}{\sqrt{2}\,3\sigma}$. This implies that the viscosity of an ideal fluid is independent of its density, i.e. of the pressure. This counterintuitive conclusion was first derived by Maxwell and its experimental verification sensibly contributed to establish in the scientific community a strong consensus on the atomistic approach of kinetic theory. It is worth stressing that such a conclusion does not hold when dealing with very dense fluids. At room temperature, diluted gases typically have η of order 10 μPa·s, while in water and blood η is of the order of few millipascal-seconds and honey at room temperature has $\eta \approx 1$ Pa·s.

It remains to consider the case when $A(z)$ is the average kinetic energy of particles $\bar{\epsilon}(z)$. At equilibrium, the energy equipartition condition yields the relation $n\bar{\epsilon}(z) = \rho\, C_V\, T(z)$, where $\rho = mn$ is the mass density of particles, C_V is the specific heat at constant volume and $T(z)$ is the temperature at height z. The net flux of kinetic energy $\Phi(\bar{\epsilon})$ can be read as the heat transported through the fluid along the z-axis,

$$\Phi(\bar{\epsilon}) = -\frac{1}{3} n\langle v\rangle\lambda \frac{\partial \bar{\epsilon}}{\partial z} = -\frac{1}{3}\, \rho\, C_V\, \langle v\rangle\, \lambda\, \frac{\partial T(z)}{\partial z} = -\kappa\, \frac{\partial T(z)}{\partial z}, \tag{1.28}$$

where the quantity $\kappa = \frac{1}{3}\, \rho\, C_V\, \langle v\rangle\, \lambda = \frac{m C_V \langle v\rangle}{\sqrt{2}\,3\sigma}$ defines the heat conductivity. Also κ is found to be independent of n. The variability of κ in real systems is less pronounced than for other kinetic coefficients: in fact, a very good conductor like silver has $\kappa \simeq 400\ \mathrm{Wm^{-1}K^{-1}}$, while for cork, an effective heating insulator, it drops down to 4×10^{-2} in the same units.

One can conclude that the transport coefficients, i.e. the diffusion constant D, the viscosity η and the heat conductivity κ, are closely related to each other and depend on a few basic properties of the particles, like their mass m, their average velocity $\langle v\rangle$, and their mean free path λ. For example, by comparing the definitions of κ and η one finds the remarkable relation

$$\frac{\kappa}{\eta} = \alpha\, C_V, \tag{1.29}$$

with $\alpha = 1$. In real systems the constant α takes different values, which depend on the presence of internal degrees of freedom (e.g., $\alpha = \frac{5}{2}$ for realistic models of monoatomic gases).

The conceptual relevance of the relation (1.29) is that it concerns quantities that originate from quite different conditions of matter. In fact, on the left-hand side we have the ratio of two transport coefficients associated with macroscopic nonequilibrium conditions, while on the right-hand side we have a typically equilibrium quantity, the specific heat at constant volume. After what has been discussed in this section, this observation is far from mysterious: by assuming that even in the presence of a macroscopic gradient of physical quantities equilibrium conditions set in locally, the kinetic theory provides a unified theoretical approach for transport and equilibrium observables.

1.4 Brownian Motion

The basic example of transport properties emerging from fluctuations close to equilibrium is Brownian motion. This phenomenon had been observed by Robert Brown in 1827: small pollen particles (typical size, 10^{-3} cm) in solution with a liquid exhibited an erratic motion, whose irregularity increased with decreasing particle size. The long scientific debate about the origin of Brownian motion lasted over almost a century and raised strong objections to the atomistic approach of the kinetic theory (see Section 1.2.1). In particular, the motion of the pollen particles apparently could not be consistent with an explanation based on collisions with the molecules of the liquid, subject to thermal fluctuations. In fact, pollen particles have exceedingly larger mass and size with respect to molecules and the amount of velocity acquired by a pollen particle in a collision with a molecule is typically 10^{-6} m/s. In Fig. 1.5 we reproduce some original data by Jean Baptiste Perrin.

The kinetic approach also implies that the number of collisions per second of the Brownian particle with liquid molecules is gigantic and random. Thermal equilibrium conditions and the isotropy of the liquid make equally probable the sign of the small velocity variations of the pollen particle. On observable time scales the effect of the very many collisions should average to zero and the pollen particle should not acquire any net displacement from its initial position in the fluid. This way of reasoning is actually wrong, as Albert Einstein argued in his theory of Brownian motion.

Einstein's methodological approach, based on the combination of simple models with basic physical principles, was very effective: he first assumed that the Brownian macroscopic particles should be considered as "big mass molecules," so that the system

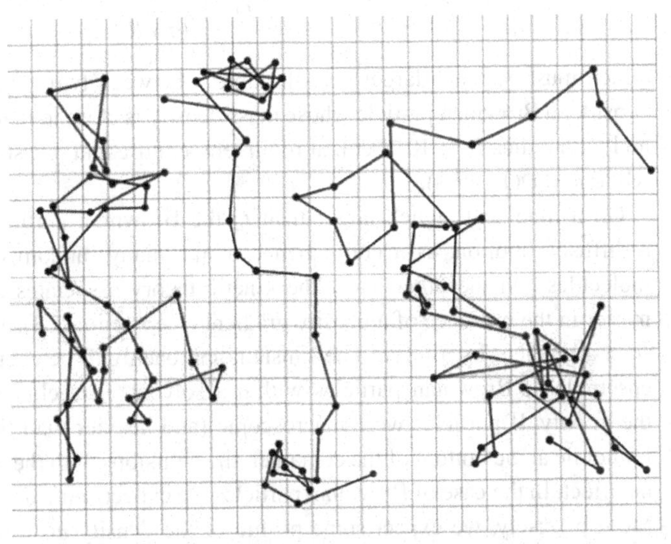

Fig. 1.5 Three tracings of the motion of colloidal particles of radius 0.53 μm, as seen under the microscope, are displayed. Successive positions every 30 s are joined by straight line segments (the mesh size is 3.1 μm). Reproduced from J. B. Perrin, *Les Atomes* (Paris: Librairie Félix Alcan, 1913). Wikimedia Commons.

composed by the Brownian particles and the solvent, in which they are suspended, can be considered a mixture of two fluids at thermal equilibrium. This implies the validity of Van t' Hoff's law,

$$P(\mathbf{x}) = Tn(\mathbf{x}), \tag{1.30}$$

where P is the osmotic pressure between the fluids, T is the equilibrium temperature of the solution and n is the volumetric density of Brownian particles (the "solute" of the binary mixture). However, according to classical kinetic theory one should not be allowed to write an equation of state like (1.30) for a collection of the mesoscopic Brownian particles; they should be better described by the laws of dynamics, rather than by those of thermodynamics. Einstein's opposite point of view can be justified by considering that, despite their dimension, the Brownian particles are not subject to standard macroscopic forces: they experience microscopic forces originated by thermal fluctuations of the fluid molecules and in this sense they are equivalent to particles in solution.

Einstein guessed that Brownian motion looks random on a much larger time scale than the one needed for dissipating the energy acquired through collisions with molecules. In practice, this amounts to assuming that the dissipation mechanism should be described by the macroscopic Stokes's law

$$\mathbf{F} = m\frac{d\mathbf{v}}{dt} = -6\pi\eta R\mathbf{v}, \tag{1.31}$$

where $\mathbf{F}$ is the friction force proportional to the velocity $\mathbf{v}$ of the Brownian particle of radius R and η is the viscosity of the solvent. Therefore, the energy dissipation process of Brownian particles has to occur on the time scale

$$t_d = \frac{m}{6\pi\eta R}, \tag{1.32}$$

which must be much larger than the time τ between two collisions; see Eq. (1.7). If we consider a Brownian particle whose mass density is close to that of the solvent (e.g., water) and with radius $R \simeq 10^{-4}$ cm, at room temperature and pressure one has $t_d = O(10^{-7}\ \text{s})$, while $\tau = O(10^{-11}\ \text{s})$.

On a time scale much larger than t_d the Brownian particles are expected to exhibit a diffusive motion, as a consequence of the many random collisions with the solvent molecules. On the other hand, the kinetic theory associates diffusion with transport of matter in the presence of a density gradient (see Section 1.3). We should therefore induce such a gradient by a force. The Einstein approach will be exemplified by considering an ensemble of Brownian particles within a solvent, in a closed container and subject to the gravity. If particles were macroscopic (and heavier than the solvent), they would sit immobile at the bottom of the container, the collisions with the atoms of the solvent having no effect. In the case of Brownian particles we expect some diffusion, hindered by gravity. More precisely, we expect some profile of the density $n(z)$ of Brownian particles, with $\partial_z n < 0$ and $n(z) \to 0$ for $z \to +\infty$. This profile is the outcome of a stationary state that we are now going to describe in terms of equilibrium between currents and equilibrium between forces.

If $F_0 = -mg$ is the force of gravity acting on a Brownian particle of mass m, its velocity satisfies the equation[5]

$$m\frac{dv}{dt} = F_0 - 6\pi\eta R v, \tag{1.33}$$

which leads to the asymptotic, sedimentation speed $v_0 = F_0/6\pi\eta R$. In the steady state, the resulting current $J_0 = n(z)v_0$ must be counterbalanced by the diffusion current, giving a vanishing total current,

$$-D\partial_z n(z) + n(z)v_0 = 0. \tag{1.34}$$

In terms of equilibrium between forces, the force related to the diffusive motion of Brownian particles is due to the osmotic pressure (1.30). Therefore, equilibrium implies

$$-\partial_z P + n(z)F_0 = 0. \tag{1.35}$$

Using the Van t' Hoff law and the explicit expression of the sedimentation speed, from Eqs. (1.34) and (1.35) we find the same equation for the density profile,

$$-\partial_z n(z) + \frac{F_0}{6\pi\eta RD}n(z) = 0 \tag{1.36}$$

$$-\partial_z n(z) + \frac{F_0}{T}n(z) = 0, \tag{1.37}$$

which implies the remarkable relation, called Einstein's formula,

$$D = \frac{T}{6\pi\eta R} \equiv \frac{T}{\tilde{\gamma}}, \tag{1.38}$$

where we have defined the friction coefficient (see Eq. (1.33)), $\tilde{\gamma} = 6\pi\eta R$. Perrin showed that this formula provides very good quantitative agreement with experimental data.

The solution of the equation for $n(z)$ gives an exponential profile for the density of Brownian particles,

$$n(z) = n(0)\exp\{(-mg/T)z\}, \tag{1.39}$$

as attested by the experimental results reported in Fig. 1.6.

In summary, Einstein's theory of Brownian motion is based on the description of the Brownian particle in both microscopic and macroscopic terms: the microscopic description is employed when we use the van t' Hoff law to describe Brownian particles as a diluted gas in a solvent, while the macroscopic description is introduced via the Stokes law.

1.4.1 The Langevin Equation for the Brownian Particle

A mechanical approach that is valid at both $t < t_d$ (ballistic regime) and $t > t_d$ (diffusive regime) was later proposed by Paul Langevin. For the first time deterministic and stochastic forces were introduced into the same equation. The deterministic component is the friction term $-\tilde{\gamma}\mathbf{v}$, acting on the Brownian particle: $\tilde{\gamma}$ is the friction coefficient, which amounts to $\tilde{\gamma} = 6\pi\eta R$ in the Stokes regime and $\mathbf{v}$ is the three-dimensional velocity vector of

[5] We can deal with relations among scalar quantities, because all of them are directed along the z-axis.

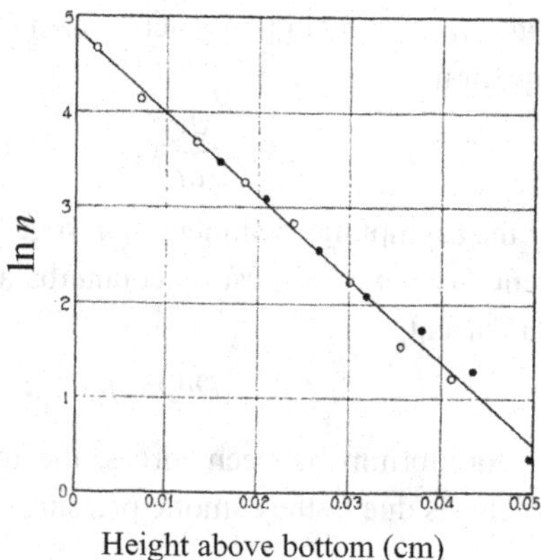

Fig. 1.6 Height distribution of the density of gold sols, showing an exponentially decreasing function. From N. Johnston and L. G. Howell, Sedimentation equilibria of colloidal particles, *Physical Review*, **35** (1930) 274-282.

the Brownian particles, whose components are denoted by v_i, $i = 1, 2, 3$. The stochastic component associated with the effect of collisions with the solvent molecules is a random, time-dependent force $\tilde{\boldsymbol{\eta}}(t)$, whose components are analogously denoted by $\tilde{\eta}_i$, $i = 1, 2, 3$. Symmetry arguments indicate that each one of these components can be assumed to be independent, isotropic, uncorrelated in time (at least for $t \gg \tau$) and Gaussian, since it results from the combination of a very large number of collisions, which can be approximately considered as independent events (see Appendix A). Accordingly, the average of the stochastic force is null and its time-correlation function can be approximated by a Dirac delta for $t \gg \tau$:

$$\langle \tilde{\eta}_i(t) \rangle = 0 \tag{1.40a}$$

$$\langle \tilde{\eta}_i(t)\tilde{\eta}_j(t') \rangle = \tilde{\Gamma}\delta_{ij}\delta(t - t'), \tag{1.40b}$$

where the brackets $\langle \;\; \rangle$ indicate the statistical average, $\tilde{\Gamma}$ is a suitable dimensional constant, and δ_{ij} is a Kronecker delta. The Langevin equation is a Newtonian equation containing the stochastic force $\tilde{\boldsymbol{\eta}}$. For each component of the three-dimensional velocity vector it reads

$$m\frac{dv_i(t)}{dt} = -\tilde{\gamma}\, v_i(t) + \tilde{\eta}_i(t), \tag{1.41}$$

which is rewritten, for the sake of simplicity, as

$$\frac{dv_i(t)}{dt} = -\gamma\, v_i(t) + \eta_i(t), \tag{1.42}$$

where we have introduced the symbol $\gamma = \tilde{\gamma}/m = 1/t_d$ and the stochastic force per unit mass, $\eta = \tilde{\eta}/m$. The latter satisfies the same equations as (1.40), with $\tilde{\Gamma}$ replaced by $\Gamma = \tilde{\Gamma}/m^2$.

Eq. (1.42) can be formally integrated, yielding the solution

$$v_i(t) = \exp(-\gamma t)\left[v_i(0) + \int_0^t d\tau \exp(\gamma\tau)\,\eta_i(\tau)\right]. \tag{1.43}$$

Accordingly, $v_i(t)$ is a stochastic function with average value

$$\langle v_i(t)\rangle = v_i(0)\exp(-\gamma t) \tag{1.44}$$

and average squared value

$$\begin{aligned}\langle v_i^2(t)\rangle &= \exp(-2\gamma t)\left[v_i^2(0) + \int_0^t d\tau \int_0^t d\tau' \exp\left(\gamma(\tau+\tau')\right)\langle \eta_i(\tau)\eta_i(\tau')\rangle\right] \\ &= \exp(-2\gamma t)\left[v_i^2(0) + \int_0^t d\tau \Gamma \exp(2\gamma\tau)\right] \\ &= v_i^2(0)\exp(-2\gamma t) + \frac{\Gamma}{2\gamma}\left[1 - \exp(-2\gamma t)\right]. \end{aligned} \tag{1.45}$$

Above we first used Eq. (1.40a) to cancel out the terms that are linear in the noise, then we explicitly used the noise correlation function (1.40b).

As time t grows the average velocity vanishes exponentially with rate γ, while the average squared velocity approaches the value

$$\lim_{t\to\infty}\langle v_i^2(t)\rangle = \frac{\Gamma}{2\gamma}. \tag{1.46}$$

In order to bridge this purely mechanical approach with Einstein theory it remains to establish a relation with thermodynamics. This is naturally obtained by attributing a thermal origin to the fluctuations of the stochastic force. In particular, if we assume that the Brownian particles and the solvent are in thermal equilibrium at temperature T, the energy equipartition principle establishes that, in the limit of very large times, the average kinetic energy per degree of freedom is proportional to the temperature,

$$\lim_{t\to\infty}\left\langle \frac{1}{2}m\,v_i^2(t)\right\rangle = \frac{1}{2}T. \tag{1.47}$$

By comparison with (1.46) one obtains the formula,

$$\Gamma = \frac{2\gamma T}{m}, \tag{1.48}$$

or equivalently

$$\tilde{\Gamma} = 2\tilde{\gamma}T, \tag{1.49}$$

which is a basic example of a fluctuation–dissipation relation.

From an experimental point of view, the fluctuations of the velocity of the particle cannot be measured because of the erratic trajectories, while it is easy to measure the mean square displacement of the Brownian particle from its initial position:

$$\begin{aligned}\langle (x_i(t) - x_i(0))^2\rangle &= \left\langle \left[\int_0^t v_i(\tau)d\tau\right]^2\right\rangle \\ &= \int_0^t d\tau \int_0^t d\tau' \langle v_i(\tau)v_i(\tau')\rangle. \end{aligned} \tag{1.50}$$

The expression between brackets in the last integral is the velocity correlation function of the Brownian particle, which can be computed from (1.43), similarly to the derivation of Eq. (1.45). We obtain

$$\langle v_i(\tau)v_i(\tau')\rangle = v_i^2(0)e^{-\gamma(\tau+\tau')} + \Gamma e^{-\gamma(\tau+\tau')}\int_0^{\tau} dt_1 \int_0^{\tau'} dt_2 e^{\gamma(t_1+t_2)}\delta(t_1-t_2). \tag{1.51}$$

The double integral depends on which variable between τ and τ' is the smallest; if $\tau < \tau'$, we must first integrate over t_2, obtaining

$$\begin{aligned}\Gamma e^{-\gamma(\tau+\tau')}\int_0^{\tau} dt_1 \int_0^{\tau'} dt_2 e^{\gamma(t_1+t_2)}\delta(t_1-t_2) &= \Gamma e^{-\gamma(\tau+\tau')}\int_0^{\min(\tau,\tau')} dt_1 e^{2\gamma t_1}\\ &= \frac{\Gamma}{2\gamma}\left(e^{-\gamma|\tau-\tau'|} - e^{-\gamma(\tau+\tau')}\right),\end{aligned}$$

a result that is valid in any case. So, Eq. (1.51) becomes

$$\langle v_i(\tau)v_i(\tau')\rangle = v_i^2(0)e^{-\gamma(\tau+\tau')} + \frac{\Gamma}{2\gamma}\left(e^{-\gamma|\tau-\tau'|} - e^{-\gamma(\tau+\tau')}\right). \tag{1.52}$$

Notice that for $\tau' = \tau$, the previous correlator simplifies to Eq. (1.45). Finally, substituting into (1.50) we obtain

$$\langle (x_i(t)-x_i(0))^2\rangle = \left(v_i^2(0) - \frac{\Gamma}{2\gamma}\right)\frac{(1-e^{-\gamma t})^2}{\gamma^2} + \frac{\Gamma}{\gamma^2}t - \frac{\Gamma}{\gamma^3}\left(1-e^{-\gamma t}\right), \tag{1.53}$$

where we have used the following relations

$$\begin{aligned}\int_0^t d\tau \int_0^t d\tau' e^{-\gamma(\tau+\tau')} &= \left(\frac{1-e^{-\gamma t}}{\gamma}\right)^2\\ \int_0^t d\tau \int_0^t d\tau' e^{-\gamma|\tau-\tau'|} &= 2\int_0^t d\tau \int_0^{\tau} d\tau' e^{-\gamma(\tau-\tau')}\\ &= \frac{2}{\gamma}t - \frac{2}{\gamma^2}\left(1-e^{-\gamma t}\right).\end{aligned} \tag{1.54}$$

We can now evaluate (1.53) in the limits $t \ll t_d$ and $t \gg t_d$. In the former case, $\gamma t \ll 1$ and $1-e^{-\gamma t} \simeq \gamma t - \gamma^2 t^2/2$; in the latter case, $\gamma t \gg 1$ and $1-e^{-\gamma t} \simeq 1$. Therefore, in the two limits the mean square displacement of the Brownian particle has the following expressions

$$\langle (x_i(t)-x_i(0))^2\rangle = \begin{cases} v_i^2(0)t^2 & t \ll t_d \quad \text{ballistic regime}\\ \dfrac{\Gamma}{\gamma^2}t & t \gg t_d \quad \text{diffusive regime.}\end{cases} \tag{1.55}$$

Using (1.48) and the Einstein relation (1.38), we obtain $\Gamma/\gamma^2 = 2T/(m\gamma) = 2D$, so that $\langle (x_i(t)-x_i(0))^2\rangle = 2Dt$ in the diffusive regime. More generally, in d spatial dimensions we have

$$\lim_{t\to\infty}\frac{\langle (\mathbf{x}(t)-\mathbf{x}(0))^2\rangle}{t} = 2dD. \tag{1.56}$$

It is straightforward to generalize the Langevin equation (1.41) to the case where the Brownian particle is subject to a conservative force $F(x) = -U'(x)$,

$$m\ddot{x}(t) = -U'(x) - \tilde{\gamma}\dot{x}(t) + \tilde{\eta}(t). \tag{1.57}$$

This problem will be reconsidered in Section 1.6.3 in the context of the Ornstein–Uhlenbeck processes and in Section 1.6.5 where detailed balance conditions are considered.

1.4.2 The Fokker–Planck Equation for the Brownian Particle

The statistical content of the Langevin approach emerges explicitly from the properties attributed to the random force in Eq. (1.40b). The statistical average $\langle \ \rangle$ can be interpreted as the result of the average over many different trajectories of the Brownian particle obtained by different realizations of the stochastic force components η_i in Eq. (1.42). By taking inspiration from Boltzmann kinetic theory one can get rid of mechanical trajectories and describe from the very beginning the evolution of Brownian particles by a distribution function $p(\mathbf{x}, t)$, such that $p(\mathbf{x}, t)d^3x$ represents the probability of finding the Brownian particle at time t in a position between $\mathbf{x}$ and $\mathbf{x} + d^3x$. The evolution is ruled by the transition rate $W(\mathbf{x}', \mathbf{x})$, which represents the probability per unit time and unit volume that the Brownian particle "jumps" from $\mathbf{x}$ to $\mathbf{x}'$.

In this approach any reference to instantaneous collisions with the solvent molecules is lost and one has to assume that relevant time scales are much larger than t_d. As proposed by Fokker and Planck, one can write the evolution equation for $p(\mathbf{x}, t)$ as a master equation, i.e. a balance equation where the variation in time of $p(\mathbf{x}, t)$ emerges as the result of two competing terms: a gain factor, due to jumps of particles from any position $\mathbf{x}'$ to $\mathbf{x}$, and a loss factor, due to jumps from $\mathbf{x}$ to any $\mathbf{x}'$. Thus

$$\frac{\partial p(\mathbf{x}, t)}{\partial t} = \int_{-\infty}^{+\infty} d^3x' \left[p(\mathbf{x}', t)W(\mathbf{x}, \mathbf{x}') - p(\mathbf{x}, t)W(\mathbf{x}', \mathbf{x}) \right], \tag{1.58}$$

where $d^3x' W(\mathbf{x}, \mathbf{x}')$ ($d^3x' W(\mathbf{x}', \mathbf{x})$) represents the probability per unit time that the particle jumps from a neighborhood of $\mathbf{x}'$ to $\mathbf{x}$ (or vice versa). If we define $\boldsymbol{\chi} = \mathbf{x}' - \mathbf{x}$ and we use the notation $W(\mathbf{x}; \boldsymbol{\chi}) = W(\mathbf{x}', \mathbf{x})$, we obtain

$$\frac{\partial p(\mathbf{x}, t)}{\partial t} = \int d^3x' \left[p(\mathbf{x}', t)W(\mathbf{x}'; -\boldsymbol{\chi}) - p(\mathbf{x}, t)W(\mathbf{x}; \boldsymbol{\chi}) \right] \tag{1.59}$$

$$= \int d^3\chi \left[p(\mathbf{x} - \boldsymbol{\chi}, t)W(\mathbf{x} - \boldsymbol{\chi}; \boldsymbol{\chi}) - p(\mathbf{x}, t)W(\mathbf{x}; \boldsymbol{\chi}) \right], \tag{1.60}$$

where in the second equality we passed from the variable $\mathbf{x}'$ to $\boldsymbol{\chi}$ and in the first integral of (1.60) we have substituted $\boldsymbol{\chi}$ with $-\boldsymbol{\chi}$.

Since large displacements of the Brownian particles are very infrequent, one can reasonably assume that the rate functions $W(\mathbf{x} - \boldsymbol{\chi}; \boldsymbol{\chi})$ and $W(\mathbf{x}; \boldsymbol{\chi})$ are significantly different from zero only for very small $\boldsymbol{\chi}$. This allows one to introduce a formal Taylor series expansion in Eq. (1.60) around $\boldsymbol{\chi} = 0$. By considering only terms up to the second order one obtains

$$\frac{\partial p(\mathbf{x},t)}{\partial t} = \int d^3\chi \left[-\nabla \left(p(\mathbf{x},t)W(\mathbf{x};\chi)\right) \cdot \chi + \frac{1}{2}\sum_{i,j} \frac{\partial^2}{\partial x_i\, \partial x_j} \left(p(\mathbf{x},t)W(\mathbf{x};\chi)\right) \chi_i \chi_j \right]. \tag{1.61}$$

Eq. (1.61) has been found with the Brownian particle in mind, but the variable $\mathbf{x}$ can be any vectorial quantity defining the state of a physical system, not only the spatial position of a particle. For this reason it has a range of application going well beyond the motion of pollen particles in a solvent and it is worth making a further step before specializing to the Brownian motion. We can formally define the quantities

$$\alpha_i(\mathbf{x}) = \int d^3\chi \; W(\mathbf{x};\chi)\; \chi_i \tag{1.62}$$

$$\beta_{ij}(\mathbf{x}) = \int d^3\chi\, W(\mathbf{x};\chi)\chi_i\chi_j \tag{1.63}$$

so that Eq. (1.61) can be written as

$$\frac{\partial p(\mathbf{x},t)}{\partial t} = -\sum_i \frac{\partial}{\partial x_i}\left(\alpha_i(\mathbf{x})p(\mathbf{x},t)\right) + \frac{1}{2}\sum_{i,j}\frac{\partial^2}{\partial x_i \partial x_j}\left(\beta_{ij}(\mathbf{x})p(\mathbf{x},t)\right). \tag{1.64}$$

This is the celebrated Fokker–Planck equation and it will be used in different contexts of this book.

Let us now come back to the Brownian motion, in which case $\mathbf{x}$ is the spatial position of the particle and the quantities $\boldsymbol{\alpha}$ and β_{ij} have a simple interpretation as average quantities. The quantity $\boldsymbol{\alpha}$ is the average displacement of a Brownian particle per unit time;[6] i.e. its average velocity is

$$\boldsymbol{\alpha} = \frac{\langle \Delta \mathbf{x}\rangle}{\Delta t}. \tag{1.65}$$

In the absence of external forces it vanishes, but in the presence of a constant force $\mathbf{F}_0$ (e.g., gravity) the average velocity is equal to the sedimentation speed $\mathbf{v}_0 = \mathbf{F}_0/\tilde{\gamma}$ (see below Eq. (1.33)), resulting from the balance of the external force acting on the Brownian particle with the viscous friction force produced by the solvent.

The quantity β_{ij} amounts to the average squared displacements per unit time,

$$\beta_{ij} = \frac{\langle \Delta x_i \Delta x_j\rangle}{\Delta t}. \tag{1.66}$$

In a homogeneous, isotropic medium it is diagonal and proportional to the diffusion constant D,

$$\beta_{ij} = 2\delta_{ij} D. \tag{1.67}$$

For the Brownian particle, we finally obtain

$$\frac{\partial p(\mathbf{x},t)}{\partial t} = -\mathbf{v}_0 \cdot \nabla p(\mathbf{x},t) + D\nabla^2 p(\mathbf{x},t). \tag{1.68}$$

[6] Note that W is a transition rate, i.e. a probability per unit time and also per unit volume.

In the absence of external forces the viscous term can be neglected and Eq. (1.68) reduces to the diffusion equation,

$$\frac{\partial p(\mathbf{x}, t)}{\partial t} = D\nabla^2 p(\mathbf{x}, t), \tag{1.69}$$

whose solution is given, for $d = 1$, by Eq. (D.11),

$$p(x, t) = \frac{1}{\sqrt{4\pi Dt}} \exp\left(-\left[\frac{(x - x_0)^2}{4Dt}\right]\right). \tag{1.70}$$

The average squared displacement is given by the variance of the Gaussian distribution, i.e.

$$\langle x^2 \rangle - \langle x \rangle^2 = 2Dt. \tag{1.71}$$

Comparing with Eq. (1.55), we can conclude that the Fokker–Planck approach predicts the same asymptotic diffusive behavior as the Langevin approach.

1.5 Discrete Time Stochastic Processes

As it happens for a Brownian particle, the evolution in time of a great many systems in nature is affected by random fluctuations, usually referred to as "noise." These fluctuations can be originated by many different physical processes, such as the interaction with thermalized particles of the solvent (as in the case of the Brownian particle), absorption and emission of radiation, chemical reactions at molecular level, different behaviors of agents in a stock market or of individuals in a population, etc. When we introduce the concept of "noise" in these phenomena we are assuming that part of their evolution is ruled by "unknown" variables, whose behavior is out of our control, i.e. they are unpredictable, in a deterministic sense. On the other hand, we can assume that such fluctuations obey general statistical rules, which still allow us to obtain the possibility of predicting the evolution of these systems in a probabilistic sense. An everyday example of such a situation is a closed bottle thrown into the sea. During its navigation the bottle is continuously subject to small-scale random fluctuations of different duration and amplitude, due to the waves interfering on the sea surface. This notwithstanding, the bottle is expected to follow over a large-scale distance some macroscopic current flow, which may eventually lead it to a populated seaside. This is at least what the castaway, who threw the bottle in the sea with a message inside, expects will eventually occur with some not entirely negligible probability.

The mathematics that has been set up for describing these kind of phenomena is the theory of stochastic processes. Many great scientists, such as Marian Smoluchowski, Louis Bachelier, Max Planck, Norbert Wiener, George Uhlenbeck, Albert Enstein, and Paul Langevin (as mentioned in the previous section), contributed to this branch of mathematics. Its foundations were established by the systematic work of the Russian mathematician Andrej Markov, who introduced the concept of the so-called Markov chains.

At variance with the previous section, here we want to introduce the reader to a more rigorous mathematical formulation of the concepts that have been described in the previous section by intuitive or heuristic arguments. First of all we shall deal with stochastic processes in discrete space and time. They are particularly useful to establish the basic mathematical tools that will be later extended to continuous space and time, into the form of stochastic differential equations. Eventually, we shall obtain a consistent formulation of the general Fokker–Planck formalism, which deals with the evolution of probability distributions.

1.5.1 Markov Chains

Markov chains are stochastic processes discrete in time and in the state space, where the value assumed by each stochastic variable depends on the value taken by the same variable at the previous instant of time. Let us translate these concepts into a mathematical language. We assume that the stochastic variable $x(t)$ takes values at each instant of time t over a set of N states, $S = \{s_1, s_2, \ldots, s_{N-1}, s_N\}$. For the sake of simplicity, we also assume that $x(t)$ is measured at equal finite time intervals, so that also time becomes a discrete variable, equivalent, in some arbitrary time unit, to the ordered sequence of natural numbers, $t = 1, 2, \ldots, n, \ldots$. The basic quantity we want to deal with is the probability, $p(x(t) = s_i)$, that $x(t)$ is in state s_i at time t. If $p(x(t) = s_i)$ does not depend on the previous history of the stochastic process, we are dealing with the simple case of a sequence of independent events, like tossing a coin. In general, one could expect that, if some time correlation (i.e. memory) is present in the evolution of $x(t)$, then $p(x(t) = s_i)$ could depend also on the previous history of the stochastic process. In this case it is useful to introduce the conditional probability

$$\Omega(x(t) = s_i \,|\, x(t-1) = s_{i_1}, x(t-2) = s_{i_2}, \ldots, x(t-n) = s_{i_n}) \tag{1.72}$$

that $x(t)$ is in the state s_i at time t, given the evolution of the stochastic process backward in time up to $t - n$. In this case we say that the stochastic process has memory n: the case $n = 1$ defines a Markov process, where

$$\Omega(x(t) = s_j \,|\, x(t-1) = s_i) \tag{1.73}$$

is the transition probability in a unit time step from s_i to s_j; see Fig. 1.7.[7]

In general, $\Omega(x(t) = s_j \,|\, x(t-1) = s_i)$ is a function of time, but here we limit ourselves to consider stationary Markov processes, where this transition rate is independent of time. In this case we can deal with many interesting problems that can be discussed and solved with basic mathematical ingredients. The time-dependent case is certainly more general

[7] An example of a non-Markovian process is a self-avoiding random walk. We can think of a random walker moving at each time step between nearby sites of a regular square lattice, choosing with equal probability any available direction. On the other hand, the walker cannot move to any site already visited in the past. The time evolution of the random process modifies the probability of future events, or said differently, the walker experiences a persistent memory effect, that typically yields long-time correlations. Such a situation is peculiar to many interesting phenomena concerning other domains of science, such as sociology or economics. For instance, the price X of a stock at time t is usually more properly represented as an outcome of a non-Markovian process, where this quantity depends on the overall economic activity $a(t)$ up to time t, i.e. $X(a(t))$ rather than simply $X(t)$.

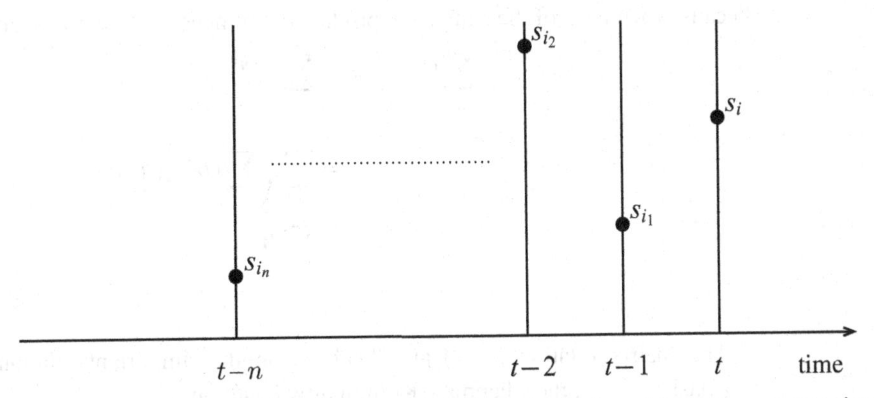

Fig. 1.7 The configuration space at time t is represented by a vertical axis crossing the time axis in t. In the most general case, the probability to be in s_i at time t depends on the previous history at times $t-1, t-2, \ldots, t-n$.

and of major interest, but even a short introduction would be hardly accessible for the audience to which this textbook is addressed.

Making use of the shorthand notations $p_i(t) \equiv p(x(t) = s_i)$ and $W_{ji} \equiv \Omega(x(t+1) = s_j \,|\, x(t) = s_i)$, these quantities must obey the following conditions:

$$p_i(t) \geq 0, \qquad \forall i, t \tag{1.74}$$

$$\sum_i p_i(t) = 1, \qquad \forall t \tag{1.75}$$

$$W_{ij} \geq 0, \qquad \forall i, j \tag{1.76}$$

$$\sum_i W_{ij} = 1, \qquad \forall j. \tag{1.77}$$

We can define the stochastic dynamical rule of the Markov chain as

$$p_j(t+1) = \sum_i W_{ji} p_i(t), \tag{1.78}$$

so one can easily realize that the W_{ij} can be viewed as the entries of an $N \times N$ matrix W, called stochastic matrix. Accordingly, Eq. (1.78) can be rewritten in vector form as

$$\mathbf{p}(t+1) = W\mathbf{p}(t), \tag{1.79}$$

where $\mathbf{p}(t) = (p_1(t), p_2(t), \ldots, p_j(t), \ldots, p_N(t))$ is the column vector of the probability. This matrix relation can be generalized to obtain

$$\mathbf{p}(t+n) = W^n \mathbf{p}(t), \tag{1.80}$$

where W^n, the nth power of W, is also a stochastic matrix, since it satisfies the same properties (1.76) and (1.77) of W. This is easily proved by induction, assuming W^n is a stochastic matrix and showing that W^{n+1} is also stochastic. In fact,

$$(W^{n+1})_{ij} = \sum_k (W^n)_{ik} W_{kj} \geq 0, \qquad \forall i, j \tag{1.81}$$

because each term of the sum is the product of nonnegative quantities, and

$$\sum_i (W^{n+1})_{ij} = \sum_{i,k} (W^n)_{ik} W_{kj} \tag{1.82}$$

$$= \sum_k \left(\sum_i (W^n)_{ik} \right) W_{kj} \tag{1.83}$$

$$= \sum_k W_{kj} \tag{1.84}$$

$$= 1. \tag{1.85}$$

The matrix relation (1.80) also leads to another important relation concerning the stochastic matrix, the Chapman–Kolmogorov equation,

$$p(t+n) = W^n p(t) = W^n W^t p(0) = W^{t+n} p(0). \tag{1.86}$$

It extends to stochastic processes the law valid for deterministic dynamical systems, where the evolution operator from time 0 to time $(t+n)$, $\mathcal{L}^{t+n}$, can be written as the composition of the evolution operator from time 0 to time t with the evolution operator from time t to time $t+n$, namely $\mathcal{L}^{t+n} = \mathcal{L}^n \circ \mathcal{L}^t$.

As usual for any $N \times N$ matrix it is useful to solve the eigenvalue problem

$$\det(W - \lambda I) = 0,$$

where I is the identity matrix and λ is a scalar quantity, whose values solving the eigenvalue equation are called the spectrum of W.

Since W is not a symmetric matrix, its eigenvalues are not necessarily real numbers.[8] Moreover, one should distinguish between right and left eigenvectors of W. We denote the right ones as $\mathbf{w}^{(\lambda)} = (w_1^{(\lambda)}, w_2^{(\lambda)}, \ldots, w_j^{(\lambda)}, \ldots, w_N^{(\lambda)})$, so that we can write

$$W\,\mathbf{w}^{(\lambda)} = \lambda\,\mathbf{w}^{(\lambda)}. \tag{1.87}$$

The spectrum of W has the following properties, whose proof is given in Appendix B:

(a) $|\lambda| \le 1$;
(b) there is at least one eigenvalue $\lambda = 1$;
(c) $\mathbf{w}^{(\lambda)}$ is either an eigenvector with eigenvalue 1, or it fulfills the condition $\sum_j w_j^{(\lambda)} = 0$.

We conclude this section by introducing some definitions:

Definition 1.1 A state s_j is accessible from a state s_i if there is a finite value of time t such that $(W^t)_{ji} > 0$.

Definition 1.2 A state s_j is persistent if the probability of returning to s_j after some finite time t is 1, while it is transient if there is a finite probability of never returning to s_j for any finite time t. As a consequence of these definitions a persistent state will be visited infinitely many times, while a transient state will be discarded by the evolution after a sufficiently long time.

[8] Notice that the asymmetry of W implies that, in general, $\sum_j W_{ij} \neq 1$, at variance with (1.77).

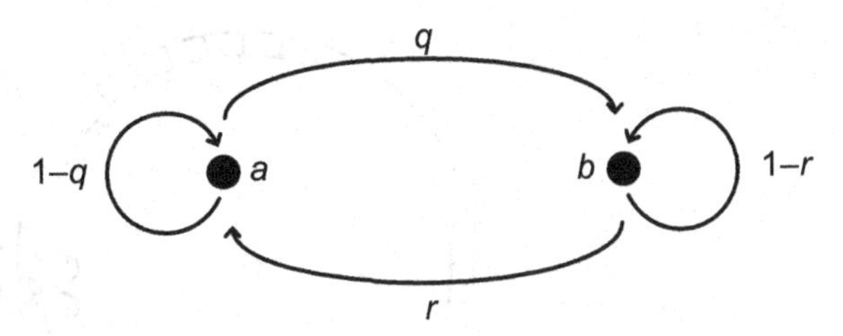

Fig. 1.8 Diagram of allowed transitions for a two-state Markov chain.

Definition 1.3 A Markov chain is irreducible when all the states are accessible from any other state.

Definition 1.4 A Markov chain is periodic when the return times T_j on a state s_j are all (integer) multiples of a period T, i.e. $(W^T)_{jj} > 0$.

1.5.2 Useful Examples of Markov Chains

The Two-State Case

We consider a Markov chain made of two states, $S = \{a, b\}$. The stochastic matrix has the form

$$W = \begin{pmatrix} 1-q & r \\ q & 1-r \end{pmatrix}, \qquad (1.88)$$

because condition (1.77) implies that the sum of elements of each column is equal to one. Accordingly, q is the probability rate per unit time of passing from a to b and $(1-q)$ of remaining in a, while r is the probability rate per unit time of passing from b to a and $(1-r)$ of remaining in b (see Fig. 1.8).

Let us denote with $p_a(t)$ the probability of observing the system in state a at time t: relation (1.75) yields $p_b(t) = 1 - p_a(t)$ at any t. By applying the stochastic evolution rule (1.78) one obtains the equation

$$p_a(t+1) = (1-q)p_a(t) + r(1-p_a(t)) = r + (1-r-q)p_a(t). \qquad (1.89)$$

This equation implies also the similar equation for $p_b(t)$, which can be obtained from (1.89) by exchanging $p_a(t)$ with $p_b(t)$ and r with q. By simple algebra[9] one can check that the explicit solution of (1.89) is

$$p_a(t) = \alpha + (1-r-q)^t\,(p_a(0) - \alpha), \qquad \alpha = \frac{r}{r+q}, \qquad (1.90)$$

where $p_a(0)$ is the initial condition, i.e. the probability of observing the state a at time 0.

There are two limiting cases: (i) $r = q = 0$, no dynamics occurs; (ii) $r = q = 1$, dynamics oscillates forever between state a and state b. In all other cases, $|1-r-q| < 1$

[9] If we start iterating Eq. (1.89), we find $p_a(1) = r + \beta p_a(0)$, $p_a(2) = r(1+\beta) + \beta^2 p_a(0)$, $p_a(3) = r(1+\beta+\beta^2) + \beta^3 p_a(0), \ldots$, with $\beta = 1-r-q$. We can make the ansatz $p_a(t) = r\sum_{\tau=0}^{t-1}\beta^\tau + \beta^t p_a(0)$. Evaluating the summation explicitly, which is equal to $(1-\beta^t)/(1-\beta)$, we find the expression given in Eq. (1.90).

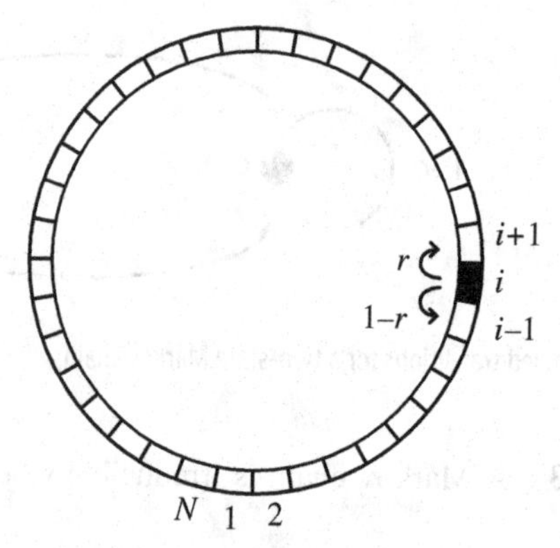

Fig. 1.9 Anisotropic diffusion on a ring, i.e., a linear segment of N sites with periodic boundary conditions.

and in the limit $t \to \infty$, $p_a \to \alpha$ and $p_b \to (1-\alpha)$. More precisely, $p_a(t)$ converges exponentially fast to α with rate $\tau = -1/\ln|r+q-1|$, i.e. the dynamics approaches exponentially fast a stationary state.[10] This simple Markov chain is irreducible and its states are accessible and persistent.

Notice that also the following relation holds

$$W_{ba}p_a(\infty) = W_{ab}p_b(\infty). \tag{1.91}$$

As we are going to discuss in more generality in Section 1.5.4, Eq. (1.91) is the detailed balance condition that establishes a sort of time reversibility of the stochastic process. Actually, Eq. (1.91) tells us that, in the stationary state, the probability of being in state a and passing in a unit time step to state b is equal to the probability of being in b and passing in a unit time step to state a.

Random Walk on a Ring

The state space is the collection of the nodes of a one-dimensional lattice, i.e. $S = \{1, 2, \dots, i, \dots, N\}$, with periodic boundary conditions: a random walker moves along the ring by jumping in a unit time step from site i to site $i+1$ with rate r or to site $i-1$ with rate $1-r$, for any i (see Fig. 1.9). This Markov chain is described by the $N \times N$ tridiagonal stochastic matrix

$$W = \begin{pmatrix} 0 & 1-r & 0 & 0 & 0 & \cdots & 0 & 0 & r \\ r & 0 & 1-r & 0 & 0 & \cdots & 0 & 0 & 0 \\ 0 & r & 0 & 1-r & 0 & \cdots & 0 & 0 & 0 \\ 0 & 0 & r & 0 & 1-r & \cdots & 0 & 0 & 0 \\ \vdots & \vdots & \vdots & \vdots & \vdots & \ddots & \vdots & \vdots & \vdots \\ 1-r & 0 & 0 & 0 & 0 & \cdots & 0 & r & 0 \end{pmatrix} \tag{1.92}$$

[10] We are allowed to speak about a stationary state, rather than an asymptotic evolution, because, after a time $t \gg \tau$, we make an exponentially small error in approximating $p_a(t)$ with α.

and Eq. (1.78) reads

$$p_i(t+1) = (1-r)p_{i+1}(t) + rp_{i-1}(t). \tag{1.93}$$

If a stationary state can be eventually attained, all probabilities should be independent of time and (1.93) simplifies to

$$p_i = (1-r)p_{i+1} + rp_{i-1}, \tag{1.94}$$

where we write p_i rather than $p_i(\infty)$. A straightforward solution of this equation is p_i =const, independently of i. Due to (1.75) we finally obtain $p_i = 1/N$. This result could be conjectured also on the basis of intuitive arguments: in the long run the random walker will have lost any memory of its initial state and, since due to the lattice symmetry all sites are equivalent, the stationary state will correspond to an equal probability of visiting any site. As in the previous example, the Markov chain is irreducible and all its states are accessible and persistent.

The spectrum of W provides us with more detailed mathematical information about this problem. Eq. (1.87) reads

$$(1-r)w_{k+1} + rw_{k-1} = \lambda w_k, \tag{1.95}$$

which has the solution $w_k = \omega^k$, where ω should be determined imposing a periodic boundary condition, which requires $w_{N+1} = w_1$, i.e. $\omega^N = 1$. We therefore have N independent solutions,

$$\omega_j = \exp\left(\frac{2\pi\, i}{N} j\right), \qquad j = 0, 1, \ldots, N-1. \tag{1.96}$$

The corresponding eigenvalues λ_j are determined by Eq. (1.95), which rewrites

$$\lambda_j = (1-r)\omega_j + r\omega_j^{-1} = \cos\left(\frac{2\pi}{N} j\right) + i(1-2r)\sin\left(\frac{2\pi}{N} j\right), \quad j = 0, 1, \ldots, N-1. \tag{1.97}$$

Notice that for $r = 1/2$ the eigenvalues λ_j are all real, at variance with the case $r \neq 1/2$, where, apart $\lambda_0 = 1$, they are complex.

The corresponding eigenvectors are

$$\mathbf{w}^{(\lambda_j)} = \frac{1}{\sqrt{N}}(1, \omega_j, \omega_j^2, \ldots, \omega_j^{N-1})^T, \tag{1.98}$$

where we have introduced the normalization factor, $1/\sqrt{N}$, such that $\mathbf{w}^{\mathbf{T}}\mathbf{w} = 1$. For $j = 0$ we have $\lambda_0 = 1$, independently of r, thus showing that the stationary solution is the same for the asymmetric and symmetric random walk on a ring. On the other hand, for $r \neq 1/2$ there is a bias for the walker to move forward ($r > 1/2$) or backward ($r < 1/2$). In fact, the walker moves with average velocity $v = r - (1-r) = 2r - 1$ and the time it takes to visit all lattice sites is $O(N/v)$.[11]

Making reference to the basic model discussed in Section 1.2.2, we can guess that for $r = 1/2$ the walker performs a diffusive motion, and the time it takes to visit all lattice

[11] This is true for v fixed and diverging N, because for N fixed the ballistic time N/v should be compared with the diffusive time, of order N^2.

sites is $O(N^2)$. In this symmetric case some additional considerations are in order. In fact, if N is an even number and at $t = 0$ the random walker is located at some even site $2j$, i.e. $p_{2j}(0) = 1$, at any odd time step, t_o, $p_{2k}(t_o) = 0$, while at any even time step, t_e, $p_{2k+1}(t_e) = 0$. A similar statement holds if the random walker at $t = 0$ is on an odd site, by exchanging odd with even times and vice versa. This does not occur if N is odd and in this case $\lim_{t\to+\infty} p_i(t) = 1/N$. Conversely, if N is even this limit does not exist, but the following limit exists

$$\lim_{t\to+\infty} \frac{1}{t}\sum_{\tau=1}^{t} p_i(\tau) = \frac{1}{N}. \tag{1.99}$$

This is the ergodic average of $p_i(t)$, that, by this definition, recovers the expected value of the stationary probability. As a final remark, we want to observe that for even N and $r = 1/2$ we have $\lambda_{N/2} = -1$, whose eigenvector is made by an alternation of 1 and -1 (see Eq. (1.98)). The existence of this peculiar eigenvalue and eigenvector can be related to the nonexistence of the limit $\lim_{t\to+\infty} p_i(t) = 1/N$.

Random Walk with Absorbing Barriers

The problem of the random walk with absorbing barriers has been widely investigated, because of its relevance for game theory and in several diffusion problems occurring in material physics. The random walker has an absorbing barrier if there is a state i such that in the corresponding Markov chain $W_{ii} = 1$ and, accordingly, $W_{ji} = 0$ for any $j \neq i$. We can sketch the problem as follows. A random walker moves on a lattice of N sites with fixed ends at states 1 and N, which are absorbing barriers (see Fig. 1.10). The stochastic dynamics on the lattice, apart at sites 1 and N, is the same as that of the random walker on a ring discussed in the previous example. The corresponding stochastic matrix has the same form of (1.92), apart from the first and the last columns, which change into $(1, 0, \ldots, 0, 0)$ and $(0, 0, \ldots, 0, 1)$, respectively, thus yielding

$$W = \begin{pmatrix} 1 & 0 & 0 & 0 & 0 & \cdots & 0 & 0 & 0 \\ 0 & 0 & 1-r & 0 & 0 & \cdots & 0 & 0 & 0 \\ 0 & r & 0 & 1-r & 0 & \cdots & 0 & 0 & 0 \\ 0 & 0 & r & 0 & 1-r & \cdots & 0 & 0 & 0 \\ \vdots & \vdots & \vdots & \vdots & \vdots & \ddots & \vdots & \vdots & \vdots \\ 0 & 0 & 0 & 0 & 0 & \cdots & 0 & 0 & 1 \end{pmatrix}. \tag{1.100}$$

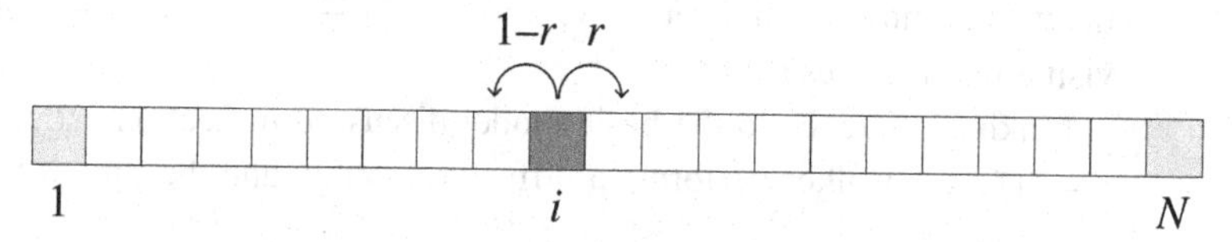

Fig. 1.10 Anisotropic diffusion on a segment with absorbing boundary conditions in $i = 1$ and $i = N$. In the main text we also use the quantity $s = (1 - r)/r$ as a parameter to characterize the asymmetry.

Let us assume that at time 0 the walker is at site j ($1 < j < N$): due to the structure of the stochastic matrix we can state that the walker will be eventually absorbed (with probability 1) either at 1 or N. We can conclude that the Markov chain is irreducible, but, apart from the persistent states 1 and N, all the other states are transient.

We can also ask the more interesting question: which is the probability that a random walker starting at j will reach 1 without being first absorbed in N? In order to answer this question we can introduce the time-independent conditional probability of being trapped at 1 starting at time 0 from site j, $\mathfrak{p}_j$. It obeys the equation ($j = 2, \ldots, N-1$)

$$\begin{aligned}\mathfrak{p}_j &= W_{j+1,j}\ \mathfrak{p}_{j+1} + W_{j-1,j}\ \mathfrak{p}_{j-1}\\ &= r\ \mathfrak{p}_{j+1} + (1-r)\ \mathfrak{p}_{j-1},\end{aligned} \tag{1.101}$$

which exhibits the same formal structure of (1.94), but with boundary conditions $\mathfrak{p}_1 = 1$ and $\mathfrak{p}_N = 0$. The first equality in (1.101) stems from the basic rule that the joint probability of independent events is the product of the probabilities of each single event: actually, the first (second) term on the right-hand side is the product of the probability rate of passing from j to $j+1$ ($j-1$) in a unit time step with the probability of being first absorbed in 1 starting from $j+1$ ($j-1$).

In order to solve Eq. (1.101) we cannot use the same solution of (1.94), because of the presence of the absorbing barriers that impose the boundary conditions $\mathfrak{p}_1 = 1$ and $\mathfrak{p}_N = 0$. However, we use the same ansatz $\mathfrak{p}_j = \psi^j$. By substituting into (1.101) one obtains

$$\psi^j = r\psi^{j+1} + (1-r)\psi^{j-1}, \tag{1.102}$$

or $1 = r\psi + (1-r)\psi^{-1}$, whose solutions are $\psi_1 = 1$, i.e. $\mathfrak{p}_j^{(1)} = 1$ and $\psi_2 = (1-r)/r \equiv s$, i.e. $\mathfrak{p}_j^{(2)} = s^j$. These equations coincide for $r = 1/2$, i.e. $s = 1$, in which case (1.101) reads

$$\mathfrak{p}_j = \frac{1}{2}(\mathfrak{p}_{j+1} + \mathfrak{p}_{j-1}). \tag{1.103}$$

The solution $\mathfrak{p}_j^{(1)} = 1$ still holds, but the second solution now reads $\mathfrak{p}_j^{(2)} = j$, as one can immediately check.

Since Eq. (1.101) is a finite-difference linear equation in the variable $\mathfrak{p}_j$, its general solution has to be a linear combination of $\mathfrak{p}_j^{(1)}$ and $\mathfrak{p}_j^{(2)}$. For $r \neq 1/2$ we have

$$\mathfrak{p}_j = As^j + B \tag{1.104}$$

with A and B constants to be determined by the boundary conditions,

$$\mathfrak{p}_1 = As + B = 1 \tag{1.105}$$

$$\mathfrak{p}_N = As^N + B = 0, \tag{1.106}$$

whose solution is

$$A = \frac{1}{s(1-s^{N-1})}, \qquad B = -\frac{s^{N-1}}{(1-s^{N-1})}. \tag{1.107}$$

Replacing in Eq. (1.104) we obtain

$$\mathfrak{p}_j = \frac{s^{j-1} - s^{N-1}}{1 - s^{N-1}}. \tag{1.108}$$

For $r = 1/2$ ($s = 1$) we have instead

$$\mathfrak{p}_j = Aj + B \tag{1.109}$$

and boundary conditions are $\mathfrak{p}_1 = A + B = 1$ and $\mathfrak{p}_N = AN + B = 0$, i.e.

$$A = -\frac{1}{N-1}, \quad B = \frac{N}{N-1}. \tag{1.110}$$

Replacing in Eq. (1.109) we obtain

$$\mathfrak{p}_j = \frac{N-j}{N-1}. \tag{1.111}$$

Let us now consider the interesting limit $N \to \infty$. For $r > 1/2$ ($s < 1$) one has $A = 1/s$, $B = 0$ and

$$\mathfrak{p}_j = s^{j-1}, \tag{1.112}$$

i.e. the probability of reaching the absorbing barrier at 1 is exponentially small with j, because the random walker has a bias toward the other absorbing barrier at infinity. For $r < 1/2$, $A = 0$, $B = 1$, and $\mathfrak{p}_j = 1$, i.e. the absorbing barrier at 1 will be eventually reached with probability 1, because the random walker has a bias to move toward 1. The same conclusion can be drawn for $r = 1/2$, because, even if this case is unbiased, any site j is at a finite distance from the absorbing barrier at 1. Therefore, we can conclude

$$\mathfrak{p}_j = 1, \qquad r \leq \frac{1}{2} \quad \text{and} \quad N \to \infty. \tag{1.113}$$

This stochastic process describes a basic problem of game theory. Imagine that you are a player with an initial capital j playing a unit of your capital on "red" or "black" for each run at a roulette table. You would like to know what the probability is that you will eventually reach the situation of being left with one single unit of capital (the absorbing barrier at 1) before reaching a capital $N > j$, that may convince you to stop playing (perhaps because your profit is adequate to your expectation). What we have discussed before allows you to conclude that if the game is biased against the player, i.e. you are in the case $r < 1/2$, you have a probability quite close to 1 of being eventually ruined. For instance, we can compute the probability $\mathfrak{p}_j$, with $j = 500$ units of capital, of eventually reaching a capital of a single unit, before having doubled your capital. Despite the unfavorable bias in playing "red" or "black" corresponding to just $r = 18/37$ and $s = 19/18$ (the zero of the roulette determines the bias favorable to the casino), $\mathfrak{p}_j \approx 1 - O(10^{-12})$; i.e. you have a probability $O(10^{-12})$ of reaching the expected capital of 1000 units, before being ruined. You can realize that this strategy of playing roulette is certainly time consuming, but practically will lead you to ruin.

1.5.3 Ergodic Markov Chains

An important class of Markov chains are the ergodic ones. Let us consider a Markov chain with a finite state space, i.e. $S = \{s_1, s_2, \ldots, s_N\}$: it is ergodic if it is irreducible, nonperiodic and all states are persistent (see Definitions 1.1–1.4 at the end of Section 1.5.1). The main property of ergodic Markov chains is that they determine a unique invariant, i.e. stationary, probability distribution. This is given by the eigenvector $\mathbf{w}^{(1)}$, which is the solution of the eigenvalue problem with $\lambda = 1$:

$$W\mathbf{w}^{(1)} = \mathbf{w}^{(1)}. \tag{1.114}$$

The spectral property (b), proved in Appendix B, guarantees that such an eigenvector exists. For an ergodic Markov chain it is also unique, because each state is accessible (i.e., the matrix is irreducible) and persistent, i.e. it will be revisited with probability 1 after a finite lapse of time. If these properties do not hold, in general one is faced with several peculiar scenarios, e.g., the reduction of the stochastic matrix into blocks, whose number equals the degeneracy of the eigenvalue $\lambda = 1$.

The stationary probability, $\mathbf{w}^{(1)}$, of an ergodic matrix will be eventually attained exponentially fast on a time scale, τ, independent of the initial conditions, namely

$$\mathbf{p}(t) \approx \mathbf{w}^{(1)} + \mathbf{A}e^{-t/\tau}, \tag{1.115}$$

where $\mathbf{A}$ is a suitable vector with constant components, that sum up to zero to fulfill condition (1.75). In the limit $t \to \infty$, $\mathbf{w}^{(1)}$ is a true dynamical state of the stochastic process, and, accordingly, it has to obey conditions (1.74) and (1.75), i.e. all its components are nonnegative and it is normalized.

Since all states of an ergodic Markov chain are persistent we would like to know the typical return time $\langle T_j \rangle$ to j (i.e. the average value of the return time T_j in the limit $t \to \infty$). The answer to this question is given by the Kac lemma, according to which

$$\langle T_j \rangle = \frac{1}{w_j^{(1)}}. \tag{1.116}$$

A simple argument to explain this result is the following. Let us denote with $T_j^{(n)}$, with $n = 1, 2, \ldots$, the nth return time to j. The total time $\mathcal{T}_M$ needed for the ergodic Markov chain to come back M times to j is given by

$$\mathcal{T}_M = \sum_{n=1}^{M} T_j^{(n)}. \tag{1.117}$$

On the other hand, this means that the fraction of time $\phi_j(\mathcal{T}_M)$ spent by the stochastic process at j in the time $\mathcal{T}_M$ is given by

$$\phi_j(\mathcal{T}_M) = \frac{M}{\mathcal{T}_M}. \tag{1.118}$$

According to the interpretation of $\mathbf{w}^{(1)}$, $\phi_j(\mathcal{T}_M)$ has to converge to $w_j^{(1)}$ in the limit $M \to \infty$ (which is equivalent to the limit $t \to \infty$), and one can write

$$\langle T_j \rangle = \lim_{M \to \infty} \frac{1}{M} \mathcal{T}_M = \lim_{M \to \infty} \frac{1}{\phi_j(\mathcal{T}_M)} = \frac{1}{w_j^{(1)}}. \tag{1.119}$$

This relation points out that ergodicity amounts to the equivalence between ensemble and time averages.

Another important result is that the spectral properties of an ergodic Markov chain determine the time scale of convergence to the stationary probability. In fact, according to the results of Appendix B, the eigenvalues $\lambda^{(j)}$ of the stochastic matrix representing an ergodic Markov chain can be ordered as

$$\lambda^{(1)} = 1 > |\lambda^{(2)}| \geq |\lambda^{(3)}| \geq \cdots \geq |\lambda^{(N)}|. \tag{1.120}$$

Let us come back to Eq. (1.115) to explain its origin. According to the projection theorem of linear algebra, any probability on the state space at time t, $\mathbf{p}(t)$, can be written as a suitable linear combination of the eigenvectors $\mathbf{w}^{(k)}$ of the stochastic matrix, which form an orthonormal basis:[12]

$$\mathbf{p}(t) = \sum_{k=1}^{N} a_k(t)\mathbf{w}^{(k)}, \qquad a_j(t) \in \mathbb{R} \tag{1.121}$$

where $a_k(t) = \mathbf{p}(t) \cdot \mathbf{w}^{(k)}$.

Let us consider the evolution of the ergodic Markov chain from time 0 to time t,

$$\mathbf{p}(t) = W^t\mathbf{p}(0) = W^t \sum_{k=1}^{N} a_k(0)\mathbf{w}^{(k)} = \sum_{k=1}^{N} a_k(0)(\lambda^{(k)})^t\mathbf{w}^{(k)} \equiv \sum_{k=1}^{N} a_k(t)\mathbf{w}^{(k)}. \tag{1.122}$$

Apart from $a_1(0)$, which does not change in time, because $\lambda^{(1)} = 1$, all the other coefficients evolve in time as

$$a_k(t) = a_k(0)(\lambda^{(k)})^t = (\pm)^t a_k(0)e^{-t/\tau_k}, \tag{1.123}$$

where $(\pm)$ is the sign of the eigenvalue $\lambda^{(k)}$ and

$$\tau_k = -\frac{1}{\ln |\lambda^{(k)}|}. \tag{1.124}$$

Therefore, for $k > 1$, $a_k(t)$ eventually vanish exponentially with rate τ_k.

Making use of property (1.120), we can conclude that the overall process of relaxation to equilibrium from a generic initial condition is dominated by the longest time scale, i.e. the one corresponding to the eigenvalue $\lambda^{(2)}$, so that in (1.115) we have

$$\tau = \tau_2 = -\frac{1}{\ln |\lambda^{(2)}|}. \tag{1.125}$$

1.5.4 Master Equation and Detailed Balance

The dynamical rule of the Markov chain, Eq. (1.78), can be rewritten as

$$p_i(t+1) = p_i(t) - p_i(t) + \sum_j p_j(t)W_{ij} = p_i(t) - p_i(t)\sum_j W_{ji} + \sum_j p_j(t)W_{ij}, \tag{1.126}$$

where we have used the condition (1.77). The previous formula can be recast in the form of a master equation,

$$p_i(t+1) - p_i(t) = \sum_{i \neq j} \left(W_{ij}p_j(t) - W_{ji}p_i(t)\right). \tag{1.127}$$

[12] For the sake of simplicity we exclude any degeneracy of the eigenvalues.

This equation tells us that the variation of the probability of being in state s_i in a unit time step can be obtained from the positive contribution of all transition processes from any state s_j to state s_i and from the negative contribution of all transition processes from state s_i to any other state s_j.

This form is particularly useful to define the conditions under which one can obtain a stationary probability, i.e. all p_i are independent of time t: in this case the left-hand side of (1.127) vanishes and the stationarity condition reads

$$\sum_{i \neq j} \left(W_{ij} p_j - W_{ji} p_i \right) = 0, \quad \forall i, \tag{1.128}$$

where the p_js are the components of the stationary probability.

Notice that (1.128) is verified if the following stronger condition holds

$$W_{ij} p_j - W_{ji} p_i = 0, \quad \forall i, j, \tag{1.129}$$

which is called the detailed balance condition. A Markov chain whose stochastic matrix elements obey (1.129) is said to be reversible and it can be shown that it is also ergodic (see Appendix C), with $\mathbf{p} \doteq \mathbf{w}^{(1)}$ representing the so-called equilibrium probability.[13]

We have already discussed in Section 1.5.2 the simple example of a two-state Markov chain obeying detailed balance. As one can immediately realize, detailed balance does not hold in the asymmetric ($r \neq 1/2$) random walk on a ring, because the stationary probability is $w_j^{(1)} = 1/N$ for all j, while $r = W_{j+1,j} \neq W_{j,j+1} = 1 - r$. In this example we are faced with an ergodic Markov chain that is not reversible. Only if the symmetry of the process is restored (i.e. $r = 1/2$) does the detailed balance condition hold and the invariant probability is an equilibrium one, although it is the same for the asymmetric case.

It is important to point out that the detailed balance condition can be used as a benchmark for the absence of an equilibrium probability in an ergodic irreversible Markov chain. Actually, in this case it is enough to find at least a pair of states for which (1.129) does not hold to conclude that the stationary probability $\mathbf{w}^{(1)}$ is not the equilibrium one. Many of the examples discussed in Section 1.5.2 belong to the class of stochastic processes that evolve to a stationary nonequilibrium probability. The difference between equilibrium and nonequilibrium steady states will be reconsidered in Section 3.3.

1.5.5 Monte Carlo Method

The Monte Carlo method is one of the most useful and widely employed applications of stochastic processes. The method aims at solving the problem of the effective statistical sampling of suitable observables by a reversible Markov chain.

More precisely, the problem we are discussing here is the following: how to estimate the equilibrium average of an observable O making use of a suitable, reversible Markov process. As in the rest of Section 1.5, we assume that the configuration space is made

[13] In the case of reversible Markov chains, the stationary probability is more properly called equilibrium probability, because it is equivalent to the concept of thermodynamic equilibrium in statistical mechanics, where the equilibrium probability is determined by the Hamiltonian functional, which engenders a time-reversible dynamics through the Hamilton equations (see Eqs. (2.27a) and (2.27b)).

by N microscopic states, $S = \{s_1, s_2, \ldots, s_{N-1}, s_N\}$, whose equilibrium probabilities are $(p_1, p_2, \ldots, p_N)$. We can write

$$\langle O \rangle_{eq} = \sum_{j=1}^{N} O_j\, p_j \tag{1.130}$$

where $O_j \doteq O(s_j)$ is the value taken by O in state s_j. The problem of computing such an equilibrium average emerges when the number of states N is exceedingly large and we want to avoid sampling O over equilibrium states, whose probabilities may be very small. This is the typical situation of many models of equilibrium statistical mechanics. For instance, the Ising model (see Section 3.2.2), originally introduced to describe the ferromagnetic phase transition, is made by locally interacting binary spin variables, $\sigma = \pm 1$, located at the nodes of a lattice. If the lattice contains L nodes, the total number of states is $N = 2^L$: already for $L = 100$ one has $N \approx 10^{30}$, quite close to an astronomical number, which makes the computation of (1.130) practically unfeasible.[14]

In order to work out the Monte Carlo strategy, we have to assume from the very beginning that we know the equilibrium probabilities $(p_1, p_2, \ldots, p_N)$. In equilibrium statistical mechanics they are given by the Gibbs weight

$$p_i = \frac{e^{-\beta E_i}}{Z}, \tag{1.131}$$

where $\beta = T^{-1}$ is the so-called inverse reduced temperature, E_i is the energy of state s_i, and $Z = \sum_{i=1}^{N} e^{-\beta E_i}$ is the partition function. One could argue that there is something contradictory in the previous statement: in order to know p_i one has also compute Z, which again is a sum over an astronomical number of states.[15] As we are going to show, the Monte Carlo procedure needs only local information, i.e. the ratios of the equilibrium probabilities $p_i/p_j = e^{-\beta(E_i - E_j)}$, which do not depend on Z.

The following step is how to define a reversible Markov chain that has the p_is as equilibrium probabilities. As we pointed out, a reversible Markov chain is also ergodic, i.e. $\langle O \rangle_{eq}$ can be approximated by a sufficiently long trajectory $(s_1, s_2, \ldots, s_n)$ in the state space of the Markov chain:

$$\frac{1}{n} \sum_{t=1}^{n} O(t) \approx \sum_{j=1}^{N} O_j\, p_j. \tag{1.132}$$

The Monte Carlo procedure is effective, because we can obtain a good estimate of $\langle O \rangle_{eq}$ also if $n \ll N$. In fact, because of ergodicity, we know (see Section 1.5.3) that the fraction of time spent by the stochastic process at state s_i is p_i for time going to infinity: the states s_k that are preferably visited by the trajectory of a stochastic process of finite length n are typically those with $p_k > 1/N$. Accordingly, the Monte Carlo method selects automatically,

[14] The value $L = 100$ is a very small number: in $d = 3$, a cube of only five nodes per side has more than 100 nodes!

[15] In statistical mechanics there are few exactly solvable models, e.g., the Ising model in $d = 1, 2$, where Z can be computed analytically in the thermodynamic limit $N \to \infty$.

after some transient time depending on the initial state, the states corresponding to higher values of the equilibrium probability.

The main point is to estimate how the quality of the approximation depends on n. In fact, any practical implementation of the Monte Carlo procedure into a numerical algorithm is subject to various problems. For instance, the rule adopted to construct the "trajectory" of the stochastic process typically maintains some correlations among the sampled values $O(1), O(2), \ldots$. In particular, we can estimate the correlation time of the Monte Carlo process by measuring how the time autocorrelation function of the observable $O(t)$ decays in time, namely $\langle O(t)O(0)\rangle - \langle O(t)\rangle^2 \sim \exp(-t/\tau)$. If the process lasts over a time n we can assume that the number of statistically independent samples is of the order n/τ. According to the central limit theorem (see Appendix A), the error we make in the approximation (1.132) is $O(\sqrt{\tau/n})$: in practice, the Monte Carlo procedure is effective only if $n \gg \tau$.

Now, let us describe how the reversible Markov chain we are looking for can be explicitly defined for the Ising model, according to the most popular algorithm, called Metropolis. The steps of the algorithm are as follows:

1. We select the initial state by sampling a random spin configuration on the lattice.
2. We select at random with uniform probability $1/L$ a spin variable, say the one at node n_k, and compute its local interaction energy with the nearby spins, E_{n_k}.
3. We flip the spin variable, i.e. $\sigma_{n_k} \to -\sigma_{n_k}$, and we compute its local interaction energy with the nearby spins in this new configuration, E'_{n_k}.
4. If $E'_{n_k} < E_{n_k}$, the next state in the Markov process is the flipped configuration.
5. If $E'_{n_k} \geq E_{n_k}$, we "accept" the new configuration with probability
$$p^* = \exp\left(-\beta(E'_{n_k} - E_{n_k})\right).$$
This means we extract a random number r, uniformly distributed in the interval $[0, 1]$: if $r < p^*$, the next state is the flipped configuration; otherwise, it is equal to the old state.
6. We iterate the process starting again from step 2.

In general terms, in equilibrium statistical mechanics the detailed balance condition is satisfied if

$$W_{ji}e^{-\beta E_i} = W_{ij}e^{-\beta E_j}, \tag{1.133}$$

where we have made explicit the equilibrium probability of occupancy of a microscopic state, given by the Gibbs weight. Therefore, the transition rates must satisfy the relation

$$\frac{W_{ji}}{W_{ij}} = e^{-\beta(E_j - E_i)}. \tag{1.134}$$

The Metropolis algorithm is defined by

$$W_{ji} = \min\left\{1, e^{-\beta(E_j - E_i)}\right\}, \tag{1.135}$$

therefore it satisfies Eq. (1.134).

1.6 Continuous Time Stochastic Processes

The stochastic processes described in Section 1.5 are defined on a discrete state space and in discrete time. On the other hand, many physical processes are better represented in continuous space and time. For instance, if we want to describe a fluid it is simpler to attribute continuous position $\mathbf{x}(t)$ and momentum $\mathbf{p}(t)$ coordinates to each fluid element (i.e. an infinitesimal portion of the fluid), rather than considering the same quantities for each particle in the fluid. A continuous time variable, on its side, allows us to take advantage of the powerful machinery of differential equations. By the way, we should keep in mind that in the previous sections we have first introduced the basic, discrete model of a diffusive random walk (Section 1.2.2), then we have passed to the continuous formulation of Brownian motion in terms of the Langevin equation (Section 1.4.1) and of the Fokker–Planck equation (Section 1.4.2). Here below we provide a continuous time version of stochastic processes, while keeping the discrete nature of the state space.

As a first step we derive the time continuous version of the master equation (1.127) making use of the Chapman–Kolmogorov relation (1.86) in continuous time t,[16]

$$W_{ij}^{t+\Delta t} = \sum_k W_{ik}^{\Delta t} W_{kj}^{t}. \tag{1.136}$$

If Δt is an infinitesimal increment of time, we can imagine that $W_{ik}^{\Delta t}$ vanishes with Δt if $i \neq k$ and that $W_{ii}^{\Delta t} \to 1$ in the same limit. We can therefore write, up to terms $O(\Delta t^2)$,

$$W_{ik}^{\Delta t} = \begin{cases} \mathcal{R}_{ik}\Delta t, & k \neq i \\ 1 - \mathcal{R}_{ii}\Delta t, & k = i. \end{cases} \tag{1.137}$$

The rates $\mathcal{R}_{ik}$ are not independent, because the normalization condition (1.77),

$$1 = \sum_i W_{ik}^{\Delta t} = 1 + \Delta t \left(-\mathcal{R}_{kk} + \sum_{i \neq k} \mathcal{R}_{ik} \right), \tag{1.138}$$

implies that

$$\mathcal{R}_{kk} = \sum_{i \neq k} \mathcal{R}_{ik}. \tag{1.139}$$

By substituting (1.137) into (1.136) and making use of (1.139) we can write

$$\begin{aligned} W_{ij}^{t+\Delta t} - W_{ij}^{t} &= \Delta t \left(\sum_{k \neq i} W_{kj}^{t} \mathcal{R}_{ik} - W_{ij}^{t} \mathcal{R}_{ii} \right) \\ &= \Delta t \left(\sum_{k \neq i} W_{kj}^{t} \mathcal{R}_{ik} - \sum_{k \neq i} W_{ij}^{t} \mathcal{R}_{ki} \right). \end{aligned} \tag{1.140}$$

[16] With respect to the case of discrete time, here W^t is not the elementary stochastic matrix raised to the power t. Rather, W^t should be understood as a transition matrix depending on the continuous parameter t.

By dividing both members by Δt and performing the limit $\Delta t \to 0$ we obtain the continuous time master equation

$$\frac{dW^t_{ij}}{dt} = \sum_{k \neq i} (W^t_{kj}\mathcal{R}_{ik} - W^t_{ij}\mathcal{R}_{ki}). \tag{1.141}$$

We can now write it in terms of the probability $p_i(t)$ the system is in state i at time t, using the relation between $p_i(t)$ and W^t_{ij},

$$p_i(t) = \sum_j W^t_{ij} p_j(0). \tag{1.142}$$

In fact, it is sufficient to derive Eq. (1.142) with respect to time,

$$\frac{dp_i(t)}{dt} = \sum_j \frac{dW^t_{ij}}{dt} p_j(0) \tag{1.143}$$

and replacing (1.141) here above, we obtain

$$\frac{dp_i(t)}{dt} = \sum_j \sum_{k \neq i} (\mathcal{R}_{ik} W^t_{kj} - \mathcal{R}_{ki} W^t_{ij}) p_j(0). \tag{1.144}$$

Finally, we can write the continuous time master equation

$$\frac{dp_i(t)}{dt} = \sum_{k \neq i} (\mathcal{R}_{ik} p_k(t) - \mathcal{R}_{ki} p_i(t)). \tag{1.145}$$

Similar to its time discrete version (1.127), this equation tells us that the variation in time of the probability of being in state s_i is obtained from the positive contribution of all transition processes from any state s_k to state s_i and from the negative contribution of all transition processes from state s_i to any other state s_k.

1.6.1 Stochastic Differential Equations

In Section 1.4.1 we introduced the Langevin equation for a Brownian particle (see Eq. (1.42)), on the basis of some heuristic arguments. The main conceptual weakness of this approach is the assumption that the components of the stochastic force, $\eta_i(t)$, acting on the Brownian particle are completely uncorrelated:

$$\langle \eta_i(t)\eta_j(t') \rangle = \Gamma \delta_{i,j} \delta(t - t'), \tag{1.146}$$

where Γ is given by the fluctuation–dissipation relation (1.48). In a physical perspective this is justified by assuming that the components of the stochastic force are independent of each other and their correlation time is practically negligible with respect to the typical time scale, $t_d = m/\tilde{\gamma} = 1/\gamma$, associated with the dynamics of the deterministic part, where m is the mass of the Brownian particle and $\tilde{\gamma}$ is the viscous friction.

Despite this assumption seeming quite plausible in physical phenomena (e.g., Brownian motion) usually modeled by stochastic processes, the mathematical structure of (1.42) is quite weird: the deterministic part is made of differentiable functions with respect to time, while $\eta_i(t)$ is intrinsically discontinuous at any time, due to (1.146). A way out of this problem is the introduction of Wiener processes, which can be considered the continuous time version of a random walk. The basic idea is quite simple: we use the integrated stochastic force to define a new stochastic process that can be made continuous in time,

$$W_i(t) = \int_0^t dt' \eta_i(t'). \tag{1.147}$$

The statistical average introduced in Section 1.4.1 has to be thought of as the average over the realizations of the stochastic process. If $\langle \eta_i(t) \rangle = 0$, then also $\langle W_i(t) \rangle = 0$, while (1.146) yields

$$\langle W_i(t) W_j(t') \rangle = \delta_{ij} \Gamma \min\{t, t'\}. \tag{1.148}$$

In particular,

$$\langle W_i^2(t) \rangle = \Gamma t, \tag{1.149}$$

meaning that the average squared amplitude of the Wiener process diffuses in time.

The Wiener process defined in (1.147) obeys the following properties:

- $W_i(t)$ is a stochastic process continuous in time and with zero average, $\langle W_i(t) \rangle = 0$.
- For any $t_1 < t_2 < t_3$ the increments $(W_i(t_2) - W_i(t_1))$ and $(W_i(t_3) - W_i(t_2))$ are independent quantities, following the same distribution.
- For any $t_1 < t_2$ the probability distribution of the increments $(W_i(t_2) - W_i(t_1))$ is a Gaussian with zero average and variance $\Gamma(t_2 - t_1)$, which is a consequence of the central limit theorem; see Appendix A.

Notice that $W_i(t)$ is not differentiable, meaning that we cannot define its time derivative, but we can define its infinitesimal increment $dW_i(t)$ for an arbitrary small time interval dt, namely,

$$dW_i(t) = W_i(t + dt) - W_i(t) = \int_t^{t+dt} dt' \eta_i(t'), \tag{1.150}$$

which implies $\langle dW_i(t) \rangle = 0$ and $\langle (dW_i(t))^2 \rangle = \Gamma dt$.

In order to simplify the notation, the infinitesimal increment of the Wiener process is usually redefined as

$$dW(t) \to \sqrt{\Gamma} \Omega(t) \sqrt{dt}, \tag{1.151}$$

where $\Omega(t)$ is a stochastic process with zero average and unit variance and $\sqrt{\Gamma}$ is extracted by the stochastic force. This relation is quite useful, because it attributes to the amplitude of the infinitesimal increment of the Wiener process a physical scale, given by the square root of the infinitesimal time increment dt.

With these ingredients we can write the general form of a well-defined stochastic differential equation,

$$dX_i(t) = a_i(X(t), t)dt + b_i(X(t), t)dW_i(t), \quad i = 1, 2 \ldots, n, \tag{1.152}$$

where $X_i(t)$, $a_i(X(t), t)$, and $b_i(X(t), t)$ are the components of generic vectors of functions in $\mathbb{R}^n$: each of these functions is differentiable with respect to its argument. In a physical perspective $a_i(X(t),t)$ is a generalized drift coefficient, while $b_i(X(t),t)$ is related to a generalized diffusion coefficient. For instance, the Langevin equation of the Brownian particle (see Section 1.4.1) is defined in $\mathbb{R}^3$ with $X_i(t) = v_i(t)$ (i.e., the ith component of the velocity of the Brownian particle), while $a_i(X(t),t) = -\gamma v_i(t)$ and $b_i(X(t),t)$ is proportional to $\sqrt{2\gamma T/m}$.

Let us further simplify the notation by overlooking the subscript i

$$dX(t) = a(X(t),t)\,dt + b(X(t),t)\,dW(t), \tag{1.153}$$

and integrate this general stochastic differential equation, continuous in both space and time. On a formal ground this task can be immediately accomplished, writing the solution of (1.153) as

$$X(t) = X(0) + \int_0^t a(X(t'),t')\,dt' + \int_0^t b(X(t'),t')\,dW(t'). \tag{1.154}$$

If we want to extract any useful information from this formal solution we have to perform its statistical average with respect to the Wiener process, as we have done for Brownian motion when passing from Eq. (1.43) to Eqs. (1.44) and (1.45) by averaging with respect to the components of the stochastic force $\eta_i(t)$.

Here, we have to face the further problem that the last integral is not uniquely defined when averaged over the stochastic process. In fact, according to basic mathematical analysis, the integral of a standard (Riemann) function $f(t)$ can be estimated by the Euler approximations

$$\int_0^t f(t')dt' \approx \sum_{i=0}^{N-1} f(t_i)\Delta t_i \approx \sum_{i=0}^{N-1} f(t_{i+1})\Delta t_i \tag{1.155}$$

where the support of t is subdivided into an ordered sequence of $N+1$ sampling times $(t_0 = 0, t_1, t_2 \ldots, t_{N-1}, t_N = t)$, with $\Delta t_i = t_{i+1} - t_i$. For the sake of simplicity let us assume a uniform sampling, i.e. $\Delta t_i = t/N$, $\forall i$. A theorem guarantees that in the limit $\Delta t \to 0$ both Euler approximations in (1.155) converge to the integral. Such a theorem does not hold for a Wiener process. The analogs of the Euler approximations for the stochastic integral are:

$$\mathcal{I}_I = \sum_{i=0}^{N-1} f(t_i)\Delta W(t_i), \tag{1.156}$$

$$\mathcal{I}_S = \sum_{i=0}^{N-1} f(t_{i+1})\Delta W(t_i). \tag{1.157}$$

Equations (1.156) and (1.157) are called the Itô and the Stratonovich[17] discretizations, respectively. For very small $\Delta W(t_i)$ (1.157) can be approximated as

$$\mathcal{I}_S = \sum_{i=0}^{N-1} [f(t_i) + \alpha\Delta W'(t_i) + O(\Delta W'^2(t_i))]\Delta W(t_i), \tag{1.158}$$

[17] After Kiyoshi Itô and Ruslan L. Stratonovich.

where $\Delta W'$ is, in general, a different realization of the Wiener process with respect to ΔW, and $\alpha = \frac{\delta f}{\delta W'}(t_i)$ is the functional derivative of $f(t)$ with respect to W' at t_i (see Appendix L.1).

By performing the limit $\Delta W \to 0$ and averaging over the Wiener process, we obtain

$$\left\langle \lim_{\Delta W \to 0} \mathcal{I}_I \right\rangle = \left\langle \int_0^t f(t')\, dW(t') \right\rangle = 0 \tag{1.159}$$

while, using (1.158),

$$\left\langle \lim_{\Delta W \to 0} \mathcal{I}_S \right\rangle = \left\langle \int_0^t f(t')\, dW(t') \right\rangle + \int_0^t \int_0^t \alpha(t'') \langle dW(t') dW(t'') \rangle = \int_0^t \alpha(t') dt' \tag{1.160}$$

where we have used the relation $\langle dW(t) dW(t') \rangle = \delta(t - t')\, dt$, which is a consequence of Eq. (1.150).

Notice that, at variance with (1.159), the last expression in (1.160) does not vanish: the Itô and the Stratonovich formulations are not equivalent. Anyway, when dealing with models based on stochastic equations, we have to choose and explicitly declare which formulation we are adopting to integrate the equations. If there are no specific reasons, in many applications it is advisable using the Itô formulation, because of property (1.159), which makes calculations simpler. However, when noise is multiplicative rather than additive, the two formulations are known to differ and the Stratonovich one is preferable. A simple (but a bit artificial) way to obtain a model with multiplicative noise is to consider the Langevin equation for the Brownian particle, with the velocity variable v replaced by its kinetic energy $E = mv^2/2$, so that Eq. (1.41) becomes

$$\frac{dE}{dt} = -2\gamma E + \sqrt{\frac{2E}{m}}\, \tilde{\eta}(t). \tag{1.161}$$

In order to verify Eq. (1.49), it is necessary to use the Stratonovich prescription.

1.6.2 General Fokker–Planck Equation

We can use the general stochastic differential equation (1.153) to derive the general Fokker–Planck equation. Let us consider a function $f(X(t), t)$ that is at least twice differentiable with respect to X. We can write its Taylor series expansion as

$$df = \frac{\partial f}{\partial t} dt + \frac{\partial f}{\partial X} dX + \frac{1}{2} \frac{\partial^2 f}{\partial X^2} dX^2 + O(dt^2, dX^3). \tag{1.162}$$

Now we also assume that $X(t)$ obeys the general stochastic differential equation (1.153), which means that $dX = adt + bdW$ (the arguments of the functions $a(X(t), t)$ and $b(X(t), t)$ have been overlooked). By substituting into (1.162) and discarding higher-order terms, we obtain the so-called Itô formula,

$$df = \left(\frac{\partial f}{\partial t} + a \frac{\partial f}{\partial X} + \frac{1}{2} b^2 \frac{\partial^2 f}{\partial X^2} \right) dt + b \frac{\partial f}{\partial X} dW(t), \tag{1.163}$$

where we have assumed $(dW)^2 = dt$. Formally, this is not correct, because $\langle (dW)^2 \rangle = dt$; however, we should consider that, in order to find the solution of (1.163), we have to integrate and average over the Wiener process. Accordingly, the previous assumption is practically correct, because of the averaging operation. Obviously, we cannot replace $dW(t)$ with its average value.

The important result is that this new stochastic equation is formally equivalent to (1.153). We can conclude that any function $f(X(t))$, at least twice differentiable, obeys the same kind of stochastic differential equation obeyed by $X(t)$. The Itô formula is quite useful, because it allows us to transform a stochastic differential equation into another one, whose solution is known by a suitable guess of the function f.

Now, let us assume that $f(t)$ does not depend explicitly on t, i.e. $\frac{\partial f}{\partial t} = 0$; we average (1.163) over the Wiener process and obtain

$$\frac{d}{dt}\langle f(X(t))\rangle = \left\langle a\,\frac{\partial f}{\partial X} + \frac{1}{2}\,b^2\,\frac{\partial^2 f}{\partial X^2}\right\rangle, \tag{1.164}$$

where we have used the condition $\langle dW(t)\rangle = 0$.

In order to specify the averaging operation we introduce explicitly the probability density $P(X, t)$, defined on the state space $\mathcal{S}$ of $X(t)$, where this stochastic variable takes the value X at time t. Given a function $f(X)$, its expectation value is

$$\langle f(X)\rangle = \int_{\mathcal{S}} dX\, P(X,t)\, f(X), \tag{1.165}$$

and we can rewrite (1.164) as

$$\begin{aligned}\frac{d}{dt}\langle f(X)\rangle &= \int_{\mathcal{S}} dX f(X)\frac{\partial\, P(X,t)}{\partial\, t}\\ &= \int_{\mathcal{S}} dX P(X,t) a\frac{\partial f}{\partial X} + \frac{1}{2}\int_{\mathcal{S}} dX P(X,t) b^2\frac{\partial^2 f}{\partial X^2}.\end{aligned} \tag{1.166}$$

Integrating by parts both integrals in the last line we obtain

$$\begin{aligned}\int_{\mathcal{S}} dX f(X)\frac{\partial\, P(X,t)}{\partial\, t} &= -\int_{\mathcal{S}} dX f(X)\,\frac{\partial\,(a(X,t)P(X,t))}{\partial\, X}\\ &+\frac{1}{2}\int_{\mathcal{S}} dX f(X)\,\frac{\partial^2 (b^2(X,t)\,P(X,t))}{\partial\, X^2},\end{aligned} \tag{1.167}$$

where it has been assumed that the probability density $P(X, t)$ vanishes at the boundary of $\mathcal{S}$. The analogous assumption for the Brownian particle is that the distribution function of its position in space vanishes at infinity.

Since (1.167) holds for an arbitrary function $f(X)$, we can write the general Fokker–Planck equation for the probability density $P(X, t)$ as

$$\frac{\partial\, P(X,t)}{\partial\, t} = -\frac{\partial}{\partial\, X}(a(X,t)P(X,t)) + \frac{1}{2}\frac{\partial^2}{\partial\, X^2}(b^2(X,t)\,P(X,t)). \tag{1.168}$$

We point out that the previous equation has been obtained from the stochastic differential equation (1.153), which describes the evolution of each component of a vector of variables (e.g., coordinates) $X_i(t)$; see Eq. (1.152). In practice, in Section 1.6.1 we have assumed that all $X_i(t)$ are independent stochastic processes, so that the general Fokker–Plank equation has the same form for any $P(X_i, t)$. On the other hand, in principle one should deal with the general case, where $a_i(X,t)$ and $b_i(X,t)$ are different for different values of i. By assuming spatial isotropy for these physical quantities, one can overlook any dependence on i in the general Fokker–Plank equation. In practice, with these assumptions, all the components of the vector variable $X_i(t)$ are described by a single general Fokker–Plank equation, where $X(t)$ is a scalar quantity.

In a mathematical perspective, finding an explicit solution of (1.168) depends on the functional forms of the generalized drift coefficient $a(X,t)$ and of the generalized diffusion coefficient[18] $b^2/2$. On the other hand, as discussed in the following section, interesting physical examples correspond to simple forms of these quantities.

1.6.3 Physical Applications of the Fokker–Planck Equation

Stationary Diffusion with Absorbing Barriers

Eq. (1.168) can be expressed in the form of a conservation law, or continuity equation,

$$\frac{\partial P(X,t)}{\partial t} + \frac{\partial J(X,t)}{\partial X} = 0, \tag{1.169}$$

where

$$J(X,t) = a(X,t)P(X,t) - \frac{1}{2}\frac{\partial}{\partial X}(b^2(X,t)\,P(X,t)) \tag{1.170}$$

can be read as a probability current.

Since $X \in \mathbb{R}$, we can consider the interval $I = [X_1, X_2]$ and define the probability, $\mathcal{P}(t)$, that at time t the stochastic process described by (1.169) is in I,

$$\mathcal{P}(t) = \int_I P(X,t)dX. \tag{1.171}$$

By integrating both sides of (1.169) over the interval I one obtains

$$\frac{\partial \mathcal{P}(t)}{\partial t} = J(X_1,t) - J(X_2,t). \tag{1.172}$$

The current is positive if the probability flows from smaller to larger values of X, and various boundary conditions can be imposed on I. For instance, the condition of reflecting barriers, i.e. no flux of probability through the boundaries of I, amounts to $J(X_1,t) = J(X_2,t) = 0$ at any time t. Accordingly, the probability of finding the walker inside I is conserved, as a straightforward consequence of (1.172). Conversely, the condition of absorbing barriers implies that the walker reaching X_1 or X_2 will never come back to I, i.e. $P(X_1,t) = P(X_2,t) = 0$ at any time t. When dealing with stochastic systems defined on a finite interval (a typical situation of numerical simulations), it may be useful to impose periodic boundary conditions that correspond to $P(X_1,t) = P(X_2,t)$ and $J(X_1,t) = J(X_2,t)$.

[18] The mathematical problem of establishing conditions for the existence and the uniqueness of the solution of Eq. (1.168) demands specific assumptions of the space and time dependence of $a(X,t)$ and $b(X,t)$. They must be differentiable functions with respect to their arguments and they must be compatible with the physical assumption that $P(X,t)$ must rapidly vanish in the limit $X \to \pm\infty$ in such a way that the normalization condition $\int_{\mathbb{R}} P(X,t)dX = 1$ holds at any time t.

In this case the probability $\mathcal{P}$ is conserved not because the flux vanishes at the borders, but because the inner and outer fluxes compensate.

Stationary solutions $P^*(X)$ of Eq. (1.169) must fulfill the condition[19]

$$\frac{d}{dX}(a(X)P^*(X)) - \frac{1}{2}\frac{d^2}{dX^2}(b^2(X)P^*(X)) = 0, \tag{1.173}$$

where we have to assume that also the functions $a(X)$ and $b(X)$ do not depend on t. For instance, in the purely diffusive case, i.e. $a(X) = 0$, $b^2(X) = 2D$, the general solution of (1.173) is

$$P^*(X) = C_1 X + C_2, \tag{1.174}$$

where C_1 and C_2 are constants to be determined by boundary and normalization conditions.

In the case of reflecting barriers at X_1 and X_2 in the interval I, there is no flux through the boundaries, so that the stationary solution must correspond to no flux in I, i.e.,

$$J^*(X) = -D\frac{dP^*(X)}{dX} = 0. \tag{1.175}$$

This condition implies $C_1 = 0$ and $P^*(X) = C_2 = |X_2 - X_1|^{-1}$, where the last expression is a straightforward consequence of the normalization condition $\int_{\mathbb{R}} P^*(X)dX = 1$. We can conclude that in this case the stationary probability of finding a walker in I is a constant, given by the inverse of the length of I.

In the case of absorbing barriers we obtain the trivial solution $P^*(X) = 0$ for $X \in I$, because for $t \to \infty$ the walker will eventually reach one of the absorbing barriers and disappear.

A more interesting case with absorbing barriers can be analyzed by assuming that at each time unit a new walker starts at site $X_0 \in I$; i.e. we consider a stationary situation where a constant flux of walkers is injected in I at X_0 (see Fig. 1.11). Due to the presence of absorbing barriers, the stationary solution of (1.173) with a source generating walkers at unit rate in X_0 is

$$P^*(X) = \begin{cases} C_1(X - X_1), & \text{for } X < X_0 \\ C_2(X_2 - X), & \text{for } X > X_0, \end{cases} \tag{1.176}$$

which fulfills the condition of absorbing barriers $P^*(X_1) = P^*(X_2) = 0$. Moreover, we have to impose the continuity of $P^*(X)$ at the source point X_0. This condition yields the relation

$$C_1(X_0 - X_1) = C_2(X_2 - X_0). \tag{1.177}$$

Because of the presence of the flux source, Eq. (1.169) should be written, more correctly, as

$$\partial_t P + \partial_x J = F\delta(X - X_0). \tag{1.178}$$

[19] Partial derivatives turn into standard derivatives, because the stationary solution is, by definition, independent of t.

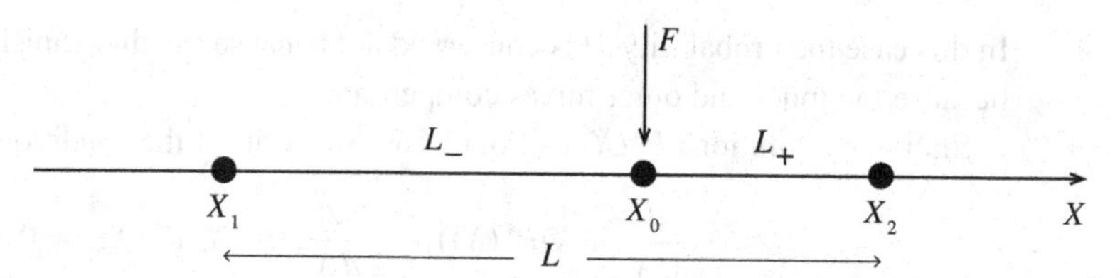

Fig. 1.11 Isotropic diffusion in a segment (X_1, X_2) of length L with absorbing boundaries at the interval extrema and a continuous injection of particles in $X = X_0$. The system relaxes to a steady state, where the stationary distribution of particle density varies linearly in each segment, because $P^*(X)$ is the solution of the equation $d^2P^*/dX^2 = 0$. Absorbing boundaries imply $P^*(X_1) = P^*(X_2) = 0$. In $X = X_0$, $P^*(X)$ is continuous, but its derivative has a jump depending on the flux F (see Eq. 1.178).

Therefore, the stationary solution must satisfy the relation

$$F = \int_{X_0^-}^{X_0^+} dX \partial_x J \tag{1.179}$$

$$= -D \left[\left. \frac{dP^*}{dX} \right|_{X_0^+} - \left. \frac{dP^*}{dX} \right|_{X_0^-} \right] \tag{1.180}$$

$$= D(C_2 + C_1). \tag{1.181}$$

The last equation, along with (1.177) gives

$$C_1 = \frac{F}{D} \frac{L_+}{L}, \qquad C_2 = \frac{F}{D} \frac{L_-}{L}, \tag{1.182}$$

where we have used the notations $L = X_2 - X_1$ for the total length of the interval and $L_- = X_0 - X_1$, $L_+ = X_2 - X_0$, for the left and right parts of the interval; see Fig. 1.11. The probability distributions and the currents are

$$P^*(X) = \begin{cases} \frac{F}{D} \frac{L_+}{L}(X - X_1) \\ \frac{F}{D} \frac{L_-}{L}(X_2 - X), \end{cases} \qquad J^*(X) = \begin{cases} -F\frac{L_+}{L} \equiv -J_-, & \text{for } X < X_0 \\ F\frac{L_-}{L} \equiv J_+, & \text{for } X > X_0, \end{cases} \tag{1.183}$$

where J_- and J_+ are the net fluxes of particles flowing in X_1 and X_2, respectively (and $J_- + J_+ = F$, because of matter conservation). Therefore, a particle deposited in X_0 has a probability

$$\Pi(X_1|X_0) = \frac{J_-}{F} = \frac{L_+}{L} = \frac{X_2 - X_0}{X_2 - X_1} \tag{1.184}$$

to be absorbed in X_1.

The same walker has an average exit time from I, $\mathcal{T}(X_0)_I$, which is equal to the number of walkers, $\mathcal{N}_w$, in the interval I divided by the total current flowing out (equal to F; see above). Since

$$\mathcal{N}_w = \int_{X_1}^{X_2} P^*(X) dX = \frac{F}{2D} L_- L_+, \tag{1.185}$$

we have

$$\mathcal{T}(X_0)_I = \frac{\mathcal{N}_w}{F} = \frac{1}{2D}(X_0 - X_1)(X_2 - X_0). \tag{1.186}$$

The reader can easily realize that we have just reconsidered in the continuous approach of the Fokker–Planck equation the problem of the random walk with absorbing barriers discussed in Section 1.5.2.

A Stochastic Particle Subject to an Elastic Force

We want to describe in the Itô formulation the motion of a stochastic particle in the presence of a conservative force and thermal fluctuations. In the case of an elastic force and of a constant diffusion coefficient, $b(X(t), t) = \sqrt{2D}$, this is known as the Ornstein–Uhlenbeck process,

$$dX(t) = -kX(t)\, dt + \sqrt{2D}\, dW(t), \tag{1.187}$$

where k is the Hook elastic constant divided by the friction coefficient $\tilde{\gamma}$.[20] This equation can be solved by introducing the function $Y(t) = X(t)\, e^{kt}$ that obeys the stochastic equation

$$dY(t) = e^{kt}(dX(t) + kXdt) = \sqrt{2D}\, e^{kt}\, dW(t), \tag{1.188}$$

which can be integrated between time 0 and time t, thus yielding

$$Y(t) = Y(0) + \sqrt{2D} \int_0^t e^{ks}\, dW(s). \tag{1.189}$$

Coming back to the variable $X(t)$ this equation becomes

$$X(t) = X(0)\, e^{-kt} + \sqrt{2D} \int_0^t e^{-k(t-s)}\, dW(s), \tag{1.190}$$

where $X(0)$ is the initial position of the particle. By averaging over the Wiener process we finally obtain

$$\langle X(t) \rangle = X(0)\, e^{-kt} \tag{1.191}$$

and

$$\langle X^2(t) \rangle = \langle X(t) \rangle^2 + 2X(0)e^{-kt}\sqrt{2D} \int_0^t e^{-k(t-s)} \langle dW(s) \rangle \tag{1.192}$$

$$+ 2D \int_0^t e^{-k(t-s)} \int_0^t e^{-k(t-s')} \langle dW(s)dW(s') \rangle. \tag{1.193}$$

Using the relations $\langle dW(s) \rangle = 0$ and $\langle dW(s)dW(s') \rangle = \delta(s - s')ds$, we find

$$\langle X^2(t) \rangle - \langle X(t) \rangle^2 = \frac{D}{k}\left(1 - e^{-2kt}\right). \tag{1.194}$$

[20] The physical interpretation of this equation is quite different from Newtonian mechanics: the presence of the Wiener process on the right-hand side points out that the overall process is the result of a balance among a friction term on the left-hand side (a force proportional to a velocity) and a conservative elastic force in the presence of thermal fluctuations. The inertia term, i.e. the one proportional to the second derivative with respect to time of $X(t)$, is negligible on time scales larger than t_d.

In the limit $t \to +\infty$ the average displacement vanishes, irrespectively of the initial condition, while the variance of the stochastic process converges to D/k. The stochastic particle diffuses around the origin and its average squared displacement is distributed according to a Gaussian with zero average, while its variance is inversely proportional to the Hook constant k and proportional to the amplitude of fluctuations, i.e. to the diffusion constant D. If we want to recover the result for the Brownian particle, we should take the limit $k \to 0$ in Eq. (1.194) before the limit $t \to \infty$, obtaining $\langle X^2(t)\rangle - \langle X(t)\rangle^2 = 2Dt$, as expected.

The Fokker–Planck equation of the Ornstein–Uhlenbeck process is

$$\frac{\partial P(X,t)}{\partial t} = k\frac{\partial}{\partial X}(XP(X,t)) + D\frac{\partial^2}{\partial X^2}P(X,t). \tag{1.195}$$

This equation can be solved via Fourier transformation and then using the method of characteristics. Here we limit to give the solution,

$$P(X,t) = \frac{1}{\sqrt{2\pi\sigma^2}}\exp\left(-\frac{(X - \langle X\rangle)^2}{2\sigma^2}\right) \tag{1.196}$$

with

$$\langle X\rangle = X_0 \exp(-kt) \tag{1.197}$$

$$\sigma^2 = \frac{D}{k}[1 - \exp(-2kt)]. \tag{1.198}$$

As discussed in Appendix D.2, $\langle X\rangle$ is the average value and σ^2 is the variance. In the limit $t \to \infty$, if $k \neq 0$ the two momenta become

$$\langle X\rangle = 0 \tag{1.199}$$

$$\langle X^2\rangle = \sigma^2 = \frac{D}{k}. \tag{1.200}$$

We can conclude that the solution of the Ornstein–Uhlenbeck process is a time-dependent Gaussian distribution, whose center and variance, in the limit $t \to +\infty$, tend to the origin and to a constant, respectively. The asymptotic constancy of σ^2 is the result of the balance between diffusion (which would tend to increase fluctuations) and the elastic restoring force (which suppresses fluctuations).

The Fokker–Planck Equation in the Presence of a Mechanical Force

Let us consider the Fokker–Planck equation for a particle subject to an external mechanical force $\mathbf{F}(\mathbf{x})$ generated by a conservative potential $U(\mathbf{x})$, i.e.,

$$\mathbf{F}(\mathbf{x}) = -\nabla U(\mathbf{x}). \tag{1.201}$$

Making use of the general formalism introduced in this section (see Eqs. (1.169) and (1.170)), we can write

$$\frac{\partial P(\mathbf{x},t)}{\partial t} = -\nabla \cdot \mathbf{J}, \tag{1.202}$$

where

$$\mathbf{J} = \frac{\mathbf{F}}{\tilde{\gamma}} P(\mathbf{x}, t) - D\nabla P(\mathbf{x}, t). \tag{1.203}$$

Here D is the usual diffusion coefficient and the parameter $\tilde{\gamma}$ is the same phenomenological quantity introduced in Section 1.4.1.

Here we avoid reporting the derivation of the solution $P(\mathbf{x}, t)$ of Eq. (1.202) and we limit our analysis to obtaining the explicit expression for the equilibrium probability distribution $P^*(\mathbf{x})$. This problem can be easily solved by considering that it is, by definition, independent of time, and Eq. (1.202) yields the relation

$$\nabla \cdot \mathbf{J} = 0. \tag{1.204}$$

On the other hand, for physical reasons, any (macroscopic) current must vanish at equilibrium, and the only acceptable solution of this equation is $\mathbf{J} = 0$. Under this condition, Eq. (1.203) reduces to

$$\frac{\nabla P^*(\mathbf{x})}{P^*(\mathbf{x})} = -\frac{\nabla U(\mathbf{x})}{\tilde{\gamma} D}, \tag{1.205}$$

or, equivalently, to

$$\nabla \ln P^* = -\frac{\nabla U}{\tilde{\gamma} D}. \tag{1.206}$$

This partial differential equation can be integrated if $U(\mathbf{x})$ diverges (and therefore $P(\mathbf{x})$ vanishes) for large $|\mathbf{x}|$,

$$P^*(\mathbf{x}) = A \exp\left(-\frac{U(\mathbf{x})}{T}\right), \tag{1.207}$$

where A is a suitable normalization constant and we have used the Einstein relation $D = T/\tilde{\gamma}$ (see Eq. (1.38)). As one should have expected on the basis of general arguments, we have obtained the equilibrium Boltzmann distribution, where T is the equilibrium temperature.

1.6.4 A Different Pathway to the Fokker–Planck Equation

In Section 1.6.2 we have derived the Fokker–Planck equation by the Itô formula (1.163) for the stochastic differential equation (1.153). A useful alternative formulation of the Fokker–Planck equation can be obtained from the fully continuous version of the Chapman–Kolmogorov equation (1.136), which is defined by the integral equation[21]

$$W(X_0, t_0|X, t + \Delta t) = \int_{\mathbb{R}} dY\, W(X_0, t_0|Y, t)\, W(Y, t|X, t + \Delta t), \tag{1.208}$$

where the transition probability $W(X, t|X', t')$ from position X at time t to position X' at time t' exhibits a continous dependence on both space and time variables.[22] As in Section 1.6.2

[21] Until now, with a discrete set of states, it has been useful to define W^t_{ij} as the transition rate between state j and state i, so as to use a matrix notation. Now, with a continuous set of states, it is simpler to use the notation given in Eq. (1.208), so that time flows from left to right and it is easier to follow the calculations.

[22] As discussed at the end of Section 1.6.2, we can assume the condition of spatial isotropy, so that Eq. (1.208) holds for any component X_i ($i = 1, \ldots, N$) of the space vector variable $\mathbf{X} \in \mathbb{R}^N$.

we can consider an arbitrary function $f(X(t))$, which has to vanish sufficiently rapidly for $X \to \pm\infty$, in such a way that

$$\int_{\mathbb{R}} dX f(X)\, \mathcal{O} W(X_0, t_0|X, t) < \infty, \tag{1.209}$$

where the operator $\mathcal{O} = \mathbb{I}, \partial_t, \partial_X, \partial_{XX}, \ldots, (\partial_X)^n, \ldots$, the last symbol being a shorthand notation for the nth-order derivative with respect to the variable X. In particular, we can define the partial derivative of $W(X_0, t_0|X, t)$ with respect to time through the relation

$$\int_{\mathbb{R}} dX f(X) \frac{\partial W(X_0, t_0|X, t)}{\partial t} = \lim_{\Delta t \to 0} \frac{1}{\Delta t} \left\{ \int_{\mathbb{R}} dX f(X) [W(X_0, t_0|X, t + \Delta t) - W(X_0, t_0|X, t)] \right\}. \tag{1.210}$$

Making use of (1.208), we can write the first integral on the right-hand side in the form

$$\int_{\mathbb{R}} dX f(X) W(X_0, t_0|X, t + \Delta t) = \int_{\mathbb{R}} dX f(X) \int_{\mathbb{R}} dY\, W(X_0, t_0|Y, t) W(Y, t|X, t + \Delta t). \tag{1.211}$$

In the limit $\Delta t \to 0$, the transition probability $W(Y, t|X, t + \Delta t)$ vanishes unless X is sufficiently close to Y, so that we can expand $f(X)$ in a Taylor series,

$$f(X) = f(Y) + f'(Y)\,(X - Y) + \frac{1}{2} f''(Y)\,(X - Y)^2 + O(X - Y)^3.$$

By substituting into (1.210) and neglecting higher-order terms, we obtain

$$\begin{aligned} \int_{\mathbb{R}} dX f(X) \frac{\partial W(X_0, t_0|X, t)}{\partial t} = \lim_{\Delta t \to 0} \frac{1}{\Delta t} \Bigg\{ \Bigg[\int_{\mathbb{R}} dX \int_{\mathbb{R}} dY \Big(f(Y) + f'(Y)\,(X - Y) \\ + \frac{1}{2} f''(Y)\,(X - Y)^2 \Big) \times W(X_0, t_0|Y, t) W(Y, t|X, t + \Delta t) \Bigg] \\ - \int_{\mathbb{R}} dX f(X) W(X_0, t_0|X, t) \Bigg\}. \end{aligned} \tag{1.212}$$

Because of the normalization condition

$$\int_{\mathbb{R}} dX\, W(Y, t|X, t') = 1, \tag{1.213}$$

the first term of the Taylor series expansion cancels with the last integral in (1.212), and we can finally rewrite (1.210) as

$$\begin{aligned} \int_{\mathbb{R}} dX f(X) \frac{\partial W(X_0, t_0|X, t)}{\partial t} = \lim_{\Delta t \to 0} \frac{1}{\Delta t} \Bigg\{ \int_{\mathbb{R}} dY \int_{\mathbb{R}} dX \Big[f'(Y)\,(X - Y) + \frac{1}{2} f''(Y)\,(X - Y)^2 \Big] \\ \times W(X_0, t_0|Y, t) W(Y, t|X, t + \Delta t) \Bigg\}. \end{aligned} \tag{1.214}$$

We can now define the quantities

$$a(Y, t) = \lim_{\Delta t \to 0} \frac{1}{\Delta t} \int_{\mathbb{R}} dX (X - Y)\, W(Y, t|X, t + \Delta t) \tag{1.215}$$

$$b^2(Y, t) = \lim_{\Delta t \to 0} \frac{1}{\Delta t} \int_{\mathbb{R}} dX (X - Y)^2\, W(Y, t|X, t + \Delta t) \tag{1.216}$$

and rewrite (1.214) as

$$\int_{\mathbb{R}} dX f(X) \frac{\partial W(X_0, t_0|X, t)}{\partial t} = \int_{\mathbb{R}} dY \left[a(Y,t) f'(Y) + \frac{1}{2} b^2(Y,t) f''(Y) \right] W(X_0, t_0|Y, t). \tag{1.217}$$

The right-hand side of this integral equation can be integrated by parts, and by renaming the integration variable $Y \to X$ we obtain

$$\int_{\mathbb{R}} dX f(X) \left[\frac{\partial W}{\partial t} + \frac{\partial}{\partial X}(a(X,t)W) - \frac{1}{2}\frac{\partial^2}{\partial X^2}(b^2(X,t)W) \right] = 0, \tag{1.218}$$

where $W \equiv W(X_0, t_0|X, t)$. Due to the arbitrariness of $f(X)$, we can finally write the Fokker–Planck equation for the transition probability $W(X_0, t_0|X, t)$,

$$\frac{\partial W(X_0, t_0|X, t)}{\partial t} = -\frac{\partial}{\partial X}(a(X,t) W(X_0, t_0|X, t)) + \frac{1}{2}\frac{\partial^2}{\partial X^2}(b^2(X,t) W(X_0, t_0|X, t)). \tag{1.219}$$

This equation is also known as the forward Kolmogorov equation. The pedagogical way this equation has been derived can be made more precise using the Kramers–Moyal expansion: in Appendix E we illustrate this technique for deriving the backward Kolmogorov equation, which describes the evolution with respect to the initial point, rather than to the final one. The backward equation reads

$$\frac{\partial W(X_0, t_0|X, t)}{\partial t_0} = -a(X_0, t_0)\frac{\partial}{\partial X_0} W(X_0, t_0|X, t) - \frac{1}{2} b^2(X_0, t_0)\frac{\partial^2}{\partial X_0^2} W(X_0, t_0|X, t). \tag{1.220}$$

Notice that this equation is not symmetric to (1.219), thus showing that the forward and backward formulations of the Kolmogorov equation are not straightforwardly related to each other. Eq. (1.220) is useful for some applications, such as the one discussed in the following section.

We want also to point out that (1.219) tells us that the transition probability $W(X_0, t_0|X, t)$ obeys a Fokker–Planck equation that is equivalent to the general Fokker–Planck equation (1.168) for the probability density $P(X, t)$. This is a consequence of the relation

$$P(X, t) = \int_{\mathbb{R}} dY P(Y, t')\, W(Y, t', |X, t), \tag{1.221}$$

which is the space–time continuous version of Eq. (1.78). In fact, by deriving the previous equation with respect to t and using Eq. (1.219), one recovers Eq. (1.168). It is worth stressing that these equivalent formulations of the Fokker–Planck equation hold because we have assumed that $W(X_0, t_0|X, t)$ vanishes in the limit $t \to t_0$ if $|X - X_0|$ remains finite. If this is not true, Eqs. (1.219) and (1.220) are not applicable, while the Chapman–Kolmogorov equation (1.208) still holds.

First Exit Time and the Arrhenius Formula

As a first interesting application of what we have discussed in the previous section we want to reconsider the problem of the escape time from an interval $I = [X_1, X_2]$. At page 42 we considered a symmetric, homogeneous diffusion process with absorbing barriers in $X_{1,2}$.

Here we will consider a more complicated process and different boundary conditions. Let us start by setting the problem in general terms.

We can use the transition probability $W(X_0, t_0|X, t)$ to define the probability $\mathbb{P}_{X_0}(t)$ of being still in I at time t after having started from $X_0 \in I$ at time t_0:

$$\mathbb{P}_{X_0}(t) = \int_{X_1}^{X_2} dY\, W(X_0, t_0|Y, t). \tag{1.222}$$

In other words, $\mathbb{P}_{X_0}(t)$ is the probability that the exit time $\mathcal{T}_I(X_0)$ is larger than t. It should be stressed that Eq. (1.222) is not valid for any boundary conditions, because if the "particle" is allowed to exit the interval and reenter, such relation between $\mathbb{P}_{X_0}(t)$ and $W(X_0, t_0|Y, t)$ is no longer true. In the following we are considering either reflecting boundary conditions or absorbing boundary conditions: in both cases, reentry is not allowed and Eq. (1.222) is valid.

Let us indicate with $\Pi(\mathcal{T})$ the probability density of the first exit time from I, starting at X_0 (to simplify the notation in the argument of $\Pi(\mathcal{T})$ we have overlooked the explicit dependence on I and X_0). By definition we have

$$\mathbb{P}_{X_0}(t) = \int_t^{+\infty} d\tau\, \Pi(\tau) \quad \rightarrow \quad \Pi(t) = -\frac{\partial \mathbb{P}_{X_0}(t)}{\partial t}. \tag{1.223}$$

The average exit time is given by the expression

$$\langle \mathcal{T}_I \rangle = \int_0^{+\infty} d\tau\, \Pi(\tau)\, \tau = -\int_0^{+\infty} d\tau \frac{\partial \mathbb{P}_{X_0}(\tau)}{\partial \tau}\, \tau = \int_0^{+\infty} d\tau \mathbb{P}_{X_0}(\tau), \tag{1.224}$$

where the last expression has been obtained integrating by parts and assuming that $\mathbb{P}_{X_0}(t)$ vanishes sufficiently rapidly for $t \rightarrow +\infty$.

Now, we consider the backward Kolmogorov equation for the transition probability $W(X_0, t_0|Y, t)$ (1.220), specialized to the diffusive case, $b^2(X, t) = 2D$, with a time-independent drift term $a(X)$, namely,

$$\frac{\partial W(X_0, t_0|Y, t)}{\partial t} = a(X_0)\frac{\partial}{\partial X_0} W(X_0, t_0|Y, t) + D\frac{\partial^2}{\partial X_0^2} W(X_0, t_0|Y, t), \tag{1.225}$$

where we have used $\partial_{t_0} W(X_0, t_0|Y, t) = -\partial_t W(X_0, t_0|Y, t)$, because time translational invariance implies that $W(X_0, t_0|Y, t)$ depends on $(t - t_0)$ only.

We can obtain an equation for $\mathbb{P}_{X_0}(t)$ by integrating both sides of (1.225) over Y, varying in the interval I:

$$\frac{\partial \mathbb{P}_{X_0}(t)}{\partial t} = a(X_0)\frac{\partial}{\partial X_0}\mathbb{P}_{X_0}(t) + D\frac{\partial^2}{\partial X_0^2}\mathbb{P}_{X_0}(t). \tag{1.226}$$

In the following we are going to replace the starting point X_0 with X, to make notations less cumbersome. Integrating further both sides of this equation over t and assuming $t_0 = 0$, we finally obtain an equation for $\langle \mathcal{T}_I(X) \rangle$:

$$a(X)\frac{\partial}{\partial X}\langle \mathcal{T}_I(X) \rangle + D\frac{\partial^2}{\partial X^2}\langle \mathcal{T}_I(X) \rangle = -1, \tag{1.227}$$

where the right-hand side comes from the conditions $\mathbb{P}_X(0) = 1$ and $\mathbb{P}_X(+\infty) = 0$.

Eq. (1.227) can be applied to the example discussed at page 43 by setting $a(X) = 0$ (no drift) and $\langle T_I(X_1)\rangle = \langle T_I(X_2)\rangle = 0$ (absorbing boundaries), yielding the solution

$$\langle T_I(X)\rangle = \frac{1}{2D}(X - X_1)(X_2 - X), \tag{1.228}$$

which is the same result reported in Eq. (1.186).

It is possible to obtain the general solution of Eq. (1.227), once we observe that defining

$$\Phi(X) = \exp\left(\frac{1}{D}\int a(X)dX\right), \tag{1.229}$$

such equation can be rewritten as

$$\frac{d}{dX}\left(\Phi(X)\frac{d\langle \mathcal{T}(X)\rangle}{dX}\right) = -\frac{1}{D}\Phi(X), \tag{1.230}$$

which can be integrated twice, first giving

$$\frac{d\langle \mathcal{T}(X)\rangle}{dX} = -\frac{1}{D\,\Phi(X)}\int^X dY\,\Phi(Y) + \frac{C_1}{\Phi(X)}, \tag{1.231}$$

then

$$\langle \mathcal{T}(X)\rangle = C_1\int^X dY\frac{1}{\Phi(Y)} - \frac{1}{D}\int^X dY\frac{1}{\Phi(Y)}\int^Y dZ\Phi(Z) + C_2, \tag{1.232}$$

with the integration constants to be determined using the appropriate boundary conditions in $X = X_{1,2}$.

Let us now suppose absorbing conditions at both extrema, $\langle \mathcal{T}(X_1)\rangle = \langle \mathcal{T}(X_2)\rangle = 0$. The former condition can be implemented by taking the lower limit of the two integrals $\int^X dY\cdots$ equal to X_1, which automatically implies $C_2 = 0$. As for the lower limit of the integral $\int^Y dZ$, it is irrelevant, because its variation leads to redefine the constant C_1, so we can take it equal to X_1 as well. Now, Eq. (1.232) is written as

$$\langle \mathcal{T}(X)\rangle = C_1\int_{X_1}^X dY\frac{1}{\Phi(Y)} - \frac{1}{D}\int_{X_1}^X dY\frac{1}{\Phi(Y)}\int_{X_1}^Y dZ\Phi(Z), \tag{1.233}$$

with the right boundary condition, $\langle \mathcal{T}(X_2)\rangle = 0$, implying

$$C_1\int_{X_1}^{X_2} dY\frac{1}{\Phi(Y)} - \frac{1}{D}\int_{X_1}^{X_2} dY\frac{1}{\Phi(Y)}\int_{X_1}^Y dZ\Phi(Z) = 0. \tag{1.234}$$

We can finally write

$$\langle \mathcal{T}(X)\rangle = \frac{A(X;X_1,X_2) - B(X;X_1,X_2)}{D\int_{X_1}^{X_2}\frac{dY}{\Phi(Y)}}, \tag{1.235}$$

where

$$A(X;X_1,X_2) = \int_{X_1}^{X_2}\frac{dY}{\Phi(Y)}\int_{X_1}^Y dZ\Phi(Z)\int_{X_1}^X\frac{dY'}{\Phi(Y')} \tag{1.236}$$

$$B(X;X_1,X_2) = \int_{X_1}^{X}\frac{dY}{\Phi(Y)}\int_{X_1}^Y dZ\Phi(Z)\int_{X_1}^{X_2}\frac{dY'}{\Phi(Y')}. \tag{1.237}$$

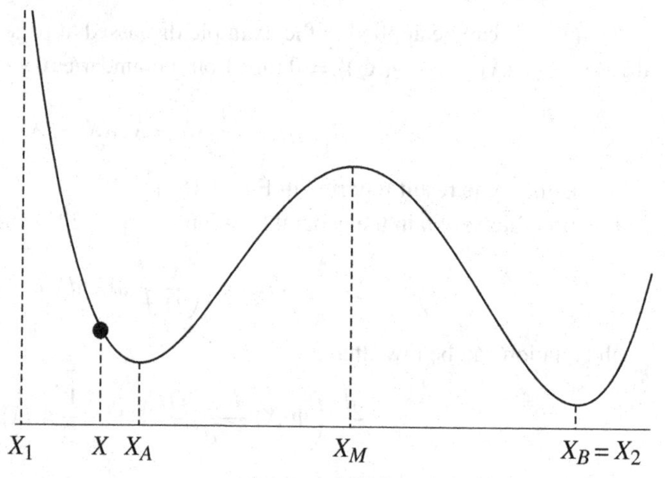

Fig. 1.12 Sketch of the double well potential $U(X)$. The value X_1 can be finite or can diverge to $-\infty$.

We conclude this section with an application of great physical relevance, the Arrhenius formula,[23] which describes the problem of the escape time of a stochastic diffusive process from an asymmetric double-well potential $U(X)$; see Fig. 1.12. In this case the generalized drift term is given by the expression $a(X) = -\frac{1}{\tilde{\gamma}}\frac{dU(X)}{dX}$, where $\tilde{\gamma}$ is the friction coefficient (see Eqs. (1.201)–(1.203)).

We want to ask the following question (see Fig. 1.12 for notations): if a diffusive walker starts at $t = 0$ close to the relative minimum X_A, what is the average time needed to overtake the barrier in X_M and to reach the absolute minimum in X_B? The question can be answered by computing the average time the particle takes to leave the interval $I = [X_1, X_2]$, with a reflecting barrier at X_1 and an absorbing barrier at $X_2 = X_B$. The left boundary condition yields

$$\lim_{X \to X_1^+} \frac{d\langle \mathcal{T}(X)\rangle}{dX} = 0 \tag{1.238}$$

and can be implemented (see Eq. (1.231)) taking $C_1 = 0$ and the lower limit of the integral equal to X_1, i.e.,

$$\frac{d\langle \mathcal{T}(X)\rangle}{dX} = -\frac{1}{D\,\Phi(X)} \int_{X_1}^{X} dY\,\Phi(Y). \tag{1.239}$$

By applying the right boundary condition, $\langle \mathcal{T}(X_B)\rangle = 0$, we obtain the final, explicit result,

$$\langle \mathcal{T}(X)\rangle = \frac{1}{D} \int_{X}^{X_B} dY e^{\frac{U(Y)}{T}} \int_{X_1}^{Y} dZ\, e^{-\frac{U(Z)}{T}}, \tag{1.240}$$

where we have used the Einstein relation to replace $\tilde{\gamma} D$ with T in the exponential functions.

[23] The Arrhenius formula has a great historical importance because it was originally formulated on the basis of heuristic arguments and successfully applied to the study of chemical reactions.

The first integral is dominated by values of the variable Y close to the maximum of $U(X)$ in X_M, while the second integral is weakly dependent on Y for $Y \sim X_M$. Therefore, we can disregard such dependence and approximate the average exit time as follows:

$$\langle \mathcal{T}(X) \rangle \approx \frac{1}{D} \int_{X_1}^{X_M} dZ\, e^{-\frac{U(Z)}{T}} \int_X^{X_B} dY e^{\frac{U(Y)}{T}}. \tag{1.241}$$

The integral on Z is now a constant and is dominated by values of $U(Z)$ close to its minimum X_A, so we can use the Taylor series expansion ($\alpha_1 = U''(X_A)$)

$$U(Z) \simeq U(X_A) + \frac{1}{2}\alpha_1 (Z - X_A)^2 \tag{1.242}$$

and obtain the approximate estimate

$$\begin{aligned} \int_{X_1}^{X_M} dZ\, e^{-\frac{U(Z)}{T}} &\approx e^{-\frac{U(X_A)}{T}} \int_{X_1}^{X_M} dZ\, e^{-\frac{\alpha_1}{2T}(Z-X_A)^2} \\ &\approx e^{-\frac{U(X_A)}{T}} \int_{-\infty}^{+\infty} dZ\, e^{-\frac{\alpha_1}{2T}(Z-X_A)^2} \\ &= \sqrt{\frac{2\pi T}{\alpha_1}}\, e^{-\frac{U(X_A)}{T}}. \end{aligned} \tag{1.243}$$

In the second passage we are allowed to extend the limits of the integral to $\pm\infty$ because we make an exponentially small error.

By applying a similar argument, we can observe that the integral over Y in Eq. (1.241) is dominated by values of Y close to X_M, and we can use the Taylor series expansion

$$U(Y) \simeq U(X_M) - \frac{1}{2}\alpha_2 (Y - X_M)^2, \tag{1.244}$$

with $\alpha_2 = -U''(X_M) > 0$, to obtain the approximate estimate

$$\begin{aligned} \int_X^{X_B} dY e^{\frac{U(Y)}{T}} &\approx e^{\frac{U(X_M)}{T}} \int_X^{X_B} dY e^{-\frac{\alpha_2}{2T}(Y-X_M)^2} \\ &\approx e^{\frac{U(X_M)}{T}} \int_{-\infty}^{+\infty} dY e^{-\frac{\alpha_2}{2T}(Y-X_M)^2} \\ &= \sqrt{\frac{2\pi T}{\alpha_2}}\, e^{\frac{U(X_M)}{T}}. \end{aligned} \tag{1.245}$$

In conclusion, the average exit time from a metastable potential well[24] is essentially independent of the starting point X, assuming it is close to X_A. The Kramers approximation for the exit time can be finally written as

$$\langle \mathcal{T} \rangle = \frac{2\pi}{\sqrt{\alpha_1 \alpha_2}} \frac{T}{D} e^{\frac{\Delta U}{T}}, \tag{1.246}$$

24 The absorbing boundary condition on the right means that on the time scale of the exit process the particle has a vanishing probability to come back. Physically, this means that the new minimum in X_B must be lower than the minimum in X_A.

where $\Delta U = U(X_M) - U(X_A)$, meaning that this quantity essentially depends on the height of the barrier separating X_A from X_B: the higher the barrier, the longer the average exit time from X_A.

The Arrhenius formula is essentially equivalent to (1.246) and provides an estimate of the reaction rate $\mathcal{R}$ of two diffusing chemical species, separated by a potential barrier ΔE, corresponding to the energy involved in the reaction process:

$$\langle \mathcal{T}(X) \rangle^{-1} \sim \mathcal{R} \propto \exp\left(-\frac{\Delta E}{T}\right). \tag{1.247}$$

1.6.5 The Langevin Equation and Detailed Balance

We want to complete this section on continuous time stochastic processes by returning to the Einstein relation (1.49), $\tilde{\Gamma} = 2\tilde{\gamma}T$, and discussing how its validity is equivalent to imposing detailed balance. Let us start with the Langevin equation for a particle in the presence of a conservative potential $U(x)$ (see Eq. (1.57)),

$$m\ddot{x}(t) = -U'(x) - \tilde{\gamma}\dot{x}(t) + \tilde{\eta}(t), \tag{1.248}$$

with $\langle \tilde{\eta}(t) \rangle = 0$ and $\langle \tilde{\eta}(t)\tilde{\eta}(t') \rangle = \tilde{\Gamma}\delta(t-t')$, as usual.

The most delicate point is the correct formulation of detailed balance when the microscopic state is labeled by a quantity, the momentum $p = m\dot{x}$, which changes sign with time reversal. In fact, until now we have tacitly assumed to make reference to spin or lattice gas systems, whose microscopic states are invariant under time reversal. For such systems, the formulation of detailed balance is simply $p_\alpha W_{\beta\alpha} = p_\beta W_{\alpha\beta}$ (see Eq. (1.129)), where $W_{\beta\alpha}$ is the transition rate between state α and state β and p_α is the probability to be in state α. Once we introduce the momentum, we should rather prove that

$$p_\alpha W_{\beta\alpha} = p_{\beta^*} W_{\alpha^*\beta^*}, \tag{1.249}$$

where the asterisk means the time-reversed state. In our case, labels read $\alpha = (x,p)$, $\beta = (x',p')$, $\alpha^* = (x,-p)$, and $\beta^* = (x',-p')$.

For a small time dt, Eq. (1.248) means

$$x(t+dt) = x(t) + \frac{p(t)}{m}dt \tag{1.250}$$

$$p(t+dt) = p(t) - U'(x(t))dt - \tilde{\gamma}\frac{p(t)}{m}dt + dW, \tag{1.251}$$

where dW is the infinitesimal increment of a Wiener process, as defined in Eq. (1.150). In practice, it is a Gaussian distributed, stochastic variable with zero average and variance $\langle (dW)^2 \rangle = \int_0^{dt} d\tau' \int_0^{dt} d\tau'' \langle \eta(\tau')\eta(\tau'') \rangle = \tilde{\Gamma}dt$. Therefore, we can write

$$W_{\beta\alpha} = \delta\left(x' - x - \frac{p}{m}dt\right)\frac{1}{\sqrt{2\pi\tilde{\Gamma}dt}}\exp\left\{-\frac{\left(p' - p + U'(x)dt + \tilde{\gamma}\frac{p}{m}dt\right)^2}{2\tilde{\Gamma}dt}\right\}$$

$$W_{\alpha^*\beta^*} = \delta\left(x - x' + \frac{p'}{m}dt\right)\frac{1}{\sqrt{2\pi\tilde{\Gamma}dt}}\exp\left\{-\frac{\left(p' - p + U'(x')dt - \tilde{\gamma}\frac{p'}{m}dt\right)^2}{2\tilde{\Gamma}dt}\right\}.$$

From Eqs. (1.250) and (1.251) we find that $(x' - x)$ is of order dt and $(p' - p)$ is of order $\sqrt{dt}$, so we can evaluate the ratio

$$\frac{W_{\beta\alpha}}{W_{\alpha^*\beta^*}} = \exp\left\{\frac{\tilde{\gamma}}{\tilde{\Gamma}}\left(\frac{p^2 - p'^2}{m} - \frac{2p}{m}U'(x)dt\right) + o(dt)\right\} \tag{1.252}$$

$$= \exp\left\{\frac{2\tilde{\gamma}}{\tilde{\Gamma}}(E(\alpha) - E(\beta^*)) + o(dt)\right\}, \tag{1.253}$$

where

$$E(\alpha) \equiv E(x,p) = \frac{p^2}{2m} + U(x). \tag{1.254}$$

Eq. (1.253) can be written as the ratio p_{β^*}/p_α, therefore attesting to the validity of the detailed balance, if and only if $2\tilde{\gamma}/\tilde{\Gamma}$ is the inverse absolute temperature, i.e.,

$$\tilde{\Gamma} = 2\tilde{\gamma}T. \tag{1.255}$$

We might repeat the same procedure in the simpler case of overdamped Langevin equation for the Brownian particle, so that the inertial term $m\ddot{x}(t)$ is negligible. Rather than having the two coupled Eqs. (1.250) and (1.251), we simply have

$$x(t+dt) = x(t) - \frac{U'(x(t))}{\tilde{\gamma}} + dW, \tag{1.256}$$

and similar calculations lead to the same result (1.255).

1.7 Generalized Random Walks

In the previous sections we have described diffusive processes ruled by Gaussian probability distributions of observables in the mathematical language of discrete as well as of continuous stochastic processes.

On the other hand, the astonishing richness of natural phenomena encourages us to explore new territories of mathematics of stochastic processes. In fact, there is a broad class of phenomena that escape standard diffusive processes and exhibit quite a different statistical description. As physical examples we can mention turbulent fluid regimes, the dynamics of ions in an optical lattice, the diffusion of light in heterogeneous media, hopping processes of molecules along polymers, etc. Instances from other fields of science are the spreading of epidemics, the foraging behavior of bacteria and animals, the statistics of earthquakes and air traffic, and the evolution of the stock market. This incomplete list is enough to point out the importance of having suitable mathematical tools for describing the statistics of such a widespread class of phenomena that have in common an anomalous diffusive behavior.

The French mathematician Paul Lévy pioneered the statistical description of anomalous diffusive stochastic processes that are usually called Lévy processes; see Fig. 1.13 for an experimental example. Just to fix the ideas from the very beginning, a symmetric

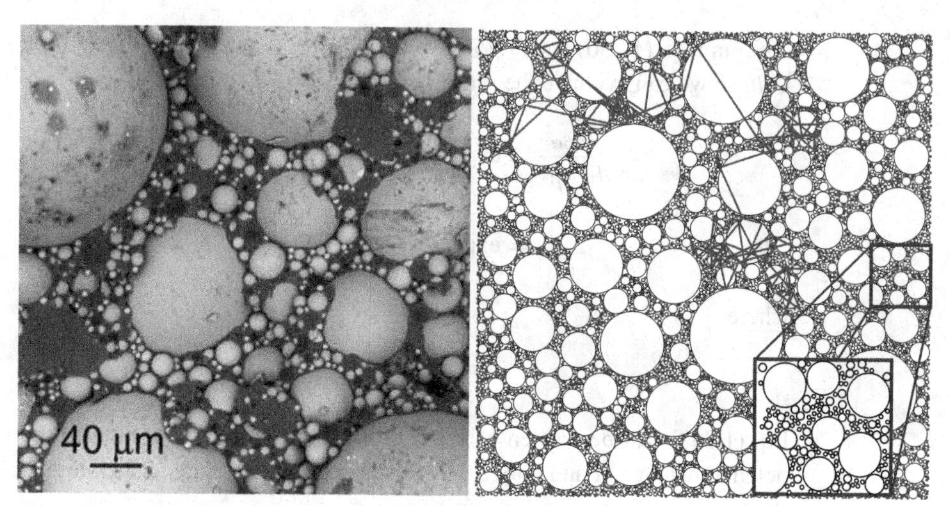

Fig. 1.13 Left: Electron micrograph of a Lévy glass. From M. Burresi et al., Weak localization of light in superdiffusive random systems, *Physical Review Letters*, **108** (2012) 110604. Right: Simulated photon walk in a two-dimensional Lévy glass, with the inset showing the scale invariance of the glass. From P. Barthelemy, J. Bertolotti and D. S. Wiersma, Lévy walk in an inhomogeneous medium, *Nature*, **453** (2008) 495–498.

stochastic process described by the variable $x(t)$ (without prejudice of generality one can take $\langle x(t)\rangle = 0$) is said to be anomalous if its variance obeys the relation

$$\langle x^2(t)\rangle \propto t^\alpha, \quad 0 < \alpha < 2. \tag{1.257}$$

For $\alpha = 1$ we recover the standard diffusive process (therefore, diffusion is not anomalous), while the cases $0 < \alpha < 1$ and $1 < \alpha < 2$ correspond to subdiffusive and superdiffusive stochastic processes, respectively. The case $\alpha = 0$ corresponds, for example, to the Ornstein–Uhlenbeck process (see page 45), while the case $\alpha = 2$ corresponds to ballistic motion.

1.7.1 Continuous Time Random Walk

The examples discussed in Section 1.5.2 describe fully discretized random walks: the walker performs a random sequence of equal-length space steps, each one in a unit time step, i.e. the walker's steps are performed at a velocity with constant modulus. This simple way of representing a random walk can be generalized by assuming that the space steps, x, are continuous, independent, identically distributed random variables, according to a probability distribution $\eta(x)$ (as we have already discussed in Section 1.2.2 for a Poisson distribution of step lengths). Similarly, we can assume that also the time intervals, t, needed to perform a step are continuous, independent, identically distributed random variables, according to a probability distribution $\omega(t)$. This implies that in this continuous time random walk (CTRW) model the walker performs elementary steps with different velocities. On the other hand, the very concept of velocity is still ill-defined, as well as for discrete random walk models, where the macroscopic motion is of diffusive type, despite

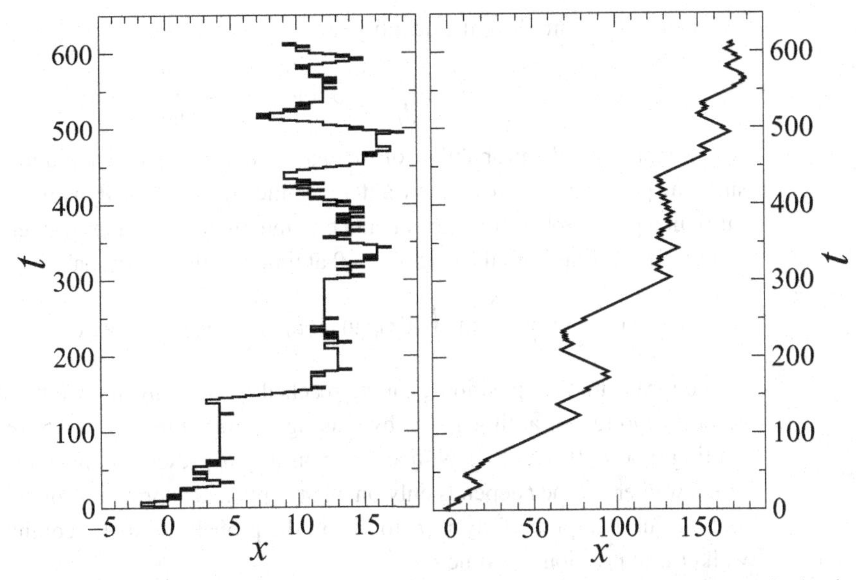

Fig. 1.14 Left: CTRW with step increment ± 1. Right: Lévy walk with velocity ± 1. Both evolutions are generated using the same time distribution $\omega(t) \sim 1/(t^{1+\alpha})$, with $\alpha = 1.5$ and the same distribution of $+/-$ signs for the direction of motion.

the unit step being performed at constant modulus of the velocity. In this perspective, it is more appropriate to interpret the continuous random variable t as a residence time of the walker at a given position in space, before performing a step in a vanishing time; see Fig. 1.14.

Let us now translate the CTRW model into a suitable mathematical language. For the sake of simplicity, we discuss here the one-dimensional case, whose extension to higher dimension does not present additional difficulties. If we assume that the space where the random walker moves is isotropic, we have to assume that a jump of length x is equally probable to a jump of length $-x$; i.e., we have to take into account a normalized symmetric distribution, $\eta(x) = \eta(-x)$ and $\int_{-\infty}^{+\infty} dx\eta(x) = 1$. Notice that, as for the standard Brownian motion, we have

$$\langle x \rangle = \int_{-\infty}^{+\infty} x\eta(x)\mathrm{d}x = 0, \tag{1.258}$$

$$\langle x^2 \rangle = \int_{-\infty}^{+\infty} x^2\eta(x)\mathrm{d}x \neq 0. \tag{1.259}$$

The normalized probability distribution for the residence time $\omega(t)$ is defined for $0 \leq t \leq \infty$, with $\int_0^\infty \omega(t)dt = 1$.[25]

[25] When dealing with probabilities depending on values of physical quantities, such as space lengths or times as in this case, one has to remember that all of these values are pure numbers. They have to be multiplied by suitable physical units to recover their physical meaning. For instance, in this way the normalization conditions for $\eta(x)$ and $\omega(t)$ do not demand that these probability distributions have the dimension of an inverse length and of an inverse time, respectively.

It is useful to introduce the quantity

$$\psi(t) = \int_t^{+\infty} \omega(\tau)d\tau = 1 - \int_0^t \omega(\tau)d\tau, \tag{1.260}$$

which represents the probability of the walker not to make a step until time t (and called survival probability). Now we want to define the random dynamics of the walker by introducing the probability $p(x, t)$ that a continuous-time random walker arrives at position x at time t, starting from the origin $x = 0$ at time $t = 0$. This probability can be written as[26]

$$p(x,t) = \int_{-\infty}^{+\infty} dy \int_0^t d\tau \eta(y)\, \omega(\tau)\, p(x-y, t-\tau) + \delta(x)\, \delta(t), \tag{1.261}$$

which means that the position x can be reached at time t by any walker that arrived at any other position $x - y$ at time $t - \tau$, by making a jump of length y $(-\infty < y < +\infty)$ after a waiting time τ $(0 \le \tau \le t)$. Notice that causality is preserved, meaning that the evolution of the walker at time t depends only on what happened at previous times. We can use $p(x, t)$ and the survival probability $\psi(t)$ to obtain the probability that a continuous-time random walker is at position x at time t,

$$\rho(x,t) = \int_0^t d\tau\, p(x, t-\tau)\psi(\tau). \tag{1.262}$$

In fact, $\rho(x, t)$ is the sum over the time intervals τ during which a walker, which arrived at x at any previous time $t - \tau$, survived until time t.

We can use the Fourier–Laplace transformation to solve the integral equations (1.261) and (1.262). Let us denote with k and s the dual variables of x and t, respectively. Eq. (1.261) is a convolution product for both variables x and t; by Fourier-transforming on x and by Laplace-transforming on t, this equation takes the simple form[27]

$$p(k,s) = \eta(k)\, \omega(s)\, p(k,s) + 1, \tag{1.263}$$

thus yielding

$$p(k,s) = \frac{1}{1 - \eta(k)\, \omega(s)}, \tag{1.264}$$

where

$$\eta(k) = \int_{-\infty}^{+\infty} dx\, e^{-ikx} \eta(x), \tag{1.265}$$

$$\omega(s) = \int_0^{+\infty} dt\, e^{-st} \omega(t). \tag{1.266}$$

Before performing the Fourier–Laplace transform of $\rho(x, t)$, it is useful to evaluate the Laplace transform of the survival probability, $\psi(t)$. Since its derivative is minus the residence time probability (see Eq. (1.260)), we can write

[26] This equation is a generalization of the case of continuous time steps of Eq. (D.36) in Appendix D.4.

[27] In the exact definition of Laplace transform the lower limit of time integration is 0^-, so the integration of $\delta(t)$ gives 1 and not 1/2. Additionally, $p(x, t)$ can be extended to $t < 0$ by assuming $p(x, t) \equiv 0$ at negative times.

$$\omega(s) = \int_0^\infty dt e^{-st} \omega(t) \tag{1.267}$$

$$= - \int_0^\infty dt e^{-st} \psi'(t) \tag{1.268}$$

$$= 1 - s\psi(s), \tag{1.269}$$

where we have integrated by parts and used the condition $\psi(0) = 1$. We can use this result in the form

$$\psi(s) = \frac{1 - \omega(s)}{s} \tag{1.270}$$

to perform the Fourier–Laplace transform of $\rho(x, t)$, thus obtaining (see Eq. (1.262))

$$\rho(k, s) = \frac{1}{1 - \eta(k)\,\omega(s)} \frac{1 - \omega(s)}{s}. \tag{1.271}$$

This is quite a useful equation, since the Fourier–Laplace transform of the density of continuous time random walkers is found to depend in a simple way on $\eta(k)$ and $\omega(s)$. Moreover, the relation between the momenta of the distribution of $\eta(x)$ and the derivatives of its Fourier transform $\eta(k)$, which is reported in Appendix D.4, applies also in this case, where discrete time steps have been substituted with continous time steps, ruled by the probability density function $\omega(t)$. More precisely, by replacing $\eta(k)$ with $\rho(k, t)$, Eq. (D.43) becomes

$$\langle x^m(t) \rangle = i^m \frac{d^m}{d\,k^m} \rho(k, t) \Big|_{k=0}. \tag{1.272}$$

For the Laplace-transformed momenta we can write

$$\langle x^m(s) \rangle = \int_0^{+\infty} dt \langle x^m(t) \rangle\, e^{-s\,t} = i^m \frac{d^m}{d\,k^m} \rho(k, s) \Big|_{k=0}. \tag{1.273}$$

In particular, for $m = 2$, we have

$$\begin{aligned} \langle x^2(s) \rangle &= - \frac{d^2}{d\,k^2} \rho(k, s) \Big|_{k=0} \\ &= - \frac{1 - \omega(s)}{s} \frac{d^2}{d\,k^2} \frac{1}{1 - \eta(k)\,\omega(s)} \Big|_{k=0} \\ &= - \frac{\omega(s)}{s\,[1 - \omega(s)]} \eta''(k = 0) - \frac{2\omega^2(s)}{s\,[1 - \omega(s)]^2} (\eta'(k = 0))^2, \end{aligned} \tag{1.274}$$

where we have used that $\eta(k = 0) = \int_{-\infty}^{+\infty} dx \eta(x) = 1$. The two derivatives appearing in Eq. (1.274) encapsulate relevant information about the distribution $\eta(x)$. More generally, it is interesting to investigate the limits $k \to 0$ and $s \to 0$, which provide information on the asymptotic behaviour of $\rho(x, t)$ at large distances and for long times. In fact, in these limits we can use the approximate expressions

$$\eta(k) \approx \int_{-\infty}^{+\infty} dx \left(1 - ikx - \frac{k^2}{2} x^2 + \cdots \right) \eta(x) = 1 - ik\langle x \rangle - \frac{k^2}{2} \langle x^2 \rangle + \cdots \tag{1.275}$$

$$\omega(s) \approx \int_0^{+\infty} dt\, (1 - st + \cdots) \omega(t) = 1 - \theta\, s + \cdots, \tag{1.276}$$

where we have considered the leading terms of the Taylor series expansion of the exponentials in Eqs. (1.265) and (1.266) and have introduced the quantities

$$\langle x \rangle = \int_{-\infty}^{+\infty} x\eta(x)dx \tag{1.277}$$

$$\langle x^2 \rangle = \int_{-\infty}^{+\infty} x^2\eta(x)dx \tag{1.278}$$

$$\theta = \langle t \rangle = \int_{0}^{+\infty} t\,\omega(t)\,dt. \tag{1.279}$$

For a symmetric distribution of spatial jumps, $\langle x \rangle = 0$ and $\sigma^2 = \langle x^2 \rangle$. Coming back to Eq. (1.274), we can replace $\eta'(k=0) = -i\langle x \rangle = 0$ and $\eta''(k=0) = -\langle x^2 \rangle = -\sigma^2$, obtaining

$$\langle x^2(s) \rangle = \frac{\omega(s)}{s[1-\omega(s)]}\sigma^2. \tag{1.280}$$

If we also replace the $s \to 0$ expansion of $\omega(s)$, we find

$$\langle x^2(s) \rangle = \frac{\sigma^2}{\theta s^2}, \tag{1.281}$$

whose Laplace antitransform is

$$\langle x^2(t) \rangle = \frac{\sigma^2}{\theta} t. \tag{1.282}$$

The last equation is proved observing that $\int_0^\infty dt e^{-st} t = 1/s^2$.

In conclusion, if θ and σ^2 are finite, we obtain standard diffusion, with

$$D = \frac{\sigma^2}{2\theta}. \tag{1.283}$$

This result can also be found by using expansions (1.275) and (1.276) in Eq. (1.271). Then, at leading order we can write $\big(1-\eta(k)\omega(s)\big) \simeq \sigma^2 k^2/2 + \theta s$ and $\big(1-\omega(s)\big) \simeq \theta s$, so that

$$\rho(k,s) \simeq \frac{1}{\left(\theta s + \frac{\sigma^2}{2}k^2\right)} \frac{\theta s}{s} \simeq \left(s + \frac{\sigma^2}{2\theta}k^2\right)^{-1}. \tag{1.284}$$

As shown in Appendix D.3, this is the Fourier–Laplace transformation of the solution of the diffusive equation

$$\frac{\partial \rho(x,t)}{\partial t} = D\frac{\partial^2 \rho(x,t)}{\partial x^2}. \tag{1.285}$$

The CTRW model provides also interesting examples of anomalous diffusion. For instance, let us consider the case where σ^2 is finite, while θ diverges. This means that $\omega(t)$ is dominated by a power law for $t \to +\infty$,

$$\omega(t) \sim \frac{\theta^\alpha}{t^{\alpha+1}}, \tag{1.286}$$

with $0 < \alpha < 1$ and θ^α having the dimension of a fractional time, $[t]^\alpha$, in such a way that ω has the correct dimension of inverse time.

Notice that, even in the presence of such an asymptotic behavior, $\omega(t)$ is a well-defined probability distribution provided it can be normalized, namely if one assumes that $\omega(t)$ is a smooth function that does not diverge for $t \to 0^+$.[28] On the other hand, the average residence time (see Eq. (1.279)) is found to diverge as $t^{1-\alpha}$ in the limit $t \to +\infty$ and the CTRW describes a much slower evolution than standard diffusion, i.e. a subdiffusive process . In this case the Fourier–Laplace transform of $\omega(t)$ reads

$$\omega(s) = 1 - \theta^\alpha s^\alpha. \tag{1.287}$$

This result is a generalization of Eq. (1.276), which is not applicable here because $\alpha < 1$ and $\langle t \rangle$ diverges. However, $\omega(t)$ is still normalizable, so it must be

$$\lim_{s\to 0} \omega(s) = 1, \tag{1.288}$$

which allows us to write

$$\begin{aligned} \omega(s) - \omega(s=0) &= \int_0^\infty dt (e^{-st} - 1)\omega(t) \\ &= \left(\int_0^{1/s} + \int_{1/s}^\infty \right) dt (e^{-st} - 1)\omega(t) \\ &\approx - \int_{1/s}^\infty dt \omega(t) + O(s) \\ &= -\theta^\alpha s^\alpha \int_1^\infty \frac{d\tau}{\tau^{1+\alpha}}, \end{aligned} \tag{1.289}$$

where we have approximated $(e^{-st} - 1)$ with 0 in the first integral $(t < 1/s)$ and with -1 in the second integral $(t > 1/s)$. In conclusion, since the definite integral in the last line is a number and θ^α has just a dimensional role, we obtain Eq. (1.287).

We can now follow the same procedure adopted for the standard diffusive case, so in the limits $k \to 0$ and $s \to 0$ we can use the approximations $1 - \eta(k)\omega(s) \simeq \sigma^2 k^2/2 + (\theta\, s)^\alpha$ and $1 - \omega(s) \simeq (\theta\, s)^\alpha$, thus obtaining at leading order the expression

$$\rho(k,s) \simeq \frac{\theta^\alpha s^{\alpha-1}}{\frac{\sigma^2 k^2}{2} + (\theta s)^\alpha} \simeq \left(s + \frac{\sigma^2 k^2}{2\theta^\alpha} s^{1-\alpha} \right)^{-1}, \tag{1.290}$$

which reduces to Eq. (1.284) for $\alpha = 1$. Using this expression of $\rho(k,s)$ we can first compute the probability of finding the walker at the origin. This is given by the expression

$$\rho(x=0,s) = \frac{1}{2\pi} \int_{-\infty}^{+\infty} dk\, \rho(k,s) \simeq \frac{\theta^\alpha s^{\alpha-1}}{2\pi} \int_{-\infty}^{+\infty} dk \left(k^2 \frac{\sigma^2}{2} + \theta^\alpha s^\alpha \right)^{-1}. \tag{1.291}$$

With the change of variable $y = k\sqrt{\sigma^2/(2\,\theta^\alpha s^\alpha)}$, we can write

$$\rho(x=0,s) \simeq \frac{1}{\pi\sqrt{2}} \frac{\theta^{\alpha/2}}{\sigma\, s^{1-\alpha/2}} \int_{-\infty}^{+\infty} dy \frac{1}{1+y^2} \simeq \frac{1}{\sqrt{2}} \frac{\theta^{\alpha/2}}{\sigma\, s^{1-\alpha/2}}. \tag{1.292}$$

[28] Heuristic arguments suggest that physically interesting processes demand a vanishing probability distribution for a zero residence time of the CTRW.

Since $\rho(x = 0, s)$ does not have a finite limit for $s \to 0$, the line of reasoning to obtain (1.289) cannot be used and a more rigorous approach, making use of the so-called Tauberian theorems, should be adopted.[29] Using such theorems, the result for the probability of finding the walker at the origin is given by

$$\rho(0,t) = \frac{1}{\sqrt{2}} \frac{\theta^{\alpha/2}}{\sigma\, \Gamma(1-\alpha/2)}\, t^{-\alpha/2}. \tag{1.293}$$

We can therefore conclude that this probability decays algebraically in time with an exponent $\alpha/2$, which for normal diffusion is $1/2$. It is important to stress that this expression provides just the asymptotic (i.e., large t) behavior of the probability density function of finding the walker at the origin.

From (1.290) we can compute all momenta of the distribution. Here we just compute the time-dependent average square displacement, using the general formula (1.274), with $\langle x\rangle = 0$, and $\omega(s)$ given by (1.287), thus yielding up to leading order in k

$$\langle x^2(s)\rangle = -\frac{\omega(s)}{s\,[1-\omega(s)]}\sigma^2 \approx \frac{1}{\theta^\alpha\, s^{\alpha+1}}\,\sigma^2. \tag{1.294}$$

By applying the results of the Tauberian theorems we obtain

$$\langle x^2(t)\rangle = \frac{1}{\Gamma(1+\alpha)}\frac{\sigma^2}{\theta^\alpha} t^\alpha = 2\,D_\alpha\, t^\alpha, \tag{1.295}$$

where D_α is a generalized diffusion coefficient.

For the sake of completeness we want to mention that (1.290) can be read as the Fourier–Laplace transform of the following, fractional partial differential equation

$$\frac{\partial^\alpha \rho(x,t)}{\partial t^\alpha} = D_\alpha \frac{\partial^2 \rho(x,t)}{\partial x^2}. \tag{1.296}$$

Although the important topic of fractional partial differential equations cannot be addressed here, we can nevertheless observe on an intuitive ground that (1.295) follows by a simple dimensional analysis of (1.296). In conclusion, Eq. (1.295) is the basic physical relation describing a subdiffusive random walk (i.e., $0 < \alpha < 1$) in continuous space and time.

Another interesting example contained in the CTRW model is the case where θ is finite, while $\eta(x)$ is a normalized symmetric probability density function, whose asymptotic behavior obeys the power law

$$\eta(x) \sim |x|^{-(1+\alpha)}, \tag{1.297}$$

with $0 < \alpha < 2$. These processes are known as Lévy flights. For symmetry reasons the first momentum of $\eta(x)$ is zero, while the second momentum is found to diverge as

[29] The Tauberian theorems allow us to determine the Laplace transform of a function $f(t)$ whose behavior for large t is of the form

$$f(t) \simeq t^{\gamma-1}\,\Phi(t) \quad \text{with} \quad 0 < \gamma < +\infty,$$

where $\Phi(t)$ is a slowly varying function of t (e.g., $\lim_{t\to+\infty}\Phi(t) = c$, where c is a constant, or even $\phi(t) = [\log(t)]^n$). The Tauberian theorems tell us that the Laplace transform of $f(t)$ is

$$f(s) = \Gamma(\gamma)\, s^{-\gamma}\, \Phi(1/s).$$

$$\langle x^2 \rangle = \lim_{R\to\infty} \int^R dx\,|x|^{1-\alpha} \approx \lim_{R\to\infty} R^{2-\alpha}. \tag{1.298}$$

The Fourier transform of $\eta(x)$ can be evaluated with the spirit of Eq. (1.289), using the fact that the distribution probability is normalizable, so $\eta(k=0)=1$:

$$\eta(k=0) - \eta(k) = \int_{-\infty}^{+\infty} dx \left(1 - e^{-ikx}\right) \eta(x) \tag{1.299}$$

$$\approx 2 \int_{1/|k|}^{\infty} dx \frac{1 - \cos(|k|x)}{x^{1+\alpha}} \tag{1.300}$$

$$= 2|k|^\alpha \int_1^\infty dy \frac{1 - \cos y}{y^{1+\alpha}}. \tag{1.301}$$

Finally, we find

$$\eta(k) \sim 1 - (\sigma\,|k|)^\alpha, \tag{1.302}$$

where σ is a suitable physical length scale.

In the limits $k \to 0$ and $s \to 0$, from Eq. (1.271) we obtain the approximate expression

$$\rho(k,s) \simeq \left(s + \frac{\sigma^\alpha}{\theta}|k|^\alpha\right)^{-1}. \tag{1.303}$$

By introducing the quantity $\mathcal{D}_\alpha = \sigma^\alpha/\theta$ we can write

$$\rho(k,t) \simeq e^{-\mathcal{D}_\alpha t\,|k|^\alpha}, \tag{1.304}$$

where $\rho(k,t)$ is the inverse Laplace transform of $\rho(k,s)$ with respect to s, because $\int_0^\infty dt e^{-st} e^{-\mathcal{D}_\alpha t|k|^\alpha} = 1/(s + \mathcal{D}_\alpha |k|^\alpha)$. The probability distribution $\rho(x,t)$ can be obtained by antitransforming $\rho(k,t)$ in the Fourier space of k, but this is not an easy task, because the exponent α in (1.304) is not an integer. Nonetheless, it is possible to derive the fractional momenta of order q of $\rho(x,t)$, as

$$\langle |x(t)|^q \rangle = \int_{-\infty}^{+\infty} \rho(x,t)|x(t)|^q dx \tag{1.305}$$

$$= \frac{1}{2\pi} \int_{-\infty}^{+\infty} dx|x|^q \int_{-\infty}^{+\infty} dk e^{ikx} e^{-\mathcal{D}_\alpha t\,|k|^\alpha} \tag{1.306}$$

$$= (\mathcal{D}_\alpha t)^{q/\alpha} \frac{1}{2\pi} \int_{-\infty}^{+\infty} dy|y|^q \int_{-\infty}^{+\infty} dp e^{ipy} e^{-|p|^\alpha} \tag{1.307}$$

$$\sim (\mathcal{D}_\alpha t)^{q/\alpha}, \tag{1.308}$$

where we have made two changes of variable, $p = (\mathcal{D}_\alpha t)^{1/\alpha} k$ and $y = x/(\mathcal{D}_\alpha t)^{1/\alpha}$, and the final integral on the right is just a number. In particular, for $q = 2$ we have

$$\langle x^2(t) \rangle \sim (\mathcal{D}_\alpha t)^{2/\alpha}. \tag{1.309}$$

As expected, for $\alpha = 2$ we recover the standard diffusive behavior, while for $\alpha < 2$ the mean squared displacement grows in time more than linearly. Notice that $\alpha = 1$ corresponds to a ballistic propagation. As we are going to discuss in the following section, a suitable description of superdiffusion in the CTRW requires consideration of finite velocities by the introduction of the Lévy walk process.

1.7.2 Lévy Walks

Lévy processes are a general class of stochastic Markov processes with independent stationary time and space increments. For example, above we studied Lévy flights, characterized by a narrow distribution of time steps and a distribution of space increments given by Eq. (1.297). The main feature of the latter distribution is that not all of its momenta are finite. As a consequence, it cannot be fully characterized by its first two momenta, i.e. the mean $\langle x \rangle$ and the variance $\langle x^2 \rangle - \langle x \rangle^2$, as is the case for Gaussian probability distributions.

The main advantage of studying Lévy flights is that they are quite simple mathematical models of anomalous diffusion. Their main drawback is that we are implicitly assuming that each step lasts over a fixed time interval, independent of the step length. Since this can be arbitrarily large, the random walker seems to move with an arbitrarily large velocity, a manifest drawback to giving a physical interpretation to such processes. Such physical interpretation can be restored by introducing Lévy walks (LWs), where the distribution of lengths is still given by

$$\eta(x) \sim \frac{1}{|x|^{1+\alpha}}, \tag{1.310}$$

but the path between the starting and the end points of a jump is assumed to be run by the walker at constant velocity $\mathbf{v}$; i.e., the time t spent in a jump is proportional to its length x, with $t = x/|\mathbf{v}|$ (see Fig. 1.14). It can be argued that a LW can be equally defined as a process where a walker moves at constant velocity in a time step t, whose asymptotic distribution is $\omega(t) \sim t^{-(1+\alpha)}$. If $\alpha < 2$ the variance of $\omega(t)$ diverges and the distances of the jumps run by the walker obey the probability distribution (1.310). In the context of LW, one can describe superdiffusive behavior.

In general terms, we can formulate the LW process as a suitable modification of the CTRW model discussed in Section 1.7.1. The probability $p(x, t)$ that a LW arrives at position x at time t, starting from the origin $x = 0$ at time $t = 0$, still obeys an equation like (1.261), namely,

$$p(x,t) = \int_{-\infty}^{+\infty} \mathrm{d}y \int_0^t \mathrm{d}\tau\, \Omega(y,\tau)\, p(x-y, t-\tau) + \delta(x)\,\delta(t), \tag{1.311}$$

where $\Omega(y, \tau)$ is the joint probability distribution of performing a step of length y in a time τ. Since a LW runs at constant velocity v, this quantity can be expressed as

$$\Omega(x,t) = \delta(|x| - vt)\,\omega(t), \tag{1.312}$$

where

$$\omega(t) \sim t^{-(1+\alpha)} \tag{1.313}$$

for large values of t. Notice that this expression admits the possibility that a walker has not yet completed a LW at time t. By Fourier-transforming on x and by Laplace-transforming on t, Eq. (1.311) yields

$$p(k,s) = \frac{1}{1 - \Omega(k,s)}. \tag{1.314}$$

If $\Omega(x,t) = \eta(x)\omega(t)$ as in Eq. (1.261), we have $\Omega(k,s) = \eta(k)\omega(s)$ and we recover Eq. (1.264). In the present case instead, the Fourier–Laplace transform of $\Omega(x,t)$ (see Eq. (1.312)) is

$$\begin{aligned}\Omega(k,s) &= \frac{1}{2}\int_{-\infty}^{+\infty} dx \int_0^{+\infty} dt e^{-ikx-st}\Big(\delta(-x-vt)+\delta(x-vt)\Big)\,\omega(t)\\ &= \frac{1}{2}\left[\int_0^{+\infty} dt e^{-(s-ivk)t}\,\omega(t) + \int_0^{+\infty} dt e^{-(s+ivk)t}\,\omega(t)\right]\\ &= \frac{1}{2}\big[\omega(s-ivk)+\omega(s+ivk)\big] \equiv \mathrm{Re}\,\omega(s+ivk). \qquad (1.315)\end{aligned}$$

For $\alpha < 1$, we can use the result $\omega(s) = 1 - \theta^\alpha s^\alpha$, obtained for the CTRW model (see Eq. (1.287)), to write

$$\Omega(k,s) \simeq 1 - \frac{1}{2}\theta^\alpha[(s-ivk)^\alpha + (s+ivk)^\alpha]. \qquad (1.316)$$

This equation must be handled with care, because we have noninteger powers of binomials. We must first take the limit of large distances ($k \to 0$), then the limit of long times ($s \to 0$); otherwise, asymptotic dynamics are limited by the artificially imposed finite size of the system. Therefore, we obtain

$$\Omega(k,s) \approx 1 - \theta^\alpha s^\alpha - \frac{\theta^\alpha v^2}{2}\alpha(1-\alpha)k^2 s^{\alpha-2}. \qquad (1.317)$$

In analogy with definition (1.262), we want to use these quantities to obtain an expression for the probability that a LW is at position x at time t,

$$\rho(x,t) = \int_{-\infty}^{+\infty} dx' \int d\tau p(x-x', t-\tau)\Psi(x',\tau), \qquad (1.318)$$

where now $\Psi(x',\tau)$ is the probability that a LW moves exactly by a distance x' in a time τ. Thus

$$\Psi(x,t) = \delta(|x|-vt)\int_t^{+\infty}\omega(t')dt', \qquad (1.319)$$

which corresponds to the motion of a walker that proceeds at constant speed v under the condition that no scattering event occurs before time t. The same kind of calculation performed to obtain $\Omega(k,s)$ yields the expression of the Fourier–Laplace transform $\Psi(k,s)$,

$$\Psi(k,s) = \mathrm{Re}\,\psi(s+ivk), \qquad (1.320)$$

where we recall that $\psi(s)$ is the Laplace transform of $\psi(t) = \int_t^\infty \omega(t')dt'$ and it is given by Eq. (1.270), which was obtained for CTRW, but still holds for LW.

By combining these results, the expression of the Fourier–Laplace transform of (1.318) is

$$\rho(k,s) = p(k,s)\,\Psi(k,s) = \frac{\Psi(k,s)}{1-\Omega(k,s)}, \qquad (1.321)$$

where the denominator can be found by Eq. (1.317). Making use of (1.320) and (1.270) the numerator can be evaluated in a similar manner, giving

$$\Psi(k,s) = \theta^\alpha s^{\alpha-1} + \frac{\theta^\alpha v^2}{2}(\alpha-1)(2-\alpha)k^2 s^{\alpha-3}. \tag{1.322}$$

We finally obtain

$$\rho(k,s) \simeq \frac{1}{s}\frac{s^\alpha - c^2 k^2 s^{\alpha-2}}{s^\alpha + b^2 k^2 s^{\alpha-2}}, \tag{1.323}$$

where $b^2 = v^2\alpha(1-\alpha)/2$ and $c^2 = v^2(1-\alpha)(2-\alpha)/2$.

By applying Eq. (1.273) to this result, after some lengthy algebra and then by Laplace antitransforming in the s variable, we finally obtain

$$\langle x^2(t)\rangle = V^2 t^2, \tag{1.324}$$

i.e. a ballistic behavior, where the characteristic spread velocity is given by $V = v\sqrt{(1-\alpha)} < v$. In the special case $\alpha = 1$, V vanishes and logarithmic corrections of the form $t^2/\ln(t)$ are expected for the mean square displacement.

For $1 < \alpha < 2$ we face quite a different situation because $\omega(t) \sim \theta^\alpha/t^{\alpha+1}$ has a finite first moment; i.e., we are dealing with a finite characteristic time,

$$\tau = \langle t\rangle = \int_0^{+\infty} dt\,\omega(t)t < \infty, \tag{1.325}$$

and we can obtain the expression of the Laplace transform of $\omega(t)$ as

$$\begin{aligned}\omega(s) - \omega(s=0) &= \int_0^\infty dt(e^{-st}-1)\omega(t)\\ &= \int_0^\infty dt(-st + \frac{1}{2}s^2t^2 + \cdots)\omega(t)\\ &= -s\tau + \int_0^{+\infty} dt\left(\frac{1}{2}s^2t^2 + \cdots\right)\omega(t),\end{aligned} \tag{1.326}$$

where the last integral still converges in $t = 0$ if we replace the large t expression, $\omega(t) \sim \theta^\alpha/t^{\alpha+1}$. So we obtain

$$\omega(s) \simeq 1 - \tau s + \theta^\alpha \int_0^{+\infty} dt \frac{\left(\frac{1}{2}s^2t^2 + \cdots\right)}{t^{\alpha+1}} \tag{1.327}$$

$$\simeq 1 - \tau s + \theta^\alpha s^\alpha \int_0^{+\infty} d\tau \frac{\left(\frac{1}{2}\tau^2 + \cdots\right)}{\tau^{\alpha+1}} \tag{1.328}$$

$$\simeq 1 - \tau s + As^\alpha, \tag{1.329}$$

where the quantity A is proportional to θ^α.

Substituting this expression into Eq. (1.315) we obtain

$$\begin{aligned}\Omega(k,s) &\simeq 1-\tau s+\frac{1}{2}A[(s-ivk)^{\alpha}+(s+ivk)^{\alpha}]\\ &\simeq 1-\tau s+A\,s^{\alpha}-\frac{A\,v^2}{2}\alpha(\alpha-1)k^2s^{\alpha-2}\\ &\simeq 1-\tau s+\frac{A\,v^2}{2}\alpha(\alpha-1)k^2s^{\alpha-2},\end{aligned}\tag{1.330}$$

where the term s^{α} has been neglected with respect to the linear term s. As for the passage from the first to the second line, we first consider (as before) the limit $k\to 0$, then $s\to 0$. Making use again of (1.320) and (1.270), we can also compute

$$\begin{aligned}\Psi(k,s) &\simeq \tau-\frac{1}{2}A\,[(s-ivk)^{\alpha-1}+(s+ivk)^{\alpha-1}]\\ &\simeq \tau-A\,s^{\alpha-1}+\frac{A\,v^2}{2}(\alpha-1)(\alpha-2)k^2s^{\alpha-3}\\ &\simeq \tau+\frac{A\,v^2}{2}(\alpha-1)(\alpha-2)k^2s^{\alpha-3},\end{aligned}\tag{1.331}$$

where $s^{\alpha-1}$ has been neglected with respect to the constant term. By substituting (1.330) and (1.331) into (1.321) we eventually obtain

$$\rho(k,s)\simeq\frac{1}{s}\frac{\tau-c_1^2\,k^2\,s^{\alpha-3}}{\tau+b_1^2\,k^2\,s^{\alpha-3}},\tag{1.332}$$

where $c_1^2=Av^2(\alpha-1)(2-\alpha)/2$ and $b_1^2=Av^2\alpha(\alpha-1)/2$. By applying Eq. (1.273) to this result, after (again) some lengthy algebra and then by Laplace antitransforming in the s variable (using the Tauberian theorems; see note 29 above), we obtain an expression for the mean square displacement,

$$\langle x^2(t)\rangle\simeq Ct^{3-\alpha},\tag{1.333}$$

where $C\propto A(\alpha-1)/\tau$. In this case we find a superdiffusive behavior that interpolates between the ballistic case ($\alpha=1$) and standard diffusion ($\alpha=2$). Also for $\alpha=2$, logarithmic corrections of the form $t\ln(t)$ have to be expected.

For $\alpha>2$ also terms of order $s^{\alpha-2}$ become negligible and $\rho(k,s)$, at leading order, recovers the form (1.284); i.e., the LW boils down to a standard diffusive process. In fact, for $\alpha>2$ the probability distribution $\omega(t)$ has finite average and variance and the asymptotic behavior of the corresponding LW can be argued to be equivalent to a standard Gaussian process equipped with the same average and variance. Summarizing, we have found that LW can be classified according to the following scheme:

$$\langle x^2(t)\rangle\propto\begin{cases}t, & \text{if } \alpha>2 \quad \text{diffusive}\\ t^{3-\alpha}, & \text{if } 1<\alpha<2 \quad \text{superdiffusive}\\ t^2, & \text{if } 0<\alpha<1 \quad \text{ballistic.}\end{cases}\tag{1.334}$$

Notice that, at variance with the CTRW, the LW model cannot include subdiffusive propagation, because such a regime is incompatible with a finite velocity of the walker.

In this perspective the subdiffusive regime described by the CTRW model is shown to be unsuitable for phenomenological applications.

1.8 Bibliographic Notes

A pedagogical introduction to the Boltzmann transport equation and related topics can be found in the book by K. Huang, *Statistical Mechanics*, 2nd ed. (Wiley, 1987). A more detailed account of the many aspects related to this fundamental equation is contained in the book by C. Cercignani, *The Boltzmann Equation and Its Applications* (Springer, 1988).

A historical perspective about kinetic theory and statistical mechanics is contained in the book by S. G. Brush, *The Kind of Motion We Call Heat* (North Holland, 1976).

Textbooks providing a basic and modern approach to equilibrium and nonequilibrium statistical mechanics are D. Chandler, *Introduction to Modern Statistical Mechanics* (Oxford University Press, 1987), and L. Peliti, *Statistical Mechanics in a Nutshell* (Princeton University Press, 2011).

The book by A. Einstein, *Investigations on the Theory of the Brownian Movement* (Dover, 1956), collects the five papers written by the great scientist between 1905 and 1908 about this central topic for the future developments of statistical mechanics. The book contains also very interesting notes by the editor, R. Fürth.

An interesting book in support of the atomic theory of matter was published in French by J. B. Perrin in 1913, and it is now available in several editions, with the English title *Atoms*.

An interesting article about the theory of Brownian motion is the original paper by G. E. Uhlenbeck and L. S. Ornstein, On the theory of the Brownian motion, *Physical Review*, **36** (1930) 823–841.

A mathematical approach to the theory of Brownian motion is contained in the book by E. Nelson, *Dynamical Theories of Brownian Motion*, 2nd ed. (Princeton University Press, 2001).

A useful reference for many questions faced in this chapter and beyond are the lecture notes by B. Derrida, *Fluctuations et grandes déviations autour du Seconde Principe* (Collège de France, 2015–2016).

A mathematical approach to the perspective of probability theory of random walk and Markov chains and processes can be found in W. Feller, *An Introduction to Probability Theory and Its Applications*, vol. 1, 3rd ed. (John Wiley & Sons, 1968).

A detailed and complete account about Monte Carlo methods and their applications can be found in K. Binder and D. W. Heermann, *Monte Carlo Simulation in Statistical Physics*, 3rd ed. (Springer, 1997).

A clear, introductory textbook on stochastic processes is D. S. Lemons, *An Introduction to Stochastic Processes in Physics* (John Hopkins University Press, 2002).

A general and complete book about stochastic processes and their applications is C. W. Gardiner, *Handbook of Stochastic Methods*, 3rd ed. (Springer, 2004).

The Fokker–Planck equation in most of its facets and applications is discussed in H. Risken, *The Fokker–Planck Equation: Methods of Solution and Applications* (Springer-Verlag, 1984).

An extended overview about recent achievements in the theory of random walks and Lévy processes is contained in J. Klafter and I. M. Sokolov, *First Steps in Random Walks: From Tools to Applications* (Oxford University Press, 2011). This book includes additional information about fractional partial differential equations.

A reference textbook of functional analysis is W. Rudin, *Functional Analysis* (McGraw-Hill, 1991), where details about Tauberian theorems can be found.

2 Linear Response Theory and Transport Phenomena

Describing linear response theory at a basic level is quite a challenging task, because it contains a great deal of physical hypotheses and mathematical difficulties. The former have to be carefully justified by heuristic arguments, while the latter have to be illustrated in detail to allow for a clear understanding of most of the physical consequences. On top of that, there is the additional difficulty of casting the main concepts into a frame, where the unexperienced reader can recognize a logical and pedagogical organization. This is why we begin this chapter showing how we can obtain the Kubo relation for a Brownian particle. Our aim is to point out from the very beginning the importance of equilibrium time correlation functions as basic ingredients for an approach to nonequilibrium phenomena. In order to explore the possibility of extending a similar approach to nonequilibrium thermodynamic quantities, we have to recognize that, on a formal ground, a thermodynamic observable can be assumed to be a fluctuating quantity, even at equilibrium (the amplitude of its relative fluctuations vanishing in the thermodnamic limit). In practice, a preliminary step in the direction of a linear response theory of nonequilibrium processes amounts to assuming that thermodyamic observables obey a fluctuation-driven evolution, described by generalized Langevin and Fokker–Planck equations, in full analogy with a Brownian particle. Although a rigorous mathematical formulation should be in order, here we justify this assumption on the basis of physical plausibility arguments.

An explicit dynamical description of the physical mechanism of fluctuations of a thermodynamic observable, as the result of the presence of an associated perturbation field switched on in the past, allows us to establish a quantitative relation with the way this thermodynamic observable relaxes to its new equilibrium value, when the perturbation field is switched off. In particular, the relaxation mechanism is found to be ruled by the decay rate of the equilibrium time correlation function of the perturbed thermodynamic observable.

Now we can come back to transport processes relying on effective theoretical tools. In fact, generalized Green–Kubo relations for the transport coefficients such as diffusivity, viscosity, and thermal conductivity of a general thermodynamic system can be expressed in terms of the equilibrium time correlation functions of the mass, momentum, and energy currents, respectively. These results are obtained by exploiting the formal equivalence of the Fokker–Planck equation with a continuity equation, thus providing a hydrodynamic basis to transport phenomena. The Onsager regression relation allows us to extend such a description to coupled transport processes.

The bridge between linear response theory and the formulation of the thermodynamics of irreversible processes can be established by considering entropy as a genuine dynamical variable: its production rate, which can be expressed as a combination of generalized thermodynamic forces (affinities) and the corresponding fluxes, is the key ingredient for characterizing transient as well as stationary nonequilibrium conditions. Combining this result with the Onsager reciprocity relations and the Onsager theorem for charged transport we can formulate a phenomenological theory of coupled transport processes, which is summarized by the Onsager matrix of generalized transport coefficients: its properties and symmetries again originate from those of equilibrium time correlation functions of pairs of thermodynamic observables. We point out that, despite entropy being treated as a dynamical variable, we assume that in a linear regime it maintains the same dependence of the equilibrium case as all the basic thermodynamic quantities, such as internal energy, pressure, and chemical potentials. This amounts to asserting that also for stationary nonequilibrium processes associated with coupled transport phenomena, local equilibrium conditions hold.

The final part of this chapter is devoted to a short illustration of linear response theory for quantum systems. Despite the formulation being not particularly involved, its application to specific examples requires the knowledge of quantum field theory. For this reason, the final Section 2.10.2 can be easily understood only by readers familiar with such topic.

2.1 Introduction

Linear response theory is a perturbative approach to nonequilibrium processes that allows us to establish a relation between equilibrium time correlation functions of fluctuating observables and measurable quantities of physical interest, such as susceptibilities, transport coefficients, relaxation rates, etc. For instance, the diffusion coefficient of a Brownian particle can be related to its velocity autocorrelation function by the Kubo relation. The basic idea is that we deal with systems whose evolution remains always close to an equilibrium state, including when some perturbation drives them out of it. At first sight, all of that may appear contradictory: how can it be that equilibrium properties are associated with nonequilibrium ones? A preliminary answer to this question has been given at the end of Section 1.3 in the framework of elementary kinetic theory of transport phenomena: in the presence of a macroscopic gradient of physical quantities, equilibrium conditions set in locally. Such an assumption may appear reasonable if the gradient is moderate, i.e., if we gently drive the system out of equilibrium, so that on local space and time scales, significantly larger than the mean free path of particles and of the equilibrium relaxation time, the system always remains very close to equilibrium conditions.

Another important consideration for proceeding in the direction of linear response theory comes from the general Fokker–Planck formulation of stochastic processes discussed in Chapter 1. Here we argue that this dynamical description can be applied to the fluctuation-driven evolution of any thermodynamic, i.e., macroscopic, observable. We should admit that this is not a straightforward statement, and, actually, it demanded the efforts of many outstanding scientists before being fully understood. The introductory level of this textbook does not allow us to enter any rigorous mathematical justification and so we are limited to providing a simple explanation based on heuristic arguments.

Once we have established that close to or at equilibrium also thermodynamic observables fluctuate, we need to identify the mechanism at the origin of these fluctuations and describe it in a suitable mathematical language. This task is accomplished making use of a Hamiltonian description, where a perturbation field may induce the variation of a thermodynamic observable. In this framework the equilibrium side of linear response theory is associated with the autocorrelation function of the thermodynamic observable averaged over the Boltzmann–Gibbs equilibrium statistical measure. A mathematical description of the linear response theory is based on the fluctuation–dissipation theorem: it concerns the analytic properties of the linear response function, which is proportional to time derivative of the equilibrium autocorrelation function. This theorem has a clear physical importance, because it allows us to establish a rigorous relation between the fluctuation and the dissipation mechanisms, which are the key ingredients of linear response theory. This can be used to construct a hydrodynamic approach to transport processes. As we have seen in Section 1.6.3, the Fokker–Planck equation can be cast into the form of a continuity equation, which can be interpreted as a hydrodynamic equation of a fluid system. Making use of constitutive relations that allow us to express density currents in terms of density gradients (as in elementary kinetic theory of transport; see Section 1.3) we can obtain general Green–Kubo relations for diffusivity, viscosity, and heat conductivity of a fluid. As in the case of Brownian particles, such relations allow us to calculate these transport coefficients from the integral of the equilibrium correlation function of the total mass, momentum, and energy currents. The practical interest of the Green–Kubo relations also relies on the possibility of obtaining numerical estimates of the equilibrium correlation functions by suitable molecular dynamics algorithms.

Linear response theory can be generalized by considering the possibility that a perturbation field acting on a given thermodynamic observable may have influence over other thermodynamic observables. This is quite a natural idea, because the presence of a perturbation field modifies the evolution of microscopic degrees of freedom of the systems, and this change, in principle, concerns any other macroscopic observable, because it is a function of the same degrees of freedom. The main reason for this approach is summarized by the Onsager regression hypothesis: when studying the evolution of a physical system around equilibrium we cannot distinguish if an instantaneous value of an observable, different from its equilibrium value, is due to a spontaneous fluctuation or to a perturbation switched off in the past. As a consequence, the regression of the instantaneous value of the observable toward its equilibrium value is related to the decay of the time correlation function. It is important to point out that this statement applies to the correlation function of any pair of fluctuating observables. Their symmetry with respect to time reversal is a crucial ingredient for characterizing the properties of the Onsager matrix of generalized transport coefficients. This is the main tool for describing coupled transport phenomena, such as thermomechanic, thermoelectric, and galvanomagnetic effects in continuous systems. In fact, a phenomenological theory of coupled transport can be constructed by combining linear response theory with a thermodynamic formulation of irreversible processes. This is based on the idea that nonequilibrium states of a thermodynamic system can be characterized by the entropy production rate, expressed as the product of generalized thermodynamic forces, usually called affinities, with the fluxes of the corresponding observable. In a phenomenological approach we can assume that fluxes and affinities are proportional to each other, and we conclude that the entries of the Onsager matrix can be expressed in terms of the static correlation functions of pairs of fluctuating observables.

Finally, we shortly discuss how linear response theory can be formulated for quantum systems. On a conceptual ground, quantum processes at variance with classical ones are characterized by the presence of intrinsic fluctuations, whose practical manifestation is that we expect to obtain statistical inferences rather than deterministic ones even at the microscopic level. In the framework of a thermodynamic description such a difference is much less relevant, although extending linear response theory to quantum systems demands the introduction of suitable tools of quantum theory, such as the density matrix formalism and time ordering, when dealing with non commuting quantum observables. This notwithstanding, classical and quantum linear response theories are quite close to each other, as shown by a few selected examples.

2.2 The Kubo Formula for the Brownian Particle

In Section 1.6.2 we introduced a general Fokker–Planck equation for a generic stochastic process $X(t)$. In order to explore in more detail the dynamical content of this equation, let us first consider the easy example of the diffusion equation (1.69) for a Brownian particle. We assume space isotropy and that $x(t)$, $y(t)$, and $z(t)$ are described by independent stochastic processes, so that we can treat the diffusion equation for just one of these variables, say $x(t)$,

$$\frac{\partial P(x,t)}{\partial t} = D\frac{\partial^2}{\partial x^2}P(x,t). \tag{2.1}$$

The explicit solution for $P(x,t)$ (see Eq. (1.70)) can be used to compute the average square displacement, given in Eq. (1.71),

$$\langle x^2\rangle - \langle x\rangle^2 = 2Dt. \tag{2.2}$$

If we assume the initial position of the Brownian particle at the origin, i.e., $\mathbf{r}(0) = 0$,

$$\langle \mathbf{r}^2\rangle = \langle x^2\rangle + \langle y^2\rangle + \langle z^2\rangle = 6Dt. \tag{2.3}$$

It is noteworthy that the same result can be obtained multiplying both sides of Eq. (2.1) by x^2 and integrating over x,

$$\frac{\partial}{\partial t}\int_{-\infty}^{+\infty} dx\, x^2\, P(x,t) = D\int_{-\infty}^{+\infty} dx\, x^2 \frac{\partial^2}{\partial x^2}P(x,t). \tag{2.4}$$

The right-hand side of this equation can be integrated twice by parts over x: since $P(x,t)$ and its derivatives with respect to x vanish for $x \to \pm\infty$ more rapidly than any power of x, one finally obtains

$$\frac{\partial\langle x^2(t)\rangle}{\partial t} = 2D, \tag{2.5}$$

whose time integration gives, again, Eq. (2.2). We should stress that the Fokker–Planck equation does not cover time scales smaller than or of the order of $t_d = m/\tilde{\gamma}$, which is the time necessary to attain equilibrium. This is the reason why Eq. (2.2) differs from (1.53), whose limiting behaviors are given in Eq. (1.55). It is therefore more correct to write

$$\lim_{t\to\infty} \frac{\langle x^2(t)\rangle}{t} = 2D. \tag{2.6}$$

We can obtain a useful relation between the diffusion constant D of the Brownian particle and its velocity autocorrelation function as follows. We start from the relation $x(t) = \int_0^t dt'\, v_x(t')$ that defines the instantaneous position of the Brownian particle $x(t)$ in terms of its velocity component in the x-direction $v_x(t)$, so we can write

$$x^2(t) = \int_0^t dt' \int_0^t dt'' v_x(t')\, v_x(t''). \tag{2.7}$$

By deriving both sides with respect to time we obtain

$$\frac{\partial x^2(t)}{\partial t} = 2\, v_x(t) \int_0^t dt'\; v_x(t') = 2 \int_0^t dt'\; v_x(t) v_x(t'). \tag{2.8}$$

We can average both sides of this equation at equilibrium and we obtain

$$\frac{\partial \langle x^2(t)\rangle}{\partial t} = 2 \int_0^t dt'\; \langle v_x(t) v_x(t')\rangle, \tag{2.9}$$

where the integral on the right-hand side contains the velocity autocorrelation function of the Brownian particle. Since at equilibrium the expectation value of any observable is invariant under time translations, we can shift the origin of time as follows

$$\langle v_x(t)\, v_x(t')\rangle = \langle v_x(t-t')\, v_x(0)\rangle. \tag{2.10}$$

By redefining the integration variable as $\tau = t - t'$, we obtain

$$\frac{\partial \langle x^2(t)\rangle}{\partial t} = 2 \int_0^t d\tau\; \langle v_x(\tau) v_x(0)\rangle \tag{2.11}$$

and using Eq. (2.6) we can finally write

$$D = \lim_{t\to\infty} \int_0^t d\tau\, \langle v_x(\tau)\, v_x(0)\rangle, \tag{2.12}$$

where the limit $t \to \infty$ has been introduced because Eq. (2.10) applies only when equilibrium has been attained.

Eq. (2.12) is the simplest form of the so-called Kubo relation. It tells us that the diffusion coefficient D of the Brownian particle can be measured by estimating the asymptotic (equilibrium) behavior of its velocity correlation function. It is important to note that we are facing the first example of a more general situation, where transport coefficients, i.e., physical quantities typically associated with nonequilibrium processes, can be obtained from equilibrium averages of correlation functions of suitable current-like observables.

In order to extend the Kubo relation to other transport coefficients of a generic thermodynamic system, such as viscosity or thermal conductivity, we must first work out the linear response theory. As we show in Section 2.5 the results of linear response theory can be used to formulate a hydrodynamic description based on the Fokker–Planck equation for fluctuating observables, associated with conserved quantities.

2.3 Generalized Brownian Motion

In Chapter 1 we extensively discussed the theory of fluctuations for a Brownian particle via the Langevin and the Fokker–Planck equations, while providing in Section 1.6 a generalization of these equations to any stochastic quantity, subject to uncorrelated fluctuations, such as the thermal fluctuations of a thermodynamic system at or close to its equilibrium state at a given temperature T. Here we want to proceed along this direction starting from a basic argument: in principle, there is no limitation to applying this mathematical description to the fluctuation-driven evolution of any thermodynamic, i.e., macroscopic, observable. However, the introductory level of this book suggests we avoid entering the awkward pathway of a mathematically rigorous justification of such an assumption. Therefore, we limit ourselves to a few simple explanatory remarks.

As well as the position and the velocity of a Brownian particle, also a thermodynamic observable $X(t)$ fluctuates in time at thermal equilibrium. Such fluctuations, as it happens for the Brownian particle, are originated by the interactions with an exceedingly large number (typically, the Avogadro number $\mathcal{N}_A \approx 10^{23}$) of microscopic degrees of freedom, present in the system. In practice, we assume that at thermal equilibrium all of these microscopic degrees of freedom evolve as independent stochastic variables and also independent of the actual value of $X(t)$. Due to their large number, relative fluctuations are expected to be quite small and eventually vanish in the thermodynamic limit for any macroscopic observable. This statement is a consequence of the central limit theorem discussed in Appendix A. Since fluctuations are irrelevant in the thermodynamic limit, the finiteness of the system we are dealing with is a necessary condition for the theory of generalized Brownian motion.

In complete analogy with this, we write the Langevin-like equation[1]

$$\tilde{\gamma} \frac{dX(t)}{dt} = \mathcal{F}(X) + \tilde{\eta}(t), \tag{2.13}$$

where $\mathcal{F}$ is a generalized force, $\tilde{\gamma}$ is a generalized friction coefficient, and $\tilde{\eta}(t)$ is the stochastic term, epitomizing the fluctuations originated by the interaction of the thermodynamic observable X with the microscopic degrees of freedom.

With the above assumptions on the origin of fluctuations, the stochastic term has to fulfill the properties

$$\langle \tilde{\eta}(t) \rangle = 0 \tag{2.14}$$

$$\langle \tilde{\eta}(t)\, \tilde{\eta}(t') \rangle = \tilde{\Gamma}\, \delta(t - t'), \tag{2.15}$$

[1] We use a notation that is in full agreement with Section 1.4.1.

where $\langle \cdot \rangle$ are equilibrium averages. The latter condition of fully time-uncorrelated stochastic fluctuations should be interpreted, in a physical perspective, as the assumption that $X(t)$ varies over much longer time scales than those typical of $\tilde{\eta}(t)$.

This problem is formally analogous to the example of a Brownian particle subject to an external force, originated by a conservative potential (see Section 1.6.3). In fact, we can formally write the Fokker–Planck equation corresponding to Eq. (2.13) as

$$\frac{\partial P(X,t)}{\partial t} = -\frac{\partial \mathcal{J}(X,t)}{\partial X}, \tag{2.16}$$

where

$$\mathcal{J} = v(X(t))P(X,t) - \frac{\tilde{\Gamma}}{2\tilde{\gamma}^2}\frac{\partial}{\partial X}P(X,t). \tag{2.17}$$

Here

$$v(X(t)) = \frac{\mathcal{F}}{\tilde{\gamma}} \tag{2.18}$$

is the analog of the limit velocity of the Brownian particle and $\tilde{\Gamma}/\tilde{\gamma}^2$ is the strength of the correlation of the noise $\tilde{f} \equiv \tilde{\eta}/\tilde{\gamma}$.

Close to thermodynamic equilibrium, the expression of $\mathcal{F}$ comes from Einstein's theory of equilibrium fluctuations, which provides us with an expression of the equilibrium probability density of the values taken by X due to fluctuations,

$$P(X) = C \exp - \left(\frac{\mathcal{U}(X)}{T}\right), \tag{2.19}$$

where $\mathcal{U}$ is the thermodynamic potential expressed as a function of X. In particular, one has

$$\mathcal{U}(X^*) = U(V,T), \tag{2.20}$$

where X^* is the equilibrium value of X and $U(V,T)$ is the internal energy of the thermodynamic system.

By assuming stationarity conditions for $P(X)$ and making use of the fluctuation–dissipation relation (1.49), $\tilde{\Gamma} = 2\tilde{\gamma}T$, we can rewrite (2.18) as

$$v(X) = \frac{\mathcal{F}}{\tilde{\gamma}} = -\frac{2T}{\tilde{\Gamma}}\frac{\partial \mathcal{U}(X)}{\partial X}, \tag{2.21}$$

where the generalized thermodynamic force $\mathcal{F}$ has been written as the derivative of the potential $\mathcal{U}$ with respect to the perturbed thermodynamic observable X. We can therefore rewrite Eq. (2.13) in the form

$$\frac{dX}{dt} = -\frac{2T}{\tilde{\Gamma}}\frac{\partial \mathcal{U}(X)}{\partial X} + \tilde{f}(t). \tag{2.22}$$

We can conclude that making use of Einstein's theory of fluctuations and of the analogy with the problem of a Brownian particle subject to a conservative force, we have derived a heuristic expression for the generalized Brownian motion in terms of the nonlinear Langevin-like equation (2.22).

We can proceed along this direction by considering that in thermodynamic equilibrium conditions $\mathcal{U}(X^*)$ is a minimum of the generalized potential $\mathcal{U}(X)$ with respect to X, i.e.,

$$\left(\frac{\partial \mathcal{U}(X)}{\partial X}\right)_{X=X^*} = 0, \tag{2.23}$$

so we can write

$$\mathcal{U}(X) = \mathrm{U}(V,T) + \frac{1}{2}\left(\frac{\partial^2 \mathcal{U}(X)}{\partial X^2}\right)_{X=X^*}(X - X^*)^2 + \cdots . \tag{2.24}$$

Notice that in a thermodynamic system made of N particles, relative fluctuations at equilibrium of any macroscopic observable, i.e., $|X - X^*|$, are of the order of $N^{-\frac{1}{2}}$, as predicted by the central limit theorem (see Appendix A): this is why higher powers can be neglected with respect to the quadratic term. Since we are interested in studying the fluctuations of X close to its equilibrium value X^*, in (2.22) we can evaluate the derivative of the potential making use of Eq. (2.24), obtaining the linearized form

$$\frac{dX}{dt} = -\frac{2T}{\tilde{\Gamma}}\,\chi_X^{-1}\,(X - X^*) + \tilde{f}(t). \tag{2.25}$$

Here the coefficient

$$\chi_X^{-1} = \left(\frac{\partial^2 \mathcal{U}(X)}{\partial X^2}\right)_{X=X^*} \tag{2.26}$$

represents the inverse of the equilibrium thermodynamic susceptibility, associated with the thermodynamic observable X, i.e., the same quantity that in a magnetic system defines how the magnetization responds to the variation of an external magnetic field. In the next section we determine the susceptibility within a linear response theory.

2.4 Linear Response to a Constant Force

In this section we discuss how a physical system responds to an external time-dependent force that is assumed to gently perturb it. We can imagine we are observers trying to get some inference about the effects of a spontaneous fluctuation of the physical system under examination, but we could equally think that we are applying some perturbation to the system in order to force it from the outside or to perform a measurement on it. All of these situations correspond to the same mathematical description, whose basic ingredients are now presented in the language of classical mechanics. As we shall shortly discuss in Section 2.9, they can be adapted by suitable modifications also to quantum mechanical systems and even to quantum field theory.

First of all we introduce the basic mathematical ingredients. We assume that the physical system under examination is made of N particles, obeying the laws of classical mechanics. Their position and momentum coordinates in a physical space of dimension d are usually denoted as $q_i(t)$ and $p_i(t)$, with $i = 1, 2, \ldots, dN$. If the system has energy E, the Hamilton functional, $\mathcal{H}(q_i(t), p_i(t)) = E$, provides all information about the deterministic dynamics of the system, through the Hamilton equations

$$\frac{dq_i(t)}{dt} = \frac{\partial \mathcal{H}}{\partial p_i(t)}, \tag{2.27a}$$

$$\frac{dp_i(t)}{dt} = -\frac{\partial \mathcal{H}}{\partial q_i(t)}. \tag{2.27b}$$

This means that, due to the time reversal symmetry of Hamilton equations, the forward and backward time evolution of the system is uniquely determined by the initial conditions $\{q_i(0), p_i(0)\}$.

On the other hand, here we are interested in describing our system in a thermodynamic language, rather than in a mechanical one. Equilibrium statistical mechanics provides us with the basic ingredient via the Boltzmann–Gibbs formula,

$$\Pi(q_i, p_i) = \frac{1}{Z}\, e^{-\beta \mathcal{H}(q_i, p_i)}, \tag{2.28}$$

where $\Pi(q_i, p_i)$ is the probability of observing the system in the microstate (q_i, p_i) and $\beta = 1/T$, with T the temperature of the thermal bath, which keeps our system at equilibrium. The partition function

$$Z = \int_{\mathcal{V}} d\mathcal{V}\; e^{-\beta \mathcal{H}(q_i, p_i)} = e^{-\beta F} \tag{2.29}$$

provides the relation with equilibrium thermodynamics, because F is the free energy, which is a function of temperature T and of thermodynamic observables, $F = F(T, \{X\})$, where X can be the volume V occupied by the system, the number of particles N present in the system, its magnetization M, etc.

In Eq. (2.29) we have also introduced the shorthand notation $\mathcal{V}$ for the phase space, whose infinitesimal volume element is[2]

$$d\mathcal{V} \equiv \frac{1}{N!} \prod_{i=1}^{dN} \frac{dq_i\, dp_i}{2\pi\hbar}, \tag{2.30}$$

where $\hbar = 1.05457 \times 10^{-34}$ Js denotes the reduced Planck constant.[3]

Now, we want to consider the situation in which we disturb the system, driving it out of its thermodynamic equilibrium state to a new state described by a perturbed Hamiltonian $\mathcal{H}'$. For a quantitative description of the response of the system we have to perform a measurement over a suitable macroscopic, i.e., thermodynamic, observable $X(q_i, p_i)$, that is also a functional defined onto phase space $\mathcal{V}$, and we assume that it is not a conserved quantity, namely, $\{\mathcal{H}(q_i, p_i), X(q_i, p_i)\} \neq 0$.

The perturbed Hamiltonian can be written as

$$\mathcal{H}' = \mathcal{H} - hX, \tag{2.31}$$

[2] The factor $1/N!$ stems from the hypothesis that identical particles are undistinguishable: it was originally introduced by Otto Sackur and Hugo Martin Tetrode for recovering the additivity of the entropy of the ideal gas.

[3] We write $2\pi\hbar$ rather than h because, here below, this symbol is used for a generalized field.

where h is the perturbation field, or generalized thermodynamic force, that allows perturbation of the system by modifying the value of X. More generally, h and X are conjugate thermodynamic variables, like pressure and volume or chemical potential and number of particles, which are related, at thermodynamic equilibrium, through basic thermodynamic relations. In fact, by replacing $\mathcal{H}$ with $\mathcal{H}'$ in Eq. (2.29), we obtain

$$Z_h = \int_{\mathcal{V}} d\mathcal{V}\ e^{-\beta(\mathcal{H}-hX(q_i,p_i))} = e^{-\beta F_h}. \tag{2.32}$$

Deriving both sides of (2.32) with respect to h, we obtain

$$\beta \int_{\mathcal{V}} d\mathcal{V} X e^{-\beta(\mathcal{H}-hX(q_i,p_i))} = -\beta Z_h \frac{\partial F_h}{\partial h}, \tag{2.33}$$

that is to say,

$$\frac{\partial F_h}{\partial h} = -\frac{1}{Z_h}\int_{\mathcal{V}} d\mathcal{V} X e^{-\beta(\mathcal{H}-hX(q_i,p_i))} \tag{2.34}$$

$$= -\langle X \rangle. \tag{2.35}$$

The above relation applies to any pair of conjugated thermodynamic variables, magnetic field and magnetization, temperature and entropy, pressure and volume, chemical potential and number of particles. In particular, this type of relation allows us to determine the static or equilibrium susceptibility χ, i.e., the response of an extensive property under variation of the conjugated intensive variable h, in the limit $h \to 0$:

$$\begin{aligned}\chi &= \left.\frac{\partial \langle X\rangle}{\partial h}\right|_{h=0} \\ &= \frac{\partial}{\partial h}\left[\frac{1}{Z_h}\int_{\mathcal{V}} d\mathcal{V} X e^{-\beta(\mathcal{H}-hX(q_i,p_i))}\right]_{h=0} \\ &= -\frac{\beta}{Z_h^2}\left[\int_{\mathcal{V}} d\mathcal{V} X e^{-\beta(\mathcal{H}-hX(q_i,p_i))}\right]^2_{h=0} + \frac{\beta}{Z_h}\int_{\mathcal{V}} d\mathcal{V} X^2 e^{-\beta(\mathcal{H}-hX(q_i,p_i))}\Big|_{h=0} \\ &= \beta[\langle X^2\rangle_0 - \langle X\rangle_0^2].\end{aligned} \tag{2.36}$$

For instance, the static magnetic susceptibility of a magnetic system, whose perturbed Hamiltonian has the form $\mathcal{H} - HM$, where M is the magnetization (the extensive quantity) conjugated to the magnetic field H (the intensive quantity), is given by

$$\chi = \left.\frac{\partial M}{\partial H}\right|_{H=0} = -\left.\frac{\partial^2 F_H}{\partial H^2}\right|_{H=0} = \beta[\langle M^2\rangle_0 - \langle M\rangle_0^2]. \tag{2.37}$$

While this section has dealt with a constant perturbation that just shifts the equilibrium, the next section concerns a time-dependent perturbation field, $h(t)$. The study of how the system responds to such perturbations leads to the celebrated fluctuation–dissipation relations.

2.4.1 Linear Response and Fluctuation–Dissipation Relations

In order to provide a simple mathematical description, it is worth posing the problem as follows: we can assume that the system was prepared, very far in the past, in an equilibrium state of the perturbed Hamiltonian $\mathcal{H}' = \mathcal{H} - h(t)X$, where $h(t) = h$ for $-\infty < t < 0$. At $t = 0$ the perturbation is switched off, $h(t) = 0$ for $t > 0$, and we let the system relax to the equilibrium state of the unperturbed Hamiltonian $\mathcal{H}$. The relaxation process is guaranteed by the presence of the thermal bath. Up to $t = 0$ the equilibrium value of X is defined by the expression

$$X^* \equiv \langle X \rangle_{eq} = \frac{1}{Z'} \int_{\mathcal{V}} d\mathcal{V}\ e^{-\beta \mathcal{H}'(q_i,p_i)} X(q_i, p_i), \tag{2.38}$$

where Z' is given by (2.29), with $\mathcal{H}'$ replacing $\mathcal{H}$. In what follows we adopt the standard notation $\langle \cdot \rangle_{eq}$ for the equilibrium average of any observable. When $h(t)$ is switched off, the evolution of the microscopic variables $\{q_i(t), p_i(t)\}$ starts to be ruled by Eqs. (2.27) (while for $t < 0$ their evolution was ruled by the same equations with $\mathcal{H}'$ replacing $\mathcal{H}$) so that, for $t > 0$, $X(t)$ is an explicit function of time: similarly to a Brownian particle, it will relax, starting from X^*, to its equilibrium value for the unperturbed Hamiltonian $\mathcal{H}$, i.e., the expression (2.38) where $\mathcal{H}'$ is replaced by $\mathcal{H}$.

In order to obtain a suitable expression for the time evolution of $X(t)$ during its relaxation, i.e., for $t > 0$, we can write

$$\langle X(t) \rangle = \frac{\int_{\mathcal{V}} d\mathcal{V}\ e^{-\beta(\mathcal{H}(q_i,p_i) - hX(0))} X(t)}{\int_{\mathcal{V}} d\mathcal{V}\ e^{-\beta(\mathcal{H}(q_i,p_i) - hX(0))}}, \tag{2.39}$$

where $X(t)$ at the numerator of the right-hand side is formally obtained by integrating Eqs. (2.27), starting from initial conditions $\{q_i(0), p_i(0)\}$, which must be weighted according to the equilibrium distribution of the perturbed Hamiltonian. This is the reason why in both the numerator and denominator is $\mathcal{H} - hX(0)$, where $X(0) = X(q_i(0), p_i(0))$ denotes the value of X when the perturbation is switched off.

Assuming that h is small, we can expand $e^{\beta h X(0)}$ at the first order in h and therefore we can approximate the right-hand side as

$$\langle X(t) \rangle \approx \frac{\int_{\mathcal{V}} d\mathcal{V}\ e^{-\beta \mathcal{H}(q_i,p_i)} (1 + \beta h X(0)) X(t)}{\int_{\mathcal{V}} d\mathcal{V}\ e^{-\beta \mathcal{H}(q_i,p_i)} (1 + \beta h X(0))}. \tag{2.40}$$

Dividing numerator and denominator by the unperturbed partition function $Z_0 = \int_{\mathcal{V}} d\mathcal{V}\ e^{-\beta \mathcal{H}(q_i,p_i)}$, we obtain

$$\langle X(t) \rangle = \frac{\langle (1 + \beta h X(0)) X(t) \rangle_0}{\langle 1 + \beta h X(0) \rangle_0}, \tag{2.41}$$

where $\langle \cdots \rangle_0$ represents the statistical average for the unperturbed system. At first order in h we finally obtain the fluctuation–dissipation relation

$$\langle X(t) \rangle - \langle X \rangle_0 = \beta h \Big(\langle X(t) X(0) \rangle_0 - \langle X \rangle_0^2 \Big), \tag{2.42}$$

where we have replaced $\langle X(t) \rangle_0$ with $\langle X \rangle_0$, because the equilibrium expectation value of X, as well as that of any other thermodynamic quantity, is independent of time.

We can conclude that the average deviation of a perturbed macroscopic observable from its unperturbed average evolution is linear in the amplitude of the perturbation field h and proportional to the equilibrium average of its connected time autocorrelation function.

The latter quantity measures the equilibrium fluctuations of X, while the former describes the relaxation process of the perturbed variable. This is why Eq. (2.42) is called the fluctuation–dissipation relation, which is the cornerstone of linear response theory. This relation has important conceptual implications for the physics of nonequilibrium processes. In practice, it tells us that the nonequilibrium evolution of a gently perturbed macroscopic thermodynamic system is essentially indistinguishable from its fluctuations around equilibrium, so that in the linear regime the nonequilibrium dynamics of a macroscopic observable can be measured by its time autocorrelation function. Without prejudice of generality, we can set to zero the equilibrium average values of X, namely $\langle X \rangle_0 = 0$, and rewrite the fluctuation–dissipation relation (2.42) in a simpler form,

$$\langle X(t) \rangle = \beta h \langle X(t) X(0) \rangle_0 . \tag{2.43}$$

A formulation of the fluctuation–dissipation relation in the presence of a general time-dependent perturbation $h(t)$ introduces the response function $\Xi(t, t')$,

$$\langle X(t) \rangle = \int dt' \, \Xi(t, t') \, h(t'), \tag{2.44}$$

which is based on the consideration that the response at time t of a perturbed observable depends on the value of the perturbation field at any other time $t' < t$. On the other hand, if we want to preserve the physical assumption of causality we have to impose that $\Xi(t, t')$ cannot depend on values of the perturbation field taken at time $t' > t$. Moreover, consistently with Eq. (2.43), $\Xi(t, t')$ has to be an equilibrium quantity; therefore, it has to be invariant under time translations, i.e., it should depend on $(t - t')$. Thus

$$\Xi(t, t') = \begin{cases} \Xi(t - t'), & \text{if } t > t' \\ 0, & \text{otherwise.} \end{cases} \tag{2.45}$$

Important properties of the response function $\Xi(t - t')$, including the Kramers–Krönig relations, are reported in Appendix F.

We can now determine the explicit expression of the response function comparing Eqs. (2.43) and (2.44). Since the former has been derived assuming

$$h(t) = \begin{cases} h, & \text{if } t < 0 \\ 0, & \text{otherwise,} \end{cases} \tag{2.46}$$

it must be

$$\int_{-\infty}^{0} dt' \; \Xi(t - t') = \beta \langle X(t) X(0) \rangle_0 , \tag{2.47}$$

or, after introducing the change of variable $\tau = (t - t')$,

$$\int_{t}^{+\infty} d\tau \; \Xi(\tau) = \beta \langle X(t) X(0) \rangle_0 . \tag{2.48}$$

Taking the time derivative of both sides and using causality, we finally obtain

$$\Xi(t) = -\beta \Theta(t) \frac{d}{dt} \langle X(t) X(0) \rangle_0 , \tag{2.49}$$

where $\Theta(t)$ is the Heaviside step distribution, which is equal to one for $t > 0$ and vanishes otherwise.

Eq. (2.49) is a more general form of the fluctuation–dissipation relation in the linear response limit, because any explicit dependence on h has disappeared and the response to any (small) perturbation $h(t)$ can be derived from Eq. (2.44). A further way of expressing the same content of Eq. (2.49) can be obtained by considering the Fourier transform of Eq. (2.49),

$$\Xi(\omega) = \int_{-\infty}^{+\infty} dt\, e^{-i\omega t}\, \Xi(t)$$
$$= -\beta \int_{-\infty}^{+\infty} dt\, e^{-i\omega t}\, \Theta(t) \frac{d}{dt} \langle X(t) X(0) \rangle_0. \tag{2.50}$$

The last expression is the Fourier transform of a product of functions, which is known to be equal to the convolution product ($\star$) of the Fourier transformed functions, according to the general relation

$$\int_{-\infty}^{+\infty} dt\, e^{-i\omega t} f(t)\, g(t) = \frac{1}{2\pi} \int_{-\infty}^{+\infty} d\omega' f(\omega - \omega')\, g(\omega') \equiv \frac{1}{2\pi} f(\omega) \star g(\omega). \tag{2.51}$$

With reference to Eq. (2.50), we have

$$f(t) = \Theta(t)$$
$$g(t) = \frac{d}{dt} \langle X(t) X(0) \rangle_0 \equiv \frac{d}{dt} C(t), \tag{2.52}$$

where we have introduced the shorthand notation $C(t)$ for $\langle X(t) X(0) \rangle_0$. Since the Fourier transform of the time derivative of a function is the Fourier transform of the function multiplied by a factor $i\omega$, we have $g(\omega) = i\omega C(\omega)$, where $C(\omega)$ is the Fourier transform of $C(t)$. Due to time translation invariance at equilibrium, this quantity must be an even function of t,[4] so that its Fourier transform must be real.[5]

In conclusion, $\Xi(\omega)$ can be written as

$$\Xi(\omega) = -\beta \int_{-\infty}^{+\infty} \frac{d\omega'}{2\pi} \Theta(\omega - \omega')(i\omega') C(\omega'). \tag{2.53}$$

This expression is well defined only in the sense of distributions, because we need to make convergent the integral expressing $\Theta(\omega)$ as follows (see also Eqs. (F.19) and (F.20) in Appendix F):

[4] In fact, at equilibrium $\langle X(t) X(0) \rangle_0$ does not change if we shift the time scale by a given amount τ: in particular, if we take $\tau = -t$ we obtain $\langle X(t) X(0) \rangle_0 = \langle X(0) X(-t) \rangle_0$.

[5] According to the Wiener–Khinchin theorem the Fourier transform of the correlation function, $C(\omega)$, of a fluctuating observable $X(t)$ represents its power spectrum. More precisely, the theorem states that the following relation holds:

$$C(\omega) = \int_{-\infty}^{+\infty} dt\, e^{-i\omega t} \langle X(t) X(0) \rangle_0 = \lim_{T \to +\infty} \frac{1}{T} \langle |X_T(\omega)|^2 \rangle$$

where

$$X_T(\omega) = \int_{-T/2}^{T/2} dt\, X(t) e^{-i\omega t}.$$

$$\Theta(\omega) = \lim_{\epsilon \to 0^+} \int_0^{+\infty} dt\, e^{-(\epsilon + i\omega)t} = \lim_{\epsilon \to 0^+} \frac{1}{\epsilon + i\omega} = \pi\delta(\omega) - i\mathrm{PV}\left(\frac{1}{\omega}\right), \tag{2.54}$$

where the PV denotes the principal value of the argument.

Making use of this relation and considering that, according to our conventions (see Appendix F), $\Xi(\omega)$ is analytic in the lower-half complex plane, we obtain

$$\Xi(\omega) = -\beta\left[\frac{(i\omega)}{2}C(\omega) + \mathrm{PV}\int_{-\infty}^{+\infty}\frac{d\omega'}{2\pi}\frac{\omega'}{\omega - \omega'}C(\omega')\right]. \tag{2.55}$$

This equation tells us that the imaginary part of the Fourier transform of the response function $\Xi^{\mathrm{I}}(\omega)$ is related to the power spectral density of thermal fluctuations of the macroscopic variable $X(t)$, more precisely,

$$\Xi^{\mathrm{I}}(\omega) = -\beta\frac{\omega}{2}C(\omega). \tag{2.56}$$

The imaginary part is also related to the real part of the same Fourier transform, $\Xi^{\mathrm{R}}(\omega)$, by the Kramers–Kronig relation (see (F.22) in Appendix F), as

$$\Xi^{\mathrm{R}}(\omega) = \mathrm{PV}\int_{-\infty}^{+\infty}\frac{d\omega'}{\pi}\frac{1}{\omega - \omega'}\,\Xi^{\mathrm{I}}(\omega'). \tag{2.57}$$

The real and imaginary parts of the response function have a well-defined parity. Since $C(\omega) = C(-\omega)$ (see note 5 above), from Eq. (2.56) it is straightforward that $\Xi^{\mathrm{I}}(\omega)$ is an odd function. Using this result and changing the sign of the integration variable in Eq. (2.57), we find that $\Xi^{\mathrm{R}}(\omega)$ is an even function.

An alternative way of writing the Kramers–Kronig relation, also discussed in Appendix F, can be found coming back to Eq. (2.53), using the central expression of Eqs. (2.54) and (2.56),

$$\Xi(\omega) = \lim_{\epsilon \to 0^+}\int_{-\infty}^{+\infty}\frac{d\omega'}{\pi}\frac{\Xi^{\mathrm{I}}(\omega')}{\omega' - \omega - i\epsilon}. \tag{2.58}$$

Eqs. (2.56)–(2.58) are the the most widely used formulations of the so-called fluctuation–dissipation theorem.

2.4.2 Work Done by a Time-Dependent Field

We want to consider the case where the time-dependent field $h(t)$, whose physical interpretation is that of a generalized thermodynamic force associated with the observable $X(t)$, is switched on and we aim at measuring the work W done by it on the system, which is given by the expression

$$W = \int dt\, h(t)\left\langle\frac{dX(t)}{dt}\right\rangle. \tag{2.59}$$

For physical reasons we assume that $h(t)$ is a sufficiently regular function of time t and, in particular, that its Fourier transform exists. For instance, we could deal with a field $h(t)$ that is a sufficiently rapidly vanishing function for $t \to \pm\infty$, or even a periodic function, as we discuss later in Section 2.4.3 for the damped harmonic oscillator. In the former case, W can be computed for $-\infty < t < +\infty$, provided the integral in (2.59) exists, while in the second case it makes sense to compute the work done on the system over a period. Let us consider here the former case, so that we can write

$$W = \int_{-\infty}^{+\infty} dt\, h(t) \left\langle \frac{dX(t)}{dt} \right\rangle. \tag{2.60}$$

We can use Eq. (2.44) and write

$$W = \int_{-\infty}^{+\infty} dt\, h(t) \int_{-\infty}^{+\infty} dt' \frac{\partial}{\partial t} \Xi(t-t')\, h(t'). \tag{2.61}$$

In the right-hand side of this equation we can substitute $\Xi(t-t')$ with its definition in terms of its Fourier antitransform and write

$$\begin{aligned} W &= \int_{-\infty}^{+\infty} dt\, h(t) \int_{-\infty}^{+\infty} dt' \frac{\partial}{\partial t} \int_{-\infty}^{+\infty} \frac{d\omega}{2\pi} \Xi(\omega) e^{i\omega(t-t')}\, h(t') \\ &= \int_{-\infty}^{+\infty} \frac{d\omega}{2\pi} (i\omega)\, \Xi(\omega) \int_{-\infty}^{+\infty} dt\, e^{i\omega t}\, h(t) \int_{-\infty}^{+\infty} dt'\, e^{-i\omega t'}\, h(t'). \end{aligned}$$

Now we can take into account that $h(t)$ is a real quantity, i.e., $h(-\omega) = h^*(\omega)$, so that $h(-\omega)h(\omega) = |h(\omega)|^2$ and we can finally write

$$W = \int_{-\infty}^{+\infty} \frac{d\omega}{2\pi} (i\omega) \Xi(\omega) |h(\omega)|^2 . \tag{2.62}$$

Due to the presence of the linear factor $i\omega$, the Fourier transform of the response function contributes to the integral on the right-hand side with its odd component, i.e., its imaginary part $\Xi^{\mathrm{I}}(\omega)$, and we can finally write

$$W = -\int_{-\infty}^{+\infty} \frac{d\omega}{2\pi} \omega\, \Xi^{\mathrm{I}}(\omega) |h(\omega)|^2 = \frac{\beta}{2} \int_{-\infty}^{+\infty} \frac{d\omega}{2\pi} \omega^2 C(\omega) |h(\omega)|^2 , \tag{2.63}$$

where we have used Eq. (2.56) to relate $\Xi^{\mathrm{I}}(\omega)$ with $C(\omega)$.

We have obtained that the work done on the system by the field $h(t)$ depends on the imaginary part of the Fourier transform of the response function. This is the reason why Eqs. (2.56)–(2.58) are named fluctuation–dissipation relations. Moreover, Eq. (2.63) tells us that the work done on the system is related through $C(\omega)$ to the power spectrum of the fluctuating observable $X(t)$ (see also note 5 above), thus implying that W, i.e., the work done on a system by a generalized thermodynamic force $h(t)$, is a positive quantity, as it should be.

2.4.3 Simple Applications of Linear Response Theory

Some basic physical examples can help in appreciating the main features of the linear response approach and the role of the response function $\Xi(t-t')$, whose properties are illustrated in Appendix F.

Susceptibility

As a first instance, let us start from Eq. (2.44). Using the property $\Xi(t,t') = \Xi(t-t')$, it can be written in Fourier space as

$$\langle X(\omega)\rangle = \Xi(\omega)h(\omega). \tag{2.64}$$

The susceptibility χ is the physical quantity measuring the response of the observable to the perturbation field (e.g., the magnetization emerging in a magnetic material when a small magnetic field is switched on). It can be defined as

$$\chi = \lim_{\omega\to 0} \Xi(\omega) = \frac{\partial \langle X\rangle(\omega)}{\partial h(\omega)}\bigg|_{\omega=0}. \tag{2.65}$$

This definition coincides with that of the static susceptibility given in Eq. (2.36). In fact, if we consider Eq. (2.50) we can write

$$\begin{aligned}
\lim_{\omega\to 0} \Xi(\omega) &= \lim_{\epsilon\to 0^+} -\beta \int_0^{+\infty} dt e^{-\epsilon t} \frac{d}{dt} \langle X(t)X(0)\rangle_0 \\
&= \lim_{\epsilon\to 0^+} \left\{\Big[-\beta e^{-\epsilon t}\langle X(t)X(0)\rangle_0\Big]_0^{+\infty} - \beta\epsilon \int_0^{+\infty} dt e^{-\epsilon t}\langle X(t)X(0)\rangle_0\right\} \\
&= \beta\langle X^2(0)\rangle_0 = \beta\langle X^2\rangle_0,
\end{aligned} \tag{2.66}$$

where the last equality holds because the equilibrium average $\langle X^2\rangle_0$ is independent of time. This proves that definition (2.65) coincides with Eq. (2.36) for $\langle X\rangle_0 = 0$.

An alternative way of expressing the static susceptibility stems from Eq. (2.58). In fact we can also write

$$\chi = \lim_{\omega\to 0}\lim_{\epsilon\to 0^+} \int_{-\infty}^{+\infty} \frac{d\omega'}{\pi} \frac{\Xi^{\mathrm{I}}(\omega')}{\omega' - \omega - i\epsilon} \tag{2.67}$$

and commuting the two limits, we find

$$\chi = \lim_{\epsilon\to 0^+} \int_{-\infty}^{+\infty} \frac{d\omega'}{\pi} \frac{\Xi^{\mathrm{I}}(\omega')}{\omega' - i\epsilon}. \tag{2.68}$$

We can interpret this equation as follows: if we know how much a system dissipates at any frequency in the presence of a perturbation field, we can determine the response of the system at zero frequency, i.e., its susceptibility, by summing over all of these contributions. This is called the thermodynamic sum rule.

The Damped Harmonic Oscillator

The dynamics of a damped harmonic oscillator under the action of a driving force $F(t)$ is described by the differential equation

$$m\ddot{x}(t) + \tilde{\gamma}\dot{x}(t) + Kx(t) = F(t). \tag{2.69}$$

Since it is a linear equation, it is not necessary that $F(t)$ is a weak perturbation and the response function $\Xi(t-t')$ is just the Green's function $G(t-t')$, which allows us to find the solution of Eq. (2.69) as a function of $F(t)$, namely,

$$x(t) = \int_{-\infty}^{+\infty} dt'\ G(t-t')\ F(t'). \tag{2.70}$$

The standard method of solving this problem amounts to expressing $G(t)$ by its Fourier antitransform $G(\omega)$

$$G(t) = \frac{1}{2\pi}\int_{-\infty}^{+\infty} d\omega e^{i\omega t} G(\omega) \tag{2.71}$$

and rewriting (2.70) as

$$x(t) = \int_{-\infty}^{+\infty} dt' \frac{1}{2\pi}\int_{-\infty}^{+\infty} d\omega e^{i\omega(t-t')} G(\omega)\ F(t'). \tag{2.72}$$

We can substitute this formal solution into (2.69) and obtain

$$\int_{-\infty}^{+\infty} \frac{d\omega}{2\pi}\int_{-\infty}^{+\infty} dt'\big(-m\omega^2 + i\tilde{\gamma}\omega + K\big)e^{i\omega(t-t')} G(\omega)\ F(t') = F(t). \tag{2.73}$$

This equation is solved if

$$G(\omega) = -\frac{1}{m\omega^2 - i\tilde{\gamma}\omega - K}, \tag{2.74}$$

because of the equality (in the sense of distributions)

$$\delta(t) = \frac{1}{2\pi}\int_{-\infty}^{+\infty} d\omega e^{i\omega t}. \tag{2.75}$$

It is common to introduce the natural frequency of the undamped oscillator, $\omega_0 = \sqrt{K/m}$, and write

$$G(\omega) = -\frac{1}{m(\omega^2 - i\gamma\omega - \omega_0^2)}, \tag{2.76}$$

where $\gamma = \tilde{\gamma}/m$.

According to (2.67) the susceptibility of the damped and forced harmonic oscillator is

$$\chi = \lim_{\omega\to 0} G(\omega) = \frac{1}{m\omega_0^2}. \tag{2.77}$$

Thus, in this zero-frequency limit the Fourier transform of Eq. (2.70) provides us with the response of the oscillator to a static (i.e., time-independent) force F

$$x = \frac{F}{m\omega_0^2}, \tag{2.78}$$

i.e., the result we expect on the basis of a straightforward visual inspection of Eq. (2.69).

In general, we can rewrite (2.76) as

$$G(\omega) = -\frac{1}{m\,(\omega - \omega_1)(\omega - \omega_2)} \tag{2.79}$$

where

$$\omega_1 = \Omega + i\frac{\gamma}{2}$$

$$\omega_2 = -\Omega + i\frac{\gamma}{2}$$

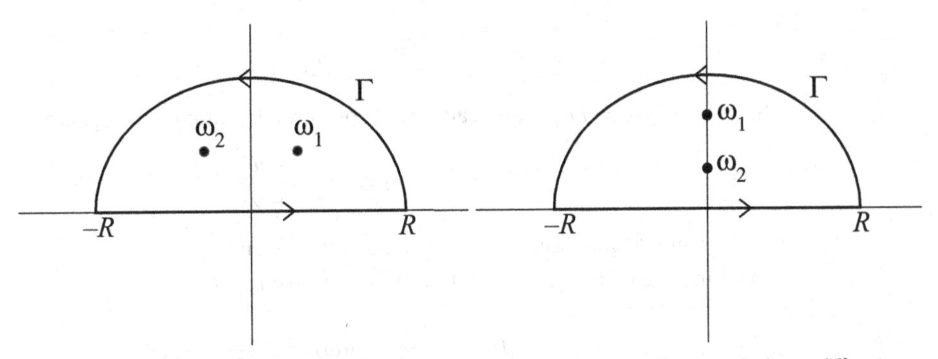

Fig. 2.1 The complex plane for calculating the integral (2.81) with the method of residues. The two poles have a different location in the underdamped case (left) and in the overdamped case (right), but they are always in the upper-half complex plane, therefore preserving causality.

are the poles of the response function in the complex plane, with

$$\Omega^2 = \omega_0^2 - \frac{\gamma^2}{4}.$$

The position of the poles $\omega_{1,2}$ in the complex plane is sketched in Fig. 2.1, in the underdamped case, i.e., $\omega_0^2 > \gamma^2/4$, on the left and in the overdamped case, i.e., $\omega_0^2 < \gamma^2/4$, on the right. In both cases the poles lay in the upper-half complex plane and causality is preserved, i.e., $G(t) = 0$ for $t < 0$ (see Appendix F).

We can obtain an explicit expression of $G(t)$ for $t > 0$ by considering the inverse Fourier transform of $G(\omega)$

$$G(t) = -\frac{1}{2\pi}\int_{-\infty}^{+\infty} d\omega e^{i\omega t}\frac{1}{m\,(\omega-\omega_1)(\omega-\omega_2)}. \tag{2.80}$$

The right-hand side of this equation can be evaluated by considering the analytic extension to the complex plane of the function to be integrated ($\omega \to z$) and by evaluating its integral over a closed contour Γ, which runs along the real axis from $-R$ to $+R$ and is closed in the upper-half complex plane by a semicircle of radius R (see Fig. 2.1),

$$G(t) = -\frac{1}{2\pi}\oint_\Gamma dz\frac{e^{izt}}{m\,(z-\omega_1)(z-\omega_2)}. \tag{2.81}$$

By performing the limit $R \to \infty$, the part of the integral over the semicircle of radius R vanishes due to Jordan's lemma, and the theorem of residues yields the expression

$$G(t) = -\frac{1}{2\pi m}2\pi i\left[\frac{e^{i\omega_1 t}}{\omega_1-\omega_2}+\frac{e^{i\omega_2 t}}{\omega_2-\omega_1}\right] \tag{2.82}$$

$$= \frac{\sin(\Omega t)}{m\Omega}\,e^{-\gamma t/2}. \tag{2.83}$$

In the strongly overdamped regime the inertial term in Eq. (2.69) can be neglected and the Fourier transform of the response function (see Eq. (2.76)) can be approximated as

$$G(\omega) \approx \frac{1}{K + i\tilde{\gamma}\omega}. \tag{2.84}$$

The pole is in $\omega = iK/\tilde{\gamma}$ and causality is preserved, while the imaginary part of $G(\omega)$ is

$$G_I(\omega) = -\frac{\tilde{\gamma}\,\omega}{K^2 + \tilde{\gamma}^2\omega^2}, \tag{2.85}$$

which is an odd function of ω, as discussed in Appendix F.

Making use of Eq. (2.56), with $X(t) \to x(t)$, we find that

$$C(\omega) = -\frac{2}{\beta\omega} G_I(\omega) = 2T\frac{\tilde{\gamma}}{K^2 + \tilde{\gamma}^2\omega^2}, \tag{2.86}$$

where $C(\omega)$ is the Fourier transform of the correlation function

$$C(t) = \lim_{T\to\infty} \frac{1}{T}\int_{-T/2}^{T/2} dt' x(t' + t)\,x(t') \equiv \langle x(t' + t)\,x(t')\rangle_{t'}. \tag{2.87}$$

By the Wiener–Khinchin theorem (see note 5 above), we know that $C(\omega)$ is proportional to the power spectrum $|x(\omega)|^2$, which, in the overdamped regime, has the form of the Lorentzian function of ω, given in the right-hand side of Eq. (2.86).

We can also compute the energy dissipated by the damped harmonic oscillator, which is the work per unit time performed by the driving force $F(t)$:

$$\begin{aligned}
\frac{dW}{dt} &= F(t)\,\dot{x}(t) \\
&= F(t)\frac{d}{dt}\int_{-\infty}^{+\infty} dt'\; G(t - t')\;F(t') \\
&= F(t)\int_{-\infty}^{+\infty} dt' \int_{-\infty}^{+\infty} \frac{d\omega}{2\pi}(i\omega)G(\omega)e^{i\omega(t-t')}\,F(t') \\
&= F(t)\int_{-\infty}^{+\infty} d\omega \frac{e^{i\omega t}}{2\pi}(i\omega)G(\omega)\int_{-\infty}^{+\infty} dt' e^{-i\omega t'}\,F(t') \\
&= \frac{F(t)}{2\pi}\int_{-\infty}^{+\infty} d\omega\,(i\omega)G(\omega)e^{i\omega t}\,F(\omega).
\end{aligned} \tag{2.88}$$

By definition the driving force $F(t)$ is a real quantity. We specialize our analysis to the case where it is a periodic function of time with frequency ν, namely, $F(t) = F_0\cos(\nu t)$. Its Fourier transform reads

$$F(\omega) = \frac{F_0}{2}\big[\delta(\omega - \nu) + \delta(\omega + \nu)\big]. \tag{2.89}$$

Substituting this expression into (2.88) we obtain

$$\frac{dW}{dt} = \frac{i}{4\pi}\,\nu\,F_0^2\cos(\nu t)\Big[G(\nu)e^{i\nu t} - G(-\nu)e^{-i\nu t}\Big]. \tag{2.90}$$

The work performed by the driving force over one period

$$\overline{W} = \int_0^{2\pi/\nu} dt\,\frac{dW}{dt} \tag{2.91}$$

is easily obtained by remarking that

$$\overline{\cos(\nu t) e^{\pm i\nu t}} = \frac{1}{2}, \tag{2.92}$$

thus yielding

$$\overline{W} = \frac{i}{4\pi} \frac{\nu F_0^2}{2} \Big[G(\nu) - G(-\nu) \Big]. \tag{2.93}$$

According to the parity properties of the Fourier transform of the response function, its real part is even and its imaginary part is odd. So, we can finally write

$$\overline{W} = -\frac{F_0^2}{4\pi} \nu \, G_I(\nu). \tag{2.94}$$

In the strong overdamped limit we can use Eq. (2.85) and write

$$\overline{W} = \frac{F_0^2}{4\pi} \frac{\tilde{\gamma}\, \nu^2}{K^2 + \tilde{\gamma}^2 \nu^2}, \tag{2.95}$$

which is, as expected, a positive quantity.

2.5 Hydrodynamics and the Green–Kubo Relation

In Section 1.6.3 we have seen that the Fokker–Planck equation for the probability of finding a stochastic particle at position $\mathbf{x}$ at time t in the presence of a conservative potential $U(\mathbf{x})$ has the formal structure of a continuity equation

$$\frac{\partial P(\mathbf{x}, t)}{\partial t} = -\nabla \cdot \mathbf{J}(\mathbf{x}, t). \tag{2.96}$$

In the absence of external forces, the current vector reads

$$\mathbf{J} = -D \nabla P(\mathbf{x}, t), \tag{2.97}$$

with D denoting the diffusion coefficient. If we multiply the probability $P(\mathbf{x}, t)$ by the total number of Brownian particles and by their mass m (we consider the simple case of identical noninteracting Brownian particles), the mass density of Brownian particles, $\rho_m(\mathbf{x}, t)$, obeys a similar equation,

$$\frac{\partial \rho_m(\mathbf{x}, t)}{\partial t} = -\nabla \cdot \mathbf{J}_m(\mathbf{x}, t), \tag{2.98}$$

where the current density is defined by the constitutive relation

$$\mathbf{J}_m(\mathbf{x}, t) = -D \nabla \rho_m(\mathbf{x}, t). \tag{2.99}$$

We want to point out that for an ensemble of identical noninteracting Brownian particles the current is just proportional to their density gradient, consistent with Eq. (1.26).

A hydrodynamic description of a system where a set of macroscopic observables is conserved (e.g., mass, momentum, energy, charge, spin, etc.) is based on continuity equations of the same kind as Eq. (2.98). More precisely, for any conserved quantity a we can write the continuity equation

$$\frac{\partial \rho_a(\mathbf{x},t)}{\partial t} = -\nabla \cdot \mathbf{J}_a(\mathbf{x},t), \tag{2.100}$$

where $\rho_a(\mathbf{x},t)$ is the density of a conserved quantity a in the system and $\mathbf{J}_a(\mathbf{x},t)$ is the correspondent density current. A constitutive relation like (2.99) can be obtained by going through the same kind of calculations performed in Section 1.3 to find an expression for the fluxes of mass, velocity, and energy as functions of the gradient of the corresponding quantities along a given direction. The following equation is just the generalization of these calculations for a density current vector:

$$\langle \mathbf{J}_a(\mathbf{x},t)\rangle = -D_a \nabla \langle \rho_a(\mathbf{x},t)\rangle. \tag{2.101}$$

It is important to observe that, while Eq. (2.100) expresses an exact, microscopic conservation law, Eq. (2.101) is a phenomenological equation that relates stationary nonequilibrium averages of fluctuating quantities. In fact, the constitutive equation holds in the presence of an applied density gradient, a setup where the role of fluctuations is intrinsic to the transport process. Since Eqs. (2.100) and (2.101) hold for any conserved quantity a, in what follows it is worthwhile overlooking the subscript a and we can write a general Fokker–Planck equation for average quantities after having substituted (2.101) into (2.100):

$$\frac{\partial \langle \rho(\mathbf{x},t)\rangle}{\partial t} = D\nabla^2 \langle \rho(\mathbf{x},t)\rangle\,. \tag{2.102}$$

On the other hand, linear response theory tells us that, for moderate applied gradients, local density fluctuations are practically indistinguishable from spontaneous relaxation to equilibrium of the same quantity (e.g., see Eq. (2.43)); more precisely, they are proportional to each other. In this case we have to consider that we are dealing with densities and current densities, i.e., quantities that depend on both space and time. Accordingly, we can formally replace in (2.102) the average nonequilibrium fluctuation[6] $\langle \rho(\mathbf{x},t)\rangle$ with the equilibrium correlation function of the density,

$$C(\mathbf{x}-\mathbf{y}, t-t') = \langle \rho(\mathbf{x},t)\,\rho(\mathbf{y},t')\rangle_0, \tag{2.103}$$

where, for the sake of simplicity, we have assumed space translation invariance, while time translational invariance is granted at equilibrium. We can finally write

$$\frac{\partial C(\mathbf{x}-\mathbf{y}, t-t')}{\partial t} = D\nabla^2 C(\mathbf{x}-\mathbf{y}, t-t'). \tag{2.104}$$

It is worth rewriting this equation for the space Fourier transform of $C(\mathbf{x}-\mathbf{y}, t-t')$,

$$C(\mathbf{k}, t-t') = \int_V d\mathbf{x}\, e^{-i\mathbf{k}\cdot(\mathbf{x}-\mathbf{y})}\, C(\mathbf{x}-\mathbf{y}, t-t'), \tag{2.105}$$

where $\int_V$ denotes the integral over the volume V of the system. Due to translation invariance this expression is independent of $\mathbf{y}$ and we can write

[6] Let us recall that in Section 2.4.1 we have finally assumed that the equilibrium average of the quantity under study vanishes, so $\langle \rho \rangle_0 = 0$.

$$C(\mathbf{k}, t-t') = \frac{1}{V}\int_V d\mathbf{y} \int_V d\mathbf{x}\, e^{-i\mathbf{k}\cdot(\mathbf{x}-\mathbf{y})}\, C(\mathbf{x}-\mathbf{y}, t-t')$$
$$= \int\int \frac{d\mathbf{x}\, d\mathbf{y}}{V}\, e^{-i\mathbf{k}\cdot(\mathbf{x}-\mathbf{y})}\, \langle \rho(\mathbf{x}, t)\, \rho(\mathbf{y}, t')\rangle_0$$
$$= \frac{1}{V}\langle \rho(\mathbf{k}, t)\, \rho(-\mathbf{k}, t')\rangle_0. \tag{2.106}$$

Because of time translation invariance of the correlation function, we can set $t > t' = 0$ and write the Fokker–Planck equation for the correlation function in the form

$$\frac{\partial\, C(\mathbf{k}, t)}{\partial\, t} = -D k^2 C(\mathbf{k}, t)\,, \tag{2.107}$$

which can be derived by taking the spatial Fourier transform of Eq. (2.104) and applying the Laplacian to both sides of Eq. (2.105). Its solution is

$$C(\mathbf{k}, t) = \exp\,(-D k^2\, t)\, C(\mathbf{k}, 0), \tag{2.108}$$

where $k^2 = |\mathbf{k}|^2$ and $t > 0$. The solution can be extended to $t < 0$ by considering that at equilibrium the correlation function is independent of the choice of the origin of time, i.e.,

$$C(\mathbf{k}, -t) = \frac{1}{V}\langle \rho(\mathbf{k}, -t) \rho(-\mathbf{k}, 0)\rangle_0 \tag{2.109}$$
$$= \frac{1}{V}\langle \rho(\mathbf{k}, 0) \rho(-\mathbf{k}, t)\rangle_0 \tag{2.110}$$
$$= C(-\mathbf{k}, t) \tag{2.111}$$
$$= C(\mathbf{k}, t), \tag{2.112}$$

where the last equality is derived from the spatial parity of the correlation function, $C(\mathbf{r}, t) = C(-\mathbf{r}, t)$. In conclusion, we can write for any t,

$$C(\mathbf{k}, t) = \exp\,(-D k^2\, |t|)\, C(\mathbf{k}, 0). \tag{2.113}$$

The time Fourier transform of $C(\mathbf{k}, t)$ is

$$C(\mathbf{k}, \omega) = \int_{-\infty}^{+\infty} dt\, e^{-i\omega t} C(\mathbf{k}, t) \tag{2.114}$$
$$= C(\mathbf{k}, 0) \int_{-\infty}^{+\infty} dt\, e^{-i\omega t} e^{-Dk^2|t|} \tag{2.115}$$
$$= C(\mathbf{k}, 0) \int_{0}^{+\infty} dt\, e^{-Dk^2 t} \left(e^{-i\omega t} + e^{i\omega t}\right) \tag{2.116}$$
$$= C(\mathbf{k}, 0) \left[\frac{1}{Dk^2 + i\omega} + \frac{1}{Dk^2 - i\omega}\right] \tag{2.117}$$
$$= C(\mathbf{k}, 0)\, \frac{2 D k^2}{\omega^2 + (Dk^2)^2}. \tag{2.118}$$

Therefore, $C(\mathbf{k},\omega)$ is a Lorentzian function centered in $\omega = 0$ and of width Dk^2. If we divide both sides by k^2 and take the limit $k \to 0$, we find

$$\lim_{k\to 0} \frac{1}{k^2} C(\mathbf{k},\omega) = \mathcal{C}\,\frac{2D}{\omega^2}, \tag{2.119}$$

where $\mathcal{C} = C\,(\mathbf{k}=0, t=0)$. So, we can obtain an expression for the generalized diffusion coefficient as

$$D = \frac{1}{2\mathcal{C}} \lim_{\omega\to 0}\lim_{\mathbf{k}\to 0} \frac{\omega^2}{k^2} C(\mathbf{k},\omega). \tag{2.120}$$

In fact, D is a quantity independent of $\mathbf{k}$ and ω and its value is obtained by taking the infinite space ($\lim_{\mathbf{k}\to 0}$) and the infinite time ($\lim_{\omega\to 0}$) limits of the density correlation function. In order to obtain the Green–Kubo relation we want to write D as a function of the density current that appears in (2.100). This equation, for space Fourier transformed variables, is

$$\frac{\partial\,\rho(\mathbf{k},t)}{\partial\,t} + i\mathbf{k}\cdot\mathbf{J}(\mathbf{k},t) = 0, \tag{2.121}$$

where

$$J_i(\mathbf{k},t) = \int d\mathbf{x}\, e^{-i\mathbf{k}\cdot\mathbf{x}} J_i(\mathbf{x},t), \quad i = x,y,z. \tag{2.122}$$

Then, we can write

$$\begin{aligned}\frac{\partial}{\partial\,t}\frac{\partial}{\partial\,t'} C(\mathbf{k},t-t') &= \left\langle \frac{1}{V}\frac{\partial\rho(\mathbf{k},t)}{\partial\,t}\frac{\partial\rho(-\mathbf{k},t')}{\partial\,t'}\right\rangle \\ &= \frac{1}{V}\sum_{i,j} k_i\,k_j\,\langle J_i(\mathbf{k},t)\,J_j(-\mathbf{k},t')\rangle_0\,. \end{aligned}\tag{2.123}$$

We can now perform the time Fourier transform of both sides. The left-hand side gives

$$\int_{-\infty}^{+\infty} d(t-t')e^{-i\omega(t-t')}\frac{\partial}{\partial\,t}\frac{\partial}{\partial\,t'}C(\mathbf{k},t-t') = \omega^2 C(\mathbf{k},\omega). \tag{2.124}$$

If we divide by k^2, for small values of $\mathbf{k}$ we can write

$$\begin{aligned}\lim_{\omega\to 0}\lim_{\mathbf{k}\to 0}\frac{\omega^2}{k^2}C(\mathbf{k},\omega) &= \lim_{\omega\to 0}\int_{-\infty}^{+\infty} d(t-t')e^{-i\omega(t-t')}\lim_{\mathbf{k}\to 0}\sum_{i,j}\frac{k_i\,k_j}{k^2}\frac{1}{V}\langle J_i(\mathbf{k},t)\,J_j(-\mathbf{k},t')\rangle_0 \\ &= \lim_{\omega\to 0}\int_{-\infty}^{+\infty} d(t-t')e^{-i\omega(t-t')}\sum_{i,j}\frac{k_i\,k_j}{k^2}\frac{1}{V}\langle J_i^T(t)\,J_j^T(t')\rangle_0,\end{aligned}$$

where $J_i^T(t)$ denotes the ith component of the total current, i.e.,

$$J_i^T(t) \equiv \int_V d\mathbf{x} J_i(\mathbf{x},t) = \lim_{\mathbf{k}\to 0} J_i(\mathbf{k},t). \tag{2.125}$$

For an isotropic system in d dimensions we have

$$\langle J_i^T(t)\,J_j^T(t')\rangle = \frac{1}{d}\,\delta_{ij}\,\langle \mathbf{J}^T(t)\cdot\mathbf{J}^T(t')\rangle_0\,. \tag{2.126}$$

We can finally rewrite (2.120) as

$$
\begin{aligned}
D &= \frac{1}{2C}\lim_{\omega\to 0}\int_{-\infty}^{+\infty} d(t-t')e^{-i\omega(t-t')}\sum_{i,j}\frac{k_i k_j}{k^2}\frac{1}{V}\frac{1}{d}\delta_{ij}\langle \mathbf{J}^T(t)\cdot\mathbf{J}^T(t')\rangle_0 \\
&= \frac{1}{dVC}\frac{1}{2}\lim_{\omega\to 0}\int_{-\infty}^{+\infty} d(t-t')e^{-i\omega(t-t')}\langle \mathbf{J}^T(t)\cdot\mathbf{J}^T(t')\rangle_0 \\
&= \frac{1}{dVC}\int_0^{+\infty} dt\langle \mathbf{J}^T(t)\cdot\mathbf{J}^T(0)\rangle_0, \qquad (2.127)
\end{aligned}
$$

where we have used the fact that the autocorrelation function is even under time reversal $t \to -t$. The previous equation can be rewritten in a mathematically more appropriate form, because the integral should be regularized as follows,

$$D = \frac{1}{dVC}\lim_{\varepsilon\to 0^+}\int_0^{+\infty} dt\, e^{-\varepsilon t}\langle \mathbf{J}^T(t)\cdot\mathbf{J}^T(0)\rangle_0. \qquad (2.128)$$

This equation is known as the conventional Green–Kubo relation, which expresses a transport coefficient D, associated with a density $\rho(\mathbf{x}, t)$, in terms of the time integral of the autocorrelation function of the total current $\mathbf{J}^T(t)$. Note that in order to determine D we also have to compute

$$C = \lim_{\mathbf{k}\to 0} C(\mathbf{k}, t = 0), \qquad (2.129)$$

which is proportional to the value of the static autocorrelation function of $\rho(\mathbf{k}, t)$ (see Eq. (2.106)).

The Green–Kubo relation is a very interesting result of linear response theory, since it allows computation of transport coefficients in terms of equilibrium averages of total current correlation functions. This is also of great practical interest, because we can compute transport coefficients by numerical simulations, which provide reliable numerical estimates of $\langle \mathbf{J}^T(t)\cdot\mathbf{J}^T(0)\rangle_0$. For instance, this method is quite useful for the progress of the theory of liquids and also for investigating anomalous transport properties in low-dimensional systems, such as carbon nanotubes and suspended graphene layers.

A widely used method to perform numerical simulations of physical models is classical molecular dynamics. Let us just sketch the basic ingredients of this method. In the simple case of a homogeneous system we deal with N equal particles of mass m, interacting among themselves, whose energy is given by a Hamiltonian $\mathcal{H}(\{\mathbf{r}_i(t), \mathbf{v}_i(t)\}_{i=1}^N)$, where $\mathbf{r}_i(t)$ and $\mathbf{v}_i(t)$ are the position and velocity coordinates of the ith particle. The equations of motion of the particles are given by the Hamilton equations and they can be integrated by suitable symplectic algorithms.[7] The system can be put in contact with reservoirs, which can be represented in the simulations by additional dynamical variables generated by deterministic or stochastic rules. The dynamical equations of the system plus reservoirs are integrated over a sufficiently long time interval to obtain reliable equilibrium averages of the total

[7] Symplectic integration algorithms are typically employed for approximating Hamiltonian dynamics into the form of a map evolving by finite time steps. In fact, the map is written in such a way to preserve any volume in phase space, i.e., to satisfy Liouville's theorem. The main practical consequence is that the approximate symplectic algorithm conserves all the invariant quantities of the original Hamiltonian.

current correlation function appearing in Eq. (2.128). In general, we can invoke ergodicity for the system under consideration, i.e., the equivalence between time and ensemble averages. In a strictly rigorous mathematical sense there are few cases where ergodicity can be proved. On the other hand, weak ergodicity[8] can be assumed to be valid for many models, where relaxation processes emerge spontaneously as a result of the interactions among particles and between particles and reservoirs.[9] The quality of numerical estimates of time correlation functions appearing in the Green–Kubo relations can be improved by averaging over many different initial conditions and realizations of the stochastic processes associated with the presence of reservoirs.

We want to conclude this section by reporting, as an example, the expressions of transport coefficients for a one-dimensional system of N particles of mass m, interacting through a nearest-neighbor confining potential $V(x_{n+1} - x_n)$, where the subscript n labels the lattice site and the variable x_n can be read as the displacement of the nth particle with respect to its equilibrium position. We assume periodic boundary conditions. In this case the Green–Kubo relations take a particularly simple form and physical interpretation. The dynamics of particles is ruled by Newton's dynamics,

$$m\ddot{x}_n = -F_n + F_{n-1}$$
$$F_n = -V'(x_{n+1} - x_n)\,,$$

where $V'(x)$ is a shorthand notation for the derivative of $V(x)$ with respect to its argument. By assuming that the velocity of the center of mass is zero,[10] one obtains the following expressions for the bulk viscosity η_B and the thermal conductivity κ,

$$\eta_B = \frac{1}{NT}\int_0^{+\infty} dt\,\langle J_p(t)J_p(0)\rangle, \tag{2.130}$$

where

$$J_p(t) = \sum_{n=1}^{N} F_n(t) \tag{2.131}$$

is the total momentum flux, and

$$\kappa = \frac{1}{NT^2}\int_0^{+\infty} dt\,\langle J_E(t)J_E(0)\rangle, \tag{2.132}$$

[8] This concept was introduced by Aleksandr Khinchin as a heuristic substitute of true ergodicity. When, starting from generic initial conditions, all quantities of physical interest typically exhibit a fast relaxation to their equilibrium values, we can assume that in practice ergodicity holds, although a true rigorous proof is unavailable. The typical test of this hypothesis amounts to checking that the time autocorrelation function of the observable $X(t)$ decays exponentially fast in time,

$$\langle X(t)\,X(0)\rangle \sim e^{-t/t_0}.$$

In general, the relaxation time scale t_0 is expected to be weakly dependent on the observable $X(t)$.

[9] When stochastic rules are at work in molecular dynamics simulations, they are usually expected to facilitate the relaxation processes by introducing in the dynamics incoherent fluctuations in the form of noise. In fact, it is well known that a purely deterministic dynamics, even in the case of large systems, may exhibit dynamical regimes where ergodic conditions cannot be achieved or might eventually show up over practically unaccessible time scales.

[10] This condition can be set in the initial condition and automatically preserved, because periodic boundary conditions allow for symmetry by translation invariance in the lattice, i.e., the conservation of momentum.

where

$$J_E(t) = \frac{1}{2} \sum_{n=1}^{N} F_n(t)(\dot{x}_{n+1} + \dot{x}_n) \tag{2.133}$$

is the total heat flux.

2.6 Generalized Linear Response Function

We have to consider that if $h(t)$ is able to perturb its conjugated macroscopic observable $X(t)$ to which it is directly coupled, it also can influence the dynamical state of other macroscopic observables. More precisely, we can generalize what was discussed in Sections 2.4 and 2.4.1 by introducing a perturbed Hamiltonian that depends on a set of thermodynamic observables $X_k(t)$ and their conjugate perturbation fields $h_k(t)$,

$$\mathcal{H}' = \mathcal{H} - \sum_k h_k(t)\, X_k, \tag{2.134}$$

where, without prejudice of generality, we can assume $\langle X_k \rangle_0 = 0\ \forall k$.

In the following we focus on the macroscopic observable $X_i(t)$ and we evaluate its response to the switching of the perturbation field $h_j(t)$, conjugate to $X_j(t)$. In analogy with (2.39), the starting point is the equation

$$\langle X_i(t) \rangle = \frac{1}{Z'} \int_{\mathcal{V}} d\mathcal{V}\ e^{-\beta \mathcal{H}'(q_i,p_i)} X_i(t). \tag{2.135}$$

In the linear approximation the response function can be evaluated by calculating the derivative of (2.135) with respect to h_j,

$$\begin{aligned} \frac{\partial \langle X_i(t) \rangle}{\partial h_j} &= \frac{\beta}{Z'} \int_{\mathcal{V}} d\mathcal{V}\ e^{-\beta \mathcal{H}'(q_i,p_i)} X_i(t) X_j(0) \\ &\quad - \frac{\beta}{(Z')^2} \int_{\mathcal{V}} d\mathcal{V}\ e^{-\beta \mathcal{H}'(q_i,p_i)} X_i(t) \int_{\mathcal{V}} d\mathcal{V}\ e^{-\beta \mathcal{H}'(q_i,p_i)} X_j(0), \end{aligned} \tag{2.136}$$

and by evaluating this expression for $h_j = 0$, so that (2.136) yields the relation (compare with (2.42))

$$\langle X_i(t) \rangle - \langle X_i(t) \rangle_0 = \beta h_j \Big(\langle X_j(0) X_i(t) \rangle_0 - \langle X_j(0) \rangle_0 \langle X_i(t) \rangle_0 \Big) + O(h_j^2). \tag{2.137}$$

Following the same procedure sketched in Section 2.4.1, we can finally obtain, in the linear approximation, an expression for the generalized response function $\Xi_{ij}(t)$,

$$\Xi_{ij}(t) = -\beta\, \Theta(t) \frac{d}{dt} \langle X_i(t)\, X_j(0) \rangle_0\,, \tag{2.138}$$

to be compared with Eq. (2.49). It is worth recalling that also in this general case we have assumed $\langle X_i \rangle_0 = \langle X_j \rangle_0 = 0$. This generalized response function satisfies the fluctuation–dissipation relation

$$\langle X_i(t) \rangle = \int dt'\ \Xi_{ij}(t - t') h_j(t'), \tag{2.139}$$

whose version in the Fourier space reads

$$\langle X_i(\omega) \rangle = \Xi_{ij}(\omega)\, h_j(\omega). \tag{2.140}$$

We can also rewrite Eq. (2.138) in the Fourier space as

$$\Xi_{ij}(\omega) = -\beta \int_{-\infty}^{+\infty} \frac{d\omega'}{2\pi} \Theta(\omega - \omega')(i\omega')\, C_{ij}(\omega'), \tag{2.141}$$

where $C_{ij}(\omega) = \langle |X_i^*(\omega) X_j(\omega)| \rangle_0$ is the Fourier transform of the time correlation function $\langle X_i(t)\, X_j(0) \rangle_0$. The previous equation generalizes Eq. (2.53).

All the properties of the response function discussed in Section 2.4.1 straightforwardly hold for $\Xi_{ij}(\omega)$. In particular, the fluctuation–dissipation theorem (generalization of Eq. (2.56)) is

$$\Xi_{ij}^I(\omega) = -\frac{\beta\omega}{2} C_{ij}(\omega), \tag{2.142}$$

while the Kramers–Kronig relation (generalization of Eq. (2.57)) now is

$$\Xi_{ij}^R(\omega) = \mathrm{PV} \int_{-\infty}^{+\infty} \frac{d\omega'}{\pi} \frac{1}{\omega' - \omega}\, \Xi_{ij}^I(\omega'), \tag{2.143}$$

where $\Xi_{ij}^R(\omega)$ and $\Xi_{ij}^I(\omega)$ denote the real and the imaginary part of the response function $\Xi_{ij}(\omega)$, respectively.

Eq. (2.142) shows that the imaginary part of the Fourier transform of the response function is related to the equilibrium fluctuations of the same pair of observables through the Fourier transform of their time correlation function.

2.6.1 Onsager Regression Relation and Time Reversal

In this section we want to illustrate the Onsager regression relation in the framework of the general linear response approach discussed in the previous section. Taking advantage of time–translation invariance at equilibrium, we can rewrite Eq. (2.138) as

$$\Xi_{ij}(t - t') = \beta\, \Theta(t - t') \frac{d}{dt'} \langle X_i(t)\, X_j(t') \rangle_0 \,. \tag{2.144}$$

Let us now consider perturbation fields whose time dependence is given by

$$h_j(t) = h\, \Theta(-t)\, e^{\epsilon t}, \tag{2.145}$$

where $0 < \epsilon \ll 1$ is a small parameter, such that $h_k(t)$ slowly increases from $-\infty$ to its maximum amplitude h for $t \to 0$, just before being switched off at $t = 0$. Then, for $t > 0$, Eq. (2.139) becomes

$$\langle X_i(t) \rangle = \beta\, h \int_{-\infty}^{0} dt'\, e^{\epsilon t'} \frac{d}{dt'} \langle X_i(t)\, X_j(t') \rangle_0 \,. \tag{2.146}$$

We can integrate by parts the right-hand side and finally obtain[11]

$$\langle X_i(t)\rangle = \beta\, h\left\{\left[e^{\epsilon t'}\langle X_i(t)X_j(t')\rangle_0\right]_{t'=-\infty}^{t'=0} - \epsilon\int_{-\infty}^{0} dt'\; e^{\epsilon t'}\langle X_i(t)X_j(t')\rangle_0\right\}$$
$$= \beta\, h\,\langle X_i(t)X_j(0)\rangle_0 + O(\epsilon). \tag{2.147}$$

This result is a generalization of Eq. (2.137) because in some sense it takes into account the switching process of the perturbation, which has now been shown to be of order ϵ. Eq. (2.147) is known as the Onsager regression relation, which expresses quite an important concept pointed out by the Norwegian physicist Lars Onsager. He argued that if we are studying a physical system during its evolution around equilibrium we cannot distinguish if an instantaneous value of an observable $\langle X_i(t)\rangle$, different from its equilibrium value $\langle X_i\rangle_0$, is due to a spontaneous fluctuation or to a perturbation switched off in the past.[12] This statement implies that the way a fluctuating observable spontaneously decays toward its equilibrium value cannot be distinguished by the way its time correlation function decays at equilibrium.

It is important to point out that the fluctuating macroscopic observables $X_i(t)$ that we consider in the linear response theory are functions of time t through the canonical coordinates $(q_k(t), p_k(t))$, because we have assumed that the physical system they belong to is represented by a Hamiltonian $\mathcal{H}(q_k(t), p_k(t))$ (see Section 2.4). As we have discussed in the previous sections, equilibrium quantities, like time correlation functions of fluctuating observables, also characterize nonequilibrium properties in the linear response theory. In a statistical sense, the equilibrium Gibbs measure is expressed in terms of the Hamiltonian, together with its symmetries. In particular, the basic one is the time-translational symmetry that is associated with the conservation of energy in the isolated system. In particular, this implies that the Hamiltonian is invariant under time reversal. For consistency reasons we also have to assume that all observables $X_i(t) \equiv X_i(q_k(t), p_k(t))$ have a definite parity with respect to time reversal, according to the way they depend on microscopic canonical coordinates.[13] For instance, observables such as particle and energy densities have even parity with respect to time reversal, because they are even functions of the particle velocities, while momentum density has odd parity, because it depends linearly on the particle velocities.

All this matter can be expressed in mathematical language by introducing a time-reversal operator $\mathcal{T}$ that applies to the phase–space points as

$$\mathcal{T}(q_k(t), p_k(t)) = (q_k(t), -p_k(t))\,, \quad \text{with } \mathcal{T}\mathcal{H} = \mathcal{H}, \tag{2.148}$$

i.e., $\mathcal{T}$ inverts the components of momenta of all particles in the system and $\mathcal{H}$ has to be an even function of momenta. This implies that if the trajectory $(q_k(t), p_k(t))$ is a solution of the Hamilton equations (2.27) with initial conditions $(q_k(0), p_k(0))$, the time-reversed trajectory $\mathcal{T}(q_k(t), p_k(t))$ is also a solution with initial condition $\mathcal{T}(q_k(0), p_k(0))$. We can

[11] We must first take the appropriate limits for t', then take the limit $\epsilon \to 0$.

[12] The average $\langle\cdot\rangle$ is performed over a statistical ensemble of perturbed systems, subject to the same perturbation field $h(t)$ as the one defined in Eq. (2.145).

[13] We are considering Hamiltonian systems in the absence of magnetic fields.

assume that all observables $X_i(t)$, as well as $\mathcal{H}$, are eigenfunctions of $\mathcal{T}$, because they have a well-defined parity. Thus

$$X_i(-t) \equiv \mathcal{T}X_i(t) = X_i(\mathcal{T}(q_k(t), p_k(t))) = X_i((q_k(t), -p_k(t))) = \tau_i X_i(t), \tag{2.149}$$

where the eigenvalues $\tau_i = \pm 1$ determine the parity of observable $X_i(t)$ with respect to time reversal.

For what we are going to discuss later on it is important to determine how $\mathcal{T}$ acts over the time correlation functions $\langle X_i(t) X_j(t') \rangle_0$. We can assume that we are in a given initial state $(q_k(0), p_k(0))$ at $t' = 0$ and apply to it the time reversal operator $\mathcal{T}(q_k(0), p_k(0)) = (q_k(0), -p_k(0))$. For instance, in a Hamiltonian system made of identical particles this operation amounts to changing the initial conditions by keeping the same space coordinates, but reversing the velocities. According to (2.149), we have $\mathcal{T}X_j(0) = \tau_j X_j(0)$. We now let the system evolve for a time $-t$ to the state $(q_k(-t), p_k(-t))$ where the corresponding value of $X_i(-t)$ has to satisfy the eigenvalue relation $\mathcal{T}X_i(-t) = \tau_i X_i(-t)$. At thermodynamic equilibrium the probability of the initial state $(q_k(0), p_k(0))$ and of the time-reversed state $(q_k(0), -p_k(0))$ are the same, because of the invariance of $\mathcal{H}$ under the action of $\mathcal{T}$. We can finally write[14]

$$\langle X_i(t) X_j(0) \rangle_0 = \langle \mathcal{T}X_i(-t)\, \mathcal{T}X_j(0) \rangle_0 = \tau_i \tau_j \langle X_i(-t) X_j(0) \rangle_0 \,. \tag{2.150}$$

We can conclude that the correlation functions of observables with the same parity with respect to time reversal are even functions of time, while those of observables with different parity are odd functions of time. This result has important consequences for the properties of the response function (2.144), that, for $t > 0$, can be rewritten

$$\Xi_{ij}(t) = -\beta \frac{d}{dt} \langle X_i(t) X_j(0) \rangle_0 = -\beta\, \tau_i\, \tau_j\, \frac{d}{dt} \langle X_i(-t) X_j(0) \rangle_0 \,. \tag{2.151}$$

Because of the invariance of the Hamiltonian dynamics under time shifts at equilibrium, we can also write

$$\Xi_{ij}(t) = -\beta\, \tau_i\, \tau_j\, \frac{d}{dt} \langle X_i(0) X_j(t) \rangle_0 = \tau_i\, \tau_j\, \Xi_{ji}(t). \tag{2.152}$$

This is an important relation that will be used in the next section to establish symmetries among kinetic coefficients in the Onsager theory of linear response.

2.7 Entropy Production, Fluxes, and Thermodynamic Forces

In this section we aim at providing a general description of nonequilibrium processes in the linear response regime based on the thermodynamics of irreversible processes. This is summarized by the second principle of thermodynamics: when a physical system is in nonequilibrium conditions its entropy, S, must increase, namely, $dS \geq 0$.[15]

[14] We recall once again that the average, equilibrium values of the various quantities are assumed to vanish, without loss of generality.

[15] Such a statement is certainly valid for physical systems where the interaction forces among the particles are short range; when long-range forces are present the situation may be much more complex.

2.7.1 Nonequilibrium Conditions between Macroscopic Systems

For pedagogical reasons we discuss the simple case where a physical system, in contact with an external reservoir, is in nonequilibrium conditions with it. We assume that in this universe the total variation of entropy $d\mathbb{S}$ can be written as

$$d\mathbb{S} = dS + dS_r, \tag{2.153}$$

i.e., it is the sum of the variation of entropy in the system, dS, and in the reservoir, dS_r. Notice that this quantity is not necessarily positive, depending on the way the reservoir is coupled to the system. For an adiabatically insulated system, $dS_r = 0$ by definition and $d\mathbb{S} \equiv dS \geq 0$. More generally, if we assume that the reservoir is at (absolute) temperature T the variation of entropy in the reservoir is proportional to the total heat supplied to the system through the Carnot–Clausius formula,

$$dS_r = -\frac{dQ}{T}, \tag{2.154}$$

where $dQ > 0$ means that the system absorbs heat from the reservoir. Since $dS \geq 0$, in nonequilibrium conditions, when the system can exchange only heat with the reservoirs, we have

$$d\mathbb{S} \geq \frac{dQ}{T}. \tag{2.155}$$

The previous equation is a way of expressing the second law of thermodynamics.

The problem of nonequilibrium thermodynamics amounts to establishing a relation between the way entropy varies in time and the irreversible phenomena that generate such a variation. A first step in this direction can be accomplished by considering that the entropy depends on a set of extensive thermodynamic observables X_i, namely,

$$S \equiv S(\{X_i\})\,. \tag{2.156}$$

Due to their extensive character, we can assume that a given value $\mathbb{X}_i$ taken in our universe by one of these observables has to be the sum of the values taken by this observable in the system, X_i, and in the reservoir, $X_i^{(r)}$, that is,

$$\mathbb{X}_i = X_i + X_i^{(r)}. \tag{2.157}$$

If fluctuations allow both X_i and $X_i^{(r)}$ to be unconstrained, we can define the thermodynamic force $\mathbb{F}_i$ associated with the extensive observable X_i as

$$\mathbb{F}_i \equiv \left(\frac{\partial \mathbb{S}}{\partial X_i}\right)\Big|_{X_i=\mathbb{X}_i} = \frac{\partial S}{\partial X_i} - \frac{\partial S_r}{\partial X_i^{(r)}} \equiv F_i - F_i^{(r)}, \tag{2.158}$$

where we have defined

$$F_i = \left(\frac{\partial S}{\partial X_i}\right) \tag{2.159}$$

$$F_i^{(r)} = \left(\frac{\partial S_r}{\partial X_i^{(r)}}\right) \tag{2.160}$$

and we have taken explicitly into account that, as a consequence of the closure condition (2.157), $dX_i = -dX_i^{(r)}$, since the value $\mathbb{X}_i$ is fixed in our universe, i.e., $d\mathbb{X}_i = 0$. Notice that the partial derivatives in (2.158) are performed while keeping constant all the other extensive thermodynamic observables. These generalized thermodynamic forces $\mathbb{F}_i$, that are also called affinities, are responsible for driving the system in out-of-equilibrium conditions. At thermodynamic equilibrium the total entropy in the universe $\mathbb{S}$ is by definition maximal, so that $\mathbb{F}_i = 0$. On the other hand, if $\mathbb{F}_i \neq 0$ an irreversible process sets in, spontaneously driving the system toward an equilibrium state.

We want to point out that an equivalent way of introducing the concept of affinities can be obtained by reinterpreting S and S_r as the entropies of two subsystems forming a closed, i.e., isolated, system whose entropy is given by $\mathbb{S}$. At a classical level the two interpretations are equivalent, because it is natural to assume that a physical system can be isolated, while the two subsystems, by definition, have the same physical properties. At a quantum level the idealization of an isolated physical system seems quite an ill-defined concept. Moreover, the presence of a thermal reservoir, which represents the complement with respect to the universe of the system under scrutiny, implies that the reservoir is not necessarily made only of the same kind of physical units (particles, atoms, molecules, etc.) that are present in the system. In this perspective, the interpretation with the reservoir seems to be more general and appropriate. On the other hand, in this case we have to postulate the existence of a physical mechanism that allows the reservoir to interact with the system. Far from being a purely philosophical argument, this is a crucial point to be taken into account, if we aim at constructing a physical model of a suitable thermal reservoir, in both the classical and quantum cases. In particular, this is of primary importance if we have to deal with numerical simulations in the presence of a thermal reservoir, and various schemes can be adopted, including stochastic as well as deterministic formulations. In spite of their intellectual and practical interest, these topics are not considered in this textbook.

We can exemplify our previous considerations about a system in out-of-equilibrium conditions by first recalling that typical examples of extensive thermodynamic observables are the internal energy U, the volume V, and the number of particles N_j of the species j contained in the system. The Gibbs relation

$$T\,dS = dU + P\,dV - \sum_j \mu_j\, dN_j \tag{2.161}$$

provides us with the functional dependence of S on these extensive observables, where T is the absolute temperature, P is the pressure, and μ_j is the chemical potential of particles of species j. Notice that this relation deals with entropy variations at close to equilibrium conditions. We assume that such a functional dependence is valid also for the above-described out-of-equilibrium conditions. For instance, we can still define the absolute temperature of the system and of the reservoir (or of the two subsystems) by the expression

$$\left(\frac{\partial S}{\partial U}\right)_{V,N_j} = \frac{1}{T} \tag{2.162}$$

and, analogously, the pressure and chemical potentials by the expressions

$$\left(\frac{\partial S}{\partial V}\right)_{U,N_j} = \frac{P}{T}, \qquad \left(\frac{\partial S}{\partial N_j}\right)_{U,V} = -\frac{\mu_j}{T}, \tag{2.163}$$

where we adopt the standard notation for partial derivatives and keep the subscript quantities constant.

With reference to Eq. (2.158), the affinities stemming from the Gibbs relation (2.161) are

$$\mathbb{F}_U = \frac{1}{T} - \frac{1}{T_r}, \quad \mathbb{F}_V = \frac{P}{T} - \frac{P_r}{T_r}, \quad \mathbb{F}_{N_j} = -\frac{\mu_j}{T} + \frac{(\mu_j)_r}{T_r}. \tag{2.164}$$

The vanishing of all affinities implies that the system and the reservoir share the same temperature, pressure, and chemical potentials; therefore, they are at equilibrium. On the other hand, if the difference of the inverse temperatures is nonzero there is a heat flow between the system and the reservoir that aims at establishing global equilibrium conditions. For instance, we can think that the system and the reservoir are in contact through a thin diathermal wall, which allows the heat to flow from the higher temperature to the lower one, until the system eventually takes the temperature of the reservoir (or the two subsystems eventually exhibit the same temperature, which results by a suitable average of the original temperatures). This is to point out that, in general, the response to an applied thermodynamic force, $\mathbb{F}_i$, should be related to the rate of change over time of the extensive observable X_i. More precisely, we can define the fluxes (rate of changes) as

$$J_i = \frac{dX_i}{dt}. \tag{2.165}$$

When the affinity vanishes, the corresponding flux also vanishes. The thermodynamics of irreversible processes aims at establishing the relation between fluxes and affinities. For this purpose it is useful to introduce an explicit dependence of the entropy on time, so we can define the entropy production rate as

$$\frac{d\mathbb{S}}{dt} = \sum_i \left(\frac{\partial S}{\partial X_i} - \frac{\partial S_r}{\partial X_i^{(r)}} \right) \frac{dX_i}{dt} = \sum_i \mathbb{F}_i J_i. \tag{2.166}$$

Since our universe is isolated, the second principle of thermodynamics states that in out-of-equilibrium conditions the total entropy has to increase, namely,

$$\frac{d\mathbb{S}}{dt} = \sum_i \mathbb{F}_i J_i \geq 0, \tag{2.167}$$

and it vanishes only when all affinities and the corresponding fluxes vanish, i.e., at thermodynamic equilibrium.[16]

2.7.2 Phenomenological Equations

After having identified the affinities, we can proceed to the main goal of the thermodynamics of irreversible processes, i.e., finding a relation between affinities and fluxes. For this purpose, within a linear nonequilibrium regime, we suppose that fluctuating macroscopic observables $X_i(t)$ evolve in time according to linear equations of the form

[16] There are peculiar situations in coupled transport processes (see Section 2.8), where, as a consequence of symmetries, a flux can be made to vanish by applying suitable, nonvanishing affinities, although the system is kept in stationary out-of-equilibrium conditions.

$$J_i = \frac{d}{dt} X_i(t) = -\sum_k \mathcal{M}_{ik} X_k(t), \tag{2.168}$$

where $\mathcal{M}_{ik}$ are phenomenological coefficients, depending both on the nature of the specific fluctuating quantities and on the specific conditions leading the system out of equilibrium. If only one quantity fluctuates, this assumption leads to an exponential relaxation toward an asymptotic, vanishing value, in line with the linear response theory, where the exponential relaxation is the result of the equilibrium correlation function.

As we are going to show later in this section, such an assumption yields a linear relation between fluxes and affinities. This relation is consistent with the assumption of a linear response to an applied thermodynamic force, meaning that, at leading order, the flux of an extensive thermodynamic observable is proportional to the affinities. The proportionality constants are the so-called (linear) kinetic coefficients.[17]

On the other hand, some physical arguments associated with the fluctuating nature of $X_i(t)$ suggest that we should reconsider the meaning we attribute to the set of differential equations (2.168). The first consideration is that we expect that Eq. (2.168) describes confidently the linear response regime over time scales sufficiently longer than the correlation time, t_c, of spontaneous fluctuations in the system: in this sense t has to be intended as a coarse-grained variable. Moreover, the flux $J_i(t)$ can be more properly defined by substituting $X_i(t)$ with its average over realizations of the fluctuation–relaxation process,[18] namely,

$$J_i = \frac{d}{dt} \langle X_i(t) \rangle = -\sum_k \mathcal{M}_{ik} \langle X_k(t) \rangle, \tag{2.169}$$

where the phenomenological coefficients are assumed to be physical quantities independent of the average over realizations.

We can use the Onsager regression relation (2.147) and rewrite (2.169) as

$$\frac{d}{dt} \langle X_i(t)\, X_j(0) \rangle_0 = -\sum_k \mathcal{M}_{ik} \langle X_k(t)\, X_j(0) \rangle_0\,, \tag{2.170}$$

with the caveat that is valid for $t > 0$. If X_i and X_j are both invariant under time reversal (i.e., $\tau_i = \tau_j = 1$), we have

$$\langle X_i(t)\, X_j(0) \rangle_0 = \langle X_j(t)\, X_i(0) \rangle_0\,. \tag{2.171}$$

By differentiating both sides of this equation with respect to time and using (2.170) we obtain

$$\sum_k \mathcal{M}_{ik} \langle X_k(t)\, X_j(0) \rangle_0 = \sum_k \mathcal{M}_{jk} \langle X_k(t)\, X_i(0) \rangle_0\,, \tag{2.172}$$

which can be evaluated for $t = 0$, yielding the relation

$$\sum_k \mathcal{M}_{ik} \langle X_k X_j \rangle_0 = \sum_k \mathcal{M}_{jk} \langle X_k X_i \rangle_0\,, \tag{2.173}$$

[17] In general, one could also consider higher-order kinetic coefficients by assuming additional nonlinear dependences of fluxes on affinities: such higher-order coefficients may become relevant in out-of-equilibrium regimes dominated by nonlinear effects.

[18] This assumption can be justified with the same arguments that we have used for replacing Eq. (2.100) with Eq. (2.102).

where $\langle X_k X_j \rangle_0$ is a shorthand notation for the equal-time equilibrium correlation. If we define the matrix L, whose elements are

$$L_{ij} = \sum_k \mathcal{M}_{ik} \langle X_k X_j \rangle_0 \,, \tag{2.174}$$

Eq. (2.173) implies that L is symmetric,

$$L_{ij} = L_{ji}. \tag{2.175}$$

This symmetry property between phenomenological transport coefficients is known as the Onsager reciprocity relation.

Now we want to show that by making use of Eqs. (2.170) we can obtain a linear relation between fluxes and affinities of the form

$$J_i = \frac{d}{dt} X_i(t) = \sum_j L_{ij} \mathbb{F}_j, \tag{2.176}$$

where the matrix elements L_{ij} are defined in Eq. (2.174). Before proving the validity of these phenomenological relations, we want to observe that they allow us to express the elements of the Onsager matrix as

$$L_{ij} = \left(\frac{\partial J_i}{\partial \mathbb{F}_j} \right)_{\mathbb{F}_k = 0}, \tag{2.177}$$

i.e., the matrix L of the linear kinetic coefficients contains elements that are computed at equilibrium conditions, i.e., when all affinities $\mathbb{F}_k$ vanish. In other words, the L_{ij} can be expressed in terms of quantities such as temperature, pressure, chemical potentials, etc., that characterize the equilibrium state of the system.[19]

As we have obtained (2.170) from (2.168), in this case we should also consider the fluctuating nature of the quantities $X_i(t)$ and $\mathbb{F}_k(t)$ appearing in (2.176). We can proceed formally multiplying both sides of (2.176) by $X_j(0)$ and considering the equilibrium averages of the products, thus obtaining

$$\frac{d}{dt} \langle X_i(t) X_j(0) \rangle_0 = \sum_k L_{ik} \langle \mathbb{F}_k(t) X_j(0) \rangle_0. \tag{2.178}$$

Taking into account (2.170) we can write

$$\sum_k \mathcal{M}_{ik} \langle X_k(t) X_j(0) \rangle_0 = -\sum_k L_{ik} \langle \mathbb{F}_k(t) X_j(0) \rangle_0. \tag{2.179}$$

In particular, this relation holds for equal-time equilibrium correlations

$$\sum_k \mathcal{M}_{ik} \langle X_k X_j \rangle_0 = -\sum_k L_{ik} \langle \mathbb{F}_k X_j \rangle_0. \tag{2.180}$$

If we are considering nonequilibrium conditions between macroscopic systems, the right-hand side of this equation can be written using (2.158) as

[19] This can be viewed as a generalization of what we found in Section 1.3, where the transport coefficients derived from the kinetic theory have been expressed as functions of particle density, mean free path, average velocity, specific heat at constant volume, etc.

$$\langle \mathbb{F}_k X_j \rangle_0 = \left\langle \frac{\partial S}{\partial X_k} X_j \right\rangle_0 - \left\langle \frac{\partial S_r}{\partial X_k^{(r)}} X_j \right\rangle_0 . \tag{2.181}$$

If the system is in contact with a reservoir, the second term on the right-hand side can be neglected, because at equilibrium its correlations between entropy fluctuations and extensive observables are irrelevant. Therefore, using Eq. (2.159) we can eventually write

$$\langle \mathbb{F}_k X_j \rangle_0 = \left\langle F_k X_j \right\rangle_0 . \tag{2.182}$$

The equilibrium correlation on the right-hand side of Eq. (2.182) can be evaluated by considering the fundamental postulate of equilibrium statistical mechanics, according to which

$$S(\{X_i\}) = \ln \Gamma, \tag{2.183}$$

where Γ is the volume in the phase space, which is compatible with the assigned values of the extensive thermodynamic variables X_i. The main consequence of this postulate is that the equilibrium value of any observable $\mathcal{O}(\{X_i\})$ can be defined as

$$\langle \mathcal{O}(\{X_i\}) \rangle_0 = \frac{1}{\mathcal{N}} \int \prod_i dX_i \mathcal{O}(\{X_i\}) e^{S(\{X_i\})}, \tag{2.184}$$

where the normalization factor is given by the expression

$$\mathcal{N} = \int \prod_i dX_i e^{S(\{X_i\})}. \tag{2.185}$$

If we take $\mathcal{O}(\{X_i\}) = \frac{\partial S}{\partial X_k} X_j$, we can write

$$\begin{aligned} \left\langle \frac{\partial S}{\partial X_k} X_j \right\rangle_0 &= \frac{1}{\mathcal{N}} \int \prod_i dX_i \frac{\partial S}{\partial X_k} X_j e^{S(\{X_i\})} \\ &= \frac{1}{\mathcal{N}} \int \prod_i dX_i \frac{\partial}{\partial X_k} \left(e^{S(\{X_i\})} \right) X_j \end{aligned} \tag{2.186}$$

and integrating by parts, we obtain

$$\langle F_k X_j \rangle_0 = \left\langle \frac{\partial S}{\partial X_k} X_j \right\rangle_0 \tag{2.187}$$

$$= -\delta_{jk} + \frac{1}{\mathcal{N}} \int \prod_{i \neq k} dX_i X_j \, e^{S(\{X_i\})} \Big|_{X_k^{\min}}^{X_k^{\max}} \tag{2.188}$$

$$= -\delta_{jk}, \tag{2.189}$$

where the second part of the integral, coming from integration by parts, can be neglected in the thermodynamic limit, while the other part reduces to the Kronecker delta δ_{jk}, because the extensive thermodynamic variables are assumed to be independent of each other.

Substituting this result into (2.180) we recover the definition (2.174), thus showing the validity of the phenomenological relations (2.176). We can conclude that such relations represent an effective approximation in a linear response regime, where local equilibrium conditions hold.

2.7.3 Variational Principle

Using Eq. (2.176) the entropy production rate, given in Eq. (2.166), can be expressed as a bilinear form of the affinities,

$$\frac{d\mathbb{S}}{dt} = \sum_{ij} L_{ij}\, \mathbb{F}_i\, \mathbb{F}_j. \tag{2.190}$$

In out-of-equilibrium conditions the right-hand side of this equation has to be a positive quantity, which must vanish at equilibrium. In particular, if we switch off all currents other than J_i, we have $\frac{d\mathbb{S}}{dt} = L_{ii}\mathbb{F}_i^2$, which implies

$$L_{ii} > 0 \quad \forall i. \tag{2.191}$$

We can also compute the time derivative of the entropy production rate,

$$\frac{d^2\mathbb{S}}{dt^2} = 2\sum_{ij}\sum_{k} \frac{\partial \mathbb{F}_i}{\partial X_k}\frac{dX_k}{dt} L_{ij}\, \mathbb{F}_j \tag{2.192}$$

$$= 2\sum_{ik} \frac{\partial \mathbb{F}_i}{\partial X_k} J_k J_i$$

$$= 2\sum_{ik} \frac{\partial^2 \mathbb{S}}{\partial X_k \partial X_i} J_k J_i, \tag{2.193}$$

where we have used, in order, the symmetry condition (2.175), the phenomenological relation (2.176), and the definition of generalized thermodynamic force in (2.158).

Since the entropy $\mathbb{S}(\{X_i\})$ is a concave function of its arguments, we can conclude that the right-hand side of Eq. (2.193) is a negative quadratic form and

$$\frac{d^2\mathbb{S}}{dt^2} \le 0. \tag{2.194}$$

Therefore, in a linear regime the entropy production rate, equal to $d\mathbb{S}/dt$, spontaneously decreases during the system evolution and there is a natural tendency of the system to minimize the entropy production rate. If the out-of-equilibrium conditions are transient, the entropy production rate will eventually vanish, while it may approach a minimum finite value when stationary out-of-equilibrium conditions are imposed.

In order to illustrate the latter scenario we take as an example a system that is maintained in nonequilibrium conditions by fixing the value of one affinity, say $\mathbb{F}_i$, while all other affinities, $\mathbb{F}_k$, are free to vary. Making use of the rate of entropy production (2.190) and of the symmetry of the matrix L, we can write

$$\left.\frac{\partial\, \dot{\mathbb{S}}}{\partial\, \mathbb{F}_k}\right|_{k\neq i} = 2\sum_j L_{kj}\, \mathbb{F}_j = 2J_k\,. \tag{2.195}$$

By this relation and (2.194), we can conclude that the stationary state, i.e., the nonequilibrium state corresponding to the vanishing of all currents J_k other than J_i, corresponds to an extremal state with respect to the varying affinities. The evaluation of the second derivative,

$$\frac{\partial^2 \dot{\mathbb{S}}}{\partial \mathbb{F}_k^2} = 2L_{kk} > 0, \tag{2.196}$$

tells us that we are dealing with a minimum entropy production rate.

Let us consider a system in contact with two reservoirs, which we name a and b. In this case Eq. (2.166) becomes

$$\frac{d\mathbb{S}}{dt} = \sum_i \left(\frac{\partial S}{\partial X_i} J_i + \frac{\partial S_a}{\partial X_i^{(a)}} J_i^{(a)} + \frac{\partial S_b}{\partial X_i^{(b)}} J_i^{(b)} \right), \tag{2.197}$$

where we have denoted with $J_i^{(a)}$ and $J_i^{(b)}$ the fluxes of the observable X_i in the reservoirs. The closure condition analogous to (2.157) in this case reads

$$\mathbb{X}_i = X_i + X_i^{(a)} + X_i^{(b)} \tag{2.198}$$

and yields the relation

$$d\mathbb{X}_i = 0 = dX_i + dX_i^{(a)} + dX_i^{(b)} \tag{2.199}$$

because the value of $\mathbb{X}_i$ is fixed in the universe. Since the system can be considered much smaller than the two reservoirs, the variations of the observable X_i can be neglected with respect to the corresponding variations in the reservoirs and we can rewrite Eq. (2.197) in the approximate form

$$\frac{d\mathbb{S}}{dt} \approx \sum_i \left(\frac{\partial S_a}{\partial X_i^{(a)}} - \frac{\partial S_b}{\partial X_i^{(b)}} \right) J_i^{(a)} \tag{2.200}$$

because $J_i^{(a)} = \frac{dX_i^{(a)}}{dt} = -\frac{dX_i^{(b)}}{dt} = -J_i^{(b)}$. We can conclude that, in stationary conditions, the flux J_i flowing through the system is approximately equal to the one flowing through the reservoirs $J_i \approx J_i^{(a)} = -J_i^{(b)}$, while the corresponding affinity is given by the approximate expression

$$\mathbb{F}_i \approx \frac{\partial S_a}{\partial X_i^{(a)}} - \frac{\partial S_b}{\partial X_i^{(b)}}, \tag{2.201}$$

which amounts to the difference of the affinities in the two reservoirs. As an explicit instance, we can think of a system in contact with two heat reservoirs at different temperatures, $T^{(a)}$ and $T^{(b)}$, with $T^{(a)} > T^{(b)}$. The only varying observable in the reservoirs is the internal energy U, and we have an average outgoing flux from the reservoir at temperature $T^{(a)}$ and an average incoming flux to the reservoir at temperature $T^{(b)}$; Eq. (2.200) becomes

$$\frac{d\mathbb{S}}{dt} \approx \left(\frac{1}{T^{(b)}} - \frac{1}{T^{(a)}} \right) |J_u^{(a)}|, \tag{2.202}$$

where $J_u^{(a)}$ denotes the energy current flowing through the system. This case can be generalized to an arbitrary number of reservoirs in contact with the system.

All together, we can state the following variational principle: the stationary state corresponds to a minimum value of the entropy production rate, compatible with the external constraints imposed by the presence of affinities yielding nonvanishing currents.

2.7.4 Nonequilibrium Conditions in a Continous System

As we have seen in Section 2.7.1, when we deal with out-of-equilibrium conditions between a system and a reservoir or between two homogeneous subsystems, the identification of the affinities $\mathbb{F}_i$ (see Eq. (2.164)) is a straightforward consequence of Gibbs relation (2.161). On the other hand, it may happen that heat, rather than flowing through a thin diathermal wall separating the two units of a system, flows along a piece of matter, where the temperature varies continuously in space, due to the presence of a temperature gradient. For instance, this is the typical situation when a heat conductor is put in contact at its ends with two thermal reservoirs at different temperatures. In cases like this, the identification of the affinity in the system is not straightforward.

The main problem is that we need a suitable definition of entropy in a continuous system driven out of equilibrium. We can overcome this conceptual obstacle by assuming that local equilibrium conditions hold in the system. This means that we can associate to a small region of the system a local entropy, whose functional dependence on the extensive observables is the same as the one given at equilibrium by Gibbs relation. Thus,

$$dS = \sum_i F_i \, dX_i, \tag{2.203}$$

which is formally equivalent to Eq. (2.159).

This is quite a crucial assumption, which is expected to hold in a continuous thermodynamic system gently driven out of equilibrium, in such a way that its entropy is still a well-defined local quantity, which maintains the same functional dependence on the extensive thermodynamic quantities. Said differently, we are assuming that in a continuous system mildly driven out of equilibrium, local equilibrium conditions spontaneously emerge on a spatial scale of the order of a few mean-free paths of the particles in the system and on a time scale of the order of a few collision/interaction times between such particles. In this perspective, one can easily realize that we are considering out-of-equilibrium conditions, where the effect of perturbations from the surrounding world is sufficiently small to remain in a linear response regime, in such a way that the system can relax locally to equilibrium conditions or, more properly, to a stationary state that is locally indistinguishable from an equilibrium one. The validity of this hypothesis can be justified a posteriori by the reliability of the predictions of the linear response theory. When strong perturbations or driving forces are applied to an out-of-equilibrium system, the linear approach fails and different descriptions have to be adopted, as happens for turbulent regimes or anomalous transport phenomena.

Now we aim at finding an explicit expression for the affinities in a continuous system. In this respect it is important to point out that, at variance with the macroscopic case discussed in Section 2.7.1, all the quantities in (2.203) should exhibit their local nature by an explicit dependence on the space coordinates, which disappears in stationary conditions, i.e., when a constant flux flows through the system. If the system was at equilibrium, these fluxes would vanish.

In order to specify the problem, we can consider an infinitely extended system in three space dimensions. In this case it is more appropriate to assume that a relation such as

(2.203) holds for the intensive quantities, i.e., the corresponding functions per unit volume $s(\{x_i\})$:

$$ds = \sum_i F_i \, dx_i \,, \tag{2.204}$$

so that we can write

$$F_i = \left(\frac{\partial s}{\partial x_i}\right). \tag{2.205}$$

As a consequence, in this formulation based on intensive quantities none of the x_i corresponds to the volume V, while all the local quantities F_i are the same functions of the extensive observables, as is the case at equilibrium. Moreover, the fluxes J_i have to be replaced by three-dimensional vectors, and Eq. (2.204) suggests the possibility of defining an entropy current density $\mathbf{j}_s$, by means of the current densities $\mathbf{j}_i$ associated with the intensive observables x_i, as

$$\mathbf{j}_s = \sum_i F_i \, \mathbf{j}_i \,. \tag{2.206}$$

In a region of a continuous system we can say that the rate of local production of entropy is given by the entropy entering/leaving this region (surface term) plus the rate of increase of entropy within this region (volume term), as

$$\frac{ds}{dt} = \frac{\partial s}{\partial t} + \nabla \cdot \mathbf{j}_s. \tag{2.207}$$

The first term on the right-hand side can be obtained from Eq. (2.204), namely,

$$\frac{\partial s}{\partial t} = \sum_i F_i \frac{\partial x_i}{\partial t}, \tag{2.208}$$

while the second one is given by the expression

$$\nabla \cdot \mathbf{j}_s = \nabla \cdot \Big(\sum_i F_i \mathbf{j}_i\Big) = \sum_i \Big(\nabla F_i \cdot \mathbf{j}_i + F_i \nabla \cdot \mathbf{j}_i\Big). \tag{2.209}$$

Therefore, we can rewrite Eq. (2.207) as

$$\frac{ds}{dt} = \sum_i F_i \frac{\partial x_i}{\partial t} + \sum_i \nabla F_i \cdot \mathbf{j}_i + \sum_i F_i \nabla \cdot \mathbf{j}_i \tag{2.210}$$

$$= \sum_i F_i \left(\frac{\partial x_i}{\partial t} + \nabla \cdot \mathbf{j}_i\right) + \sum_i \nabla F_i \cdot \mathbf{j}_i. \tag{2.211}$$

Here the quantities in parentheses vanish if the extensive observables are conserved quantities, like the energy (in the absence of chemical reactions) and the mole numbers, because in this case the continuity equations,

$$\frac{\partial x_i}{\partial t} + \nabla \cdot \mathbf{j}_i = 0, \tag{2.212}$$

must hold. In conclusion, we finally obtain

$$\frac{ds}{dt} = \sum_i \nabla F_i \cdot \mathbf{j}_i. \tag{2.213}$$

This relation indicates that the affinities $\mathbb{F}_i$, introduced for macroscopic systems as the difference of thermodynamic quantities F_i (see Eq. (2.158)), in the continuous case transform into the gradient of the same quantities, i.e.,

$$\mathbb{F}_i = F_i - F_i^{(r)} \quad \rightarrow \quad \mathbf{F}_i = \nabla F_i(\mathbf{r}). \tag{2.214}$$

For instance, if we consider the x-component of the energy current density $(j_u)_x$ flowing in the continuous system, the quantity multiplying it in (2.213) is given by the gradient of the inverse temperature along the x-direction, namely $\partial_x(1/T)$. In fact, it is worth recalling that in the nonequilibrium case the temperature T, as well as the entropy density and the current densities, are functions of the space coordinates. In a similar way, if we consider the x-component of the ith mole number current density $(j_{n_i})_x$, the associated affinity is given by $-\partial_x(\mu_i/T)$, where μ_i is also a function of the space coordinates.

2.8 Physical Applications of Onsager Reciprocity Relations

Up to now in this section we have illustrated the basic mathematical and theoretical aspects of the thermodynamics of irreversible processes, based on the linear response approach and on the local equilibrium hypothesis for continuous systems. In what follows we describe some examples and applications that should help the reader to appreciate the physical content of this approach and to understand how to use such theoretical tools for studying transport processes in nonequilibrium phenomena. Since the case of charged particles in the presence of a magnetic field requires a specific description, we first discuss some examples of coupled transport of neutral particles, then the case of charged particles.

2.8.1 Coupled Transport of Neutral Particles

Mechanothermal and Thermomechanical Effects

As a first pedagogical example we consider a system made of two volumes containing the same gas that are in contact through a wall, where a small hole allows for a slow exchange of particles. The exchanged particles also carry their (internal) energy from one gas to the other. In this sense, this geometry represents the simplest example of coupled transport. Making reference to the conditions described in Section 2.7.1, we can consider one of the two gases as the reservoir and the other as the thermodynamic system. The affinities (see Eq. (2.164)) in this simple example are

$$\mathbb{F}_U = \frac{1}{T} - \frac{1}{T_r} \tag{2.215}$$

$$\mathbb{F}_N = -\frac{\mu}{T} + \frac{\mu_r}{T_r}, \tag{2.216}$$

where T, μ and T_r, μ_r are the temperatures and chemical potentials in the system and in the reservoir, respectively. If U is the internal energy and N is the number of particles of the system, the corresponding fluxes are

$$J_U = \frac{dU}{dt} \quad \text{and} \quad J_N = \frac{dN}{dt} \tag{2.217}$$

and the phenomenological equations (2.176) are

$$J_U = L_{UU}\mathbb{F}_U + L_{UN}\mathbb{F}_N, \tag{2.218}$$

$$J_N = L_{NU}\mathbb{F}_U + L_{NN}\mathbb{F}_N. \tag{2.219}$$

Since both observables are even with respect to the time reversal operator $\mathcal{T}$, $\tau_U = \tau_N = +1$, the following conditions hold:

$$L_{UN} = L_{NU}, \qquad L_{UU} > 0, \qquad L_{NN} > 0, \qquad L_{UU}L_{NN} - L_{UN}^2 = \det L > 0. \tag{2.220}$$

The first condition comes from the the symmetry of the Onsager matrix (see Eq. (2.175)), and the following inequalities derive from the condition

$$\frac{d\mathbb{S}}{dt} = \sum_{ij} L_{ij}\,\mathbb{F}_i\,\mathbb{F}_j \geq 0, \tag{2.221}$$

which must hold for any $\mathbb{F}_i$ and $\mathbb{F}_j$.[20]

If the gases in the two volumes are at the same temperature T (thermal equilibrium), $\mathbb{F}_U = 0$ and the energy flux is given by

$$J_U = L_{UN}\left(-\frac{\mu}{T} + \frac{\mu_r}{T}\right), \tag{2.222}$$

while the particle flux is

$$J_N = L_{NN}\left(-\frac{\mu}{T} + \frac{\mu_r}{T}\right), \tag{2.223}$$

thus yielding the relation

$$\frac{J_U}{J_N} = \frac{L_{UN}}{L_{NN}}. \tag{2.224}$$

In this situation we have coupled transport of particles and heat by a purely mechanical effect even in thermal equilibrium conditions.

Another interesting situation is when the stationary flux of particles vanishes $J_N = 0$, and we let just the energy flux survive. In this case the coupled transport yields a thermomechanical effect. The phenomenological relation (2.219) reduces to

$$\frac{\mathbb{F}_N}{\mathbb{F}_U} = -\frac{L_{NU}}{L_{NN}} = -\frac{L_{UN}}{L_{NN}}, \tag{2.225}$$

where the last equality is a consequence of the Onsager reciprocity relation (2.175). We can also conclude that the thermomechanical effect is characterized by the presence of a stationary energy current that is a combined effect of both affinities, because temperatures and chemical potentials are different both in the system and in the reservoir. In particular, we find that

$$J_U = \frac{\det L}{L_{NN}}\mathbb{F}_U. \tag{2.226}$$

[20] More precisely, if we choose $\mathbb{F}_N = 0$ or $\mathbb{F}_U = 0$, we find the positivity of the diagonal terms, and if we choose the minima $\frac{\mathbb{F}_U}{\mathbb{F}_N} = -\frac{L_{UN}}{L_{UU}}$ or $\frac{\mathbb{F}_U}{\mathbb{F}_N} = -\frac{L_{NN}}{L_{UN}}$, we find the positivity of the determinant.

Notice also that the ratio between the fluxes in the first case (i.e., the mechanothermal effect) is equal to minus the ratio of affinities in the second case (i.e., the thermomechanical effect). This shows that Onsager relations establish interesting phenomenological similarities among quite different physical phenomena.

Coupled Transport in Linear Continuous Systems

As discussed in Section 2.7.2 the linear approach to transport phenomena, yielding a proportionality between fluxes and affinities (see Eq. (2.176)), stems from phenomenological considerations. For instance, in many thermal conductors it has been found experimentally that, in the absence of a matter current density, the energy current density $\mathbf{j}_u$ is proportional to the applied temperature gradient, according to Fourier's law

$$\mathbf{j}_u = -\kappa \nabla T(\mathbf{r}) = \kappa T^2 \nabla \left(\frac{1}{T}\right), \tag{2.227}$$

where κ is the thermal conductivity (see Section 1.3). For the sake of simplicity, in the examples hereafter discussed we assume we are dealing with homogeneous isotropic systems, so that all phenomenological transport coefficients do not depend on the space directions. In fact, the last expression in (2.227) indicates that the energy current density $\mathbf{j}_u$ depends linearly on the gradient of the affinity, $\nabla(1/T)$ (see Section 2.7.4), with a proportionality coefficient κT^2, where κ is a scalar quantity. On a physical ground it is worth stressing that Fourier's law is expected to hold for small temperature gradients, while a dependence on higher-order terms, such as $[\nabla(1/T)]^2$ and $[\nabla(1/T)]^3$, is expected to emerge for values of the applied temperature gradient sufficiently large to yield strong nonlinear effects.[21]

Another well-known example of a linear phenomenological law is Fick's law of diffusion, which tells us that, in the absence of an energy current density, the matter current density can be written as

$$\mathbf{j}_\rho = -D \nabla \rho(\mathbf{r}), \tag{2.228}$$

where D is the diffusion coefficient and ρ is the matter density field.[22] In a linear regime at fixed temperature T, ρ is proportional to the chemical potential (per unit volume) μ and (2.228) can be rewritten as

$$\mathbf{j}_\rho = D \nabla \left(\frac{-\mu(\mathbf{r})}{T}\right). \tag{2.229}$$

Also in this case we are facing linear conditions, which correspond to sufficiently moderate matter density gradients, in such a way that the matter current density is found to be proportional to the gradient of the corresponding affinity.

[21] Deviations from the phenomenological Fourier's law have been predicted and experimentally observed in low-dimensional models, such as carbon nanotubes and graphene layers. In such cases, the deviations are due to specific anomalous diffusive behaviors (like the superdiffusive regime described in Section 1.7), which are typical of low-dimensional fluids and nanostructures in the absence of on-site potentials.

[22] It is a number density, not a mass density.

Now we want to consider the situation of a continuous thermodynamic system made of a single species of particles contained in a fixed volume V, where stationary energy and matter current densities are present at the same time. In this case the Gibbs relation for intensive observables (2.204) specializes to

$$du = Tds + \mu d\rho, \tag{2.230}$$

where u, s, and ρ are the densities of internal energy, entropy, and number of particles, respectively. Taking into account the previous phenomenological considerations, we can write the linear coupled transport equations (2.176) in the form

$$\mathbf{j}_u = L_{uu}\, \nabla\Big(\frac{1}{T}\Big) + L_{u\rho}\, \nabla\Big(\frac{-\mu}{T}\Big) \tag{2.231}$$

$$\mathbf{j}_\rho = L_{\rho u}\, \nabla\Big(\frac{1}{T}\Big) + L_{\rho\rho}\, \nabla\Big(\frac{-\mu}{T}\Big), \tag{2.232}$$

where $\mathbf{j}_u$ and $\mathbf{j}_\rho$ are the current densities. Note that, even if not explicitly indicated, all the physical quantities appearing in these equations are functions of the coordinate vector $\mathbf{r}$. The symmetric structure of Onsager matrix of linear coefficients (see Eq. (2.175)) yields the relation $L_{u\rho} = L_{\rho u}$, while, according to Eq. (2.214), the affinities are

$$\mathbf{F}_u = \nabla\Big(\frac{1}{T}\Big)$$

$$\mathbf{F}_\rho = \nabla\Big(\frac{-\mu}{T}\Big). \tag{2.233}$$

The entropy density production rate in the system is given by the expression

$$\frac{ds}{dt} = \mathbf{j}_u \cdot \nabla\Big(\frac{1}{T}\Big) + \mathbf{j}_\rho \cdot \nabla\Big(\frac{-\mu}{T}\Big), \tag{2.234}$$

which exemplifies Eqs. (2.213) and (2.214) for an out-of-equilibrium continuous system in the presence of irreversible processes, produced by thermodynamic forces (in the form of applied gradients) determining energy and mass current densities flowing through the system.

We want to point out that (2.234) is obtained as a consequence of the continuity equations,

$$\frac{\partial u}{\partial t} + \nabla \cdot \mathbf{j}_u = 0 \tag{2.235}$$

$$\frac{\partial \rho}{\partial t} + \nabla \cdot \mathbf{j}_\rho = 0, \tag{2.236}$$

which express the conservation of energy and matter densities in the system. In general, stationary conditions imply divergence-free currents; i.e., in this case we have $\nabla \cdot \mathbf{j}_u = \nabla \cdot \mathbf{j}_\rho = 0$. Notice that this is not the case for the entropy density current, because stationary conditions are associated with finite currents and with a finite entropy production rate. Combining (2.234) and the Onsager relations (2.231) and (2.232) we can also write

$$\frac{ds}{dt} = \sum_{\alpha,\beta} L_{\alpha\beta} \mathbf{F}_\alpha \cdot \mathbf{F}_\beta, \tag{2.237}$$

with indices α and β running over the subscripts u and ρ. The latter relation is the analog of Eq. (2.190) in the case of a continuous system.

2.8.2 Onsager Theorem and Transport of Charged Particles

Another phenomenological example of a linear transport phenomenon is Ohm's law, where the electric current (i.e., the flux) is found to be proportional to the electric potential difference (i.e., the affinity) applied at the ends of a standard electric conductor (e.g., a copper or a silver wire), whose ohmic resistance (the inverse of the transport coefficient named conductivity) is the proportionality constant.

On the other hand, when dealing with transport phenomena including electromagnetic effects the Onsager reciprocity relation (2.175) has to be reconsidered in the light of Onsager theorem, which can be formulated as

$$L_{ij}(\mathbf{B}) = L_{ji}(-\mathbf{B}), \tag{2.238}$$

where we have indicated explicitly only the dependence of the linear kinetic coefficients on the applied external magnetic field **B**, although they have to be intended as functions of the other intensive parameters also. This theorem states that the value of the linear kinetic coefficient L_{ij} measured in the presence of an applied external magnetic field **B** is equal to the value of the linear kinetic coefficient L_{ji} measured in the reversed magnetic field $-\mathbf{B}$. Said differently, in the presence of an external magnetic field **B** the Onsager reciprocity relation is maintained by changing the sign of **B**. We do not report here the proof of this theorem, which is related to the time reversal properties of the magnetic field.

In what follows we illustrate the Onsager theory in the case of transport of charged particles, in particular electrons in metals or semiconductor materials. This theory allows us to explain many interesting coupled transport phenomena that are quite common in our everyday life, for example, when we experience the cooling in a refrigerator or the heat produced by a lamp or by the electronic circuits of a computer. In particular, we first deal with thermoelectric effects in one-dimensional conductors. This analysis exemplifies how the Onsager theory can be used to establish relations between phenomenological coefficients as a result of the relations between Onsager kinetic coefficients. The situation is just more involved, but it works the same way when the same approach is applied to the study of thermomagnetic effects in three–dimensional conductors. This is the second set of cases that we discuss hereafter, as they are the standard framework for the application of Onsager theorem, concerning the effects associated with the presence of an external magnetic field on charged particle currents.

Thermoelectric Effects

These physical phenomena concern coupled transport of electric and heat currents in the absence of a magnetic field. They result from the mutual interference of heat flow and electric current. We can cast them in the language of Onsager theory by considering, for the sake of simplicity, a conductor where the electric current and the heat current flow in one dimension and the electric current is carried by electrons only, as happens in a metal wire.[23] We adopt here the description of the conductor as a continuous system and we specialize the Gibbs relation (2.204) to

$$ds = \frac{1}{T}du - \frac{\mu}{T}dn, \tag{2.239}$$

where T is the temperature, s is the local entropy density, u is the local energy density, n is the number of electrons per unit volume, and, accordingly, μ is the electrochemical potential per particle (not per mole, as usual). In principle, we should add to the right-hand side the contribution of other components, $-\sum_i(\mu_i/T)dn_i$, like the atomic nuclei, that form the structure of the conductor. However, their role in thermoelectric effects is immaterial, because in a linear regime components other than electrons do not contribute to the electric current flowing through the conductor. In fact, Eq. (2.206) in this case becomes

$$j_s = \frac{1}{T}j_u - \frac{\mu}{T}j_n, \tag{2.240}$$

where j_s, j_u, and j_n are the one-dimensional current densities of entropy, energy, and number of electrons, respectively. Since we are dealing with a continuous system we can also write the entropy density production rate (see (2.213)) as

$$\frac{ds}{dt} = \partial_x\left(\frac{1}{T}\right)j_u - \partial_x\left(\frac{\mu}{T}\right)j_n, \tag{2.241}$$

where we have denoted with x the direction along the one-dimensional conductor.

Given the negative sign of the elementary charge, it is useful to consider the (positive) flux $-j_n$, which can be used, along with j_u, to write the linear coupled transport equations (2.176) in the form

$$-j_n = L'_{nn}\,\partial_x\frac{\mu}{T} + L'_{nu}\partial_x\frac{1}{T} \tag{2.242}$$

$$j_u = L'_{un}\,\partial_x\frac{\mu}{T} + L'_{uu}\partial_x\frac{1}{T}, \tag{2.243}$$

where the use of the prime symbol for the Onsager coefficients is made clearer by the following considerations. In fact, for physical reasons it is better to describe thermoelectric effects by replacing the current density of the total internal energy with the current density of heat. Making reference to the thermodynamic relation $dQ = TdS$ we can define a heat current density as

$$j_q = Tj_s, \tag{2.244}$$

where we maintain the lowercase subscripts to denote the intensive quantities. Using Eq. (2.240), we can write

$$j_q = j_u - \mu j_n. \tag{2.245}$$

This result can be interpreted by considering that μ amounts to the potential energy per particle, so that μj_n represents the potential energy current density. In this sense subtracting

[23] Despite these limiting hypotheses the theory, in principle, allows one to deal with problems where charge is transported also by other kinds of particles, such as protons and charged nuclei in plasmas or ions in chemical reactions.

μj_n from the total energy current density allows us to think of the heat current density j_q as a sort of kinetic energy current density. We can now rewrite (2.241) in the form

$$\frac{ds}{dt} = \left(\partial_x \frac{1}{T}\right) j_q - \frac{1}{T}\left(\partial_x \mu\right) j_n , \tag{2.246}$$

so that, in the heat representation, the affinities associated with j_q and $-j_n$ become $\partial_x(1/T)$ and $(1/T)\partial_x \mu$, respectively. Therefore, the linear coupled transport equations (2.176) for the heat current and the number of electrons current can be rewritten as

$$-j_n = L_{nn} \frac{1}{T} \partial_x \mu + L_{nq} \partial_x \frac{1}{T} \tag{2.247}$$

$$j_q = L_{qn} \frac{1}{T} \partial_x \mu + L_{qq} \partial_x \frac{1}{T}. \tag{2.248}$$

These relations replace Eqs. (2.242) and (2.243) in the heat representation.

Now, we want to find a physical interpretation of above linear kinetic coefficients L_{ij}. First of all, we observe that $L_{nq} = L_{qn}$, because no external magnetic field is applied. Then, we remark that the electrochemical potential μ is made of two contributions,

$$\mu = \mu_e + \mu_c, \tag{2.249}$$

which correspond to the electrical (μ_e) and chemical (μ_c) contributions to the potential, respectively. In particular, we can write $\mu_e = e\,\Phi$, where e is the (negative) electron charge and Φ is the electrostatic potential. On the other hand, μ_c is a function of T and of the electronic concentration. In other words, the electrochemical potential per unit charge $(1/e)\mu$ is such that its gradient $(1/e)\partial_x \mu$ is the sum of the electric field $(1/e)\partial_x \mu_e$ and of an effective driving force $(1/e)\partial_x \mu_c$ associated with the presence of a concentration gradient of electrons. If we assume that the conductor is made of a homogeneous isothermal material, $\partial_x \mu_c = 0$ and $\partial_x \mu = \partial_x \mu_e$, while the electric conductivity σ is defined as the electric current density $e j_n$ per unit potential gradient $(1/e)\partial_x \mu$, i.e., the electromotive force. Thus for $\partial_x T = 0$ (isothermal condition),

$$\sigma = -\frac{e^2 j_n}{\partial_x \mu}. \tag{2.250}$$

Eq. (2.247), for spatially constant T, yields the relation

$$L_{nn} = \frac{\sigma T}{e^2} , \tag{2.251}$$

which allows us to attribute an explicit physical interpretation to the Onsager coefficient L_{nn} in the case of thermoelectric transport.

The heat conductivity, on its side, when $j_n = 0$, i.e., no electric current flows through the conductor, is given by Fourier's law (see Eq. (2.227)),

$$\kappa = -\frac{j_q}{\partial_x T}. \tag{2.252}$$

Using the condition $j_n = 0$, from Eq. (2.247) it is possible to deduce the relation

$$\partial_x \mu = -T \frac{L_{nq}}{L_{nn}} \partial_x \frac{1}{T} = \frac{L_{nq}}{T L_{nn}} \partial_x T \tag{2.253}$$

that, once replaced in (2.248), gives the result

$$\kappa = \frac{\det L}{T^2 L_{nn}}. \tag{2.254}$$

This equation, Eq. (2.251), and the symmetry relation $L_{nq} = L_{qn}$, put together, imply that only one of the Onsager coefficients remains to be determined. In the examples that we discuss hereafter we describe, in specific thermoelectric geometries, how we can find an additional, physical relation allowing us to determine all the Onsager coefficients.

The Seebeck Effect

This effect was discovered by the Estonian physicist Thomas Johannes Seebeck in 1821. He observed that a circuit made by two different conductors, whose junctions are taken at different temperatures, produced an electric current that was signaled by the deflection of a magnetic needle. Later the Danish scientist Hans Christian Ørsted realized that such a setup points out the manifestation of a thermoelectric effect: the difference of temperature at the junctions and the different electronic mobility in the two materials produces an effective *thermoelectric power*, which amounts to an electromotive force (emf) making a current flow through the circuit. In order to interpret this phenomenon according to the Onsager kinetic theory, we can adopt a different setup, i.e., a thermocouple (see Fig. 2.2): the junctions connecting the two conductors A and B (e.g., one copper and one iron wire or a couple of differently doped semiconductors) are kept at different temperatures, $T_1 < T_2$, and a voltmeter is inserted along one of the two arms of the thermocouple, allowing for the passage of the heat flow, but not of the electric current. In practice, the voltmeter measures the emf, which is produced by the heat flow, while $j_n = 0$. We assume that the voltmeter is at a point along the arm at some temperature intermediate between T_1 and T_2 and that it is sufficiently small and made by a good heat conductor, in such a way as to not perturb significantly the heat flow.

Since $j_n = 0$, we can use Eq. (2.253) to determine the relation between $\partial_x \mu$ and $\partial_x T$, which can be written more transparently as

$$d\mu = \frac{L_{nq}}{T L_{nn}} dT, \tag{2.255}$$

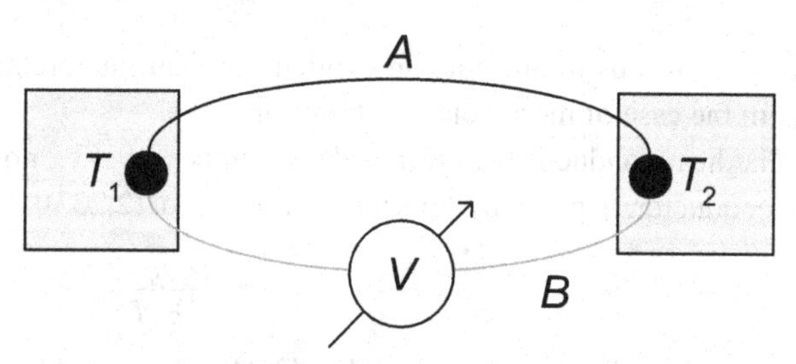

Fig. 2.2 Sketch of the experimental apparatus to observe the Seebeck effect. Two different conducting (or semiconducting) wires, A and B, are joined at their ends (black dots), which are kept at different temperatures T_1, T_2. This temperature gradient induces a difference of electric potential (electromotive force) at the inputs of the voltmeter V.

i.e., directly linking the variation of temperature with the variation of chemical potential. This equation allows us to compute the difference of electrochemical potential between the junctions along the two arms of the thermocouple as

$$\mu_2 - \mu_1 = \int_1^2 \frac{L_{nq}^A}{TL_{nn}^A} dT \tag{2.256}$$

$$\mu_2 - \mu_r = \int_r^2 \frac{L_{nq}^B}{TL_{nn}^B} dT \tag{2.257}$$

$$\mu_\ell - \mu_1 = \int_1^\ell \frac{L_{nq}^B}{TL_{nn}^B} dT, \tag{2.258}$$

where μ_r and μ_ℓ are the values of the electrochemical potential measured at the right and left electric contacts of the voltmeter, respectively. The previous equations yield the relation

$$\mu_r - \mu_\ell = \int_1^2 \left(\frac{L_{nq}^A}{TL_{nn}^A} - \frac{L_{nq}^B}{TL_{nn}^B} \right) dT. \tag{2.259}$$

We have assumed that the voltmeter is at a fixed value of the local temperature and the voltage that it measures is given by the expression

$$V = \frac{1}{e}(\mu_r - \mu_\ell) = \int_1^2 \left(\frac{L_{nq}^A}{eTL_{nn}^A} - \frac{L_{nq}^B}{eTL_{nn}^B} \right) dT. \tag{2.260}$$

The thermoelectric power of a thermocouple, ε_{AB}, is defined as the increment of voltage per unit temperature difference, while its sign is conventionally said to be positive if the increment of voltage drives the current from A to B at the hot junction, T_2. In practice, if $T_1 = T$ and $T_2 = T + \Delta T$, in the limit of small ΔT

$$\varepsilon_{AB} = \frac{V}{\Delta T} \equiv \varepsilon_B - \varepsilon_A, \tag{2.261}$$

where

$$\varepsilon_X = -\frac{L_{nq}^X}{eTL_{nn}^X}, \quad X = A, B, \tag{2.262}$$

defines the absolute thermoelectric power of a single electric conductor.

As announced after Eq. (2.254), this is the first example of an additional relation between Onsager coefficients and a physical, measurable quantity, the thermoelectric power of a conductor. We can now express all coefficients in terms of σ, κ, and ε: L_{nn} is given by (2.251), then $L_{nq} = L_{qn} = -(T^2\sigma\varepsilon)/e$ are derivable from (2.262), and finally $L_{qq} = (T^3\sigma\varepsilon^2 + T^2\kappa)$ can be found by (2.254). Inserting these expressions in Eqs. (2.247-2.248), we obtain

$$-j_n = \left(\frac{\sigma}{e^2}\right)\partial_x \mu - \left(\frac{T^2\sigma\varepsilon}{e}\right)\partial_x \frac{1}{T} \tag{2.263}$$

$$j_q = -\left(\frac{T\sigma\varepsilon}{e}\right)\partial_x \mu + (T^3\sigma\varepsilon^2 + T^2\kappa)\partial_x \frac{1}{T}. \tag{2.264}$$

By multiplying (2.263) by the factor $T\varepsilon e$ and summing it to (2.264) we find the relation

$$j_q = T\varepsilon\, e j_n + T^2 \kappa\, \partial_x \frac{1}{T}. \tag{2.265}$$

Recalling that $j_s = j_q/T$, we finally obtain

$$j_s = \varepsilon\, e j_n + T\kappa\, \partial_x \frac{1}{T}. \tag{2.266}$$

This formula tells us that each electron contributes to the entropy current as a charge carrier (first term on the right-hand side) and as a heat carrier (second term on the right-hand side). In particular, its contribution as a charge carrier amounts to an entropy εe, which allows us to interpret the thermoelectric power as the entropy transported per unit charge by the electronic current.

The Peltier Effect

The Peltier effect was discovered in 1834 by the French physicist Jean Peltier. He observed that when a current flows through the junction between two different conductors, heat is exchanged between the junction and the environment; see Fig. 2.3. One can easily realize that such a device may work as a heater or as a cooler, according to the direction of the electric current. The Peltier effect is therefore different from the Joule effect in two (related) respects: it is due to the presence of a junction and it may produce heating as well as cooling. The Peltier effect is usually described as the reverse of the Seebeck effect: in the former a heat current is induced by an electric current, in the latter it is the opposite.

The junction between the conductors A and B is assumed to be kept in isothermal conditions. The Onsager theory predicts that a certain amount of heat has to be supplied or removed from the junction in order to keep the electronic current stationary. The total energy current is discontinuous through the junction. In fact, in each conductor it is given by the expression (see Eq. (2.245))

$$j_u = j_q + \mu j_n. \tag{2.267}$$

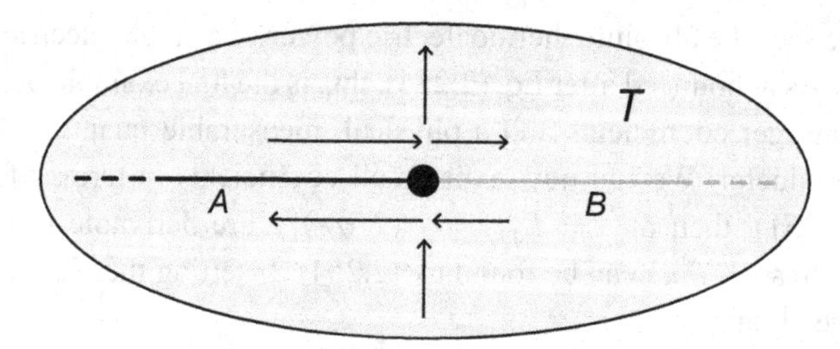

Fig. 2.3 The diagram of the Peltier effect. A junction (black dot) of two different conductors A and B, when kept at constant temperature T, must exchange heat with the reservoir in order to sustain a constant electric current. The horizontal arrows correspond to j_q^A and j_q^B: their difference amounts to the heat exchanged with the environment at temperature T, which may be either positive or negative; see the two cases above and below the junction.

On the other hand, at the isothermal junction μ and j_n must be continuous in order to have a stationary current, so that the discontinuity in j_u amounts to

$$j_u^A - j_u^B = j_q^A - j_q^B . \tag{2.268}$$

The condition of maintaining the setup at constant temperature implies that no thermal gradient is present and Eqs. (2.263) and (2.264) simplify to

$$j_n = -\Big(\frac{\sigma}{e^2}\Big)\partial_x \mu \tag{2.269}$$

$$j_q = -\Big(\frac{T\sigma\,\varepsilon}{e}\Big)\partial_x \mu, \tag{2.270}$$

which can be combined to obtain the relation

$$j_q = T\varepsilon\,(ej_n). \tag{2.271}$$

From this equation we can write

$$j_q^B - j_q^A = T(\varepsilon_B - \varepsilon_A)(ej_n), \tag{2.272}$$

thus showing that the discontinuity of the heat current at the junction is due to the different thermoelectric power of the conductors and is proportional to the electric current, ej_n. The proportionality constant defines the Peltier coefficient of the junction,

$$\Pi_{AB} = \frac{(j_q^B - j_q^A)}{ej_n} = T(\varepsilon_B - \varepsilon_A)\,, \tag{2.273}$$

which can be interpreted as the heat to be supplied to the junction per unit of electric current flowing from A to B. Eq. (2.273) proves the relation between the Peltier and the Seebeck effects, showing that the two effects are simply proportional to one another.

The Thomson–Joule Effect

The Thomson–Joule effect appears when an electric current flows through a conductor with an applied temperature gradient. We consider first a conductor through which a heat current flows in the absence of an electric one. This implies that a temperature gradient sets in the conductor, where the temperature field $T(x)$ along the conductor is determined by the dependence on temperature of its kinetic coefficients. We can now perform an idealized setup, where the conductor is put in contact at each point x with a heat reservoir at temperature $T(x)$. In such conditions there is no heat exchanged between the conductor and the reservoirs. Then, we switch on a stationary electric current j_n flowing through the conductor, thus producing a heat exchange with the reservoirs, in such a way that any variation of the energy current through the conductor has to be supplied by the reservoirs. Before switching on the electric current we know that the energy current must be conserved, i.e., $\partial_x j_u = 0$. After the electric current is switched on, according to (2.267) we can write

$$\partial_x j_u = \partial_x j_q + j_n \partial_x \mu, \tag{2.274}$$

because j_n is a constant current. Using (2.265) and (2.263) we can rewrite this equation as

$$\partial_x j_u = \partial_x\Big(T\varepsilon e j_n + T^2\,\kappa\,\partial_x\frac{1}{T}\Big) + \Big(-\frac{e^2}{\sigma}j_n + T^2\varepsilon e\partial_x\frac{1}{T}\Big)j_n. \tag{2.275}$$

In the adopted setup the only quantities[24] that depend on the space coordinate x are T and ε, and we can simplify (2.275), obtaining

$$\partial_x j_u = T\Big(\partial_x\varepsilon\Big)e j_n - \kappa\,\partial_{xx}T - \frac{e^2}{\sigma}j_n^2. \tag{2.276}$$

Since in the absence of an electric current, i.e., $j_n = 0$, we must have $\partial_x j_u = 0$; the temperature field must be such that the second addendum on the right-hand side must vanish, i.e.,

$$\kappa\,\partial_{xx}T = 0. \tag{2.277}$$

This means that the temperature field $T(x)$ is expected to exhibit a linear dependence on the space coordinate x, consistently with the Fourier's law. Thus, assuming as a first approximation that the temperature profile does not change when $j_n \neq 0$, we can simplify Eq. (2.276) as

$$\partial_x j_u = T\Big(\partial_x\,\varepsilon\Big)\,e j_n - \frac{e^2}{\sigma}j_n^2. \tag{2.278}$$

On the other hand, the thermoelectric power depends on x because it is a function of temperature, so we can write

$$\partial_x\,\varepsilon = \frac{d\varepsilon}{dT}\,\partial_x\,T, \tag{2.279}$$

from which

$$\partial_x j_u = T\frac{d\varepsilon}{dT}\,\partial_x\,T e j_n - \frac{e^2}{\sigma}j_n^2. \tag{2.280}$$

The second term on the right-hand side is the so-called Joule heat, which is produced even in the absence of a temperature gradient. The first term is the so-called Thomson heat, which has to be absorbed by the reservoirs to maintain the temperature gradient, $\partial_x T$, when the electric current flows through the conductor. We can define the Thomson coefficient, τ, as the amount of Thomson heat absorbed per unit electric current ($e j_n$) and per unit temperature gradient ($\partial_x T$), obtaining

$$\tau = T\frac{d\varepsilon}{dT}. \tag{2.281}$$

Making use of this definition and of Eq. (2.273) we can establish a relation between the Peltier and Thomson coefficients with the thermoelectric power,

$$\frac{d\,\Pi_{AB}}{dT} = (\tau_B - \tau_A) + (\varepsilon_B - \varepsilon_A)\,, \tag{2.282}$$

which can be interpreted as a consequence of the energy conservation. In fact, the thermoelectric power of a junction is the result of the contributions of the heat per unit

[24] In principle κ should also depend on T and, accordingly, on x, but for sufficiently small temperature gradients, it can be assumed to be constant.

temperature and per unit electric current supplied to the junction by the Peltier and by the Thomson effects.

Thermomagnetic and Galvanomagnetic Effects

We now consider a three-dimensional conductor in which heat current and electric current can flow simultaneously in the presence of an external magnetic field **B**. For the sake of simplicity we assume that **B** is directed along the z-axis, while currents and affinities (i.e., gradients) depend only on the coordinates x and y. In this configuration we are assuming a planar symmetry, meaning that the observed effects are independent of z.[25] The basic formulas are essentially the same employed for studying the thermoelectric effects, extended to the case $d = 2$. Eqs. (2.240), (2.244), and (2.245) become

$$\mathbf{j}_s = \frac{1}{T}\mathbf{j}_u - \frac{\mu}{T}\mathbf{j}_n \tag{2.283}$$

$$\mathbf{j}_q = T\mathbf{j}_s \tag{2.284}$$

$$\mathbf{j}_q = \mathbf{j}_u - \mu\,\mathbf{j}_n, \tag{2.285}$$

respectively, while (2.246) turns into

$$\begin{aligned}\frac{ds}{dt} &= \nabla\frac{1}{T}\cdot\mathbf{j}_q - \frac{1}{T}\nabla\mu\cdot\mathbf{j}_n\\ &= \partial_x\frac{1}{T}(j_q)_x - \frac{1}{T}\partial_x\mu\,(j_n)_x + \partial_y\frac{1}{T}(j_q)_y - \frac{1}{T}\partial_y\mu\,(j_n)_y.\end{aligned} \tag{2.286}$$

Following the same procedure adopted for thermoelectric effects, we can write the set of equations analogous to (2.247) and (2.248) in the form

$$-(j_n)_x = L_{11}\frac{1}{T}\partial_x\mu + L_{12}\partial_x\frac{1}{T} + L_{13}\frac{1}{T}\partial_y\mu + L_{14}\partial_y\frac{1}{T} \tag{2.287}$$

$$(j_q)_x = L_{21}\frac{1}{T}\partial_x\mu + L_{22}\partial_x\frac{1}{T} + L_{23}\frac{1}{T}\partial_y\mu + L_{24}\partial_y\frac{1}{T} \tag{2.288}$$

$$-(j_n)_y = L_{31}\frac{1}{T}\partial_x\mu + L_{32}\partial_x\frac{1}{T} + L_{33}\frac{1}{T}\partial_y\mu + L_{34}\partial_y\frac{1}{T} \tag{2.289}$$

$$(j_q)_y = L_{41}\frac{1}{T}\partial_x\mu + L_{42}\partial_x\frac{1}{T} + L_{43}\frac{1}{T}\partial_y\mu + L_{44}\partial_y\frac{1}{T}. \tag{2.290}$$

Space isotropy in the (x, y)-plane implies, in particular, the invariance under a $(\pi/2)$-rotation, corresponding to the transformation $x \to y$ and $y \to -x$. Considering that under this transformation, $(j_{n,q})_x \to (j_{n,q})_y$, this invariance allows us to find the following relations among coefficients:

$$\begin{aligned}L_{31} &= -L_{13} & L_{32} &= -L_{14} & L_{33} &= L_{11} & L_{34} &= L_{12}\\ L_{41} &= -L_{23} & L_{42} &= -L_{24} & L_{43} &= L_{21} & L_{44} &= L_{22}.\end{aligned}$$

We now consider a reflection of the x-axis, which must be accompanied by the inversion of the magnetic field. Invariance under this transformation implies that L_{11}, L_{12}, L_{21}, and L_{22}, are even functions of **B**, while L_{13}, L_{14}, L_{23}, and L_{24}, are odd functions of **B**. Finally,

[25] If the system extends over a finite size along z, we assume that surface corrections are negligible.

using the general relations (2.238) between Onsager coefficients when we reverse the sign of **B**, we find the additional relations $L_{23} = L_{14}$ and $L_{21} = L_{12}$. In conclusion, the previous set of equations can be rewritten in such a way to contain only six independent kinetic coefficients:

$$-(j_n)_x = L_{11}\frac{1}{T}\partial_x\mu + L_{12}\partial_x\frac{1}{T} + L_{13}\frac{1}{T}\partial_y\mu + L_{14}\partial_y\frac{1}{T} \tag{2.291}$$

$$(j_q)_x = L_{12}\frac{1}{T}\partial_x\mu + L_{22}\partial_x\frac{1}{T} + L_{14}\frac{1}{T}\partial_y\mu + L_{24}\partial_y\frac{1}{T} \tag{2.292}$$

$$-(j_n)_y = -L_{13}\frac{1}{T}\partial_x\mu - L_{14}\partial_x\frac{1}{T} + L_{11}\frac{1}{T}\partial_y\mu + L_{12}\partial_y\frac{1}{T} \tag{2.293}$$

$$(j_q)_y = -L_{14}\frac{1}{T}\partial_x\mu - L_{24}\partial_x\frac{1}{T} + L_{12}\frac{1}{T}\partial_y\mu + L_{22}\partial_y\frac{1}{T}. \tag{2.294}$$

With this result at hand, we can make use of the definitions of the phenomenological transport coefficients already introduced to describe the thermoelectric effects, to provide an analogous description of magnetoelectric and galvanometric effects.

Isothermal and Adiabatic Electric Conductivities

For the sake of simplicity we assume that an electric current flows in the x-direction, i.e., $(j_n)_x \neq 0$ and $(j_n)_y = 0$. The electric conductivity of the material is defined by the relation

$$\sigma = -\frac{e^2\,(j_n)_x}{\partial_x\,\mu}. \tag{2.295}$$

The isothermal transport conditions correspond to the absence of thermal gradients, i.e., $\partial_x T = \partial_y T = 0$. We can use Eq. (2.293) to find $\partial_y\mu = (L_{13}/L_{11})\partial_x\mu$, then replace in (2.291) to obtain an expression of the isothermal conductivity $\sigma_{\rm I}$ in terms of the Onsager kinetic coefficients,

$$\sigma_{\rm I} = \frac{e^2}{TL_{11}}\left(L_{11}^2 + L_{13}^2\right). \tag{2.296}$$

The adiabatic transport conditions correspond to a situation where a gradient of temperature is allowed only along the y-direction, transverse to the current $(j_n)_x$, while no transverse current can flow, i.e., $\partial_x T = (j_n)_y = (j_q)_y = 0$. Now we need three equations, (2.291), (2.293), and (2.294), for expressing the adiabatic electric conductivity in terms of the Onsager kinetic coefficients, thus obtaining[26]

$$\sigma_{\rm A} = \frac{e^2 D_3}{TD_2}, \tag{2.297}$$

where D_2 and D_3 are the determinants of the reduced matrices:

$$D_2 = \begin{vmatrix} L_{11} & L_{12} \\ L_{12} & L_{22} \end{vmatrix}, \qquad D_3 = \begin{vmatrix} L_{11} & L_{13} & L_{14} \\ -L_{13} & L_{11} & L_{12} \\ -L_{14} & L_{12} & L_{22} \end{vmatrix}. \tag{2.298}$$

[26] The mathematics behind these calculations is the standard procedure of solving linear systems of equations.

Isothermal and Adiabatic Heat Conductivities

Similar to what we did for the electric conductivities, we now assume that a thermal gradient is present only in the x-direction, i.e., $\partial_x T \neq 0$ and $\partial_y T = 0$. The heat conductivity of the material is defined by the relation

$$\kappa = -\frac{(j_q)_x}{\partial_x T}. \tag{2.299}$$

In this case the isothermal transport conditions correspond to the absence of electric currents, namely, $(j_n)_x = (j_n)_y = 0$. In this case we have to use Eqs. (2.291)–(2.293) for expressing the isothermal heat conductivity in terms of the Onsager kinetic ones, thus obtaining

$$\kappa_{\mathrm{I}} = \frac{1}{T^2} \frac{D_3}{L_{11}^2 + L_{13}^2}. \tag{2.300}$$

The appearance of the determinant D_3, the same introduced in the expression of the adiabatic electric conductivity (2.297), at the numerator of κ_I is again a manifestation of the symmetries of the Onsager matrix.

The conditions for adiabatic heat transport correspond to the absence of electronic current and to the confinement of the heat current to the x-direction, namely, $(j_n)_x = (j_n)_y = (j_q)_y = 0$. Imposing these conditions in Eqs. (2.291), (2.293), and (2.294), the fourth equation (2.292) yields

$$\kappa_{\mathrm{A}} = \frac{D_4}{T^2 D_3}, \tag{2.301}$$

where

$$D_4 = \begin{vmatrix} L_{11} & L_{12} & L_{13} & L_{14} \\ L_{12} & L_{22} & L_{14} & L_{24} \\ -L_{13} & -L_{14} & L_{11} & L_{12} \\ -L_{14} & -L_{24} & L_{12} & L_{22} \end{vmatrix} \tag{2.302}$$

is the determinant of the complete matrix of Onsager kinetic coefficients.

Isothermal Hall and Nernst Effects

This case corresponds to a situation where the presence of the external magnetic field $\mathbf{B}$ in the z-direction yields an electric current flowing along the x-direction, while a transverse emf is applied along the y-direction. The effect is essentially due to the Lorentz force acting on electrons (or, more generally, on charge carriers). The Hall coefficient is defined as

$$R_{\mathrm{H}} = \frac{\partial_y \mu}{e^2 |\mathbf{B}| \, (j_n)_x}. \tag{2.303}$$

The isothermal conditions correspond to a null transverse electric current and to the absence of thermal gradients, i.e., $(j_n)_y = \partial_x T = \partial_y T = 0$. An expression of R_{H} as a function of Onsager kinetic coefficients can be obtained using Eqs. (2.291) and (2.293),

$$R_{\mathrm{H}} = -\frac{T}{e^2 |\mathbf{B}|} \frac{L_{11}}{L_{11}^2 + L_{13}^2}. \tag{2.304}$$

The adiabatic case corresponds to $\partial_y T \neq 0$ and we have no way to compute a similar expression for the Hall coefficient from Onsager relations. In fact, the presence of a transverse thermal gradient does not allow us to identify the expression $(1/e)\partial_y \mu$ as the actual emf acting on charge carriers, and the problem lacks a unique solution.

We face a similar situation with the Nernst effect, which amounts to producing a thermal gradient along the x-direction, while an emf is applied in the y-direction in the absence of electronic currents, i.e., $(j_n)_x = (j_n)_y = 0$. The Nernst coefficient is defined as

$$N = -\frac{\partial_y \mu}{e|\mathbf{B}|\partial_x T}. \tag{2.305}$$

The isothermal condition refers just to the absence of a temperature gradient along the y-direction, i.e., $\partial_y T = 0$: making use again of Eqs. (2.291) and (2.293) we can express the isothermal Nernst coefficient in terms of the Onsager kinetic coefficients as

$$N_I = \frac{1}{e|\mathbf{B}|T} \frac{L_{11}L_{14} - L_{12}L_{13}}{L_{11}^2 + L_{13}^2}. \tag{2.306}$$

Ettingshausen and Righi–Leduc Effects

In this case we deal with a situation where both currents in the y-direction are absent, i.e., $(j_n)_y = (j_q)_y = 0$, while in the x-direction there is no thermal gradient, $\partial_x T = 0$ (Ettingshausen effect), or no electric current, $(j_n)_x = 0$ (Righi–Leduc effect). The Ettingshausen coefficient is defined as

$$P = \frac{\partial_y T}{e|\mathbf{B}|(j_n)_x} \tag{2.307}$$

and making use of Eqs. (2.291), (2.293), and (2.294) it can be expressed as

$$P = \frac{T^2}{e|\mathbf{B}|} \frac{L_{11}L_{14} - L_{12}L_{13}}{D_3}. \tag{2.308}$$

The Righi–Leduc coefficient is defined as

$$A = \frac{\partial_y T}{|\mathbf{B}|\partial_x T} \tag{2.309}$$

and making use again of Eqs. (2.291), (2.293), and (2.294) it can be expressed as

$$A = \frac{D_3'}{|\mathbf{B}|D_3}, \tag{2.310}$$

where

$$D_3' = \begin{vmatrix} L_{11} & L_{13} & -L_{12} \\ -L_{13} & L_{11} & L_{14} \\ -L_{14} & L_{12} & L_{24} \end{vmatrix}. \tag{2.311}$$

Summarizing, we have obtained eight different phenomenological coefficients as a function of six independent Onsager kinetic coefficients. This means that there must be two relations among the phenomenological coefficients. One, pointed out by the American physicist Percy Williams Bridgman, is

$$TN_{\mathrm{I}} = \kappa_{\mathrm{I}} P, \tag{2.312}$$

which can be immediately checked. Obtaining the second relation is definitely more involved because of the presence of third- and fourth-order powers of the Onsager kinetic coefficients. It is important to point out that, as a consequence of the parity properties with respect to **B** of the Onsager coefficients, all the phenomenological coefficients of the various thermomagnetic and galvanomagnetic effects are even functions of **B**, as we can argue by the physical consideration that the measurement of these coefficients cannot depend on the switching from **B** to −**B**.

2.9 Linear Response in Quantum Systems

The linear response formalism that we have described for classical systems can be extended to quantum systems by suitable adjustments. First of all, it is useful to work in the Heisenberg picture, where quantum operators are time dependent. The reason is that we want to take into account the response of the quantum system to an external time-dependent source of perturbation. If the unperturbed quantum system is described by the Hamiltonian operator $\hat{H}$, we can introduce a perturbation by considering the perturbed Hamiltonian operator

$$\hat{H}'(t) = \hat{H} - \hat{H}_s \tag{2.313}$$

$$\hat{H}_s = \sum_k \eta_k(t)\,\hat{X}_k(t), \tag{2.314}$$

where $\hat{X}_k(t)$ are Hermitian operators corresponding to physical observables of the quantum system and $\eta_k(t)$ are the associated source terms. This Hamiltonian is the quantum counterpart of (2.134), where the classical perturbation fields $h_k(t)$, which represent generalized thermodynamic forces, and the classical observables X_k have been replaced by the source terms $\eta_k(t)$ and by the quantum observables $\hat{X}_k(t)$, respectively.

In analogy with the classical case, we assume that fluctuations of any observable depend linearly on the sources of perturbation through a response function, thus

$$\delta\langle\hat{X}_k(t)\rangle = \int dt' \sum_l \Xi_{kl}(t,t')\,\eta_l(t'), \tag{2.315}$$

where $\delta\langle\hat{X}_k(t)\rangle = \langle\hat{X}_k(t)\rangle - \langle\hat{X}_k\rangle_0$ is the difference between the expectation values of $\hat{X}_k(t)$ in the perturbed and in the unperturbed systems, while $\Xi_{kl}(t,t')$ is the response function of the observable $\hat{X}_k(t)$ to the perturbation source $\eta_l(t)$. Also in the quantum case we can assume the relation $\Xi_{kl}(t,t') \equiv \Xi_{kl}(t-t')$, as a consequence of the time-translation invariance of the quantum system. Therefore we can rewrite (2.315) as

$$\delta\langle\hat{X}_k(t)\rangle = \int dt' \sum_l \Xi_{kl}(t-t')\,\eta_l(t'), \tag{2.316}$$

and by transforming in Fourier space we obtain

$$\begin{aligned}\delta\langle \hat{X}_k(\omega)\rangle &= \int dt' \int dt\, e^{-i\omega t} \sum_l \Xi_{kl}(t-t')\, \eta_l(t') \\ &= \sum_l \int dt' \int dt\, e^{-i\omega(t-t')}\, \Xi_{kl}(t-t')\, e^{-i\omega t'}\, \eta_l(t') \\ &= \sum_l \Xi_{kl}(\omega)\, \eta_l(\omega),\end{aligned} \tag{2.317}$$

which is the analog of (2.140) for classical observables. In fact, as in classical systems, linear response theory implies that the response to a perturbation source of frequency ω is local in the frequency space; i.e., the fluctuations of the corresponding quantum observable exhibit the same frequency of the perturbation source. Moreover, if we assume that the perturbation sources $\eta_l(t)$ are real quantities and the operator $\hat{X}_k(\omega)$ is Hermitian, i.e., $\langle \hat{X}_k(t)\rangle$ is real, then also the response function $\Xi_{kl}(t-t')$ has to be real, and its Fourier transform $\Xi_{kl}(\omega)$ is such that $\Xi^*_{kl}(\omega) = \Xi_{kl}(-\omega)$. All considerations concerning causality and the analytic properties of $\Xi_{kl}(\omega)$, including the Kramers–Krönig relations (see Sections 2.4.1, 2.6, and Appendix F) also apply to the quantum case. In particular, the susceptibility can be defined again (see Eq. (2.67)) as the zero-frequency limit of the response function, namely,

$$\chi_{kl} = \lim_{\omega\to 0} \Xi_{kl}(\omega) = \lim_{\epsilon\to 0^+} \int_{-\infty}^{+\infty} \frac{d\omega'}{\pi} \frac{\Xi^I_{kl}(\omega')}{\omega' - i\epsilon}, \tag{2.318}$$

where $\Xi^I_{kl}(\omega)$ is the imaginary part of the Fourier transform of the response function.

Now, we want to obtain an explicit expression of the response function, in terms of sources and observables. For this purpose we have to introduce the evolution operator associated with the source Hamiltonian $\hat{H}_s$ in (2.314),

$$\hat{U}(t, t_0) = \exp\Big(-\frac{i}{\hbar} \int_{t_0}^{t} dt'\, \hat{H}_s(t')\Big). \tag{2.319}$$

The evolution operator is defined to obey the differential operator equation

$$\frac{d\hat{U}}{dt} = -\frac{i}{\hbar} \hat{H}_s\, \hat{U}. \tag{2.320}$$

In the interaction representation, specified by the subscript I, time evolution is determined by the unperturbed Hamiltonian $\hat{H}$ and $\hat{U}(t, t_0)$ maps the quantum state of the system at time t_0 to the quantum state at time t,

$$|\psi(t)\rangle_I = \hat{U}(t, t_0)\, |\psi(t_0)\rangle_I. \tag{2.321}$$

In general, we can assume that the ensemble of states of the quantum system is described by a density matrix $\hat{\rho}(t)$. We can think that far in the past, $t_0 \to -\infty$, before the perturbation is switched on, the density matrix is given by $\hat{\rho}_0$. Accordingly, at any time t we can write

$$\hat{\rho}(t) = \hat{U}(t)\, \hat{\rho}_0\, \hat{U}^{-1}(t)\,, \tag{2.322}$$

where $\hat{U}(t) = \hat{U}(t, t_0 \to -\infty)$. Having at our disposal an expression of the density matrix at time t, we can compute the expectation value of a perturbed observable as

$$\langle \hat{X}_k(t) \rangle = \mathrm{Tr}\ \hat{\rho}(t)\, \hat{X}_k(t) = \mathrm{Tr}\ \hat{\rho}_0 \hat{U}^{-1}(t)\, \hat{X}_k(t)\, \hat{U}(t), \tag{2.323}$$

where we have used the cyclic property of the trace. If we assume that the sources $\eta_k(t)$ are small quantities, we can take advantage of quantum perturbation theory and write the approximate formula

$$\begin{aligned} \langle \hat{X}_k(t) \rangle &\approx \mathrm{Tr}\ \hat{\rho}_0 \Big(\hat{X}_k(t) + \frac{i}{\hbar} \int_{-\infty}^{t} dt'\ [\hat{H}_s(t'), \hat{X}_k(t)] + \cdots \Big) \\ &= \langle \hat{X}_k(t) \rangle_0 + \frac{i}{\hbar} \int_{-\infty}^{t} dt'\ \langle [\hat{H}_s(t'), \hat{X}_k(t)] \rangle_0 + \cdots , \end{aligned} \tag{2.324}$$

where $\langle \hat{X}_k \rangle_0 = \mathrm{Tr}\ \hat{\rho}_0 \hat{X}_k$ is the expectation value of the operator $\hat{X}_k$ in the unperturbed system, while $\langle [\hat{A}(t'), \hat{B}(t)] \rangle_0$ is the same kind of expectation value for the commutator between the unspecified operators $\hat{A}(t')$ and $\hat{B}(t)$. The dots in the previous equation stand for higher-order terms of the perturbative expansion.

Using the definition (2.314) and the Heaviside distribution $\Theta(t - t')$, which allows us to extend the upper limit of the integral to $+\infty$, we can rewrite (2.324) as the quantum version of the fluctuation–dissipation relation:

$$\delta \langle \hat{X}_k(t) \rangle = \frac{i}{\hbar} \int_{-\infty}^{+\infty} dt'\ \Theta(t - t') \sum_l \langle [\hat{X}_l(t'), \hat{X}_k(t)] \rangle_0\, \eta_l(t') \,. \tag{2.325}$$

Finally, by comparing (2.325) with (2.316) we have

$$\Xi_{kl}(t - t') = -\frac{i}{\hbar}\ \Theta(t - t') \langle [\hat{X}_k(t), \hat{X}_l(t')] \rangle_0, \tag{2.326}$$

which is the quantum version of Eq. (2.138): the time derivative of the correlation function of classical observables is replaced by the commutator of the quantum observables at different times. The inverse temperature β does not appear explicitly in this formula, because it has been derived making use of a density matrix for zero temperature quantum states. In fact, up to here we have derived information concerning the dynamics of linear response theory in the quantum case. As we discuss hereafter, in order to make β come into play we have to take into account averages performed by the density matrix of a finite temperature thermodynamic state, $\hat{\rho} = \exp(-\beta \hat{H})$, which corresponds to the Boltzmann–Gibbs measure in the canonical ensemble.

As we have seen in Section 2.6 the simplest formulation of the fluctuation–dissipation theorem in the classical case is Eq. (2.142): the dissipative component of the response function, more precisely the imaginary part of its Fourier transform, is related to the fluctuations of the observables of the system through the Fourier transform of their correlation function. In what follows we want to find a quantum version of the fluctuation–dissipation theorem. Since $\Xi_{kl}(t - t')$ is invariant under time translation, we can set $t' = 0$ in Eq. (2.326) and we define the quantity

$$\Xi''_{kl}(t) = -\frac{i}{2} [\Xi_{kl}(t) - \Xi_{lk}(-t)]. \tag{2.327}$$

As we are going to show, the Fourier transform of this quantity is proportional to the anti-Hermitian part of the Fourier transform of the response function: the prefactor $-\frac{i}{2}$ makes the Fourier transform a real quantity. In fact,

$$\begin{aligned}
\Xi''_{kl}(\omega) &= -\frac{i}{2}[\Xi_{kl}(\omega) - \Xi^*_{lk}(\omega)] \\
&= -\frac{i}{2}\int_{-\infty}^{+\infty} dt\,[\Xi_{kl}(t)e^{-i\omega t} - \Xi_{lk}(t)e^{i\omega t}] \\
&= -\frac{i}{2}\int_{-\infty}^{+\infty} dt\, e^{-i\omega t}[\Xi_{kl}(t) - \Xi_{lk}(-t)] \qquad (2.328) \\
&\equiv \int_{-\infty}^{+\infty} dt e^{-i\omega t}\,\Xi''_{kl}(t), \qquad (2.329)
\end{aligned}$$

where we have used the fact that $\Xi_{kl}(t)$ is a real function.

By substituting (2.326) into (2.327) we obtain

$$\begin{aligned}
\Xi''_{kl}(t) &= -\frac{1}{2\hbar}\Theta(t)\big[\langle \hat{X}_k(t)\,\hat{X}_l(0)\rangle_0 - \langle \hat{X}_l(0)\,\hat{X}_k(t)\rangle_0\big] \\
&\quad + \frac{1}{2\hbar}\Theta(-t)\big[\langle \hat{X}_l(-t)\,\hat{X}_k(0)\rangle_0 - \langle \hat{X}_k(0)\,\hat{X}_l(-t)\rangle_0\big] \\
&= -\frac{1}{2\hbar}\langle \hat{X}_k(t)\,\hat{X}_l(0)\rangle_0 + \frac{1}{2\hbar}\langle \hat{X}_l(-t)\,\hat{X}_k(0)\rangle_0, \qquad (2.330)
\end{aligned}$$

where we have used translation invariance, i.e., $\langle \hat{X}_l(0)\,\hat{X}_k(t)\rangle_0 = \langle \hat{X}_l(-t)\,\hat{X}_k(0)\rangle_0$, and the identity $\Theta(t) + \Theta(-t) \equiv 1$. We can reorder the operators in the last term by considering that the average operation is performed in the canonical ensemble with a density matrix given by the Boltzmann weight $\hat{\rho} = \exp(-\beta\hat{H})$:

$$\begin{aligned}
\langle \hat{X}_l(-t)\,\hat{X}_k(0)\rangle_0 &= Z^{-1}\mathrm{Tr}\, e^{-\beta\hat{H}}\hat{X}_l(-t)\,\hat{X}_k(0) \\
&= Z^{-1}\mathrm{Tr}\, e^{-\beta\hat{H}}\hat{X}_l(-t)\, e^{\beta\hat{H}}e^{-\beta\hat{H}}\hat{X}_k(0) \\
&= Z^{-1}\mathrm{Tr}\, e^{-\beta\hat{H}}\hat{X}_k(0)\,\hat{X}_l(-t-i\hbar\beta) \\
&= \langle \hat{X}_k(t+i\hbar\beta)\,\hat{X}_l(0)\rangle_0, \qquad (2.331)
\end{aligned}$$

where $Z = \mathrm{Tr}\hat{\rho}$. The third equality above is obtained in a formal way, by interpreting the Boltzmann weight as a sort of evolution operator in the imaginary time $i\hbar\beta$, while the last one stems again from time-translation invariance in the complex plane. The formal trick of introducing an imaginary time is far from a rigorous procedure, but it provides us a shortcut through more complex calculations (i.e., solving the complete spectral problem) to obtain the quantum formulation of the fluctuation–dissipation theorem. In fact, coming back to (2.330) we can now write

$$\Xi''_{kl}(t)) = -\frac{1}{2\hbar}\big[\langle \hat{X}_k(t)\,\hat{X}_l(0)\rangle_0 - \langle \hat{X}_k(t+i\hbar\beta)\,\hat{X}_l(0)\rangle_0\big]. \qquad (2.332)$$

We can finally compute the Fourier transform of both sides and obtain

$$\Xi''_{kl}(\omega) = -\frac{1}{2\hbar}\big[1 - e^{-\beta\hbar\omega}\big]\, C_{kl}(\omega), \qquad (2.333)$$

where

$$C_{kl}(\omega) = \int_{-\infty}^{+\infty} dt e^{-i\omega t} C_{kl}(t), \tag{2.334}$$

with

$$C_{kl}(t) = \langle \hat{X}_k(t) \hat{X}_l(0) \rangle_0. \tag{2.335}$$

Eq. (2.333) represents the quantum version of the fluctuation–dissipation theorem (2.142). A physical interpretation of (2.333) emerges if we rewrite this equation in the form

$$C_{kl}(\omega) = -2\hbar \big[n_B(\omega) + 1\big] \, \Xi''_{kl}(\omega), \tag{2.336}$$

where $n_B(\omega) = 1/(\exp(\beta\hbar\omega) - 1)$ is the Bose–Einstein distribution function. We can conclude that, in the quantum case, the contributions to fluctuations come not only from thermal effects, i.e., from $n_B(\omega)$, but also from intrinsic quantum fluctuations, i.e., from the "+1" term in the square brackets. In the limit $\hbar \to 0$, or (more physically) in the limit $\beta \to 0$, $[n_B(\omega) + 1] \approx 1/(\beta\hbar\omega)$ and we recover the classical equation Eq. (2.142)

$$C_{kl}(\omega) = -\frac{2}{\beta\omega} \, \Xi^I_{kl}(\omega), \tag{2.337}$$

provided we consider that the classical counterpart of $\Xi''_{kl}(\omega)$ is $\Xi^I_{kl}(\omega)$.

2.10 Examples of Linear Response in Quantum Systems

The discussion in the previous section indicates a clear analogy between the classical and the quantum formulation of linear response theory. In fact, apart from some formal aspects to be carefully taken into account, also in the quantum case we have found that time correlation functions of observables are the main ingredients that allow us to identify quantities of physical interest. Due to space limitations in what follows we cannot provide an exhaustive discussion of the many applications of linear response theory to quantum systems. We sketch just a few selected examples that summarize conceptual similarities as well as technical peculiarities of the quantum case with respect to the classical one.

2.10.1 Power Dissipated by a Perturbation Field

In Section 2.4.2 we have discussed how the response function is related to the power dissipated by the external work made on a classical system by a perturbation field (see Eqs. (2.59) and (2.60)), and we have computed this quantity for the damped harmonic oscillator in the presence of a periodic driving force (see Eq. (2.93)).

Now we can study the problem of dissipated power for a general quantum system, represented by the Hamiltonian operator $\hat{H}'$. Its general expression amounts to the rate of change in time of the energy in the quantum system, namely,

$$W(t) = \frac{d}{dt}\,\mathrm{Tr}(\hat{\rho}\hat{H}') = \mathrm{Tr}\Big(\frac{d\hat{\rho}}{dt}\,\hat{H}' + \hat{\rho}\,\frac{d\hat{H}'}{dt}\Big), \tag{2.338}$$

where both operators $\hat{\rho}$ and $\hat{H}'$ are functions of time t. Evaluating a physical observable, such as $W(t)$ is independent of the chosen representation. In this case it is more convenient working in the Schrödinger one, where the density matrix evolves according to the equation

$$i\frac{d\hat{\rho}}{dt} = \big[\hat{H}', \hat{\rho}\big], \tag{2.339}$$

so that the first term on the right-hand side of (2.338) vanishes because of the cyclic property of the trace. On the other hand, we assume that $\hat{H}'$ varies in time due to the presence of a perturbation introduced by the source term (2.314), where, for the sake of simplicity, we assume that only one source term is active, namely,

$$\hat{H}_s = \eta(t)\,\hat{X}. \tag{2.340}$$

In fact, in the Schrödinger representation the dependence on time is limited to the perturbation field η, while the observable $\hat{X}$ is a time-independent quantity, so that

$$\frac{d\hat{H}'}{dt} = \hat{X}\frac{d\eta(t)}{dt}. \tag{2.341}$$

Substituting into (2.338) we obtain

$$W(t) = \mathrm{Tr}\Big(\hat{\rho}\,\frac{d\hat{H}'}{dt}\Big) = \langle\hat{X}\rangle\frac{d\eta}{dt} = \Big(\langle\hat{X}\rangle_0 + \delta\langle\hat{X}\rangle\Big)\frac{d\eta}{dt}. \tag{2.342}$$

If we assume that we are dealing with a periodic forcing of period $2\pi/\Omega$, the source term has the form

$$\eta(t) = \mathrm{Re}\left(\eta_0\, e^{i\Omega t}\right) = \frac{1}{2}\left(\eta_0\, e^{i\Omega t} + \eta_0^*\, e^{-i\Omega t}\right) \tag{2.343}$$

and we can express the average dissipated power as

$$\overline{W} = \frac{\Omega}{2\pi}\int_0^{2\pi/\Omega} dt\, W(t) = \frac{\Omega}{2\pi}\int_0^{2\pi/\Omega} dt\Big(\langle\hat{X}\rangle_0 + \delta\langle\hat{X}\rangle\Big)\frac{d\eta}{dt}. \tag{2.344}$$

The contribution from the term $\langle\hat{X}\rangle_{\eta=0}$ cancels when integrated over a full period and we have

$$\begin{aligned}
\overline{W} &= \frac{\Omega}{2\pi}\int_0^{2\pi/\Omega} dt \int_{-\infty}^{+\infty} dt'\; \Xi(t-t')\eta(t')\,\frac{d\eta}{dt} \\
&= \frac{\Omega}{2\pi}\int_0^{2\pi/\Omega} dt \int_{-\infty}^{+\infty} dt'\; \frac{1}{2\pi}\int_{-\infty}^{+\infty} d\omega\; \Xi(\omega)e^{i\omega(t-t')} \\
&\quad\times\frac{1}{4}\Big[\eta_0\, e^{i\Omega t'} + \eta_0^*\, e^{-i\Omega t'}\Big]\Big[i\Omega\eta_0\, e^{i\Omega t} - i\Omega\eta_0^*\, e^{-i\Omega t}\Big] \\
&= \frac{i\Omega}{4}\frac{\Omega}{2\pi}\int_0^{2\pi/\Omega} dt \int_{-\infty}^{+\infty} d\tau\; \frac{1}{2\pi}\int_{-\infty}^{+\infty} d\omega\; \Xi(\omega)e^{-i\omega\tau} \\
&\quad\times\Big[\eta_0^2 e^{i\Omega(2t+\tau)} - \eta_0^{*2} e^{-i\Omega(2t+\tau)} + |\eta_0|^2 e^{-i\Omega\tau} - |\eta_0|^2 e^{i\Omega\tau}\Big],
\end{aligned} \tag{2.345}$$

where we have made a change of variable from t' to $\tau = t' - t$. We can now observe that the contributions from terms ${\eta_0}^2$ and η_0^{*2} vanish when integrated over t, while the terms proportional to $|\eta_0|^2$ are independent of t, and, when integrated over τ, they yield Dirac delta distributions $\delta(\omega - \Omega)$ and $\delta(\omega + \Omega)$. We therefore obtain

$$\overline{W} = \frac{i\Omega}{4}\Big[\Xi(-\Omega) - \Xi(\Omega)\Big]\,|\eta_0|^2, \tag{2.346}$$

which can be further simplified observing that $\Xi^{\mathrm{I}}(\omega)$ is the odd component of $\Xi(\omega)$ (see Appendix F),

$$\overline{W} = \frac{\Omega}{2}\Xi^{\mathrm{I}}(\Omega)\,|\eta_0|^2. \tag{2.347}$$

If we perform work on a system, its energy has to increase and, accordingly, the quantity $\Omega\,\Xi^{\mathrm{I}}(\Omega)$ has to be positive. These results can be extended to the general case described by the perturbed Hamiltonian operator (2.313).

2.10.2 Linear Response in Quantum Field Theory

In order to describe transport problems in a quantum framework we need to consider a quantum field theory (QFT) formulation, where we assume that the physical observables are quantum Hermitian operators that depend on the space coordinates $\mathbf{r}$ and on time t, i.e., $\hat{X}(\mathbf{r}, t)$. In this case the perturbative component of the Hamiltonian (see Eq. (2.314)) is

$$\hat{H}_s = \int d\mathbf{r} \sum_k \eta_k(\mathbf{r}, t)\,\hat{X}_k(\mathbf{r}, t), \tag{2.348}$$

while the response function $\Xi_{kl}(\mathbf{r}, t; \mathbf{r}', t')$ (see Eq. (2.315)) is defined by the relation

$$\delta\langle\hat{X}_k(\mathbf{r}, t)\rangle = \int d\mathbf{r}'dt' \sum_l \Xi_{kl}(\mathbf{r}, t; \mathbf{r}', t')\;\eta_l(\mathbf{r}', t'). \tag{2.349}$$

Again, all the considerations concerning causality and the analytic properties of the response function that hold in the classical as well in the quantum case straightforwardly extend to the quantum field theory formulation. In particular, we can write the analog of Eq. (2.326) in the form

$$\Xi_{kl}(\mathbf{r}, \mathbf{r}'; t - t') = -\frac{i}{\hbar}\,\Theta(t - t')\langle[\hat{X}_k(\mathbf{r}, t), \hat{X}_l(\mathbf{r}', t')]\rangle_0, \tag{2.350}$$

where we have introduced explicitly the condition that the response function is invariant under time translations. For systems that are also translationally invariant in space, the response function depends on $\mathbf{r} - \mathbf{r}'$, rather than separately on both coordinate vectors, and, passing to Fourier-transformed quantities in space and time, Eq. (2.349) becomes

$$\delta\langle\hat{X}_k(\mathbf{k}, \omega)\rangle = \sum_l \Xi_{kl}(\mathbf{k}, \omega)\,\eta_l(\mathbf{k}, \omega)\,. \tag{2.351}$$

Electric Conductivity

In order to describe electric conductivity we have to consider a QFT with global $U(1)$ symmetry, associated with a gauge field, which is the electromagnetic four-potential. According to Noether's theorem, there is an associated conserved four-current $j^\mu = (j^0, j^i)$ (with $i = x, y, z$), which is a conserved quantity, i.e., $\partial_\mu j^\mu = 0$. We assume that the current j^μ is coupled to the background, electromagnetic gauge field $A_\mu(\mathbf{r})$, yielding the perturbation Hamiltonian operator

$$H_s = \int d\mathbf{r} A_\mu j^\mu. \tag{2.352}$$

For the sake of simplicity, here we assume that A_μ is a fixed perturbation source under our control. On the other hand, the conserved current in the presence of a background field is given by

$$j^\mu = ie[\phi^\dagger \partial^\mu \phi - (\partial^\mu \phi^\dagger)\phi] - e^2 A^\mu \phi^\dagger \phi, \tag{2.353}$$

where e is the electric charge and ϕ is a free, relativistic, complex scalar field.

As a first step we can apply Eq. (2.349) to the case where the perturbation is (2.352), thus obtaining

$$\begin{aligned} \delta\langle j_\mu \rangle &= \langle j_\mu \rangle - \langle j_\mu \rangle_0 \\ &= -\frac{i}{\hbar} \sum_\nu \int d\mathbf{r}' \int_{-\infty}^{t} dt' \, \langle [j_\mu(\mathbf{r}, t), j_\nu(\mathbf{r}', t')] \rangle_0 \, A_\nu(\mathbf{r}', t'), \end{aligned} \tag{2.354}$$

where the limits of the second integral on the right-hand side are imposed by the causality condition; see Eq. (2.350).

Let us first consider the contribution coming from the term $\langle j_\mu \rangle_0$. Even if the unperturbed situation has no currents, the second term in the right-hand side of (2.353) provides a contribution

$$\langle j_\mu \rangle_0 = -e^2 A_\mu \langle \phi^\dagger \phi \rangle_0 = -e A_\mu \rho, \tag{2.355}$$

where ρ is the background charge density that is evaluated in the unperturbed quantum state.

As a physical example we want to describe the problem of electric conductivity, i.e., Ohm's law, in the language of QFT. This amounts to studying linear response in the presence of a background electric field $\mathbf{E} = (E_x, E_y, E_z)$. In order to evaluate the right-hand side of Eq. (2.354), we can decide to fix the gauge $A_0 = 0$, so that the space components of the electric field are given by the expression

$$E_l = -\frac{dA_l}{dt} \rightarrow A_l(\omega) = -\frac{E_l(\omega)}{i\omega}. \tag{2.356}$$

In the absence of an external magnetic field, the system is invariant under rotation and parity transformation, so that the current has to be parallel to the applied electric field and the phenomenological equation in Fourier space describing Ohm's law reads

$$\langle j(\mathbf{k}, \omega) \rangle = \sigma(\mathbf{k}, \omega) \, E(\mathbf{k}, \omega), \tag{2.357}$$

where $E(\mathbf{k}, \omega)$ is the background electric field in Fourier space and σ is the conductivity. Note that in this simple setup the dependence on the subscript indices can be omitted and Eq. (2.354) simplifies to

$$\begin{aligned}\delta\langle j\rangle &= \langle j\rangle - \langle j\rangle_0 \\ &= -\frac{i}{\hbar}\int d\mathbf{r}'\int_{-\infty}^{t} dt'\,\langle[j(\mathbf{r},t), j(\mathbf{r}',t')]\rangle_0\, A(\mathbf{r}',t') \\ &= \int d\mathbf{r}'\int_{-\infty}^{+\infty} dt'\,\Xi(\mathbf{r}-\mathbf{r}', t-t')A(\mathbf{r}',t')\,. \end{aligned} \tag{2.358}$$

Now we can Fourier-transform this equation to cast it in the form of Eq. (2.357) and we obtain that the conductivity has two contributions,

$$\sigma(\mathbf{k}, \omega) = -i\frac{e\rho}{\omega} + i\frac{\Xi(\mathbf{k}, \omega)}{\omega}, \tag{2.359}$$

where the first one stems from the background charge density, while the second one is associated with the Fourier transform of the response function,

$$\Xi(\mathbf{k}, \omega) = -\frac{i}{\hbar}\int_{-\infty}^{+\infty} dt\, d\mathbf{r}\, \Theta(t)\, e^{-i(\omega t + \mathbf{k}\cdot\mathbf{r})}\,\langle[j(\mathbf{r},t), j(\mathbf{0},0)]\rangle_0\,. \tag{2.360}$$

This equation can be interpreted as the Kubo formula for electric conductivity in QFT.

Viscosity

The procedure that we have just described to obtain an expression for the conductivity is quite instructive, because it can be followed to obtain, by suitable adjustments, other expressions for transport coefficients from linear response theory applied to QFT. As an example, here we want to sketch how to obtain an expression for the viscosity η in QFT. From the classical case (see Section 1.3) we know that viscosity is the phenomenological coefficient associated with the transport of momentum, which, in the hydrodynamic representation, is a conserved quantity (as electric charge is a conserved quantity in the previous example). Moreover, we have to consider a QFT that is invariant under space and time translations: in this case Noether's theorem yields four currents that are associated with energy and momentum conservation. They are represented by the stress-energy tensor $T^{\mu\nu}$ and the conservation condition reads $\partial_\mu T^{\mu\nu} = 0$. We can proceed as for electric conductivity by replacing the electric current with the momentum current. In order to simplify the problem, we can assume a setup where momentum in the x-direction is transported along the z-direction, thus making immaterial any dependence on the y-direction (this setup is the same adopted in the classical case discussed in Section 1.3). In this case we have to deal with a single, relevant component of the stress-energy tensor, namely, T^{xz}, which, in this case, plays the same role as the electric current in the previous example. Accordingly, the expression that we have obtained for the electric conductivity tensor can be formally turned into an expression for the viscosity η. Anyway, in this case there are two main differences with respect to electric conductivity. First of all, there is no background charge density to be taken into account, because momentum current is a mechanical observable. The second difference is that viscosity, as in the classical setup, has to be computed in the presence of a constant driving force. This implies that we have to consider the limits $\omega \to 0$ and $\mathbf{k} \to 0$ in the expression analogous to Eq. (2.359). Thus we can write

$$\eta = \lim_{\omega \to 0} \lim_{\mathbf{k} \to 0} \frac{\Xi_{xz,xz}(\mathbf{k}, \omega)}{i\omega}, \tag{2.361}$$

where

$$\Xi_{xz,xz}(\mathbf{k}, \omega) = -\frac{i}{\hbar} \int_{-\infty}^{+\infty} dt \, d\mathbf{r} \, \Theta(t) \, e^{-i(\omega t + \mathbf{k}\cdot\mathbf{r})} \, \langle [T_{xz}(\mathbf{r}, t), T_{xz}(\mathbf{0}, 0)] \rangle_0. \tag{2.362}$$

Eq. (2.361) can be interpreted as the Kubo formula for viscosity in QFT.

2.11 Bibliographic Notes

A classical textbook about linear response theory, the fluctuation–dissipation theorem, and their applications to nonequilibrium statistical mechanics is R. Kubo, M. Toda, and N. Hashitsume, *Statistical Physics II: Nonequilibrium Statistical Mechanics*, 3rd ed. (Springer, 1998). In particular, this book contains a rigorous extension of the formalism of stochastic differential equations to macroscopic observables by the so-called projection formalism.

A classical textbook on nonequilibrium thermodynamics containing a detailed presentation of the Onsager theory of transport processes is S. R. de Groot and P. Mazur, *Non-Equilibrium Thermodynamics* (Dover, 1984).

The derivation of the hydrodynamic equations from the Boltzmann model of the ideal gas as a consequence of the conservation of mass, momentum, and energy densities is contained in the book by K. Huang, *Statistical Mechanics*, 2nd ed. (Wiley, 1987).

A pedagogical introduction to entropy is the book by D. S. Lemons, *A Student's Guide to Entropy* (Cambridge University Press, 2013).

An advanced texbook on nonequilibrium statistical physics with applications to condensed matter problems is G. F. Mazenko, *Nonequilibrium Statistical Mechanics* (Wiley-VCH, 2006).

Many applications of nonequilibrium many-body theory to condensed matter physics are contained in the book by P. M. Chaikin and T. C. Lubensky, *Principles of Condensed Matter Physics* (Cambridge University Press, 2000).

An interesting and modern book about nonequilibrium statistical mechanics of turbulence, transport processes. and reaction–diffusion processes is J. Cardy, G. Falkovich, and K. Gawedzki, *Non-equilibrium Statistical Mechanics and Turbulence* (Cambridge University Press, 2008). This book contains the lectures given by the authors at a summer school. This notwithstanding, these lectures concern frontier topics in nonequlibrium statistical mechanics that may not be immediately accessible to undergraduate students.

A comprehensive and up-to-date account on anomalous transport phenomena in low-dimensional systems can be found in the book by S. Lepri (ed.), *Thermal Transport in*

Low Dimensions, Lecture Notes in Physics (Springer, 2016). Some of the contributions contained in this book illustrate experimental results.

The peculiar thermodynamic properties of equilibrium and out-of-equilibrium systems with long-range interactions are discussed in the book by A. Campa, T. Dauxois, D. Fanelli, and S. Ruffo, *Physics of Long-Range Interacting Systems* (Oxford University Press, 2014).

A mathematical textbook about analytic functions and their Fourier and Laplace transform, together with functional analysis, is W. Rudin, *Functional Analysis* (McGraw-Hill, 1991).

Many interesting aspects concerning the applications of linear response theory and Kubo formalism to linear irreversible thermodynamics are contained in the book by D. J. Evans and G. P. Morris, *Statistical Mechanics of Nonequilibrium Liquids*, 2nd ed. (ANU Press, 2007). This book contains also an instructive part concerning the use of computer algorithms useful for performing numerical simulations of fluids.

A short survey about models of thermal reservoirs useful for numerical simulations is discussed in the review paper by S. Lepri, R. Livi, and A. Politi, Thermal conduction in classical low-dimensional lattices, *Physics Reports*, **377** (2003) 1–80.

3 From Equilibrium to Out-of-Equilibrium Phase Transitions

This chapter is a bridge between equilibrium and nonequilibrium phase transitions, making use of concepts the reader should know: control parameter and order parameter, critical exponents, phenomenological scaling theory, and mean-field theory. The first part of the chapter is a not so short guide to equilibrium phase transitions; rather than a reminder, it is where we introduce the above concepts, making use of familiar physical phenomena, the liquid–gas and the ferro-paramagnetic phase transitions.

However, when dealing with nonequilibrium phase transitions we cannot take for granted that the reader is familiar with any of these. For this reason we make use of two equilibrium systems that are easily generalized to be brought out of equilibrium. The first system is the lattice gas to which we apply a driving force so as to favor the hop of a particle in a given direction and unfavor the hop in the opposite direction. With suitable boundary conditions, such driving breaks the detailed balance. The same sort of driving may be imposed on the spreading of wetting during a percolation process, where neighboring sites of a lattice may wet each other if the corresponding bond is an open pore, which occurs with some assigned probability p. If spreading is isotropic, we get the standard, equilibrium percolation transition, but we may force the spreading to take place in a given direction only, giving rise to the so-called directed percolation. In this case the spreading direction can be interpreted as a time arrow and the percolation in d spatial dimension is now a time-dependent process in $(d-1)$ spatial dimensions.

The feature that time is a sort of additional dimension will be seen to be true also for phenomena where this interpretation is not as trivial as for directed percolation. In general (and somewhat abstract) terms, this fact may help us to understand why the same type of $d = 1$ model (either Ising or percolation) gives a trivial equilibrium phase transition and an entirely nontrivial nonequilibrium phase transition.

3.1 Introduction

Phase transitions are physical phenomena that are present in our everyday life: liquid water cools into ice and water vapor condensates on the surface of a cold glass; mixtures of various components separate into different phases; a piece of magnetized iron loses its magnetization by heating; traffic flow collapses into a jam; and so on. Phase transitions provide us with the main example of how matter can go through changes of state and properties under the action of varying physical conditions. It is astonishing that phase transitions occur at the atomic and molecular levels as well as at cosmological scales.

Moreover, independent of the actual physical scales, they exhibit universal properties, which stem from the basic symmetries contained in the interaction rules between the constituents of matter.

All of these features have been widely investigated in many different realms of experimental science over the last centuries. Statistical mechanics has provided us with the theoretical framework by which these phenomena can be interpreted and predicted. This is certainly one of its major breakthroughs, although we have to point out that our present knowledge about equilibrium phase transitions relies on a much more solid ground than the one we have about nonequilibrium ones. We can say that many concepts and methods that have been worked out for describing equilibrium phase transitions can be borrowed and adapted to nonequilibrium phase transitions, relying on the power of analogies in the physical world.

Anyway, in general this is not enough, because there are some crucial differences between equilibrium and nonequlibrium phenomena. For instance, at equilibrium we can deal with statistical properties associated with static conditions, while in out-of-equilibrium conditions we have to deal with an explicit dynamical description, where time enters as a basic ingredient and takes the form of an additional dimension: if in equilibrium statistical physics spatial fluctuations are important and let mean-field theory fail in low dimension, in out-of-equilibrium systems we must also consider temporal fluctuations. A second, important difference is that the free energy, which has a very relevant role at equilibrium, generally has no out-of-equilibrium counterpart. This means it is not even obvious how to define a nonequilibrium phase transition.

In this chapter we first offer a concise review of equilibrium phase transitions, focusing on those basic concepts that are useful for nonequilibrium ones. Then we discuss nonequibrium steady states (NESSs), because a feature of nonequilibrium phase transitions is the transition between different NESSs when tuning a suitable parameter. NESSs are not equilibrium states because they do not satisfy detailed balance; an extreme example of a NESS is an absorbing state, i.e., a microscopic state from which it is impossible to escape (so, dynamics stops once the absorbing state is attained).

More generally, NESS can be settled under the action of weak, stationary external forces. The force can be applied by an external field or by imposing a sustained gradient of intensive thermodynamic parameters (e.g., a constant temperature gradient) across the system. The basic feature common to most NESSs is the presence of some stationary current(s) flowing through the system. This is also the case in some kinetic models, where a driving force is applied (see Section 3.4) or where there is a supply of injected particles from the environment, balanced by a restitution mechanism to the environment (see Section 4.5). In the case of directed percolation (see Section 3.5.1), which is the basic example of a model exhibiting a nonequilibrium phase transition in the presence of an absorbing state, the NESS in the active phase can be thought of as a stationary current of particles flowing through a porous medium under the action of gravity.

Systems with one or more absorbing states play an important role in the study of nonequilibrium phase transitions. In this chapter we limit our study to the already mentioned directed percolation (DP), which is the prototypical example (universality class) of a phase transition in the presence of one single absorbing state. Different models are studied in the next chapter, where we also consider the possibility to have equivalent NESSs, i.e., NESSs that are connected to one another by a symmetry operation, similar to what happens in equilibrium phase transitions, when the low temperature phase has equivalent ground states: in this case, both at equilibrium and at nonequilibrium, the phase transition is accompanied by a symmetry breaking.

3.2 Basic Concepts and Tools of Equilibrium Phase Transitions

Phase transitions in physical systems at thermodynamic equilibrium occur because some symmetry of the system at hand breaks when a control parameter, e.g., temperature, pressure, or volume, is varied.

In first-order phase transitions this mechanism is associated with the exchange of a finite amount of energy (heat) with the environment, as happens when passing from a gas to a liquid phase for $T < T_c$ (see the following section for the van der Waals model). In this case one observes a discontinuity in the fluid density $\rho = N/V$, when the pressure P or the temperature T are varied by an infinitesimal amount. The heat supplied by the environment when passing from the liquid to the gas phase is what is needed to overtake the interaction energy between the atoms in the liquid phase.

In continuous, i.e., second-order, phase transitions, such as the one occurring along the critical isothermal line for a van der Waals gas, or as happens in the Ising model of a ferromagnet in the absence of a magnetic field (see Section 3.2.2), the symmetry breaks spontaneously, because of microscopic fluctuations. These continuous phase transitions are characterized by the presence of long-range correlations at the transition point, which impose the typical scale-invariant structure of critical behavior. The main consequence is the presence of critical exponents, associated with the power-law behavior of physical quantities, which may either vanish or diverge at the critical point. Dimensional analysis as well as renormalization group methods have usually been employed to predict these exponents approximately and, if possible, exactly.

3.2.1 Phase Transitions and Thermodynamics

In 1873 the Dutch scientist Johannes Diderik van der Waals realized that the changes of state observed experimentally in real gases could not be accounted for by the equation of state of the ideal gas,

$$PV = \mathcal{N}RT, \tag{3.1}$$

where the product of pressure P and volume V is proportional to the temperature T and to the number of moles $\mathcal{N}$ contained in the gas, with a proportionality constant equal to the so-called constant of gases, $R = 8.314$ J mol^{-1} K^{-1}. The above equation can also be written as

$$Pv = T, \tag{3.2}$$

where $v = V/N$ is the volume per molecule and the Boltzmann constant has been set equal to one, as everywhere in this book (otherwise, on the right-hand side it would be $K_B = R/N_A$, with $N_a = 6.022140857 \times 10^{23}$, the Avogadro number).

In fact, van der Waals proposed to modify (3.2) by taking into account two basic facts. At variance with the ideal gas model, molecules in a real gas are not "point-like" particles but occupy a finite volume. One can easily realize that this is a crucial feature to be taken into account, when the gas is compressed to make it change to a liquid or a solid. Accordingly, one should subtract from the volume of the container V a phenomenological quantity Nb (the so-called covolume), so that $V - Nb$ represents the effective volume available to the molecules.

Moreover, molecules in a real gas are subject to an attractive interaction, which reduces the pressure exerted on the walls of the volume containing the real gas. By combining these two heuristic ingredients (see Appendix G for more details) van der Waals obtained the following equation of state

$$P = \frac{T}{v - b} - \frac{a}{v^2}, \tag{3.3}$$

which is characterized by a critical value $T_c = 8a/(27b)$ below which isothermal curves $P(v)$ are no longer a decreasing function. At $T = T_c$ there is an inflexion point that allows us to define a critical volume (per particle) $v_c = 3b$ and a critical pressure $P_c = P(v_c, T_c) = a/(27b^2)$. If we rescale thermodynamical variables with respect to their critical values, we obtain the parameter-free equation of state given in Eq. (G.12),

$$P^* = \frac{8}{3}\frac{T^*}{v^* - \frac{1}{3}} - \frac{3}{(v^*)^2}. \tag{3.4}$$

Despite its heuristic character, Eq. (3.3) has the main merit of pointing out the concept of universality. In fact, any van der Waals gas, independent of its atomic or molecular constituents (i.e., independent of the specific values of a and b) is found to obey the same equation of state given in Eq. (3.4) and plotted for different values of T^* in Fig. 3.1. For $T^* > 1$ (upper curve) we are in the gas phase and $P(v)$ is a monotonously decreasing function. At $T^* = 1$ (the critical temperature, middle curve) comes an inflection point, and for $T^* < 1$ (lower curve and inset) is a region where P would increase with v. This unphysical piece of isothermal curve should be replaced by a horizontal line according to the Maxwell construction and joining the two phases in equilibrium, the liquid A and the gas B. The reason why the van der Waals equation cannot account for this process is that it cannot describe a nonhomogeneous system.

The passage from state A to state B along one of the isobaric and isothermal segments corresponds to a first-order phase transition, where a finite amount of heat has to be supplied to the fluid. Since temperature and pressure remain constant, the supplied heat serves to break the interaction energy between particles in the liquid phase.[1] At $T = T_c$ the phase transition from the liquid to the gas phase occurs continuously, without heat exchange with the environment. According to the classification of Paul Ehrenfest, this scenario corresponds to a second-order phase transition.

[1] In the reversed transformation, $B \rightarrow A$, the same amount of (latent) heat is returned to the environment. This is what happens with the condensation process that comes before a rainfall.

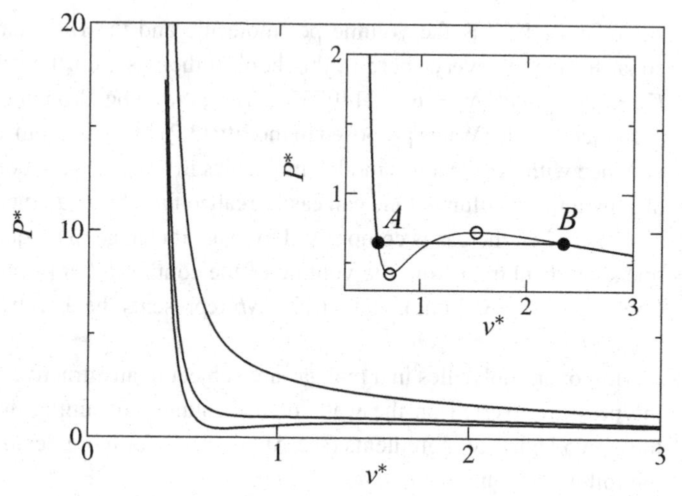

Fig. 3.1 Plot of the rescaled van der Waals equation (3.4), for $T^* = 1.5$ (gas phase, upper curve), $T^* = 1$ (critical point, middle curve), and $T^* = 0.9$ (phase coexistence, lower curve). In the inset we focus on $T^* < 1$, where there is coexistence between the liquid phase *A* and the gas phase *B*. In between A and B the equilibrium curve is the horizontal segment. The curved portions between solid and open circles correspond to metastable states, and the one between open circles corresponds to unstable states (with the pressure unphysically increasing with the volume at constant temperature).

The van der Waals equation points out another distinctive feature of equilibrium second-order phase transitions, i.e., the singular behavior of thermodynamic observables at the critical temperature T_c. For instance, the isothermal compressibility χ of a van der Waals gas at the critical temperature diverges as

$$\chi = -\left(\frac{\partial V}{\partial P}\right)_T \sim |T - T_c|^{-\gamma}, \tag{3.5}$$

where γ is the corresponding critical exponent. This result can be easily obtained by differentiating (3.4) with respect to P^* while keeping T^* constant, thus obtaining

$$-\left(\frac{\partial v^*}{\partial P^*}\right)_{T^*} = \left(\frac{8T^*}{3\left(v^* - \frac{1}{3}\right)^2} - \frac{6}{(v^*)^3}\right)^{-1}. \tag{3.6}$$

For $v = v_c$ (i.e., $v^* = 1$), the above equation simplifies to

$$-\left(\frac{\partial v^*}{\partial P^*}\right)_{T^*} = \frac{1}{6}(T^* - 1)^{-1}, \tag{3.7}$$

which gives[2] (see Eq. (3.5)) $\gamma = 1$. We should also point out that experimental measurements yield $\gamma \simeq 1.25$.

The singular behavior at the critical point of various observables other than the isothermal compressibility is the typical signature of critical phenomena. For instance, the specific heat at constant volume of real gases, c_V, has been found to diverge at T_c as

[2] Coming back to unreduced variables, we get $\chi = -\left(\frac{\partial V}{\partial P}\right)_T = \frac{N v_c T_c}{6 P_c}(T - T_c)^{-1} = 4b^2(T - T_c)^{-1}$.

$$c_V \sim |T - T_c|^{-\alpha}, \tag{3.8}$$

where the experimental estimate of the critical exponent α is close to 0.12. Also in this case, the van der Waals theory fails, since it predicts just a discontinuity of c_V at $T = T_c$ (see Appendix G). These discrepancies are a consequence of the mean-field nature of the van der Waals phenomenological equation, which underestimates the strongly correlated nature of microscopic fluctuations at the critical point. This scenario can be understood in the light of Landau theory of critical phenomena (see Section 3.2.3).

3.2.2 Phase Transitions and Statistical Mechanics

A microscopic theory of critical phenomena associated with second-order phase transitions has been successfully worked out in equilibrium statistical mechanics. This has been made possible by studying simple models that share with many other physical systems two basic universal features, namely, symmetry and dimensionality. For instance, binary mixtures and lattice gas models have been found to be equivalent to the Ising model of ferromagnetism. Hereafter we report a short description of the latter model, which can be considered as the test bed for the main achievements in this field.

The Ising model of ferromagnetism is represented by the Hamiltonian

$$\mathcal{H}[\{\sigma_i\}] = -\sum_{\langle ij \rangle} J_{ij}\, \sigma_i \sigma_j - \sum_i H_i \sigma_i. \tag{3.9}$$

The first sum is performed over all distinct and unordered pairs of nearest-neighbor sites of a regular lattice, where the discrete spin-like variables $\sigma_i = \pm 1$ are located. The interaction coupling constant J_{ij}, called the exchange constant, is positive in order to favor spin alignment for minimizing the local energy, while H_i is a local magnetic field favoring the alignment of σ_i with its sign. If we assume homogeneity of the physical parameters, we can deal with the simplified version of the Ising model where $J_{ij} = J$ and $H_i = H$, independent of the lattice sites i and j:

$$\mathcal{H}[\{\sigma_i\}] = -J \sum_{\langle ij \rangle} \sigma_i \sigma_j - H \sum_i \sigma_i. \tag{3.10}$$

In the absence of the external magnetic field, i.e., $H = 0$, the Ising Hamiltonian is symmetric with respect to the global "spin-flip" transformation $\sigma_i \to -\sigma_i$ (i.e., it exhibits the discrete $\mathbb{Z}_2$ symmetry).

In equilibrium statistical mechanics we assume that the system is in contact with a thermostat at temperature T and that the probability $p(\{\sigma_i\})$ of a spin configuration $\{\sigma_i\}$ is given by the expression

$$p(\{\sigma_i\}) = \frac{1}{Z} \exp\left(- \frac{\mathcal{H}[\{\sigma_i\}]}{T} \right), \tag{3.11}$$

where

$$Z(V, T, h) = \sum_{\{\sigma_i\}} \exp\left(- \frac{\mathcal{H}[\{\sigma_i\}]}{T} \right) \tag{3.12}$$

is the partition function, obtained by summing the so-called Boltzmann weights over the collection of all possible spin configurations. The relation with thermodynamics can be established by the following equation,

$$F(V,T,H) = -T \ln Z, \tag{3.13}$$

where $F(V,T,H)$ is the thermodynamic free energy, which depends on the volume of the system V,[3] on the temperature T and on the external magnetic field H.

Eq. (3.13) can be justified by rewriting it in the form

$$\sum_{\{\sigma_i\}} \exp\left(\frac{(F(V,T,H) - \mathcal{H}[\{\sigma_i\}]}{T}\right) = 1 \tag{3.14}$$

and by differentiating both sides with respect to T, thus yielding the relation

$$\sum_{\{\sigma_i\}} \exp\left(\frac{(F(V,T,H) - \mathcal{H}[\{\sigma_i\}]}{T}\right)\left[-\frac{1}{T^2}\left(F - \mathcal{H}[\{\sigma_i\}] - T\frac{\partial F}{\partial T}\right)\right] = 0. \tag{3.15}$$

Since the equilibrium value of any observable $\mathcal{O}[\{\sigma_i\}]$ is given by the expression

$$\langle \mathcal{O} \rangle = \frac{1}{Z} \sum_{\{\sigma_i\}} \mathcal{O}[\{\sigma_i\}] \exp\left(-\frac{\mathcal{H}[\{\sigma_i\}]}{T}\right), \tag{3.16}$$

multiplying both sides of Eq. (3.15) by $-T^2$, we obtain

$$F - \langle \mathcal{H} \rangle - T\frac{\partial F}{\partial T} = F - U - T\frac{\partial F}{\partial T} = 0, \tag{3.17}$$

where $U \equiv \langle \mathcal{H} \rangle$ is the internal energy and the last equation can be written in the form

$$F = U - TS, \tag{3.18}$$

which is the thermodynamic definition of the free energy F, where $S = -\frac{\partial F}{\partial T}$ is the entropy.[4]

The ferromagnetic phase transition characterizing the homogeneous Ising model (3.10) can be described by studying the dependence of the magnetization M on the temperature T. These quantities are usually called the order parameter and the control parameter, respectively. While T is fixed by the thermostat, M is defined as the equilibrium average of the sum of all spin variables over the lattice volume V, namely,

$$\begin{aligned} M(V,T,H) &= \left\langle \sum_i \sigma_i \right\rangle \\ &= \frac{1}{Z} \sum_{\{\sigma_i\}} \left(\sum_i \sigma_i\right) \exp\left(-\frac{\mathcal{H}[\{\sigma_i\}]}{T}\right) \\ &= T\frac{\partial \ln Z}{\partial H}. \end{aligned} \tag{3.19}$$

[3] This dependence stems from the summation over all spin configuration of a system that occupies a volume V, which is proportional to the total number of spins.

[4] Therefore, the relation (3.13) between the free energy and the partition function has been justified by showing that the well-known thermodynamic relation (3.18) can be derived from it. Alternatively, we might start from Eq. (3.18) and derive Eq. (3.13).

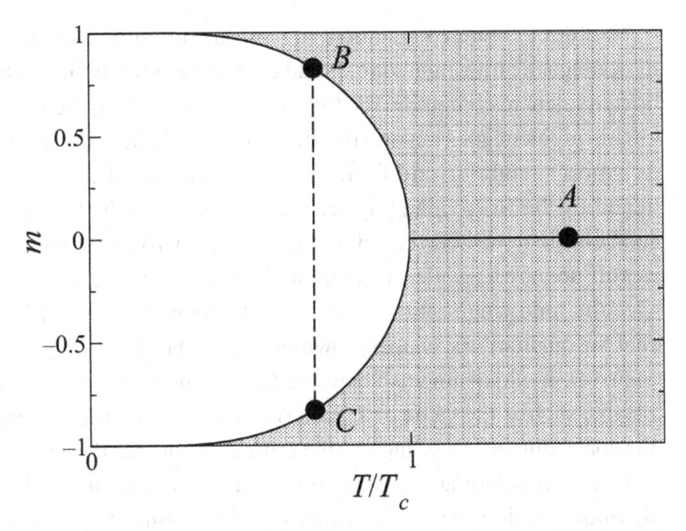

Fig. 3.2 Schematic plot of the temperature dependence of the order parameter density, $m = M/V$, in an Ising-like system. A, B, and C are equilibrium states at zero magnetic field. A is a disordered state at $T > T_c$, while B, C are a pair of equivalent, ordered states at $T < T_c$. The shaded region corresponds to equilibrium states for $H \neq 0$.

In the absence of the external magnetic field, i.e., $H = 0$, and for[5] $d \geq 2$ a qualitative description is summarized by the phase diagram shown in Fig. 3.2. In the high temperature phase (point A) M vanishes, because entropic fluctuations cancel the ordering effect of the ferromagnetic coupling J. By cooling the system through the critical temperature T_c it acquires a spontaneous magnetization, either positive or negative, along one of the two branches bifurcating at T_c. This phenomenon occurs because the magnetic susceptibility of the Ising ferromagnet,

$$\chi(V,T) = \frac{1}{V} \lim_{H\to 0} \frac{\partial M(V,T,H)}{\partial H}, \tag{3.20}$$

diverges at T_c, so that even a random microscopic fluctuation is able to produce a macroscopic effect. In a statistical perspective, we can say that, if we repeat the cooling procedure several times through T_c, positive or negative spontaneous magnetization is eventually observed with equal probability, as in a toss-a-coin process. Let us assume that one cooling procedure has selected a positively magnetized state B, and, now, we want to perform a thermodynamic transformation making the system pass from B, to its symmetric, negatively magnetized state C. Such a thermodynamic transformation is a first-order phase transition, because it can be performed by switching on a negative external magnetic field that has to supply a sufficient amount of magnetic energy to make a macroscopic portion of spins flip. If, after flipping the magnetization the external field H is switched off, we obtain state C. In general terms, the shaded region corresponds to equilibrium states in the presence of a nonvanishing field, either positive ($M > 0$) or negative ($M < 0$). The white region, corresponding to nonequilibrium states, will be studied in Chapter 6.

[5] For $d = 1$ the ferromagnetic transition of the Ising model is found to occur at $T = 0$; see Appendix H.1.

In order to obtain a quantitative rigorous description of the phase diagram of the Ising ferromagnetic transition, one should compute explicitly the partition function (3.12), which is equivalent to computing the free energy $F(V, T, H)$ of the Ising model. A detailed account of this problem goes beyond the aim of this book; here we just mention that this calculation is almost straightforward for $d = 1$ (see Appendix H.1), while for $d = 2$ the solution was found by the Norwegian physicist Lars Onsager in the last century, making use of refined mathematical methods based on the transfer matrix approach. The exact solution in $d = 3$ is still unknown, while in the limit $d \to \infty$ it can be shown that a mean-field approach (see the following section) allows one to compute $F(V, T, H)$. Numerical studies confirm that the qualitative scenario shown in Fig. 3.2 holds for $d \geq 2$. From a quantitative point of view we should distinguish between (irrelevant) properties that depend on the microscopic details, such as the value of T_c, and (relevant) properties that depend on general features as the space dimension d, such as the critical exponents (see Section 3.2.4).

Before concluding this section we want to comment on the crucial role played by the thermodynamic limit in the theory of equilibrium phase transitions. In fact, a thermodynamic system is assumed to be made up of an extremely large number of particles, such as the Avogadro number of atoms or molecules (of order 10^{23}) contained in a mole of a gas. This physical assumption can be translated into a rigorous mathematical formulation by a precise definition: the thermodynamic limit is obtained by making the number of microscopic variables N (e.g., the number of spins of the Ising model) go to infinity, while their density $\rho = N/V$ is kept constant.

We have mentioned above that the magnetic susceptibility $\chi(V, T)$ defined in (3.20) diverges at T_c. This indicates the presence of a singular (nonanalytic) behavior of χ that can be traced back to M and Z through Eqs. (3.19) and (3.20). On the other hand, Z is defined as a sum of analytic functions (see Eq. (3.12)), and a nonanalyticity in one of its derivatives, such as χ, may appear only if this sum is made by an infinite number of nonvanishing contributions. This simple mathematical argument shows that, in order to reproduce in the theory the diverging behavior of χ at T_c, it is necessary to sum Z over an infinite volume V (i.e., over an infinite number of spin configurations), while keeping ρ constant. Since a rigorous mathematical theory of phase transitions has to take into account from the very beginning the thermodynamic limit as a basic ingredient, it is more appropriate to deal with intensive quantities rather than extensive ones. In particular, we have to redefine the magnetic susceptibility as

$$\chi(\rho, T) = \lim_{H\to 0} \frac{\partial m(\rho, T, H)}{\partial H}, \tag{3.21}$$

where

$$m(\rho, T, H) = \lim_{V\to\infty} \frac{M(V, T, H)}{V} \tag{3.22}$$

is the magnetization density. Consistently, we can define the free energy density as

$$f(\rho, T, H) = \lim_{V\to\infty} \frac{F(V, T, H)}{V}. \tag{3.23}$$

3.2.3 Landau Theory of Critical Phenomena

A general phenomenological theory of phase transitions was proposed in the last century by Lev Davidovich Landau. The basic idea amounts to describing a phase transition close to the critical point by introducing an effective free energy density $f(m, T, H)$ that depends on the temperature T (or any equivalent thermodynamic quantity playing the role of a control parameter), on a coarse-grained order parameter density $m(\mathbf{x})$, and on its conjugate field $H(\mathbf{x})$, where $\mathbf{x}$ is the position vector in a space of dimension d. The attribute "coarse-grained" means that $m(\mathbf{x})$ is averaged over a suitable scale length in such a way that it is insensitive to fluctuations on the atomic scale. For instance, in a magnetic system such as those described by the Ising model (see previous section), $m(\mathbf{x})$ is the magnetization density and $H(\mathbf{x})$ is an external magnetic field.

The basic assumption of Landau theory is that, close to the transition point denoted by the critical value of the temperature T_c, $m(\mathbf{x})$ is small and $f(m, T, H)$ can be expanded at lowest orders in powers of $m(\mathbf{x})$ and of its derivatives as

$$f(m, T, H) = \frac{1}{2}|\nabla m(\mathbf{x})|^2 - H(\mathbf{x})\, m(\mathbf{x}) + \frac{1}{2} a\, m^2(x) + b\, m^3(\mathbf{x}) + u\, m^4(\mathbf{x}) + \cdots . \quad (3.24)$$

In Appendix I we provide an explicit derivation of the exact form of $f(m, T, 0)$ from an Ising-like Hamiltonian. However, in the spirit of Landau approach, the above expression can be thought as a general, small-m and small-H expansion, which is valid close to the critical point, which corresponds to $T = T_c$ and $H = 0$.

The first term on the right-hand side (whose prefactor can be set equal to $\frac{1}{2}$ by a suitable rescaling of the energy; see Eq. (I.18)) accounts for spatial variations of the order parameter, which are not negligible at the critical point. The second term accounts for a linear coupling with $H(\mathbf{x})$, which is justified by assuming that both $m(\mathbf{x})$ and $H(\mathbf{x})$ are small. The following terms represent a perturbative expansion with respect to $m(\mathbf{x})$, with a, b, and u being phenomenological parameters. In particular (see Appendix I), a is assumed to be proportional to the so-called reduced temperature,

$$a = a_0\, t, \quad \text{with} \quad t = \frac{(T - T_c)}{T_c}, \quad (3.25)$$

where a_0 is a positive constant and also $u > 0$ in order to guarantee a meaningful free energy, because the thermodynamic equilibrium states correspond to the absolute minima of the free energy density $f(m, T, H)$. As for b, it should vanish in systems invariant under change of sign of $m(\mathbf{x})$, as in the case of the ferromagnetic Ising model, because we require that $f(-m, T, -H) = f(m, T, H)$. This is no longer true, and $b \neq 0$, if the system does not hold the Ising symmetry, as occurs in the first-order van der Waals transition; see Appendix G.

The homogeneous mean-field solution of the Landau theory of critical phenomena is obtained by neglecting spatial fluctuations, so we deal with the mean-field free energy density,

$$f(m, T, H) = \frac{1}{2} a m^2 + u m^4 - H m. \quad (3.26)$$

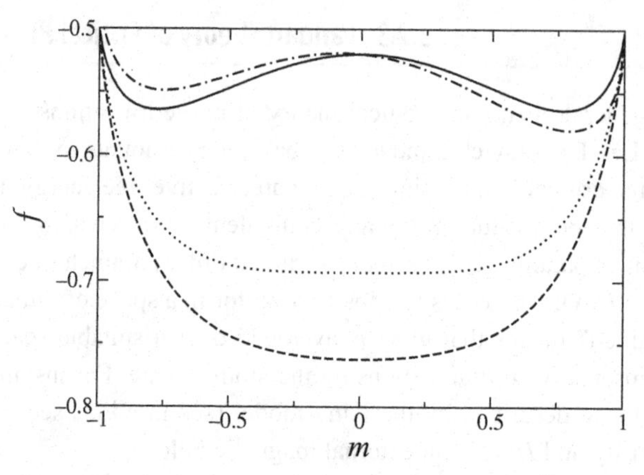

Fig. 3.3 The Ginzburg–Landau free energy (3.26) for $H = 0$, above T_c (dashed line), at T_c (dotted line), and below T_c (solid line). A field $H > 0$ modifies the double well potential as indicated by the dot-dashed line.

The thermodynamic equilibrium value of m is obtained by solving the equation $\partial f/\partial m = 0$, i.e.,

$$am + 4um^3 - H = 0, \tag{3.27}$$

and by identifying the absolute minimum of $f(m, T, H)$. In the case of $H = 0$ the absolute minima of $f(m, T, H)$ are

$$m = \begin{cases} 0, & \text{for } T > T_c \\ \pm\left(\frac{a_0}{4u}\right)^{\frac{1}{2}} |t|^{\frac{1}{2}}, & \text{for } T < T_c. \end{cases} \tag{3.28}$$

Accordingly, the mean-field theory predicts that the magnetization vanishes when T approaches T_c from below as $m \sim |t|^\beta$, with $\beta = 1/2$. See Fig. 3.3 for a pictorial representation of $f(m, T, 0)$ above, at, and below the critical temperature T_c. This result provides an a posteriori explanation of the definition (3.25) of the phenomenological parameter a. In fact, for $T > T_c$, a is positive and the minimum of $f(m, T, 0)$ is at $m = 0$, while for $T < T_c$, a is negative and the minima of $f(m, T, 0)$ correspond to a finite magnetization.

Beyond the magnetization, other thermodynamic observables exhibit power-law behavior close to or at the critical point. In particular, the equilibrium values of the magnetic susceptibility χ and of the specific heat at constant volume c_V are expected to diverge at T_c as

$$\chi = \left.\frac{\partial m}{\partial H}\right|_{H=0} \sim |t|^{-\gamma} \tag{3.29}$$

$$c_V = -T\left.\frac{\partial^2 f}{\partial T^2}\right|_{H=0} \sim |t|^{-\alpha}, \tag{3.30}$$

where both critical exponents α and γ are positive numbers. An explicit expression of the equilibrium value of χ in the mean-field approximation can be obtained by differentiating (3.27),

$$(a + 12\,um^2)dm = dH, \tag{3.31}$$

and making use of (3.25) and (3.28), thus yielding the expression

$$\chi = \begin{cases} \frac{1}{a_0}|t|^{-1}, & \text{for } T > T_c \\ \frac{1}{2a_0}|t|^{-1}, & \text{for } T < T_c \end{cases} \tag{3.32}$$

so that $\gamma = 1$. In order to obtain c_V we can write the equilibrium value of the mean-field free energy for $H = 0$ by combining (3.26), (3.27), and (3.28):

$$f(m, T, 0) = \begin{cases} 0, & \text{for } T > T_c \\ -\left(\frac{a_0^2}{16u}\right) t^2, & \text{for } T < T_c. \end{cases} \tag{3.33}$$

By applying definition (3.30) one obtains

$$c_V = \begin{cases} 0, & \text{for } T > T_c \\ \frac{T_c a_0^2}{8u}, & \text{for } T < T_c, \end{cases} \tag{3.34}$$

which predicts a finite discontinuity of c_V at T_c, so that $\alpha = 0$.

At the critical point $T = T_c$ (i.e., $a = 0$) the magnetization m is expected to exhibit a power-law dependence on the external magnetic field H of the form $m \sim H^{1/\delta}$. From Eq. (3.27) we obtain

$$m(T_c, H) = \left(\frac{H}{4u}\right)^{\frac{1}{3}}, \tag{3.35}$$

thus yielding $\delta = 3$.

The mean-field solution provides an approximate description of critical phenomena, since it completely disregards the role of fluctuations that are known to play an important role close to the critical point, as mentioned in the previous section. We can take into account fluctuations by studying the connected autocorrelation function,

$$C(\mathbf{x}) = \langle m(\mathbf{x})m(0)\rangle - \langle m(\mathbf{x})\rangle\langle m(0)\rangle = \langle m(\mathbf{x})m(0)\rangle - \langle m(0)\rangle^2, \tag{3.36}$$

where the brackets $\langle \cdots \rangle$ denote the equilibrium average, and the second equality stems from the assumption of translation invariance. For simplicity, let us consider the case $H = 0$ and $T > T_c$, i.e., $\langle m(0)\rangle = 0$. By introducing the Fourier transform of $m(\mathbf{x})$

$$m(\mathbf{k}) = \int d^d\mathbf{x}\, e^{-i\mathbf{k}\cdot\mathbf{x}}\, m(\mathbf{x}) \tag{3.37}$$

the Fourier transform of $C(\mathbf{x})$ takes the simple form

$$C(\mathbf{k}) = \frac{1}{(2\pi)^d}\langle |m(\mathbf{k})|^2\rangle. \tag{3.38}$$

We can now determine $C(\mathbf{k})$ because $|m(\mathbf{k})|^2$ is related to the quadratic part of the free energy so that its average value, which appears in Eq. (3.38), can be easily evaluated using the equipartition theorem. In the paramagnetic region and if $H = 0$ we can consider only the quadratic part of the free energy,

$$f(m(\mathbf{x}), T, 0) \approx \frac{1}{2}|\nabla m(\mathbf{x})|^2 + \frac{1}{2}a\, m^2(\mathbf{x}), \tag{3.39}$$

whose space integral can be evaluated in the Fourier space,

$$\begin{aligned}\int d\mathbf{x} f(m(\mathbf{x}),T,0) &= \frac{1}{2}\int d\mathbf{x}\frac{1}{(2\pi)^{2d}}\int d\mathbf{k}_1\int d\mathbf{k}_2(a-\mathbf{k}_1\cdot\mathbf{k}_2)e^{i\mathbf{k}_1\cdot\mathbf{x}}e^{i\mathbf{k}_2\cdot\mathbf{x}}m(\mathbf{k}_1)m(\mathbf{k}_2)\\ &= \frac{1}{2}\int d\mathbf{k}_1\int d\mathbf{k}_2(a-\mathbf{k}_1\cdot\mathbf{k}_2)\frac{1}{(2\pi)^d}m(\mathbf{k}_1)m(\mathbf{k}_2)\delta(\mathbf{k}_1+\mathbf{k}_2)\\ &= \frac{1}{2}\frac{1}{(2\pi)^d}\int d\mathbf{k}(a+\mathbf{k}^2)|m(\mathbf{k})|^2\\ &\equiv \int d\mathbf{k} f(m(\mathbf{k}),T,0).\end{aligned}$$

In the first line above we have expressed $m(\mathbf{x})$ and $\nabla m(\mathbf{x})$ in terms of $m(\mathbf{k})$; in the second line we have spatially integrated the exponents, obtaining a Dirac delta; and in the third line we have used it to pass to a single integral in the $\mathbf{k}$-space.

We can now use the equipartition theorem, writing

$$T = \langle f(m(\mathbf{k}),T,0)\rangle = \frac{1}{2}\frac{1}{(2\pi)^d}(\mathbf{k}^2+a)\langle|m(\mathbf{k})|^2\rangle \tag{3.40}$$

and finally obtaining

$$C(\mathbf{k}) = \frac{1}{(2\pi)^d}\langle|m(\mathbf{k})|^2\rangle = \frac{2T}{\mathbf{k}^2+a}. \tag{3.41}$$

By antitransforming this expression we can write the connected autocorrelation function in the so-called Ornstein–Zernike form,

$$C(\mathbf{x}) \sim |\mathbf{x}|^{2-d}e^{-|\mathbf{x}|/\xi}, \tag{3.42}$$

where $\xi = a^{-1/2}$ is the correlation length, i.e., the range of fluctuation correlations for $T > T_c$. By recalling (3.25) we can see that ξ diverges for $T \to T_c^+$ as

$$\xi \sim t^{-\nu}, \quad \text{with} \quad \nu = \frac{1}{2}. \tag{3.43}$$

We conclude this part by evaluating the relevance of fluctuations, which have been disregarded in the previous mean-field approach. We can assess it through the ratio of the spatial fluctuations of the order parameter over a distance ξ (the only relevant scale length, if the system is close to criticality) and the average value of the order parameter itself. This means we are tacitly assuming to be in the ordered region, $T < T_c$. We must therefore consider the ratio

$$\left.\frac{C(\xi)}{m^2}\right|_{T\to T_c^-} = \frac{\xi^{2-d}}{\frac{a_0}{4u}|t|}, \tag{3.44}$$

where the numerator has been determined by Eq. (3.42) and the denominator by Eq. (3.28).

Since $\xi = (a_0|t|)^{-1/2}$, this ratio can be simplified as

$$\left.\frac{C(\xi)}{m^2}\right|_{T\to T_c^-} \sim 4u(a_0|t|)^{(d-4)/2}, \tag{3.45}$$

and the requirement that fluctuations are negligible in the limit $t \to 0$ is equivalent to saying that such as ratio is much smaller than one, which is true, according to the previous equation, if $d \geq 4$. This is known as the Ginzburg criterion, which points out the crucial role played by space dimension d in critical phenomena, identifying the upper critical dimension $d_c^u = 4$. In fact, we can say that for $d \geq d_c^u$ the mean-field solution and its critical exponent are correct,[6] while for $d < d_c^u$ the mean-field solution fails.

3.2.4 Critical Exponents and Scaling Hypothesis

In the previous section we have sketched the main elements of the Landau theory of phase transitions, which is particularly useful for describing critical phenomena. On the other hand, this theory is based on many approximations, which are unable to capture some crucial aspects of critical phenomena associated with the peculiar role of dimensionality and fluctuations. For instance, the critical exponents predicted by Landau theory, called classical exponents, are different from those obtained in experiments or from exact/numerical solutions of specific models, thus proving the failure of this approximate theory. In Table 3.1 it is possible to compare the critical exponents for the Ising model with the exact values in $d = 2$, the numerical estimates in $d = 3$, and the classical (mean-field) values, which are valid for $d \geq 4$.

A better description of critical phenomena can be made by introducing the scaling hypothesis, which is partially inspired by heuristic arguments. In fact, it has been observed that the universality of critical phenomena stems from the property of scale invariance of thermodynamic observables at the critical point. The considerations that follow are based on two main ideas concerning the form of the correlation function close to the critical point and the "singular part" of the free energy density.

According to the Landau theory, the connected autocorrelation function has the Ornstein–Zernike form (3.42). If we generalize the exponent of the power term, this form has a validity that goes beyond the mean-field theory. Therefore, close to the critical point and in the absence of external sources (i.e., $H(\mathbf{x}) = 0$), the connected autocorrelation function $C(\mathbf{x})$, defined in (3.36), is assumed to have the universal functional form

Table 3.1 Exact values of the critical exponents for the Ising model in $d = 2$, numerics in $d = 3$, and mean-field values

Exponent	Ising ($d = 2$)	Ising ($d = 3$)	Mean-Field
α	0	0.110	0
β	1/8	0.326	1/2
γ	7/4	1.237	1
δ	15	4.790	3
ν	1	0.630	1/2
η	1/4	0.0363	0

[6] The case $d = 4$ deserves a special attention, because subleading logarithmic corrections to the scaling behavior at the critical point should be taken into account.

$$C(\mathbf{x}) \sim \frac{e^{-|\mathbf{x}|/\xi}}{|\mathbf{x}|^p} \tag{3.46}$$

with

$$p = d - 2 + \eta \tag{3.47}$$

$$\xi \propto |t|^{-\nu}. \tag{3.48}$$

Above d is the space dimension and η is an exponent that has been found (in experiments and models) to take values different from zero, at variance with Landau theory, which predicts $\eta = 0$; see Eq. (3.42). The functional form of $C(\mathbf{x})$ implies that close to the critical temperature, i.e., $T \to T_c$, the correlation length diverges, i.e., $\xi \to \infty$. Said differently, the only physical length-scale that matters close to T_c is ξ, and at $T = T_c$, $C(\mathbf{x})$ has a power-law behavior.

The second idea concerns the free energy density f, which is nonanalytic at T_c, because quantities derived from it are singular. Since the nonanalyticity comes from the thermodynamic limit, the dependence of the singular component of f on the reduced temperature $t = (T - T_c)/T_c$ is assumed to be fixed by dimensional considerations: since the free energy density has the dimension of the inverse volume $V^{-1} = L^{-d}$ (see Eq. (3.23)), we simply write

$$f \sim \xi^{-d} \sim |t|^{d\nu}. \tag{3.49}$$

In the following we start from Eqs. (3.46) and (3.49) and using standard thermodynamic and statistical relations we derive the singular behavior of various physical quantities, finding that all critical exponents introduced in the previous section are not independent of each other, because some relations among them, called scaling laws, hold. In particular, once the form (3.46) of the autocorrelation function has been fixed, all critical exponents can be expressed by simple algebraic relations as functions of η and ν, together with the space dimension d.

A first relation can be obtained by observing that, close to T_c, the specific heat at constant volume c_V, Eq. (3.30), can be written as

$$c_V = -T_c \frac{\partial^2 f}{\partial T^2}\bigg|_{H=0} = -\frac{1}{T_c}\frac{\partial^2 f}{\partial t^2}\bigg|_{H=0} \sim |t|^{d\nu - 2}, \tag{3.50}$$

where the last relation is a consequence of (3.49). By comparing this result with the expected scaling relation reported in (3.30) we have

$$\boxed{d\nu = 2 - \alpha} \tag{3.51}$$

which is known as the Josephson scaling law.

A second relation can be derived by first rewriting (3.46) as

$$C(\mathbf{y}, \xi) \sim \xi^{-p}\,\mathcal{C}(\mathbf{y}), \tag{3.52}$$

where $\mathbf{y} = \mathbf{x}/\xi$ is a nondimensional variable, as well as the scaling function $\mathcal{C}(\mathbf{y}) = |\mathbf{y}|^{-p}e^{-|\mathbf{y}|}$. Accordingly, the physical scale of C is determined only by the factor ξ^{-p}. Making reference to definition (3.36), dimensional analysis allows us to write

$$m \sim C^{1/2} \sim \xi^{-p/2} = |t|^{p\nu/2}. \tag{3.53}$$

Since the magnetization is expected to vanish as $m \sim |t|^{\beta}$, we can write the scaling relation

$$\beta = \nu(d - 2 + \eta)/2, \tag{3.54}$$

which will be used below.

A third scaling relation can be obtained by the thermodynamic relation between the magnetic susceptibility χ and the correlation function $C(\mathbf{x})$,[7]

$$\chi = \frac{1}{T}\int d\mathbf{x}\, C(\mathbf{x}). \tag{3.55}$$

Dimensional analysis of this relation implies that the scaling dependence of χ on the scale length ξ close to T_c has to be of the form

$$\chi \sim \xi^{d}\,\xi^{-p} = \xi^{2-\eta} = |t|^{-\nu(2-\eta)}, \tag{3.56}$$

where the factor ξ^{d} is just due to the measure of the space integral in d dimensions. Considering that χ is expected to diverge at T_c as in Eq. (3.29), we can obtain the scaling relation

$$\boxed{\gamma = \nu(2 - \eta)} \tag{3.57}$$

which is known as the Fisher scaling law.

The fourth final relation can be derived by considering that the magnetization density m is defined by the relation

$$m = -\left.\frac{\partial f}{\partial H}\right|_{H=0}. \tag{3.58}$$

Dimensional analysis of this equation (see (3.49) and (3.53)) yields

$$H \sim \xi^{-d}\,\xi^{p/2} = \xi^{(\eta-d-2)/2} \sim |t|^{\nu(d+2-\eta)/2}. \tag{3.59}$$

Phenomenological considerations indicate that at T_c the equation of state of a ferromagnetic system is given by the relation

$$m \sim H^{1/\delta}, \tag{3.60}$$

which establishes how the magnetization density m depends on the amplitude of the perturbation produced by an external magnetic field H. Since m is expected to scale at T_c as $|t|^{\beta}$, we can conclude that H has to scale as

$$H \sim |t|^{\beta\delta}, \tag{3.61}$$

thus yielding the relation

$$\beta\delta = \nu(d + 2 - \eta)/2. \tag{3.62}$$

By subtracting (3.54) to (3.62) and taking into account (3.57), we can write

$$\boxed{\beta(\delta - 1) = \gamma} \tag{3.63}$$

[7] This relation can be derived using Eqs. (3.20) and (3.19).

which is known as the Widom scaling law. Finally, by summing (3.54) and (3.62) and taking into account (3.51), we obtain $\beta(\delta + 1) = d\nu = 2 - \alpha$. Since, according to (3.63), $\beta\delta = \beta + \gamma$, we can write

$$\boxed{\alpha + 2\beta + \gamma = 2} \tag{3.64}$$

which is known as the Rushbrooke scaling law.

In summary, the four phenomenological critical exponents α, β, γ and δ, together with η and ν obey the four scaling laws, (3.51), (3.57), (3.63), and (3.64), thus proving that only two of them are actually independent and sufficient to describe the critical behavior associated with a continuous equilibrium phase transition. Notice that also the classical exponents obtained in the previous section obey these four scaling laws, because their validity is independent of the way one computes f.

3.2.5 Phenomenological Scaling Theory

The analysis of the previous section is based on the observation that close to the critical point the correlation length ξ dominates over all other scales and the behavior of physical quantities is derived using thermodynamic and statistical relations. In this section we offer a more elegant and transparent way to formulate a phenomenological scaling theory of equilibrium phase transitions, which will be useful to study nonequilibrium phase transitions as well (see Section 3.6.2).

In fact, we can assume that all observables exhibit scaling properties close to the critical point; i.e., they must be homogeneous functions of some scaling parameters Λ_i, which have to be eventually related to the correlation length $\xi \sim |t|^{-\nu}$ or, equivalently, to t. In this respect, we point out that in the Landau theory of critical phenomena the free energy density can be assumed to depend on two physical variables, namely, t and H. Accordingly, any other thermodynamic observable has to depend on the same variables. For instance, close to the critical point, i.e., for $t, H \to 0$, the scaling hypothesis amounts to imposing the condition that the magnetization density $m(t, H)$ has to be invariant by suitably rescaling itself and its arguments, thus

$$m(t, H) = \Lambda_m m(\Lambda_t t, \Lambda_H H). \tag{3.65}$$

For $H = 0$ we know that the magnetization density vanishes as $m \sim |t|^\beta$. We are free to choose $\Lambda_t = \Lambda = |t|^{-1}$; this is quite a natural choice, because we are assuming that our reference scaling parameter Λ amounts to the inverse of the rescaled distance from the critical temperature T_c. We can specialize the previous formula to the relation

$$\Lambda_m m(1, 0) = m(t, 0) \sim |t|^\beta \quad \rightarrow \quad \Lambda_m = |t|^\beta = \Lambda^{-\beta}. \tag{3.66}$$

Applying the same argument to $m(0, H) = \Lambda_m m(0, \Lambda_H H)$ and taking into account the phenomenological scaling relation (3.60), we obtain $H^{1/\delta} = \Lambda_m (\Lambda_H H)^{1/\delta}$. We can therefore express the scaling parameter Λ_H in terms of Λ:

$$\Lambda_H = \Lambda_m^{-\delta} = \Lambda^{\beta\delta}. \tag{3.67}$$

A scaling relation analogous to (3.65) holds for the free energy density,

$$f(t,H) = \Lambda_f f(\Lambda t, \Lambda^{\beta\delta} H). \tag{3.68}$$

In order to obtain Λ_f we can use the scaling law for c_V (see Eq. (3.30)),

$$c_V \sim \Lambda^\alpha \sim \left.\frac{\partial^2 f(t,H)}{\partial t^2}\right|_{H=0} = \Lambda_f \Lambda^2 \left.\frac{\partial^2 f(t,H)}{\partial t^2}\right|_{H=0,t=1}, \tag{3.69}$$

yielding $\Lambda_f = \Lambda^{\alpha-2}$, i.e.,

$$f(t,H) = \Lambda^{\alpha-2} f(\Lambda t, \Lambda^{\beta\delta} H). \tag{3.70}$$

This scaling formula can be used for establishing relations among the different scaling exponents. According to the arguments reported in the previous section (see Eq. (3.49)), simple dimensional analysis imposes that the free energy density scales as

$$f(t,0) \sim \Lambda^{-d\nu}. \tag{3.71}$$

Comparison of Eqs. (3.70) and (3.71) allows us to obtain the Josephson law,

$$d\nu = 2 - \alpha. \tag{3.72}$$

Moreover, by deriving both sides of (3.70) with respect to H, and recalling that $m = -\frac{\partial f(t,H)}{\partial H}$, we obtain

$$m(t,H) = \Lambda^{\alpha-2}\Lambda^{\beta\delta} m(\Lambda t, \Lambda^{\beta\delta} H). \tag{3.73}$$

Comparing this relation with Eq. (3.65), we obtain $\Lambda_m = \Lambda^{\alpha-2+\beta\delta}$. Since $\Lambda_m = \Lambda^{-\beta}$, we find

$$\beta = 2 - \alpha - \beta\delta. \tag{3.74}$$

If we now derive both sides of (3.73) with respect to H and recall that $\chi = \frac{\partial m}{\partial H}$, we obtain

$$\chi(t,H) = \Lambda^{\alpha-2}\Lambda^{2\beta\delta}\chi(\Lambda t, \Lambda^{\beta\delta} H). \tag{3.75}$$

Making use of the phenomenological scaling relation $\chi(t,0) \sim \Lambda^\gamma$ (see Eq. (3.29)), we can write

$$\gamma = 2\beta\delta + \alpha - 2. \tag{3.76}$$

Summing Eqs. (3.74) and (3.76), we obtain the Widom scaling law

$$\beta(\delta - 1) = \gamma, \tag{3.77}$$

while summing twice (3.74) to (3.76) we obtain the Rushbrooke scaling law

$$\alpha + 2\beta + \gamma = 2. \tag{3.78}$$

Other quantities that exhibit scale invariance (i.e., homogeneity) properties close to the critical point are the correlation length ξ (see Eqs. (3.42) and (3.43)),

$$\xi(t,H) = \Lambda^\nu \xi(\Lambda t, \Lambda^{\beta\delta} H), \tag{3.79}$$

and the autocorrelation function,

$$C(\mathbf{x}, t, H) = \Lambda_c C(\Lambda_{\mathbf{x}} \mathbf{x}, \Lambda t, \Lambda^{\beta\delta} H). \tag{3.80}$$

Notice that the dependence of C on the position variable $\mathbf{x}$ requires us to introduce a scaling parameter $\Lambda_{\mathbf{x}}$. As a direct consequence of Eq. (3.46) we can conclude that

$$\Lambda_{\mathbf{x}} = \Lambda^{\nu}, \tag{3.81}$$

because the ratio $|\mathbf{x}|/\xi$ should not rescale, while

$$\Lambda_c = \Lambda_{\mathbf{x}}^{-p} = \Lambda^{-p\nu}. \tag{3.82}$$

Taking into account the fluctuation–dissipation relation (3.55), we can write

$$\chi \sim \Lambda_{\mathbf{x}}^{d} \Lambda_c = \Lambda^{\nu(2-\eta)}, \tag{3.83}$$

thus recovering the Fisher scaling law,

$$\gamma = \nu(2 - \eta). \tag{3.84}$$

3.2.6 Scale Invariance and Renormalization Group

Taking inspiration from what we have discussed in the previous section, we can reformulate the scaling hypothesis in a more general form. In fact, if f describes a system that exhibits a continuous phase transition, it must be a homogeneous function of its variables, i.e., the control parameter t and the source field H, in the vicinity of the critcal point $t = 0$. Thus

$$f(t, H) = \Lambda^{-d} f(\Lambda_t t, \Lambda_H H), \tag{3.85}$$

where Λ is a scale length that close to the critical point coincides with the correlation length ξ (see Eq. (3.48)),

$$\Lambda \sim \xi \sim t^{-\nu}, \tag{3.86}$$

and we have explicitly taken into account that $f(t, H)$ has the physical dimension of an inverse volume in a space of dimension d.

We can express the two scaling parameters Λ_t and Λ_H as

$$\Lambda_t = \Lambda^{D_t} \tag{3.87}$$

$$\Lambda_H = \Lambda^{D_H}, \tag{3.88}$$

where D_t and D_H are the scaling exponents, or dimensions, associated with the variables t and H, respectively. As discussed at the end of Section 3.2.4 the knowledge of these two exponents is enough to determine all the other phenomenological exponents by the four scaling laws therein derived. Since t is a dimensionless quantity we can fix $D_t = 1/\nu$, so that $\Lambda_t = \xi^{1/\nu} = |t|^{-1}$, and $D_H = \beta\delta/\nu$, so that $\Lambda_H = \xi^{D_H} = |t|^{-\beta\delta}$ (see Eq. (3.61)). This choice corresponds to the phenomenological scaling theory described in the previous section.

On the other hand, it is possible to compute D_t and D_H explicitly by applying the scale invariance hypothesis directly to the Hamiltonian $\mathcal{H}(\phi(\mathbf{x}), t, H)$ of the model at hand, which depends on some state variable $\phi(\mathbf{x})$, $\mathbf{x}$ being the space coordinate in d dimensions, and on the physical parameters t and H. This is the basic idea of the renormalization group method. In order to illustrate this approach, we have to recall first the definition of the partition function and its relation with the free energy density (see Section 3.2.2):

$$Z(t, H, V) = e^{-F(t,H,V)} = \int_V d\phi(\mathbf{x}) e^{-\mathcal{H}(\phi(\mathbf{x}),t,H)} \tag{3.89}$$

$$f(t, H) = \frac{F(t, H, V)}{V}, \tag{3.90}$$

where both $F(t, H, V)$ and $\mathcal{H}(\phi(\mathbf{x}), t, H)$ are measured in units of T. This standard choice allows us to simplify the notation and is equivalent to rescaling the physical parameters of the model, e.g., $\frac{J}{T} \to J$ and $\frac{H}{T} \to H$. In order to apply the scale-invariance hypothesis to $\mathcal{H}(\phi(\mathbf{x}), t, H)$ we have to introduce a scale length explicitly. A way to accomplish this task amounts to modifying the dependence of the state variable ϕ from a continuous space coordinate vector $\mathbf{x}$ to a discrete lattice coordinate vector,

$$\phi(\mathbf{x}) \to \phi_i, \tag{3.91}$$

where i is a set of d integer coordinates $i_1, i_2, \ldots, i_d$, that identify a site on a d-dimensional cubic lattice at distance a from its nearest-neighbor sites. For instance, this is the case of the Ising model introduced in Section 3.2.2. Without prejudice of generality we can set the lattice spacing $a = 1$.

We can subdivide the lattice volume into blocks of size ℓ^d, $\ell > 1$, and we can associate with each block a new collective block variable

$$\phi'_j = \sum_{i \in j} \phi_i, \tag{3.92}$$

where j is the coordinate vector of the block and i those of the sites belonging to it. Here we adopt the rule of defining the collective block variable by summing over all the original variables in the block, but one can adopt different rules, provided the new block-transformed Hamiltonian preserves the symmetries of the original one. For instance, in the case of the Ising model, where $\phi_i = \sigma_i = \pm 1$, we can define the block transformation as

$$\sigma'_j = \begin{cases} +1, & \text{if } \sum_{i \in j} \sigma_i > 0 \\ -1, & \text{if } \sum_{i \in j} \sigma_i < 0 \\ \sigma_k, & \text{if } \sum_{i \in j} \sigma_i = 0, \end{cases} \tag{3.93}$$

where σ_k is the value of a spin chosen at random in the block. This is known as the majority rule, illustrated in Fig. 3.4.

We have to assume that close to the critical point this block transformation $\mathcal{R}$ (or any other transformation that amounts to rescaling the length of the lattice spacing) does not change the form of the original Hamiltonian, but it rescales the values of its variables and parameters.[8] The previous assumption implies that[9]

[8] The validity of the renormalization group method relies on the assumption that its results are independent of the adopted method for the lattice rescaling transformation.

[9] Since performing a renormalization transformation involves the trace over a subset of state variables ϕ_i (see the examples discussed in Appendixes H.1 and H.2), $\mathcal{H}'(\phi'_i, t', H')$ also contains some additional constant $C(t, H)$.

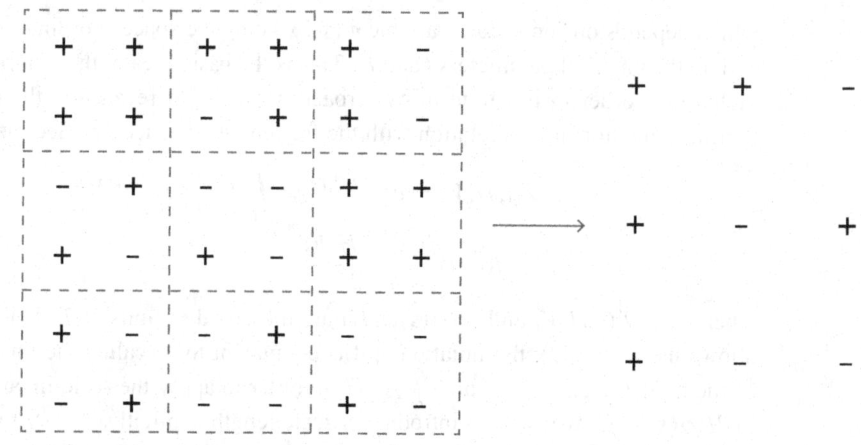

Fig. 3.4 Starting from an Ising model of unitary lattice constant, we perform a coarse graining by replacing each separate set of four spins with a single spin, following the majority rule. In case of a tie, the sign of the spin is chosen randomly. The new lattice constant is $\ell = 2$.

$$\mathcal{R}\Big[\mathcal{H}(\phi_i, t, H)\Big] = \mathcal{H}'(\phi_i', t', H') \tag{3.94}$$

and also

$$Z(t, H, V) = Z(t', H', V') \tag{3.95}$$

$$f(t, H) = \ell^{-d} f(t', H'), \tag{3.96}$$

where $Z(t', H', V')$ and $f(t', H')$ are functionals of the renormalized Hamiltonian $\mathcal{H}'(\phi_i', t', H')$.

For the moment we do not need to specify $\mathcal{R}$. We can observe that the homogeneity property (3.85) is recovered by assuming that

$$t' = \ell^{D_t}\, t \tag{3.97}$$

$$H' = \ell^{D_H}\, H. \tag{3.98}$$

Close to the critical point we can expect that both t and H are very small as well as the renormalized variables t' and H' in such a way that at leading order $\mathcal{R}$ is a linear transformation of the form

$$t' = \tau\, t + O(t^2) \tag{3.99}$$

$$H' = \zeta\, H + O(H^2), \tag{3.100}$$

thus yielding the relations

$$\frac{1}{\nu} = D_t = \frac{\ln \tau}{\ln \ell} \tag{3.101}$$

$$\frac{\beta\,\delta}{\nu} = D_H = \frac{\ln \zeta}{\ln \ell}, \tag{3.102}$$

which could allow us to obtain the scaling exponents from the knowledge of τ and ζ.

Despite the heuristic value of this formulation of the renormalization group method worked out in real space, mainly due to the intuition of the American physicist Leo Philip Kadanoff, many of the simplifying assumptions that have been introduced need to be properly reconsidered if we want to translate the formal renormalization transformation $\mathcal{R}$ into an explicit algorithm. In order to make the reader appreciate the difficulties inherent in such a procedure, we discuss some examples in Appendixes H.1 and H.2. However, describing in full generality the renormalization group transformations goes beyond the aims of this book.

For what we are going to discuss in Section 5.7.4, it is more useful to sketch the renormalization group approach in Fourier space, which was originally proposed by the American physicist Kenneth Geddes Wilson. At variance with the real space renormalization group procedure, which typically needs to be worked out in an infinite-dimensional parameter space (see Appendix H.2), we can consider a Hamiltonian that contains the minimum number of terms necessary to determine a critical behavior. For pedagogical reasons it is worth illustrating this method by an example, considering the case of the Landau theory at the lowest order, whose Hamiltonian, in d space dimensions, is[10]

$$\mathcal{H}[m(\mathbf{x})] = \frac{1}{2}\int d\mathbf{x}\left(|\nabla m(\mathbf{x})|^2 + a\, m^2(\mathbf{x})\right). \qquad (3.103)$$

Using Fourier-transformed variables,

$$m(\mathbf{k}) = \int_{-\infty}^{+\infty} d\mathbf{x}\, e^{-i\mathbf{k}\cdot\mathbf{x}}\, m(\mathbf{x}), \qquad (3.104)$$

and following the calculations at page 148, we can rewrite (3.103) as

$$\mathcal{H}[m(\mathbf{k})] = \frac{1}{2}\frac{1}{(2\pi)^d}\int d\mathbf{k}\,(k^2 + a)|m(\mathbf{k})|^2, \qquad (3.105)$$

where $k^2 = |\mathbf{k}|^2$. We can introduce a cutoff Λ in $\mathbf{k}$-space that fixes a wavelength, above which we assume a coarse-graining of the model; as in the case of renormalization transformations in the space of coordinates where fluctuations below a certain scale a_0 are ignored, here we should ignore fluctuations of wavelength larger than $\Lambda = 2\pi/a_0$.

We can write the partition function of the Landau model as

$$Z = \prod_{|\mathbf{k}|<\Lambda}\int dm(\mathbf{k})\, dm^*(\mathbf{k}) e^{-\beta\mathcal{H}[m(\mathbf{k})]}. \qquad (3.106)$$

A renormalization transformation can be performed by defining a new energy $\mathcal{H}'[m(\mathbf{k})]$ that is obtained by summing over all the wavelengths whose modulus is between Λ/b and Λ, with $b > 1$,

$$e^{-\mathcal{H}'[m(\mathbf{k})]} \equiv e^{-F_\Lambda} \prod_{\Lambda/b<|\mathbf{k}|<\Lambda}\int dm(\mathbf{k})\, dm^*(\mathbf{k}) e^{-\beta\mathcal{H}[m(\mathbf{k})]}, \qquad (3.107)$$

where F_Λ is a function of the cutoff Λ and of the coupling constant a. Once we have integrated over all degrees of freedom, F_Λ converges to the free energy F.

[10] It is enough to fix a single parameter a to fix the scale of m.

By this definition we have obtained a new Hamiltonian that depends on $m(\mathbf{k})$, which is now defined only for $|\mathbf{k}| < \Lambda/b$. On the other hand, $\mathcal{H}'[m(\mathbf{k})]$ takes the same form of $\mathcal{H}[m(\mathbf{k})]$, but it depends on a new coupling a',

$$\mathcal{H}'[m(\mathbf{k})] = \frac{1}{2}\frac{1}{(2\pi)^d}\int d\mathbf{k}\,(k^2 + a')|m(\mathbf{k})|^2. \tag{3.108}$$

In terms of the new Hamiltonian $\mathcal{H}'$, the partition function can be rewritten as

$$Z = e^{-F_\Lambda}\prod_{|\mathbf{k}|<\Lambda/b}\int dm(\mathbf{k})\,dm^*(\mathbf{k})e^{-\beta\mathcal{H}'[m(\mathbf{k})]}. \tag{3.109}$$

We can restore the original cutoff Λ by performing a suitable change of variable in the definition of $\mathcal{H}'[m(\mathbf{k})]$, which amounts to rescaling $\mathbf{k}$ by the factor b, i.e., $\mathbf{k}' = b\mathbf{k}$,

$$\mathcal{H}'[m(\mathbf{k}')] = \frac{1}{2}\frac{1}{(2\pi)^d}\,b^{-d}\int d\mathbf{k}'\left(\frac{1}{b^2}k'^2 + a\right)|m(\mathbf{k}'/b)|^2. \tag{3.110}$$

The last step of the renormalization procedure in $\mathbf{k}$-space amounts to restoring the original normalization of the order parameter; i.e., we want the coefficient of the term k'^2 to again be 1. We can obtain this result by the replacement

$$m(\mathbf{k}'/b) = \sqrt{b^{d+2}}m'(\mathbf{k}'). \tag{3.111}$$

This operation is the analog of the Kadanoff block-spin transformation. The renormalized Hamiltonian eventually is

$$\mathcal{H}'[m(\mathbf{k}')] = \frac{1}{2}\frac{1}{(2\pi)^d}\int d\mathbf{k}'\left(k'^2 + a'\right)|m'(\mathbf{k}')|^2, \tag{3.112}$$

where

$$a' = b^2 a \tag{3.113}$$

is the relation that defines the renormalization of the coupling constant a.

The partition function can be finally rewritten as

$$Z = e^{-F_\Lambda}\prod_{|\mathbf{k}'|<\Lambda}\int dm'(\mathbf{k}')\,dm'^*(\mathbf{k}')e^{-\beta\mathcal{H}'[m'(\mathbf{k}')]}. \tag{3.114}$$

As we anticipated, the renormalization group (RG) method is definitely simpler in $\mathbf{k}$-space than in real space. The procedure that we have sketched here for the Landau theory at lowest order can be extended, with additional technical difficulties, to the case where we consider higher-order terms, e.g., the quartic one. Treating in detail this problem in view of what we discuss in the following sections is unnecessary. We just want to conclude this section by noting that, since in the Landau theory we deal with continuous variables, the rescaling parameter b can be taken as close as we want to 1, and since we are interested in determining just the rate of change of the coupling constants with b, the renormalization transformation can be finally cast in the form of a set of differential equations, as many as the number of coupling parameters. Determining their fixed points and the corresponding stable and unstable manifolds (as in a standard dynamical model) allows us to obtain a complete description of the renormalization group transformation in $\mathbf{k}$-space and to obtain a clear inference on its physical implications.

3.3 Equilibrium States versus Stationary States

The first part of this chapter has dealt with transitions between equilibrium states. Before discussing nonequilibrium phase transitions it is necessary to introduce the concept of nonequilibrium steady state (NESS). Both equilibrium states and NESSs are time-independent, macroscopic states, whose difference is easily highlighted by making use of the master equation, derived in Chapter 1:

$$\frac{\partial p(s,t)}{\partial t} = \sum_{s'} \left[p(s',t) w_{s,s'} - p(s,t) w_{s',s} \right]. \tag{3.115}$$

In the above equation, $p(s,t)$ is the probability that the system is in the microstate s at time t, while $w_{s,s'}$ is the transition rate between microstate s' and microstate s. Eq. (3.115) has been written with a discrete notation for states s, which is applicable to the present chapter.

In general, a stationary (or steady) state is a macrostate characterized by occupation probabilities p that are independent of time, $p(s,t) = p(s)$, which implies

$$\sum_{s'} \left[p(s') w_{s,s'} - p(s) w_{s',s} \right] = 0 \quad \forall s \qquad \Longleftrightarrow \qquad \text{steady state.} \tag{3.116}$$

This relation means that the net flow of probability between each microstate s and all other microstates vanishes. It is well known that an equilibrium state satisfies the condition of detailed balance, which corresponds to separately vanishing each term of the above summation,

$$p^{\text{eq}}(s') w_{s,s'} = p^{\text{eq}}(s) w_{s',s} \quad \forall s, s' \qquad \Longleftrightarrow \qquad \text{equilibrium state;} \tag{3.117}$$

i.e., it corresponds to vanishing the net flow of probability between each pair of states, s, s'.

A NESS satisfies Eq. (3.116), but it does not satisfy Eq. (3.117). The two equations differ not only because (3.117) is a stronger condition than (3.116); they also have a different use. The ensemble theory of equilibrium statistical mechanics provides the expression of $p^{\text{eq}}(s)$. For instance, the occupation probability of each microstate s in a system described by the canonical ensemble is given by the Boltzmann distribution,

$$p^{\text{eq}}(s) = \frac{e^{-\beta E(s)}}{Z} \qquad \text{(canonical ensemble),} \tag{3.118}$$

where Z, the partition function, is the normalization factor. Therefore, Eq. (3.117) can be used to impose a relation that must be satisfied by transition rates,

$$\frac{w_{s',s}}{w_{s,s'}} = e^{-\beta(E(s') - E(s))}. \tag{3.119}$$

Focusing on a Monte Carlo simulation (see Section 1.5.5) can be of some help. The Monte Carlo method creates a trajectory in phase space that does not correspond to real dynamics (as is the case for molecular dynamics), but the resulting visiting probability of each microstate must equal the ensemble probability. The fictitious trajectory is defined assigning the transition rates, which must satisfy Eq. (3.119). For instance, according to the popular Metropolis algorithm (see Section 1.5.5), we can adopt the transition rates

$$w_{s',s} = \min\left\{1, e^{-\beta(E(s')-E(s))}\right\}. \tag{3.120}$$

We might say that $p^{\text{eq}}(s)$ is the building block of equilibrium statistical mechanics and $w_{s',s}$ are derived quantities (in such a way as to satisfy detailed balance). In nonequilibrium theory, there is no general principle such as the Boltzmann weight and we do not know a priori the probability distribution for the nonequilibrium steady states. Rather, the model is defined by assigning the rates $w_{s',s}$. In this sense, Eqs. (3.117) and (3.116) have a different use: for an equilibrium state, we know $p^{\text{eq}}(s)$ and (3.117) provides a relation that must be satisfied by the transition rates; for a nonequilibrium steady state, we define the transitions $w_{s',s}$ and (3.116) is a relation that must be satisfied by the probabilities of state occupation.

Now imagine that someone provides us with a set of transition rates, $w^*_{s',s}$, asking if they would lead the system to an equilibrium state or not. The question is equivalent to asking if they satisfy detailed balance or not, but we do not know what (if any) is the Hamiltonian governing the system (in some simple cases we might guess it from the knowledge of w^*, but in general we cannot). In other words, can we test detailed balance from the plain knowledge of the transition rates? The answer is positive and it is proved in the following theorem (see also Fig. 3.5).

Theorem 3.1 The transition rates $w_{s',s}$ satisfy the detailed balance condition

$$\frac{w_{s',s}}{w_{s,s'}} = \frac{p^{\text{eq}}(s')}{p^{\text{eq}}(s)} \qquad \forall s, s', \tag{3.121}$$

if and only if they satisfy the condition

$$\prod_{i=1}^{N} w_{s_{i+1},s_i} = \prod_{i=1}^{N} w_{s_{i-1},s_i} \tag{3.122}$$

for any set $(s_1, \ldots, s_N)$ of N microstates (with the definition $s_0 = s_N$ and $s_{N+1} = s_1$).

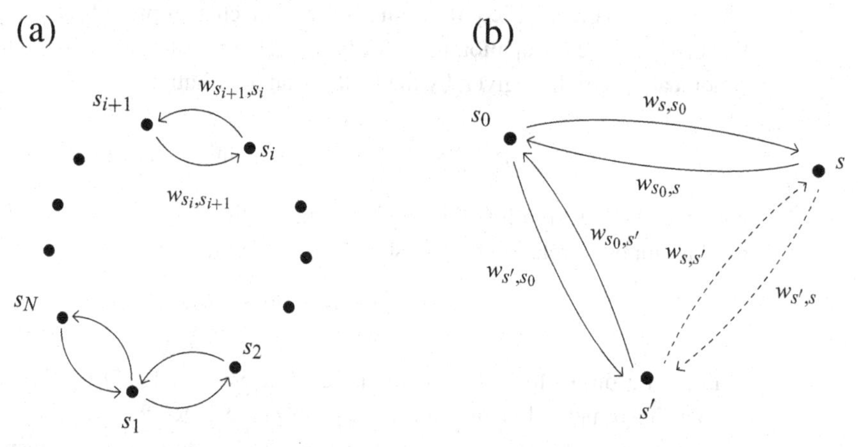

Fig. 3.5 Graphical representation of the transition rates appearing in the proof of Theorem 3.1: (a) Eq. (3.122); (b) Eqs. (3.124) and (3.126).

The first part of the theorem, detailed balance (3.121) implies Eq. (3.122), is trivial. As a matter of fact (see Fig. 3.5(a)),

$$\prod_{i=1}^{N} \frac{w_{s_{i+1},s_i}}{w_{s_{i-1},s_i}} = \prod_{i=1}^{N} \frac{p^{\text{eq}}(s_{i+1})}{p^{\text{eq}}(s_i)} = 1, \tag{3.123}$$

which is equivalent to Eq. (3.122).

The second part of the theorem requires us to define $p^{\text{eq}}(s)$ from the transition rates. For this purpose, let's take a microstate as a reference state and label it s_0 (see Fig. 3.5(b)). For any state s we define

$$p^{\text{eq}}(s) \doteq p^{\text{eq}}(s_0)\frac{w_{s,s_0}}{w_{s_0,s}}, \qquad \forall s, \tag{3.124}$$

where $p^{\text{eq}}(s_0)$ has to be determined by the normalization of occupation probabilities,

$$\sum_s p^{\text{eq}}(s) = 1. \tag{3.125}$$

In simple words, we define $p^{\text{eq}}(s)$ in such a way that any state satisfies detailed balance with respect to a specified microstate s_0. This does not guarantee that detailed balance is satisfied for any pair of states s, s', and in general it is not. However, if Eq. (3.122) is valid, then it does, as we are going to prove.

Let us start from the relation $w_{s,s_0}w_{s',s}w_{s_0,s'} = w_{s',s_0}w_{s,s'}w_{s_0,s}$, which must be satisfied because Eq. (3.122) is assumed to be true. We can rewrite it as

$$\left(\frac{w_{s,s_0}}{w_{s_0,s}}\right) w_{s',s} = \left(\frac{w_{s',s_0}}{w_{s_0,s'}}\right) w_{s,s'}, \tag{3.126}$$

and let us evaluate the two fractions between parentheses using Eq. (3.124):

$$\frac{p^{\text{eq}}(s)}{p^{\text{eq}}(s_0)} w_{s',s} = \frac{p^{\text{eq}}(s')}{p^{\text{eq}}(s_0)} w_{s,s'}. \tag{3.127}$$

Once we cancel $p^{\text{eq}}(s_0)$ from both sides, the above equation proves that detailed balance between s and s' is valid.

In conclusion, we have given an alternative formulation of detailed balance that involves transition rates only, so we can test if detailed balance is broken or not by checking Eq. (3.122).

3.4 The Standard Model

The Ising model, with binary variables and nearest-neighbor interactions, is the simplest model we can think of, and, in fact, it has a central role in equilibrium statistical physics. Therefore, it should not be surprising that an important role in nonequilibrium statistical physics is played by a model that is directly derived from it.

With a suitable change of variables the Ising model is known to describe a lattice gas (see Appendix H), which can be brought out of equilibrium by imposing a macroscopic current of mass. This was done by Sheldon Katz, Joel Lebowitz, and Herbert Spohn, who imagined driving the lattice gas in a given direction $\hat{x}$. In practice, this means favoring particles hopping in the $+\hat{x}$-direction and disadvantaging particles hopping in the opposite ($-\hat{x}$) direction, which can be obtained by modifying the original transition rates. Let us refer to the Metropolis algorithm (see Section 1.5.5 and Appendix J), according to which

$$w^{\mathrm{eq}}_{s',s} = A_{s',s}\,\omega\left(\beta\Delta\mathcal{H}^{\mathrm{LG}}\right), \tag{3.128}$$

where $A_{s',s}$ is the adjacency matrix, $\omega(x) \equiv \min\{1, e^{-x}\}$, and $\Delta\mathcal{H}^{\mathrm{LG}} = \mathcal{H}^{\mathrm{LG}}(s') - \mathcal{H}^{\mathrm{LG}}(s)$, where $\mathcal{H}^{\mathrm{LG}} = -\epsilon_0 \sum_{\langle ij\rangle} n_i n_j$ is the lattice gas energy; see Eq. (H.7). The (symmetric) matrix $A_{s',s} = 0, 1$ lists what states s' can be attained from an initial state s. Due to matter conservation and for the sake of simplicity we may limit our choice to states s' differing from s just for the position of one particle, which has moved to a nearest-neighbor site.

It is now necessary to implement the driving force, favoring the process $(10) \to (01)$ with respect to the inverse one, where the notation $(n_i n_{i+1})$ refers to two neigboring sites $i, i+1$ along the $\hat{x}$-axis. This can be done defining the following rates for the driven lattice gas,

$$w^{\mathrm{DLG}}_{s',s} = A_{s',s}\,\omega\left(\beta(\Delta\mathcal{H}^{\mathrm{LG}} - qE)\right), \tag{3.129}$$

where $E > 0$ measures the strength of the driving and $q = 0, \pm 1$, according to the move of the tagged particle: $q = +1$ if the particle moves to the right, along the $\hat{x}$-axis; $q = -1$ if the particle moves to the left, along the $\hat{x}$-axis; $q = 0$ if the particle moves perpendicularly to the $\hat{x}$-axis.

If we define the potential energy associated with the field E,

$$\mathcal{H}^{\mathrm{E}} = -E\sum_i n_i x_i, \tag{3.130}$$

where x_i is the $\hat{x}$-coordinate of the ith site, it appears that

$$\Delta\mathcal{H}^{\mathrm{LG}} - qE = \Delta\mathcal{H}^{\mathrm{LG}} + \Delta\mathcal{H}^{\mathrm{E}}. \tag{3.131}$$

Therefore, it seems that driving the lattice gas has only the effect of modifying the total Hamiltonian,

$$\mathcal{H}^{\mathrm{DLG}} = \mathcal{H}^{\mathrm{LG}} + \mathcal{H}^{\mathrm{E}}, \tag{3.132}$$

which does not lead the system out of equilibrium, since transition rates still satisfy detailed balance.

However, the above statement is true only if the system is virtually infinite, otherwise driving does break detailed balance through boundary conditions. The simplest possible proof involves only one particle, hopping in the $\pm\hat{x}$-direction, along which periodic boundary conditions apply. Microstates are simply labeled by the particle position i along $\hat{x}$ and application of Eq. (3.129) simply gives $w^{\mathrm{DLG}}_{i+1,i} = 1$, while $w^{\mathrm{DLG}}_{i-1,i} = \exp(-\beta E)$. We can finally make use of Theorem 3.1 and evaluate the products

$$P_r = \prod_{i=1,L} w^{\mathrm{DLG}}_{i+1,i} \tag{3.133}$$

$$P_l = \prod_{i=1,L} w^{\mathrm{DLG}}_{i-1,i}\,, \tag{3.134}$$

assuming periodic boundary conditions (state "0" is equivalent to state "L" and state "$L+1$" is equivalent to state "1"). $P_{r,l}$ correspond to a sequence of L hoppings on the right/left, so that the particle is back to the initial position. We immediately obtain $P_r = 1$ and $P_l = \exp(-L\beta E)$, so that $P_r \neq P_l$ and detailed balance is broken.

Before moving on to truly nonequilibrium phase transitions, we can ask what the effect is of the driving on the equilibrium second-order phase transition of the lattice gas. For a vanishing field, $E = 0$, and in dimension $d > 1$ the system undergoes a continuous phase transition between a gas, disordered, high temperature phase and a condensed, low temperature phase. The ground state corresponds to one single cluster of particles, whose shape minimizes the energy cost of its border, therefore depending on boundary conditions. For periodic boundary conditions, the ground state cluster is a stripe along the horizontal or vertical axis. If we are at a low temperature and we switch on the driving field E, it has different effects depending on the stripe orientation (see Fig. 3.6): if the stripe is perpendicular to E, the field promotes the breaking of the bonds, starting from the right edge; if the stripe is parallel to the field, the exclusion constraint hinders such breaking.

Therefore, there is a geometry of the condensed phase that is stable with respect to the breaking process and the phase transition is not destroyed by the driving. But what about

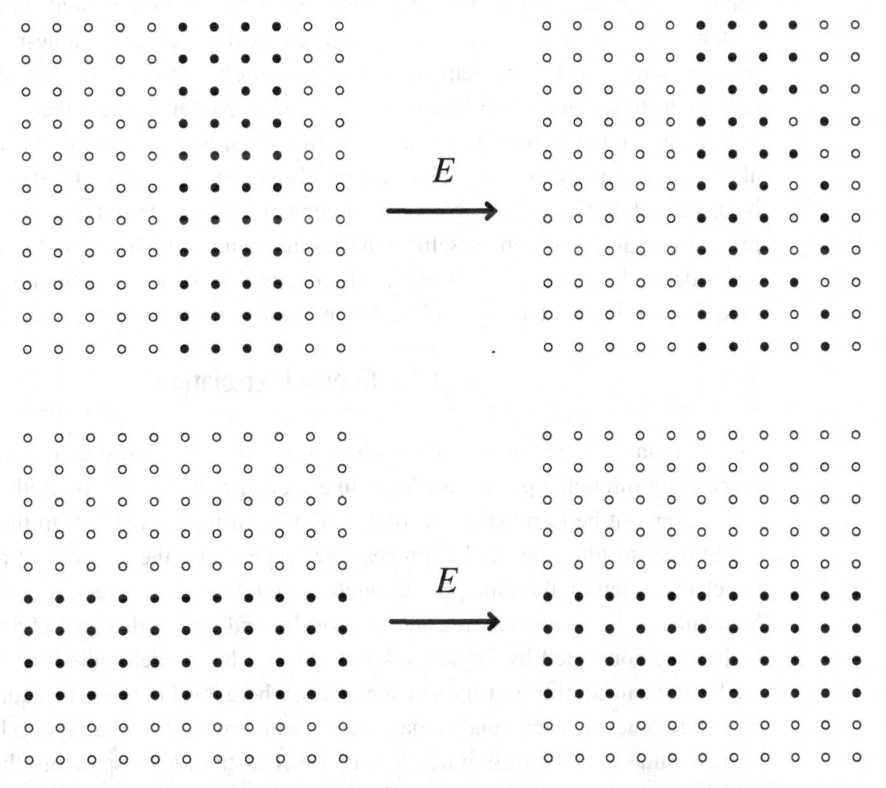

Fig. 3.6 Effect of a field E on the ground state cluster in the standard model, which is a stripe, closed on itself (because of periodic boundary conditions). If the stripe is orthogonal to E, the field promotes the hopping of particles, while in the parallel case, the field is immaterial.

the critical temperature $T_c(E)$? Contrary to naive expectations, the critical temperature increases with the field. In fact, let us consider a cluster of two particles. Their bond can be broken by thermal effects by moving one of the two particles to any neighboring, empty site. If we switch on the field, some of these moves (which break the bond) have a larger probability of occurring, while others have a reduced probability, so it is not straightforward to evaluate what the global effect of the field is.

A possible, simple explanation of the increasing character of $T_c(E)$ makes reference to the equilibrium free energy, $F = U - TS$, and to the effect of the field, which is irrelevant on U, because of periodic boundary conditions, while it determines a reduction of S, because of the mechanism described in Fig. 3.6: at a given temperature $T < T_c$ only stripes parallel to E survive. Therefore, a higher T is necessary in order to destabilize the condensed phase.

3.5 Phase Transitions in Systems with Absorbing States

As discussed in Section 3.2.1, equilibrium second-order phase transitions can be grouped into universality classes, characterized by the values of their critical exponents. From a semi-quantitative point of view relevant parameters are the spatial dimension d of the system, the symmetry of the order parameter, and the possibly long-range character of the interaction. In two dimensions, more rigorously, a complete classification of phase transitions relies on the solid theoretical ground of conformal field theory.

In addition, nonequilibrium phase transitions can be grouped into universality classes, although an equally effective and general classification is presently unavailable. As will be discussed in this section, many microscopic models are found to share the same critical properties that identify universality. On the other hand, field theoretical approaches are typically derived making use of heuristic arguments, which grasp the main symmetries of the model, as well as the basic role played by noise in a stochastic mean-field-like formulation.

3.5.1 Directed Percolation

The first class of models we deal with is inspired by the physical mechanism of a fluid percolating through a porous medium, like ground coffee paper, or cloth. In the first case our goal might be to make a cup of coffee using a flip coffee pot; in the other cases we might want to filter the fluid, a process that depends on the network of active pores, i.e. the channels letting the fluid pass through. A simple model of such process is represented by a lattice where each site is either wet or dry and a wet site can wet a neighboring site if they are connected by an active bond (pore). This model, called bond percolation, is well studied in equilibrium statistical physics, where it is known to be equivalent to a Potts model. In practice, each bond is made active with probability p, and for p larger than some critical value there is an infinite cluster of wet (active) sites spanning the entire system, which means that a fluid may percolate through it.[11] Alternatively, we may start from an

[11] A similar model is called site percolation: in this case sites rather than bonds are made active with probability p. The critical values for the formation of an infinite cluster are different in the two models.

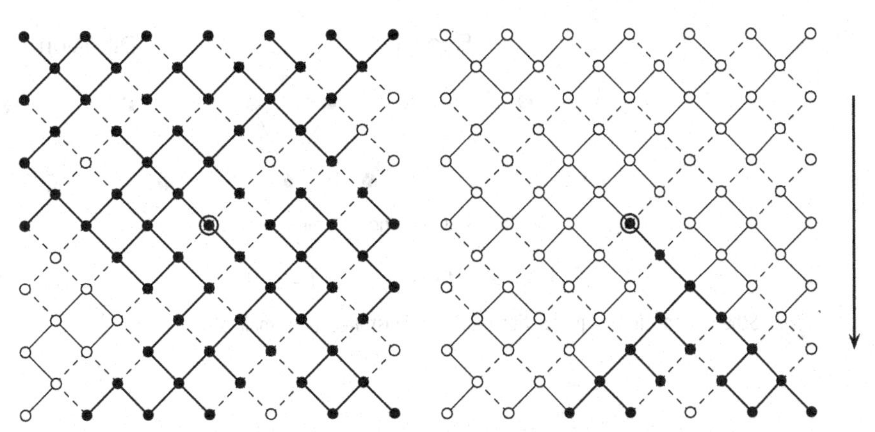

Fig. 3.7 Schematic of isotropic (left) and directed (right) bond percolation. The circled point in the center is the site where percolation originates; i.e., it is the only wet site when dynamics starts. Each bond may be active (solid line) or inactive (dashed line) and a site becomes wet if it is linked to a wet site via an active bond. Active bonds between dry sites (open circles) are thin, while active bonds between wet sites are thick. In isotropic percolation (left) the spread of wet sites proceeds isotropically, while in directed percolation (right) wetting proceeds only in the direction of the arrow.

active site and proceed to wet the system according to the above rule: does the wetting process stop ($p < p_c$) or does it spread ($p > p_c$)? This question may correspond to a specific experimental setup, e.g., to a horizontal sheet of paper. However, if we tilt the system (or, in general, if the fluid can flow in the vertical direction), it is clear that gravity plays a role, because the fluid will tend to flow in a certain direction. The two cases are called, respectively, isotropic (bond) percolation and directed (bond) percolation and are illustrated in Fig. 3.7.

The reason why we have introduced directed percolation (DP) is that the flow direction can be interpreted as the arrow of time, so that DP in $d + 1$ dimensions corresponds to a nonequilibrium process in d spatial dimension ($d = 1$ in Fig. 3.7). The example of DP is therefore specially suited to illustrate a statement made in the prologue of this chapter, i.e., that in nonequilibrium phase transitions time takes the form of an additional dimension. The universality class of DP is generally characterized by a phase transition from a fluctuating active phase to an inactive phase where the system is eventually trapped into an absorbing state. In the specific model of DP the (only) absorbing state corresponds to all dry (inactive) sites.[12]

In Fig. 3.8 we illustrate the possible processes of the DP dynamics, with an active site that can die, give rise to a newborn active site, coalesce with a neighboring active site, or diffuse. All these processes depend on the probability p that a bond is active and they define a reaction–diffusion system among particles where an active site corresponds to a particle A and an inactive site corresponds to an empty site $\emptyset$. The three reaction processes are depicted, in symbols, as $A \to \emptyset$ (death), $A \to 2A$ (offspring), and $2A \to A$ (coalescence): this language allows an easier generalization to different physical processes.

An operational definition of DP dynamics is better illustrated in Fig. 3.9, where the system is distorted so that the one-dimensional lattice has a simpler geometry and labeling.

12 Which makes the terminology unfortunate.

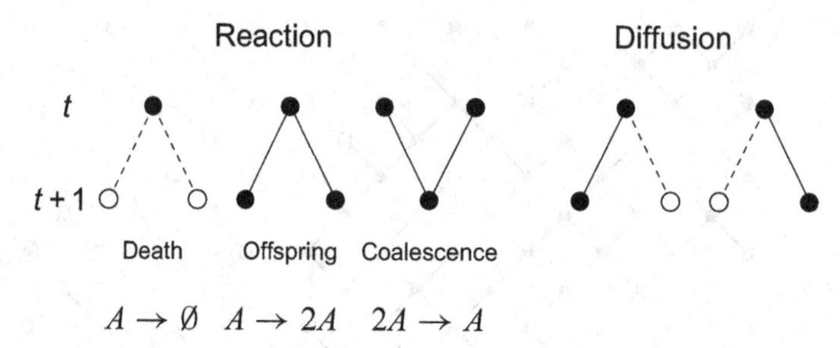

Fig. 3.8 Schematic of directed percolation processes, illustrated in the main text.

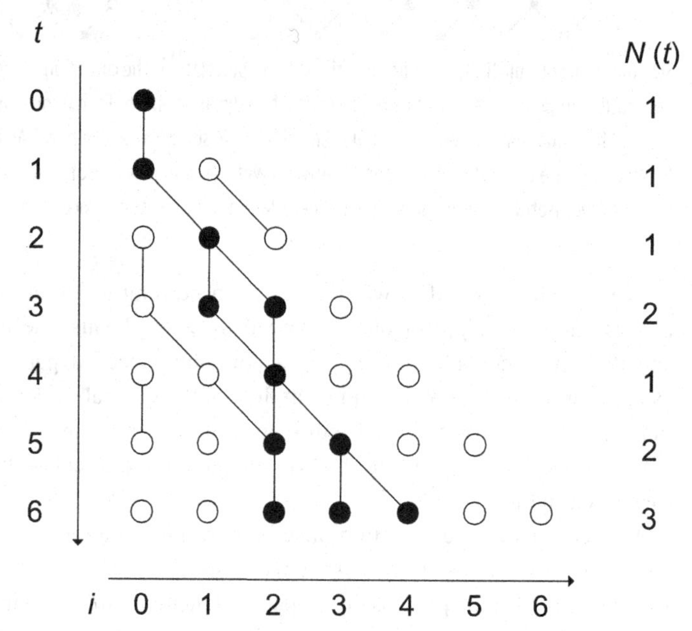

Fig. 3.9 In order to understand the implementation of directed percolation it is useful to deform, as shown here, the lattice of Fig. 3.7 on the right. For the sake of simplicity, inactive bonds are not indicated. The quantity $N(t)$ counts the number of wet sites at time t.

At time $t = 0$ only the site $i = 0$ is active; if site k is active at time t, at time $t+1$ it may wet with probability p the ("left") site k and with the same probability the ("right") site $k+1$. Therefore at time t only sites $k = 0, \ldots, t$ need to be considered. We can now introduce the variable $s(i,t) = 0, 1$ to specify if site i is active ($s = 1$) or inactive ($s = 0$) at time t. If we initialize all sites at time $t+1$ as inactive ($s(i, t+1) = 0\ \forall i$), a simple evolution rule that can be implemented in a numerical program is the following: we span the lattice at time t locating all active sites k, $s(k,t) = 1$, then with probability p we set $s(k, t+1) = 1$ and, independently, with probability p we set $s(k+1, t+1) = 1$. One can easily realize that the total number of active states at time t, $N(t) = \sum_{i=0}^{t} s(i,t)$, is an important observable.

It is straightforward that the fully inactive (dry) state plays the role of the absorbing state: if the system eventually falls into such a state, it will be trapped there forever, since no active sites can be produced.[13] It is just as simple to understand that the fully active (wet) state is not a stationary state, because there is a finite probability $(1-p)^2$ that a site dies.

We want to point out that the very existence of an absorbing state in the dynamical evolution rule of DP processes tells us that they are nonequilibrium phenomena, since detailed balance is violated. Actually, the absorbing state can be entered from any previously active configuration with some finite probability, while the time-reversed process has probability zero. As a consequence, phase transitions in such models cannot be described in terms of equilibrium ensemble theory.

The phase transition between the active and the inactive phases in bond-directed percolation is essentially due to the competition between the birth and the death processes and it occurs at a critical value p_c of the percolation probability p. In the subcritical (inactive) phase, $p < p_c$, clusters generated by a single initial active site have a finite lifetime and eventually die. In the supercritical (active) phase, $p > p_c$, there is a finite probability that these clusters extend up to infinity, spreading inside a cone, whose opening angle depends on the value of $|p - p_c|$, i.e., on the distance of the actual probability from its critical value.

At $p = p_c$ one can obtain active clusters of all sizes, which exhibit the typical sparse structure of fractal, i.e., self-similar, objects. This is a pictorial indication that at criticality many physical observables exhibit characteristic scale-free, i.e., power-law, behavior in analogy with equilibrium critical phenomena. As we shall discuss later (see Section 3.6.1) the main difference is that such power laws emerge in space–time correlation functions. On the other hand, the spatial and the temporal critical exponents are found to be universal signatures of all models belonging to the class of DP processes. Another analogy with equilibrium phase transitions is that the value of p_c is model dependent; i.e., it is not a universal property. For instance, in directed bond percolation in 1+1 (one space and one time) dimensions the best numerical estimate is $p_c \simeq 0.645$.[14]

A quantitative description of the DP nonequilibrium phase transition can be obtained by defining a suitable order parameter. The one adopted for DP processes generated by a single active site is the average number of active sites at time t, $\langle N(t)\rangle$. The average operation $\langle\cdots\rangle$ has to be performed over many "stories" of the percolation process. These stories correspond to the ensemble of percolation clusters, where channels are open with a given probability p.[15] In particular, numerical analysis shows that for $p < p_c$ (inactive phase), $\langle N(t)\rangle$ eventually exhibits an exponential decay to zero, while for $p > p_c$, $\langle N(t)\rangle$ grows rapidly and eventually crosses over to a linear increase in time, so as to obtain a finite density of active sites, $\langle N(t)\rangle / t$. In both cases the crossover time depends on $|p - p_c|$, i.e., the distance from the critical probability.

[13] As discussed later, in Section 3.5.2, there are other reaction–diffusion processes with more than one single absorbing state.

[14] In the realm of nonequilibrium phase transitions, the simple definition and simulation of DP processes makes this model as paradigmatic as the Ising model of ferromagnetism for equilibrium phase transitions. On the other hand, Onsager was able to find an exact solution of the Ising model in $d = 2$, while, despite many efforts over decades, no exact solution of the critical properties of DP processes is presently known.

[15] The comparison among different stories shows that the values of $N(t)$ may exhibit wild fluctuations, as expected on single realizations of stochastic processes. This notwithstanding, the stationarity of the process guarantees that such fluctuations are progressively smoothed out by increasing the number of stories in the ensemble.

At $p = p_c$ the average number of active sites is found to increase asymptotically according to a power law, namely, $\langle N(t) \rangle \sim t^{\theta}$. Numerical analysis provides the following estimate of the critical exponent, $\theta \simeq 0.302$. It is worth mentioning that methods for obtaining reliable numerical estimates of critical exponents in DP-like processes demand specialized techniques and considerable computational effort. As in equilibrium critical phenomena, theoretical arguments predict that this is due to logarithmic corrections to scaling, which vanish only in the thermodynamic limit. At variance with equilibrium critical phenomena, such a limit has to be intended as a suitable combination of the limits where the lattice dimension L diverges together with time. Actually, we are in the presence of a dynamical phenomenon that is ruled by a diffusive process at the microscopic level. In this sense, one can argue that in the active phase one should perform the limits in such a way that $\lim_{t\to\infty} \lim_{L\to\infty} L \cdot t^{-1/2}$ goes to a constant, which represents the diffusion coefficient of the DP process. For $p < p_c$, it is enough to assume that $\lim_{t\to\infty} \lim_{L\to\infty} L \cdot t^{-\theta} \simeq$ constant.[16]

3.5.2 The Domany–Kinzel Model of Cellular Automata

Many basic models of nonequilibrium phase transitions can be formulated as dynamical processes of interacting particles moving on a lattice. When a lattice site can be either occupied by a single particle or empty (exclusion process) and the evolution rule is local, synchronous, and Markovian, the model at hand is just a cellular automaton (CA). "Local" means that the evolution rule of a site depends only on neighboring sites; "synchronous" means that sites are updated in parallel;[17] "Markovian" means that the configuration at time $t + 1$ depends on the configuration at time t only.

Deterministic CA were introduced in the last century by the great mathematician John von Neumann to show that even a deterministic, discrete system evolving in discrete time with a local rule may display an unpredictable evolution.[18] Some decades later, Eytan Domany and Wolfgang Kinzel introduced a very simple stochastic CA model that epitomizes several "contact processes" including the bond DP model described in the previous section. In this perspective, the Domany-Kinzel (DK) model has played a very important role in the scientific literature, as a general pedagogical example for different stochastic models of nonequilibrium phase transitions.

The DK model is defined on a tilted square lattice, as shown in Fig. 3.10, which evolves by parallel updates. Let us consider a given lattice configuration at time t, which can be

[16] In principle there is a finite but exponentially vanishing probability that evolution grows ballistically, in which case we should take the limit L/t as constant. Additional details about the correct way to take the L and t limits are given in note 21, below.

[17] A synchronous evolution rule could be thought of as a time-rescaled sequential one, where the time unit has been replaced by the number of lattice sites.

[18] Since on a finite lattice of size L the number of possible dynamical configurations in a deterministic boolean CA is finite and certainly bounded from above by 2^L, its deterministic evolution must be periodic. On the other hand, the unpredictability of certain deterministic CA rules emerges when the thermodynamic limit is performed and the transient time needed to reach a periodic state, once averaged over random initial conditions, is typically found to grow exponentially with L. CA are a milestone of modern mathematics because of their conceptual consequences and implications for the theory of computation and, more generally, in many other fields of science including computer science, physics, biology, chemistry, etc.

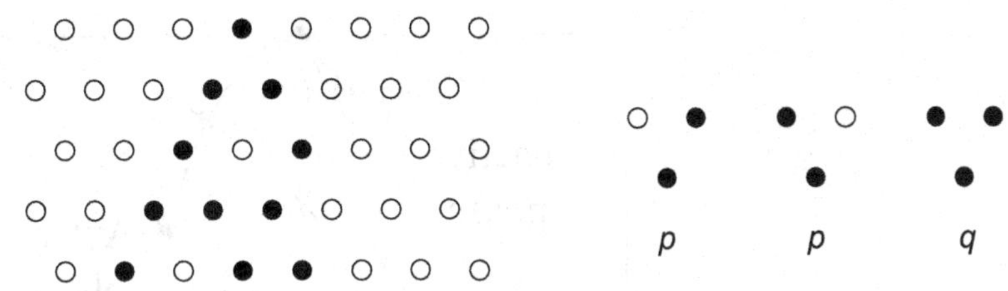

Fig. 3.10 The Domany-Kinzel CA model. Left: Possible evolution of a system where only one site is active at time t. Right: Probabilistic evolution rules.

coded as a sequence of binary symbols $s_i(t) \in \{0, 1\}$: if $s_i(t) = 0\,(1)$ at time t we have an empty/inactive (occupied/active) lattice site at position i. The evolution rule is defined in terms of conditional probabilities involving the two neighbors of each lattice site, $P(s_i(t+1)|s_{i-1}(t), s_{i+1}(t))$,

$$\begin{aligned} P(1|0,0) &= 0 \\ P(1|0,1) &= P(1|1,0) = p \\ P(1|1,1) &= q, \end{aligned} \tag{3.135}$$

with $P(0|\cdot,\cdot) = 1 - P(1|\cdot,\cdot)$. One can easily realize that due to the first rule in Eq. (3.135), a fully empty configuration corresponds to the "absorbing state." Moreover, a site can be occupied at time $t+1$ with probability p if one of its neighbor sites was occupied at time t or with probability q if both neighbor sites were occupied at time t.[19]

The phase diagram of DK model can be represented in the plane of parameters p and q; see Fig. 3.11. The transition line $p_c(q)$ separates the absorbing phase (on its left) from the active one (on its right). As it happens for bond DP, in the absorbing phase, clusters of active sites eventually vanish exponentially fast in time, while in the active phase one observes a fluctuating steady state with an average fraction of active sites depending on the choice of parameter values of the DK model.

There are some special cases in this phase diagram. (i) The line $q = p(2-p)$ corresponds to the bond DP process described in the previous section. In fact, if one identifies with p the probability of an open channel, in the bond DP the coalescence process of two active sites occurs if at least one channel is open, namely, with probability $q = 1-(1-p)^2 = p(2-p)$. (ii) The line $p = q$ defines the so-called site DP process: in this case all bonds are open, but sites are "permeable" (i.e., they can be wet) with probability p. (iii) For $q = 0$, two active sites always annihilate (this process usually occurs with probability $1-q$).[20] (iv) For $q = 1$, two active sites always coalesce, which implies a symmetry between particles and holes (or, equivalently, between active and inactive sites), because the dynamics is now invariant under the transformation of one type of site into the other. Because of this symmetry, $p_c(q = 1) = \frac{1}{2}$ (see below for more details).

[19] We should make explicit a feature of CA models, which may be easily overlooked. The "tilted" graphical representation of the evolution (see Fig. 3.9) actually means that even and odd sites are represented in following rows, because they are updated in turn. So, the real time unit to update all sites is $dt = 2$, not $dt = 1$.

[20] This situation is a stochastic variant of the rule called "W18" according to the Wolfram classification of CA.

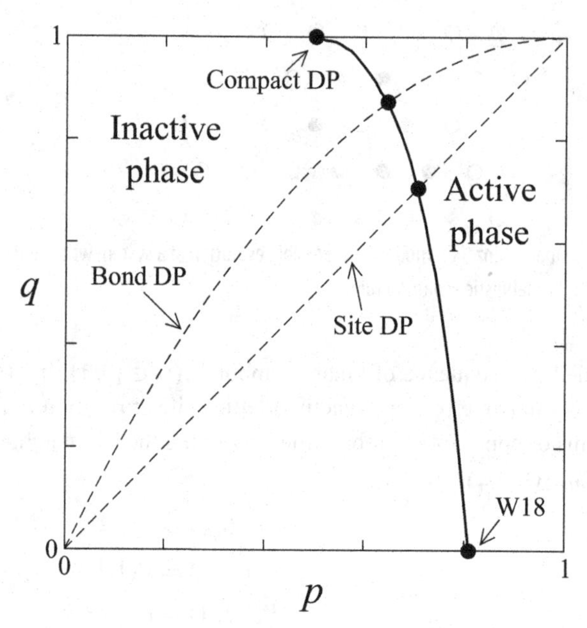

Fig. 3.11 Phase diagram of the DK model. The thick solid line is the curve $p_c(q)$, separating the absorbing (inactive) phase from the active one. The dashed lines correspond to the bond DP ($q = p(2 - p)$) and to the site DP percolation ($q = p$). Original data courtesy of Haye Hinrichsen. Haye Hinrichsen, Non-equilibrium critical phenomena and phase transitions into absorbing states, *Advances in Physics*, **49** (2000) 815–958.

The first, qualitative feature of the phase diagram (3.11) is that the value of the critical point, $p_c(q)$, is a decreasing function of q. This is not surprising because a larger q means that coalescence prevails over annihilation, contributing to an increase in the number of active sites, therefore favoring the transition from the inactive to the active phase. The most important feature of the phase diagram concerns, however, the nature of the phase transition along the transition line, i.e., for different values of q. An exact solution is lacking, but careful numerical studies provide strong evidence that all versions of the DK model belong to the same universality class of bond DP, except for $q = 1$, whose additional symmetry particle $\leftrightarrow$ hole changes the critical behavior of the system. In short, with the caveat that $q < 1$, q is an irrelevant parameter of the DK model.

Belonging to the same universality class means that the critical behavior along the transition line exhibits the same kind of long-range correlations, namely, it is characterized by the same critical exponents. More precisely, in the limits $L \to +\infty$ and $t \to +\infty$[21]

[21] It is important to point out that the order in which these limits are performed is quite a crucial point. In general they do not commute with each other, and, accordingly, their exchange may yield different conclusions on the asymptotic behavior of evolution rules. In the absence of any prior knowledge about the dynamical features of the problem at hand the physically suitable procedure amounts to first making the limit $L \to +\infty$ and then the limit $t \to +\infty$, in order to include any possible propagation process emerging from the adopted evolution rule. In many cases, however, we can use some preliminary information to define a proper way of performing these limits: for instance, if we are in the presence of ballistic propagation phenomena with a limit upper velocity $v_{\max}$, the two limits can be performed in such a way that $\lim_{t\to+\infty} L/t \geq v_{\max}$; in the case of a generalized diffusive behavior, we can impose the condition $\lim_{t\to+\infty} L/t^{\alpha} \geq D_{\max}$, where $\alpha < 1$ and $D_{\max}$ is an upper estimate of the corresponding diffusion parameter.

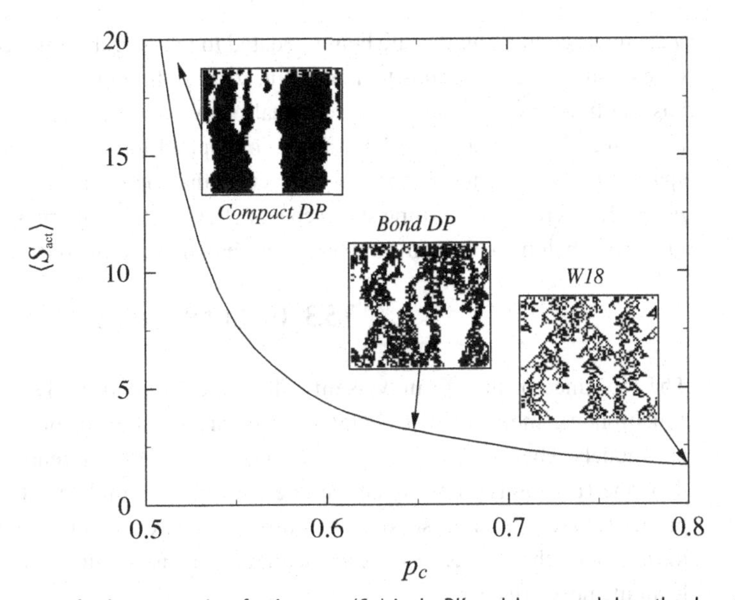

Fig. 3.12 Numerical estimates for the average size of active spots $\langle S_{act} \rangle$ in the DK model measured along the phase transition line. The insets show typical clusters for three special cases discussed in the text. Original data courtesy of Haye Hinrichsen. Haye Hinrichsen, Non-equilibrium critical phenomena and phase transitions into absorbing states, *Advances in Physics*, **49** (2000) 815–958.

the percolation clusters at $p_c(q)$ can be mapped onto each other by a suitable scale transformation, which amounts to a renormalization procedure analogous to that employed for equilibrium critical phenomena, although in this case space and time coordinates play different roles and yield different scaling properties.

We should also stress that, at variance with long-range correlations, short-range correlations at $p_c(q)$ are found to depend on q. This can be easily checked by measuring the average size of active sites, $\langle S_{act} \rangle$, in a critical percolating cluster, for different values of q. This quantity (see Fig. 3.12) is found to increase when decreasing $p_c(q)$, i.e., when passing from the stochastic W18 rule ($q = 0$) to the special case $q = 1$ through the bond DP case. As shown in the insets of the figure, an increasing value of $\langle S_{act} \rangle$ means that critical clusters become increasingly dense. In fact the model with $q = 1$ is also called compact DP (CDP), and it will be treated in more detail in the next chapter, Section 4.2.1, because it belongs to a different universality class.

The DK cellular automata can be generalized to $d \geq 2$ space dimensions making use of suitable conditional probabilities to define the evolution rule as follows:

$$P(1|n) = \begin{cases} 0, & n = 0 \\ p_n, & n \geq 1, \end{cases} \tag{3.136}$$

where $0 \leq n \leq 2d$ is the number of occupied sites in the neighborhood of the evolving site. The overall evolution rule depends on the $2d$ parameters p_n, while the absorbing state remains a fully unoccupied lattice. The generalization of bond DP in d dimensions is obtained by taking $p_n = 1 - (1 - p)^n$, which is the probability that at least one of the

n channels connecting the site being updated to the neighboring occupied site is open. The case of site DP corresponds to a constant, $p_n = p$. In more than one space dimension, one has the freedom of choosing a great deal of evolution rules, for instance, by combining bond-like DP with site-like DP, which correspond to hypersurfaces in the $2d$ parameter space. On the other hand, one can conjecture that, apart from the special case $P(1|2d) = p_{2d} = 1$, where two absorbing states coexist, the critical properties associated with any evolution rule belong to the same universality class, namely, the one of DP in $d+1$ dimensions.

3.5.3 Contact Processes

The relevance of the DP universality class goes beyond the DK CA, whose evolution is synchronous, meaning that all lattice sites are updated in one time step. In fact, there are models where DP critical properties emerge from sequential update rules, meaning that one (randomly chosen) lattice site at a time is updated. However, even if critical properties are the same, short-range correlations are expected to be highly influenced by the adopted scheme, because the sequential update is certainly more effective in destroying local memory effects.

The simplest sequential evolution rule reproducing DP scaling properties at criticality is the contact process, which was introduced as a model of epidemic spreading without immunization. The binary occupied/empty state at each lattice site adopted for describing percolating fluids in a porous medium can be turned to the binary infected/healthy state in the language of epidemic propagation in a population whose individuals are in contact with a finite neighborhood of other individuals. Each infected individual has two possibilities: either heal itself or infect one of its neighbors according to some assigned probability rates. The possibility of the infection to propagate depends on the choice of these probability rates.

The contact process can be defined on a d-dimensional square lattice, whose sites are labeled by the integer index i. At any time t, on each lattice site i we can find either an infected ($s_i(t) = 1$) or a healthy individual ($s_i(t) = 0$). The state of this population is updated at each time step by choosing at random one site and by assigning to it a new state $s_i(t+1) = 0, 1$. The outcome depends on $s_i(t)$, on the number of infected sites in its neighborhood, $n_i(t) = \sum_{j\in\{i\}} s_j(t)$,[22] and on certain transition rates $w(s_i(t) \to s_i(t+1), n_i(t))$.

A customary way to define w is to take a probability of infection that is barely linear with the fraction of infected neighbors, while the probability of recovery is some value r independent of the neighborhood.[23] Thus,

$$\begin{aligned} w(1 \to 0, n) &= r \\ w(0 \to 1, n) &= \lambda \frac{n}{2d}. \end{aligned} \tag{3.137}$$

Numerical investigations based on Monte Carlo methods and series expansions indicate that for the contact process in $1 + 1$ dimensions one finds a phase transition in the universality class of DP at a critical value of the ratio $(\lambda/r)_c \simeq 3.29785$.

[22] The symbol $\{i\}$ identifies the set of sites j that are neighbors of i, so that $0 \le n_i(t) \le 2d$.
[23] It would be more correct to define w as transition rate per unit time, which requires passing to a continuous time. We prefer to keep using a discrete time.

3.6 The Phase Transition in DP-like Systems

3.6.1 Control Parameters, Order Parameters, and Critical Exponents

The DP class of nonequilibrium phase transitions to one single absorbing state is characterized by typical scaling properties at the transition point that are widely reminiscent of critical phenomena in the equilibrium case (see Section 3.2). For instance, the ferromagnetic transition in the Ising model is found to occur at a critical temperature T_c, where the magnetization vanishes as $M \sim (T_c - T)^\beta$. The temperature T and the magnetization M are the control and the order parameter of this phase transition, respectively. The divergence of the correlation length as $\xi \sim |T - T_c|^{-\nu}$ implies that very close to T_c there is no typical macroscopic length scale; i.e., the physical system is invariant under scale transformations.

From the discussion in the previous section, one can easily infer that in the DP class the natural control parameter, analogous to the temperature T in the equilibrium case, is some probability p, e.g., the probability of an open channel in bond DP, the parameter $p(q)$ in the DK model or the ratio (λ/r) in the contact process. As in equilibrium phenomena, the critical value of the control parameter, p_c, is a model-dependent quantity.

As for the order parameter, in DP-like systems there are two possible ways to define it: (i) we can count the active sites and evaluate asymptotically in time their number (or their density), which must vanish in the inactive phase or (ii) we can evaluate the probability of not yet having reached the absorbing phase at time t (and, again, taking the limit $t \to \infty$). More precisely, the first choice depends on the initial conditions, which may be characterized by a vanishing or a finite density of active sites. In the former case (think of the limiting case of one single active site at $t = 0$) the morphology does not scale with the size of the system and we must simply count the number of active sites,

$$N(t) = \left\langle \sum_i s_i(t) \right\rangle, \tag{3.138}$$

where the ensemble average $\langle \bullet \rangle$ is performed over many realizations of the stochastic evolution. For homogeneous initial conditions, instead, we should use the density of active sites,[24]

$$\rho(t) = \frac{1}{L} \sum_i \langle s_i(t) \rangle = \frac{N(t)}{L}, \tag{3.139}$$

where L is the total number of sites.

The second choice for the order parameter is the survival probability for a trajectory whose initial state has one active site only, $s_i(0) = \delta_{i,k}$. It can be formally defined as

$$P(t) = \left\langle 1 - \prod_i (1 - s_i(t)) \right\rangle, \tag{3.140}$$

[24] The density of active sites can actually be evaluated even for a single, initial active site if we normalize $N(t)$ with respect to t, which is the maximal possible number of active sites at time t, starting from a single active site.

where $P(t)$ is the ensemble average of an observable that is equal to 1 until an active site is present, while it vanishes only when the system evolves into the fully inactive absorbing state. In other words, $P(t)$ is the fraction of the ensemble of stochastic evolutions that at time t have not yet reached the absorbing state.

In the limit $t \to \infty$, these order parameters can be associated to critical exponents as

$$\rho(\infty) \sim (p - p_c)^{\beta}, \tag{3.141}$$

$$P(\infty) \sim (p - p_c)^{\beta'}. \tag{3.142}$$

These relations indicate that when approaching the critical point p_c from the active phase, $p > p_c$, both order parameters vanish according to a power law. It is customary to associate the survival probability with the annihilation process and the density of active sites with the creation process, so $\rho(t)$ and $P(t)$ are sometimes referred to as creation and annihilation order parameters, respectively.

It is not obvious to say a priori if β and β' should be equal or not. A primary criterion seems to be related to the number of absorbing states, as shown by the DK model: for $q < 1$ the DK model belongs to the bond DP universality class, has only one absorbing state, and, as proved below, $\beta = \beta' \simeq 0.276$; for $q = 1$ there are two absorbing states and $\beta \neq \beta'$, as proved in Section 4.2.1.

The equivalence between the exponents β and β' in DP is a consequence of a special symmetry of this model, which amounts to a sort of time-reversal symmetry. A heuristic explanation can be given in the case of bond DP, considering a configuration of open and closed bonds as the one shown in Fig. 3.13. The quantity $P(t)$ is evaluated in direct time (left panel) activating a single site and determining if there is a directed path through active bonds, leading to the opposite (bottom) side. If we are in the thermodynamic limit, the average over disorder realizations can be replaced, using self-averaging, by an average over the starting site, so $P(t)$ is just the fraction of initial sites that are connected to the opposite side. Now we revert the time arrow, as shown in the right panel of the same figure: we start from all active sites and evaluate the density of active sites at time t, $\rho(t)$. It is clear that a "top" site is now active if and only if there is a directed path connecting it to the bottom side, which is exactly the condition because, in direct time, it contributes to $P(t)$. In conclusion, if we average over disorder or we take the limit $L \to \infty$, we expect that

$$P(t) = \rho(t). \tag{3.143}$$

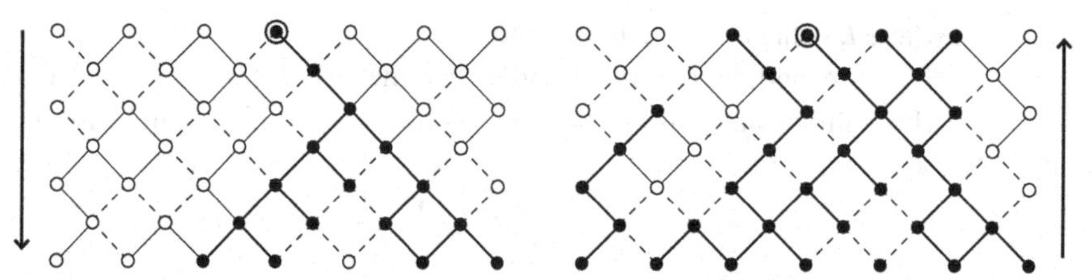

Fig. 3.13 Left: Directed percolation process starting from a single active site. Right: Time-reversed process of the left one, starting from a fully active state. If there is one directed path from top to bottom (left), there must be a directed path from bottom to top (right).

Accordingly, in the active phase both order parameters $P(\infty)$ and $\rho(\infty)$ have to saturate to the same value and have to exhibit the same critical behavior at p_c; i.e., β and β' have to be the same critical exponent. This equivalence holds also for other kinds of DP processes such as site DP or the contact process, although the same argument based on the exact time-reversal symmetry typical of bond DP does not apply. Nonetheless, there is numerical evidence that this symmetry still holds asymptotically, while $P(t)$ and $\rho(t)$ become proportional to each other in the long time limit and $\beta = \beta'$.

The main feature that makes nonequilibrium critical phenomena different from equilibrium ones is the presence of independent spatial ($\xi_\perp$) and time ($\xi_\parallel$) correlation lengths, where the $\perp$ and $\parallel$ symbols refer to the time arrow. These quantities are associated with the asymptotic behavior of the space and (positive) time correlation functions

$$c_{|i-j|} = \left\langle \lim_{t\to\infty} \frac{1}{t} \sum_{\tau=0}^{t} (s_i(\tau) - \bar{s})(s_j(\tau) - \bar{s}) \right\rangle \sim e^{-|i-j|/\xi_\perp} \tag{3.144}$$

$$c(t) = \left\langle \lim_{L\to\infty} \frac{1}{L} \sum_{i=1}^{L} (s_i(0) - \bar{s})(s_i(t) - \bar{s}) \right\rangle \sim e^{-t/\xi_\parallel}, \tag{3.145}$$

where the average value

$$\bar{s} = \lim_{t\to\infty} \frac{1}{t} \sum_{\tau=0}^{t} \langle s_i(\tau) \rangle \tag{3.146}$$

does not depend on the site index.

The physical interpretation of $\xi_\perp$ and $\xi_\parallel$ is illustrated in Fig. 3.14. In the inactive phase, $p < p_c$, the clusters of active sites generated by a single initially active site (panel (a)) typically extend in space up to a size $\xi_\perp$ and in time up to $\xi_\parallel$, before eventually being absorbed. In the active phase, $p > p_c$, and for the same kind of initial condition (panel (b)), the surviving clusters grow within a "cone" whose opening angle is characterized by the ratio $\xi_\perp/\xi_\parallel$. The correlation lengths can be identified also when using homogenous initial conditions. In fact, for $p < p_c$ (panel (c)), $\xi_\parallel$ amounts to the typical decay time of active clusters, while in the stationary state of the active phase (panel (d)), $\xi_\perp$ identifies the typical size of inactive islands and $\xi_\parallel$ identifies their duration.

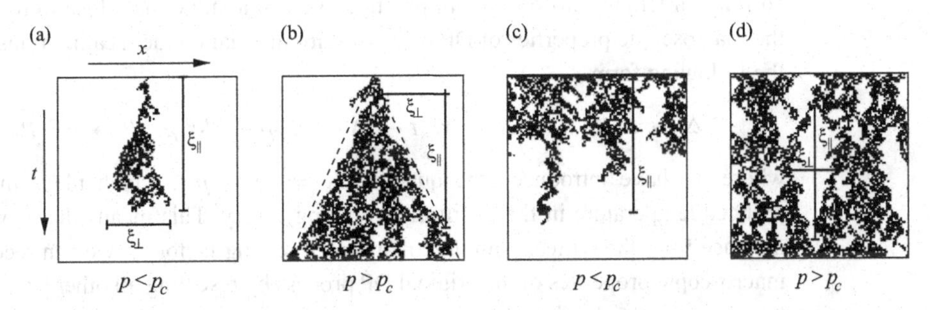

Fig. 3.14 Pictorial description of the correlation lengths $\xi_\parallel$ and $\xi_\perp$ in a DP process for different initial conditions, below and above criticality. The explanation for part labels (a)–(d) are discussed in the text. From Haye Hinrichsen, Non-equilibrium critical phenomena and phase transitions into absorbing states, *Advances in Physics*, **49** (2000) 815–958.

Very close to p_c both correlation lengths are found to diverge as

$$\xi_\perp \sim |p - p_c|^{-\nu_\perp}, \tag{3.147}$$

$$\xi_\parallel \sim |p - p_c|^{-\nu_\parallel}. \tag{3.148}$$

At variance with the exponents β, β' associated with the creation and annihilation order parameters, these two scaling exponents are naturally expected to be different, and their ratio, $z = \nu_\parallel/\nu_\perp$, is called the dynamical exponent.

From a practical point of view, the most efficient order parameter for determining the critical point of DP processes by numerical simulation is the average cluster mass (3.138), which is found to grow algebraically at p_c as

$$N(t)|_{p=p_c} \sim t^\theta. \tag{3.149}$$

Instead, for $p < p_c$, $N(t)$ asymptotically decreases and vanishes after an initial increase, while for $p > p_c$, $N(t)$ grows exponentially. Therefore, plotting $N(t)$ versus t in a log-log scale, we obtain a linear behavior at criticality, a positive curvature above criticality, and a negative curvature below criticality.

All these critical exponents, characterizing nonequilibrium phase transitions in systems with absorbing states, are not independent of each other. In equilibrium critical phenomena we could use well-known thermodynamic and statistical mechanical relations to derive the Josephson, Fisher, Widom, and Rushbrooke relations. In the present, nonequilibrium context the machinery required to get the known hyperscaling relation is more complex and we just write it,

$$\theta = \frac{d\nu_\perp - \beta - \beta'}{\nu_\parallel}, \tag{3.150}$$

where d is the number of space dimensions.

3.6.2 Phenomenological Scaling Theory

As in equilibrium critical phenomena, we can work out a phenomenological scaling theory by making explicit the scale invariance engendered by the divergence of space and time correlation lengths close to p_c. In practice, we can assume that close to the critical point the macroscopic properties of DP processes are invariant under scaling transformations of the following form

$$\Delta \to \Lambda\,\Delta, \;\; x \to \Lambda^{-\nu_\perp} x, \;\; t \to \Lambda^{-\nu_\parallel} t, \;\; \rho \to \Lambda^{\beta} \rho, \;\; P \to \Lambda^{\beta'} P, \tag{3.151}$$

where we have introduced the quantity $\Delta = |p - p_c|$, which plays the role of the reduced temperature in the equilibrium scaling theory. This means that if we rescale the distance from the critical point by an arbitrary scaling factor Λ, we can recover the same macroscopic properties of the original DP process by rescaling all other physical quantities (space x, time t, density of active sites ρ, and survival probability P) making use of the relations (3.141), (3.142), and (3.144)–(3.148).

Anyway, there is a degree of arbitrariness in choosing the quantity to be rescaled, similar to the equilibrium case, where we used either the rescaled temperature (Section 3.2.5) or the correlation length (Section 3.2.6) as building blocks of a scaling theory. For instance, we could impose an equivalent assumption of scale invariance by first rescaling the space variable x and all the other quantities accordingly:

$$x \to \Lambda x, \quad \Delta \to \Lambda^{-1/\nu_\perp} \Delta, \quad t \to \Lambda^z t, \quad \rho \to \Lambda^{-\beta/\nu_\perp} \rho, \quad P \to \Lambda^{-\beta'/\nu_\perp} P. \tag{3.152}$$

One can easily realize that this set of scale transformations is the same as (3.151), modulo a redefinition of the scale parameter $\Lambda \to \Lambda^{-1/\nu_\perp}$.

These scaling relations allow us to determine the dependence on time of the order parameters close to the critical point. For instance, the average density of active sites $\rho(t)$ starting from a homogeneous initial condition has to be scale invariant at the critical point. Using Eq. (3.151) we obtain

$$\rho(\Lambda^{-\nu_\parallel} t) = \Lambda^\beta \rho(t). \tag{3.153}$$

We can choose Λ such that $\Lambda^{-\nu_\parallel} t = 1$ and we immediately obtain how this quantity decays in time close to the critical point,

$$\rho(t) = t^{-\beta/\nu_\parallel} \rho(1) \sim t^{-\delta}. \tag{3.154}$$

In a similar way, we can obtain the decay of the survival probability $P(t)$,

$$P(t) \sim t^{-\delta'}, \tag{3.155}$$

with $\delta' = \beta'/\nu_\parallel$. Since in DP processes $\beta = \beta'$, we have also $\delta = \delta'$ and the hyperscaling relation (3.150) simplifies to

$$\theta = \frac{d\nu_\perp - 2\beta}{\nu_\parallel}. \tag{3.156}$$

It is important to point out that the knowledge of the critical exponents provides relevant information about the behavior of the order parameters close to the critical point and in a finite-size system. Actually, in these conditions ρ and P should depend on three parameters, namely, the time t, the distance from the critical point Δ, and the system size V (equal to L, in $d = 1$). On the other hand, the property of scale invariance implies that one of these parameters can be expressed in terms of the others, thus yielding the expressions

$$\rho(t, \Delta, V) \sim t^{-\beta/\nu_\parallel} f(\Delta t^{1/\nu_\parallel}, t^{-d/z} V) \tag{3.157}$$

$$P(t, \Delta, V) \sim t^{-\beta'/\nu_\parallel} g(\Delta t^{1/\nu_\parallel}, t^{-d/z} V), \tag{3.158}$$

where f and g are suitable scaling functions whose explicit expression is unknown.

In Table 3.2 we list the known values of DP critical exponents in different space dimensions. The mean-field theory, corresponding to the infinite-dimensional case ($d = \infty$), but whose exponents are known to be exact above the upper critical dimension, will be considered in the next section.

3.6.3 Mean-Field Theory

In analogy with equilibrium phase transitions, it has been conjectured that the notion of universality applies also to continuous nonequilibrium phase transitions. This means

Table 3.2 Numerical and mean-field critical exponents for the DP universality class

Exponent	$d = 1$	$d = 2$	$d = 3$	Mean Field
β	0.276	0.583	0.813	1
$\nu_\perp$	1.097	0.733	0.584	$\frac{1}{2}$
$\nu_\parallel$	1.734	1.295	1.110	1
θ	0.314	0.229	0.114	[a]

[a]The hyperscaling relation is not valid in the mean-field regime.

that scaling properties characterizing the behavior close to the critical point depend on basic properties only and are independent of the details of the model at hand. It is worth pointing out that such a conjecture has been successfully checked in many cases by careful numerical studies, but a rigorous mathematical proof is still unknown, because the dynamical renormalization group method encounters more serious technical difficulties to be worked out than the static one.

For DP processes, they are characterized by a transition from a fluctuating active phase to a unique absorbing state and the order parameter is a nonnegative scalar quantity, while the evolution rule involves short-range interactions (like those between nearest-neighbor sites in the DK model). No other symmetries or conservation laws should be present in the model. In fact, models sharing such basic properties have been found to belong to the DP universality class.

By further extending the analogy with equilibrium phase transitions, we can wonder if there is a general mean-field-like approach to DP processes, analogous to the field-theoretic formulation of critical phenomena provided by the Landau theory. Since we want to deal with nonequilibrium processes, where time plays a crucial role, we certainly need a sort of mean-field dynamical equation. In the previous chapters we have widely discussed the Langevin equation as an effective continuous-time, coarse-grained formulation of microscopic stochastic processes (e.g., the random-walk model of diffusion). We can proceed in a similar way for DP or contact processes by introducing a phenomenological Langevin equation for the density of active sites at position $\mathbf{x}$ at time t, $\rho(\mathbf{x}, t)$. In full generality, we assume here that $\mathbf{x}$ is a vector in d space dimensions. Due to the hypothesis of scaling invariance at the critical point, we have to think about the quantity $\rho(\mathbf{x}, t)$ as a space–time coarse-grained average of the number of active sites in the microscopic configuration. Rather than following a rigorous mathematical procedure based on the master equation of the contact process, here we prefer to derive this equation by heuristic arguments. Let us start by making reference to the microscopic mechanism of contact processes, because this allows us to set aside for a moment the diffusion processes. According to Eqs. (3.137), r is the recovery rate of an infected site, while the infection rate of a site is equal to $\lambda(n/2d)$, therefore being proportional to the fraction of infected neighbor sites. In a mean-field formulation the infection process occurs at a rate $\lambda\rho(1 - \rho)$, because it requires the site in question to be inactive (which occurs with probability $1 - \rho$) and the fraction of infected neighbors is the density ρ itself. Even more simply, the recovery process occurs at a rate $r\rho$. We can sum up the different contributions and obtain the equation

$$\frac{\partial \rho}{\partial t} = \lambda\rho(1-\rho) - r\rho \equiv a\rho - \lambda\rho^2, \tag{3.159}$$

where $a = \lambda - r$. If $\lambda < r$, $\partial_t \rho < 0$, and the only steady solution is $\rho_1^* = 0$, corresponding to all inactive sites (the absorbing state). If $\lambda > r$, the absorbing state is unstable because $\partial_t \rho > 0$ for a small density of active sites, and a new steady solution appears, $\rho_2^* = a/\lambda = 1 - r/\lambda$, which is stable. Therefore, this simple mean-field picture provides a transition between an absorbing and an active phase at the critical threshold of the control parameter, $(\lambda/r)_c = 1$, which should be compared with the exact value $(\lambda/r)_c = 3.298$, determined numerically.

In the above formulation we have put aside diffusion with the justification that we are focusing on the contact process rather than on DP. In fact, a more formal derivation of the time evolution of ρ should take into account that the sites neighboring $\mathbf{x}$ are one lattice constant far from $\mathbf{x}$, so the density evaluated there should be Taylor expanded, which would give a diffusion-like term on the right-hand side of Eq. (3.159). Such a term would also be produced directly from microscopic diffusion processes in the case of DP.

Therefore, below we are going to modify Eq. (3.159) so as to take into account diffusion, but also the key ingredient of stochastic processes, responsible for the possible failure of a mean-field approach: noise, i.e., fluctuations with respect to the local, average value of $\rho(\mathbf{x}, t)$. The noise term is less trivial than we can expect on the basis of the Langevin equations we have studied until now. In fact, the dynamics of an absorbing state is a trivial, deterministic dynamics that simply reproduces the same state at all times: therefore, the noise term must be switched off in proximity to an absorbing state. This is possible because for contact/DP/DK models noise is not due to some "external" source as in the case of a Brownian particle; instead, it is due to the density itself of the active sites and must vanish when $\rho(\mathbf{x}, t) = 0$. We can now write a phenomenological Langevin equation for contact processes as

$$\frac{\partial \rho(\mathbf{x}, t)}{\partial t} = a\,\rho(\mathbf{x}, t) - \lambda\rho^2(\mathbf{x}, t) + D\nabla^2\rho(\mathbf{x}, t) + \eta(\mathbf{x}, t), \tag{3.160}$$

where D is a diffusion constant and the stochastic field $\eta(\mathbf{x}, t)$ amounts to a zero average process with a correlation function proportional to $\rho(\mathbf{x}, t)$,

$$\langle \eta(\mathbf{x}, t)\rangle = 0 \tag{3.161}$$

$$\langle \eta(\mathbf{x}, t)\eta(\mathbf{x}', t')\rangle = \Gamma\rho(\mathbf{x}, t)\delta(\mathbf{x} - \mathbf{x}')\delta(t - t'), \tag{3.162}$$

where Γ has the physical dimension of inverse time.

The last relation indicates that the stochastic field $\eta(\mathbf{x}, t)$ is proportional[26] to $\sqrt{\rho(\mathbf{x}, t)}$. This square-root dependence can be argued to be a consequence of the central limit theorem. More precisely, since we are assuming that $\rho(\mathbf{x}, t)$ is a mesoscopic representation of the density of active sites, the large number of individual, independent noise contributions emerging from the microscopic stochastic process on a coarse-grained scale sum up to a Gaussian distribution, with a variance proportional to the number of active sites in this mesoscopic region.

[26] For this reason the noise is sometimes written as $\eta(\mathbf{x}, t) = \sqrt{\rho(\mathbf{x}, t)}\bar{\eta}(\mathbf{x}, t)$, where $\bar{\eta}(\mathbf{x}, t)$ is delta-correlated.

The stochastic field $\eta(\mathbf{x}, t)$ is an example of multiplicative noise, where the effective amplitude of fluctuations is modulated by the density field $\rho(\mathbf{x}, t)$ itself: a peculiar situation where one of the competing "ground states" of the phase transition is set at zero temperature. This feature modifies significantly the nature of the Langevin equation with respect to the standard case of a purely additive noise (as in the model of generalized Brownian motion), and it has a crucial influence on the critical properties of contact processes.

The next step is to apply the scale transformations (3.151) to Eq. (3.160), taking into account that the distance Δ from the critical point is now equal to $a = \lambda - r$ because the critical point is defined precisely by the condition $a = 0$. With the condition $\Lambda = a$ in mind, the rescaled equation (3.160) is

$$\Lambda^{\beta+\nu_\parallel}\frac{\partial\rho(\mathbf{x},t)}{\partial t} = \Lambda^{\beta+1}\rho(\mathbf{x},t) - \lambda\Lambda^{2\beta}\rho^2(\mathbf{x},t) + D\Lambda^{\beta+2\nu_\perp}\nabla^2\rho(\mathbf{x},t) + \Lambda^{\gamma}\eta(\mathbf{x},t), \quad (3.163)$$

where the exponent $\gamma = (\beta + d\nu_\perp + \nu_\parallel)/2$ stems from Eq. (3.162), while taking into account the property of the Dirac delta distribution $\delta(cx) = \frac{1}{|c|}\delta(x)$.

Dividing all terms by $\Lambda^{\beta+\nu_\parallel}$ we obtain

$$\frac{\partial\rho(\mathbf{x},t)}{\partial t} = \Lambda^{1-\nu_\parallel}\rho(\mathbf{x},t) - \lambda\Lambda^{\beta-\nu_\parallel}\rho^2(\mathbf{x},t) + D\Lambda^{2\nu_\perp-\nu_\parallel}\nabla^2\rho(\mathbf{x},t) + \Lambda^{\gamma-\beta-\nu_\parallel}\eta(\mathbf{x},t). \quad (3.164)$$

If we impose that the deterministic part of the Langevin equation is invariant under scale transformations, we must impose the vanishing of the relative scaling exponents, which allows us to obtain

$$\beta = 1, \quad \nu_\perp = \frac{1}{2}, \quad \nu_\parallel = 1. \quad (3.165)$$

These values define the mean-field critical exponents for the DP universality class. Their validity depends on the possibility of disregarding fluctuations, which in turn depends on the exponent of the term renormalizing the noise. In fact, we expect noise is irrelevant if

$$\gamma - \beta - \nu_\parallel > 0, \quad (3.166)$$

i.e.,

$$d > \frac{\beta + \nu_\parallel}{\nu_\perp}. \quad (3.167)$$

Using the mean-field values (3.165) we obtain the condition $d > 4$.

In conclusion, dimensional analysis indicates that noise is irrelevant for $d > 4$ (i.e., approaching the critical point it scales to zero as a positive power of Λ); it is marginal for $d = 4$; and it is relevant for $d < 4$. Since, as in equilibrium phase transitions, fluctuations determine the critical behavior, these considerations amount to obtaining the Ginzburg criterion for identifying the upper critical dimension, $d_c = 4$, for contact processes. Accordingly, we expect that the mean-field critical exponents become exact only for $d > 4$, while for lower dimension they are expected to depart significantly from exact ones, as shown in Table 3.2. For $d = 4$, we expect subleading logarithmic corrections to mean-field predictions.

Above we have not mentioned the exponent θ because it should be determined by the hyperscaling relation (3.150), but the latter is not valid in the mean-field regime. In practice, such a relation can be used with mean-field exponents for $d = 4$ only. In this case, we obtain $\theta = (d-4)/2 = 0$, which is in agreement with the observed decrease of $\theta(d)$ for $d < 4$.

3.7 Bibliographic Notes

An extensive and basic illustration of critical phenomena, Landau theory, mean-field approximations, and renormalization group can be found in K. Huang, *Statistical Mechanics*, 2nd ed. (Wiley, 1987). A deeper and more specialized insight about renormalization group methods is contained in D. Amit, *Field Theory, the Renormalization Group, and Critical Phenomena*, 2nd ed. (World Scientific Publishing, 1984), while a historical introduction to critical phenomena is given in the book by C. Domb, *The Critical Point* (CRC Press, 1996).

A complete classification of phase transitions in two space dimensions relies on the solid theoretical ground of conformal field theory. Details can be found in the book by M. Henkel, *Conformal Invariance and Critical Phenomena* (Springer, 1999). See also the pioneering papers by A. A. Belavin, A. M. Polyakov, and A. B. Zamolodchikov, Infinite conformal symmetry in two-dimensional quantum field-theory, *Nuclear Physics B*, **241** (1984) 333–380, and J. L. Cardy, Conformal invariance and universality in finite-size scaling, *Journal of Physics A: Mathematical and General*, **17** (1984) L385–L388.

A more recent reference book about nonequilibrium phase transitions is M. Henkel, H. Hinrichsen, and S. Lübeck, *Non-Equilibrium Phase Transitions,* volume 1*: Absorbing Phase Transitions* (Springer, 2008).

The so-called DP conjecture about the basic ingredients characterizing the universality class of DP is contained in the papers by H.-K. Janssen, On the nonequilibrium phase transition in reaction–diffusion systems with an absorbing stationary state, *Zeitschrift für Physik B*, **42** (1981) 151–154, and by P. Grassberger, On phase transitions in Schlögl's second model, *Zeitschrift für Physik B*, **47** (1982) 365–374.

An account about the many facets concerning driven lattice gases, reaction–diffusion, catalysis, and contact processes is contained in the book by J. Marro and R. Dickman, *Nonequilibrium Phase Transitions in Lattice Models* (Cambridge University Press, 1999).

The computational problems concerning the study of the critical properties of directed percolation are illustrated in the paper by I. Jensen, Low-density series expansions for directed percolation: I. A new efficient algorithm with applications to the square lattice, *Journal of Physics A: Mathematical and General*, **32** (1999) 5233–5250.

A detailed overview about percolation processes in the framework of a field theory approach can be found in H. K. Janssen and U. C. Täuber, *Annals of Physics*, **315** (2005) 147–192.

Field theoretical arguments about nonequilibrium hyperscaling relations are discussed in the paper by J. F. F. Mendes, R. Dickman, M. Henkel, and M. C. Marques, Generalized scaling for models with multiple absorbing states, *Journal of Physics A: Mathematical and Theoretical*, **27** (1994) 3019–3028.

For what concerns the determination of the value of the critical point in the contact process, see the papers by R. Dickman and I. Jensen, Time-dependent perturbation theory for nonequilibrium lattice models, *Physical Review Letters*, **67** (1991) 2391–2394, and R. Dickman and J. K. da Silva, Moment ratios for absorbing–state phase transitions, *Physical Review E*, **58** (1998) 4266–4270.

A review on deterministic cellular automata with selected papers can be found in S. Wolfram, *Theory and Applications of Cellular Automata* (World Scientific Publishing, 1986). A more recent and comprehensive illustration of this field of mathematics, including probabilistic models, can be found in A. Ilachinski, *Cellular Automata: A Discrete Universe* (World Scientific Publishing, 2001).

4 Out-of-Equilibrium Critical Phenomena

In the previous chapter we have emphasized the similarities between equilibrium and nonequilibrium phase transitions, stressing common ideas and concepts. Equilibrium phenomena have been described using the Ising model and its continuum counterpart, based on the Landau Hamiltonian. Nonequilibrium phenomena have then been illustrated using directed percolation. In this chapter we go beyond DP, discussing what parameters are relevant or irrelevant in the sense of universality classes.

Therefore, the first part of this chapter focuses on absorbing phase transitions, resuming DP and extending it to different universality classes, characterized by a larger number of absorbing states or by additional symmetries. The final part instead starts from the classical model of Katz, Lebowitz, and Spohn to discuss a clear nonequilibrium symmetry breaking, the bridge model, where two classes of particles are driven in opposite directions with the same dynamics and one current of particles can finally prevail over the other, similar to what occurs in the Ising model, where one class of spins (either up or down) may prevail over the other and induce a symmetry breaking. The central part of this chapter discusses self-organized criticality (SOC): on the one hand, it has a great interest by itself; on the other hand, it may be considered a sort of bridge between absorbing and driven phase transitions, because SOC models share something of both features.

Since nonequilibrium phase transitions are a relatively recent research domain, it is not possible to present it systematically and we do not pretend to offer a common guiding thread among all sections of this chapter. Nonetheless, we tried to make connections and show similarities whenever possible.

4.1 Introduction

The directed percolation (DP) process represents an example of nonequilibrium phase transition of the utmost importance. It is no exaggeration to say it plays a role similar to the Ising model in equilibrium phase transitions, both for historical reasons and because of its full generality: any absorbing phase transition with just one absorbing state (or more but inequivalent absorbing states) is expected to fall into DP universality class. There is, however, one relevant difference: the Ising model has been solved exactly in both one and two spatial dimensions, while DP has not been even in $d = 1$, which also explains the continuing current interest for DP. These reasons explain why we focused the nonequilibrium part of the previous chapter on DP. The Domany–Kinzel cellular automata support the generality of the DP universality class because of the variety of different physical interpretations we can give to the evolution rules and because the parameter q is found to be irrelevant except for $q = 1$, when the model has two equivalent absorbing states. This fact suggests that the number of absorbing states is expected to be a relevant parameter.

We are therefore encouraged to ask an ambitious question: is it possible to classify nonequilibrium phase transitions, similar to equilibrium ones? Actually not, but the first step toward some classification is certainly to identify which features of a given model are relevant and which are not. The physical dimension of the space is certainly a relevant one and it is tempting to think that the number of equivalent absorbing states might play a role similar to the dimension of the order parameter in $O(n)$ equilibrium models. However, no spontaneous symmetry breaking occurs in absorbing phase transitions, and an external symmetry breaking (e.g., induced by a field) removes the equilibrium phase transition, while it restores the universality class, which is typical of a single absorbing state, in nonequilibrium. Examples of nonequilibrium symmetry breaking exist indeed and we will discuss in detail one example, using a simplified version of a driven lattice gas.

A further, relevant feature when studying nonequilibrium phase transitions (and, more generally, nonequilibrium phenomena) is the possible existence of conservation laws. This is in contrast to equilibrium phase transitions; just think of the Ising model and its interpretation as a magnet (with a nonconserved order parameter, the magnetization) and a lattice gas (with a conserved order parameter, the density): the equilibrium universality class is the same. We provide a simple example that in nonequilibrium, things work differently, but more convincing pieces of evidence of the importance of conservation laws in nonequilibrium systems can be found in the following chapters.

Absorbing and driven systems share the relevant feature (also proper to equilibrium systems) that it is necessary to tune a suitable control parameter in order to obtain criticality. Instead, in self-organized critical systems it is the dynamics itself leading the system to the critical point, where the relevant quantities characterizing the nonequilibrium steady state (NESS) attained by the system present self-similarity and power-law behavior. We will recognize these features in the next chapter as well, when studying kinetic roughening.

4.2 Beyond the DP Universality Class

In the previous chapter we focused on the model of DP because it can be formulated in simple terms and because many other models share the same long-range properties, therefore the same critical exponents: the Domany–Kinzel cellular automata and the contact processes are two relevant examples and all these models belong to the DP universality class. We now want to discuss other universality classes, but in order to go beyond DP it is necessary to introduce some relevant changes. One possibility has already been mentioned in Section 3.5.2, where we have shown that for $q = 1$ the DK model passes from one to two absorbing states. A second possibility is to introduce a conservation law. These two new universality classes are discussed in Sections 4.2.1 and 4.2.2, respectively.

4.2.1 More Absorbing States

Compact Directed Percolation

The DK cellular automata trivially have one absorbing state, where all sites are inactive (∘), because ∘∘ → ∘. However, for $q = 1$ the same process with active sites occurs too, •• → •, implying that the fully active state is an absorbing state as well. Even more important, there is a symmetry between the fully inactive and the fully active state, because dynamics is invariant under the exchange of particles and holes if $p \to 1 - p$. This symmetry implies that it must be $p_c(q = 1) = 1 - p_c(q = 1)$, i.e., $p_c(1) = \frac{1}{2}$.

The simplest way to show that the case $q = 1$ (called compact directed percolation, CDP) lies in a universality class other than DP is to prove that $\beta(q = 1) \neq \beta'(q = 1)$. The CDP model can be rephrased in terms of domain walls between active and inactive regions. In fact, new walls cannot be created, because this would require the formation of inactive (active) sites in the middle of an active (inactive) region, which is forbidden by the absorbing character of both states: this is the reason why this is called compact directed percolation. Since the creation of domain walls is prohibited, their number can only decrease through annihilation processes, which occur when an isolated active or inactive site disappears.

The exponent β' can be easily evaluated by considering the evolution of a single active site, which produces a compact cluster of width $L(t)$, and we wonder what the probability $P(p)$ is that it grows to infinity rather than dying out. In close proximity to p_c such probability has the expression $P(p) \approx (p - p_c)^{\beta'}$, which defines the sought-after exponent β'. Using the evolution rules of DK cellular automata for $q = 1$ (see Fig. 3.10), we can say that a cluster of L active sites at time t: (i) increases its size at time $t + 1$ if both processes occurring at the bounds of the cluster produce active sites, ∘• → • and •∘ → •, (ii) decreases its size if both such processes produce inactive sites, and (iii) maintains its size if one process produces an active site and the other process produces an inactive site. Thus,

$$L(t+1) = \begin{cases} L(t) + 1, & \text{with probability } p^2 \\ L(t) - 1, & \text{with probability } (1-p)^2 \\ L(t), & \text{with probability } 2p(1-p). \end{cases} \tag{4.1}$$

We therefore have that $L(t)$ performs an asymmetric random walk with a probability $r = p^2/(p^2 + (1-p)^2)$ to move to the right and a probability $1 - r$ to move to the left,[1] see Fig. 1.10. We wonder what the probability is that $L(t = \infty) = \infty$ knowing that $L(0) = 1$ and that the random walk has an absorbing barrier for $L = 0$. The answer is given in Section 1.5.2, Eq. (1.112),

$$P(p) \equiv 1 - \mathfrak{p}_2 = 1 - \left(\frac{1-r}{r}\right) = \frac{2}{p^2}\left(p - \frac{1}{2}\right). \tag{4.2}$$

Close to the critical value $p_c = \frac{1}{2}$ we have $P(p) \simeq \frac{2}{p_c^2}(p - p_c)$, so that $\beta' = 1$.

[1] The probability to stay the same value of L does not affect the evaluation of $P(p)$; see Eq. (4.2).

The exponent β instead refers to the asymptotic, time-independent fraction of active sites when we start from a homogeneous state with a finite density of active sites, $\rho(p) \simeq (p - p_c)^\beta$. In the language of domain walls, at $t = 0$ we have an ensemble of particles separating active and inactive regions, and for $p > p_c$ they diffuse so as to favor the active phase with respect to the inactive one: the final state is fully active, so $\rho(p > p_c) = 1$ and a discontinuity appears, which means $\beta = 0$. In conclusion, $\beta = 0$ and $\beta' = 1$, proving that CDP does not belong to the DP universality class, whose critical exponents are given in Table 3.2 (in particular, $\beta(\text{DP}) = \beta'(\text{DP}) \simeq 0.276$ in $d = 1$).

Also the other critical exponents can be found by using the properties of random walkers. For simplicity we are confined to an initial condition corresponding to a single active site, whose dynamics is summarized in Eq. (4.1). The exponent θ rules how the size L of an active region increases in time at criticality ($p = p_c$), $L(t) \sim t^\theta$. In Appendix D.5 we determine $L(t)$ in a continuum picture, using the equivalence of the CDP model with a random walker on the line, starting at position $x_0 = 1$ (because $L(0) = 1$) and with a trap in $x = 0$ (because the cluster dies). At criticality, where the walker diffuses symmetrically (see Eqs. (4.1)), the result is that the average value of $L(t)$, corresponding to the average position of the walker, is constant in time. Therefore, $\theta = 0$.

The exponents $\nu_\parallel$ and $\nu_\perp$ describe the divergence of the correlation lengths. Below criticality they are defined as $\xi_{\parallel,\perp} \sim (p_c - p)^{-\nu_{\parallel,\perp}}$ and the quantity $\xi_\parallel$ is nothing but the average lifetime of the walker, while $\xi_\perp$ is the average maximal distance attained before the process dies out; see Fig. 3.14 for a graphical representation. In Appendix D.6, Eq. (D.71), we find that in the limit of small drift v the trapping time is inversely proportional to v. The drift is proportional to the asymmetry δ,[2] defined as the normalized difference between the rate of hopping to the right and the rate of hopping to the left (see Eq. (4.1)),

$$\delta = \frac{p^2 - (1-p)^2}{p^2 + (1-p)^2} = -\frac{1 - 2p}{1 + 2p(1-p)}. \tag{4.3}$$

In proximity to the critical value $p_c = \frac{1}{2}$, $\delta \simeq -(p_c - p)$. Summarizing, $\xi_\parallel \sim v^{-1} \sim |\delta|^{-1} \sim (p_c - p)^{-1}$, so $\nu_\parallel = 1$. As for the orthogonal correlation length, $\xi_\perp$, it is sufficient to note that during the time $\xi_\parallel$ the size of the cluster performs a standard random walk, therefore attaining a maximal size of the order $\xi_\perp \sim \xi_\parallel^{1/2} \sim (p_c - p)^{-1/2}$, so $\nu_\perp = \frac{1}{2}$.

In conclusion, for the CDP model we have the following critical exponents: $\beta = 0$, $\beta' = 1$, $\theta = 0$, $\nu_\parallel = 1$, and $\nu_\perp = \frac{1}{2}$. Comparing with Table 3.2 we see that most of them are equal to mean-field values, with the notable exception that $\beta \neq \beta'$ for CDP.

More Inactive States

The CDP model has the value of being exactly solvable, but the feature of compact clusters limits its generality. For this reason we mention another change of universality class induced by having more than one absorbing state. If we pass from DP to CDP making the fully active state an absorbing state, a different scenario is to have more than one inactive state (which is absorbing by definition). We may think to have two inactive states

[2] See the problem of random walk on a ring in Section 1.5.2.

(I_1 and I_2) and one active state (A). As in DP, I_k sites can reproduce only themselves ($P(I_k|I_k, I_k) = 1$) and an (active, inactive) pair can produce an active site ($P(A|I_k, A) = p$) or the *same* inactive site ($P(I_k|I_k, A) = 1 - p$). Furthermore, two active sites can reproduce ($P(A|A, A) = q$) or give rise to an inactive one ($P(I_1|A, A) = P(I_2|A, A) = (1 - q)/2$). These rules are the natural generalization of DK evolution rules (see Eq. (3.135)), but we also need to specify what the result is of a pair of different inactive sites. The model DP2 corresponds to implementing a sort of interfacial noise, with $P(A|I_1, I_2) = P(A|I_2, I_1) = 1$.

It is worth mentioning that the two inactive (absorbing) states are equivalent and this is what makes DP2 a universality class different from DP. If we create an asymmetry between I_1 and I_2, for example, by imposing $P(I_1|A, A) > P(I_2|A, A)$, this asymmetry would induce a preference for I_1 that would rule over I_2 and the model would fall in the DP universality class.

Contact Processes

In the previous chapter we have shown that stochastic processes like DP can be also interpreted as contact processes modeling, e.g., an epidemic spreading: the activation of a site in a lattice by an active neighboring looks quite similar to the mechanism of infection transmission by contact from an infected individual to a healthy one. In fact, the probability per unit time of activating a neighboring site can be read as the infection rate of other individuals. Moreover, the probability per unit time that an active site turns into an inactive state is analogous to the rate of immunization of infected individuals. On the basis of these simple considerations one can appreciate the strong similarity of DP to epidemic spreading processes. On the other hand, epidemic spreading mechanisms, in reality, depend on additional ingredients. For instance, real epidemics spread in a highly disordered environment (rather than in a regular lattice), where individuals may have different responses to infection, according to their attitude of adopting immunization strategies, like vaccination. Moreover, infection can be transmitted by short-range as well as long-range interactions. It is well known that many epidemics may propagate through international travel via the transport of pathogens to geographical areas different from their origin.

In general, epidemic processes without immunization belong to the universality class of DP. As soon as immunization is introduced, the overall mechanism changes. In fact, we can assume that there are two different probabilities, p_1 and p_2 for the first and the second infection, respectively. According to common experience, the condition $p_1 > p_2$ is usually assumed. As soon as we introduce immunization we obtain an infinite number of absorbing states, because any combination of healthy individuals (either immunized or not) cannot evolve.

Perfect immunization corresponds to the case $p_2 = 0$, where previously infected sites, once recovered, cannot be infected any longer. This process is called dynamical percolation (DyP) and we can exemplify this spreading process by considering an initial state where all sites are equally susceptible to infection and inserting a single infected site (the so-called patient zero) at the origin: how does the epidemic spread through the lattice? Since previously infected sites become immune, the infection front propagates and leaves behind a cluster of immune sites, but the precise form of this cluster depends on the infection probability p_1. In the supercritical phase ($p_1 > p_{1c}$) there is a finite probability that the infected front propagates to infinity, while in the subcritical phase the infection process eventually stops after a finite time. The DyP universality class is robust with respect to a small value of the reinfection rate $p_2 > 0$.

4.2.2 Conservation Laws

A different way to modify a DP model in order to enter a new universality class is to add a symmetry. The simplest example makes use of the following representation of DP: an active site can activate neighboring sites ($A \to A + A$), it can deactivate ($A + A \to A$), and it can diffuse ($A0 \to 0A$). We can modify this dynamics in order to conserve the parity of the number of active sites, replacing previous rules with the following: ($A \to A + A + A$), ($A + A \to 0$), and ($A0 \to 0A$). An alternative, operational definition in terms of particle dynamics can also be given.

Particles are located on a regular one-dimensional lattice and one particle at a time is randomly chosen. The particle can either diffuse to a nearby site with probability p or generate two offspring at the neighboring sites with probability $1-p$. In the former case, the particle moves with equal probability to the left site or to the right one: if the selected site is occupied by another particle they annihilate with probability r; otherwise, the particle remains in its site. In the latter case, the two offspring occupy the two nearest-neighbor sites. If one or both sites are occupied by particles, they annihilate with probability r; otherwise, no new birth occurs.

As we have defined the model with two offspring we might also define it with just one offspring (or more than two). This is not irrelevant because two offspring preserve the parity of the total number N of particles, $N \to N \pm 2$, while this is not true for just one offspring.

The parity conserving (PC) model has been found to define a new universality class. In order to prove that the change of universality class is really due to the additional symmetry, it is enough to define the above particle model with just one offspring and verify that it falls in the standard DP universality class.

Another effect of the conservation law is that a real absorbing state, the empty lattice due to the complete annihilation of particles, may occur only if N is even. In this case, by tuning the control parameters p and r it is possible to pass from the absorbing state to an active fluctuating phase, characterized by a finite, constant density of particles. Instead, if the initial number of particles is odd, the absorbing phase is replaced by a time-dependent phase, where the density of particles vanishes in the thermodynamic limit.

Numerical estimates of the critical exponents in 1+1 dimensions indicate that this PC class is quite different from the DP class of nonequilibrium critical phenomena, as shown by the values

$$\beta = \beta' \approx 0.92, \quad \nu_\parallel \approx 3.22, \quad \nu_\perp \approx 1.83. \tag{4.4}$$

The exponents θ and δ, which are defined at the critical point, depend on the initial condition. For instance, if we start with one single particle (which can never be destroyed), $\delta = 0$ (so the scaling relation $\delta = \beta/\nu_\parallel$ does not hold) and $\theta \approx 0.285$. On the other hand, if one starts with two initially active sites, quite surprisingly the roles of δ and θ are exchanged. Moreover, the relaxation properties in the inactive phase have been found to exhibit a power-law dependence in time ($\rho(t) \sim 1/\sqrt{t}$) rather than an exponential one, at variance with the DP class, where $\rho(t) \sim e^{-t/\xi_\parallel}$.

4.3 Self-Organized Critical Models

The critical phenomena studied up to now both of equilibrium and of nonequilibrium had the common feature that the fine tuning of a suitable order parameter was necessary in order to get a critical behavior, characterized by scale invariance. There are, however, driven-dissipative systems that spontaneously evolve toward a critical dynamics, characterized by a power-law distribution of relaxation events, and this phenomenon does not require the tuning of any parameter, hence it is named self-organized criticality (SOC).

This phenomenon was discovered in some seminal papers by Per Bak and coworkers at the end of the 1980s. Since then, several models have been proposed to provide a mathematical theory for a class of phenomena that seems to be ubiquitous in nature. In fact, SOC phenomena have been identified in cosmology, evolutionary biology, plasma physics, sociology, neurobiology, etc. For instance, already in the 1950s Beno Gutenberg and Charles Francis Richter had found an empirical law according to which the probability to have (in any given region and time period) an earthquake of at least magnitude E obeys a universal power-law behavior,

$$P(E) \sim E^{-\gamma}, \tag{4.5}$$

over quite a large range of values of E, with $\gamma \approx 1$.

The main consequence of this behavior is that the magnitude distribution of earthquakes is not characterized by a typical size, which obeys the Gaussian statistics imposed by the central limit theorem (see Appendix A), as happens in random independent events. We must therefore conclude that earthquakes are not random independent events and their occurrence is correlated with previous events through a sort of memory effect. The bad news is that a power-law distribution implies a nonnegligible probability that an earthquake magnitude is sensibly large.

On a physical ground, we can observe that the basic mechanism for producing SOC behavior is the presence of a slow driving process allowing for the accumulation of energy or matter, which is suddenly released through an avalanche-type process and which leads toward an absorbing state. Therefore, a typical feature of SOC is the wide separation between the driving and the relaxation time scales. Once we inject new energy or matter, the whole process restarts and leads to a new absorbing state. For this reason it has been argued that SOC is closely related to nonequilibrium phase transitions into infinitely many absorbing states. In this sense, SOC can be viewed as an extreme case of the theory of nonequilibrium critical phenomena.

Before going on to describe a few systems displaying SOC it is worth starting with a model without driving nor dissipation, i.e., a model of fixed energy or mass density ϵ,

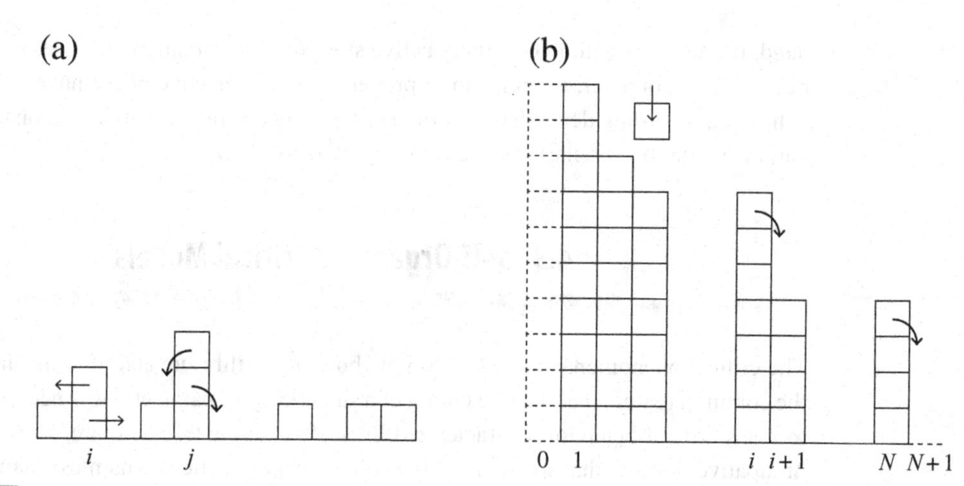

Fig. 4.1 (a) Sketch of the model, without driving nor dissipation, introduced in Section 4.3 and preparatory to SOC models. We distribute N bricks on a lattice of fixed size and dimension d ($d = 1$ in the figure). Sites whose height h is larger than or equal to a given value h^* ($h^* = 2$ in the figure) are active sites. An active site is chosen randomly and its height is reduced by two bricks, which are distributed one per neighboring site. In the specific case, there are two active sites, i and j, and such redistributions make active the sites $i - 1$ and $j - 1$. (b) The Bak-Tang-Wiesenfield (BTW) model, with the three basic processes occurring during its dynamics: the toppling of a new sand grain (driving) on a column; the move of a sand grain from site i to site $i + 1$, if $z_i = h_i - h_{i+1} > Z$ ($Z = 2$ in the figure); the removal of a sand grain (dissipation) from site N, if $h_N > Z$. Dashed lines depict boundary conditions, introducing fictitious sites in $i = 0$ (with $h_0 = h_1$) and in $i = N + 1$ (with $h_{N+1} = 0$).

which acts as control parameter. Let us consider a d-dimensional square lattice of linear size L, where we distribute N bricks on the L^d sites. At the end of the process, each site has a height h_i and is defined to be an active site if $h_i \geq h^*$ (see Fig. 4.1(a)). Dynamics proceeds as follows: an active site is randomly chosen and its height is reduced by the quantity h^*, which is distributed among neighboring sites (periodic boundary conditions are usually assumed). The simplest rule is to choose $h^* = 2d$ and distribute the $2d$ bricks, one for each neighboring site.

For small $\epsilon = N/L^d$ we expect either a frozen dynamics (no active site exists) or that dynamics lasts for a very short time (a vanishing time, in the thermodynamic limit). For large ϵ instead, the cascade following the process $h_i \to h_i - h^*$ has a finite probability to activate new sites, producing a self-sustaining process with continuous cascades. Therefore, we have a usual nonequilibrium phase transition from an absorbing phase to an active phase, at a critical value ϵ_c. We can now switch on dissipation, allowing bricks to leave the system through its boundaries. If $\epsilon > \epsilon_c$, because of dissipation ϵ is going to decrease until $\epsilon < \epsilon_c$, when the system attains an absorbing state and dynamics stops. However, if we also switch on driving, i.e., the inflow of new energy or matter, ϵ can increase, exceed the critical value, and be back in the active phase.

When both inflow and outflow are allowed, the system places itself at the critical point, regardless of $\epsilon(t = 0)$ is, and dynamics can be interpreted as a continuous passage through the critical point, in both directions. Such ability to attain the critical point independently is the most relevant feature of SOC. However, it should be stressed again that this is due to the perfect separation of time scales, with dissipation much faster than driving (i.e., accumulation of energy or mass).

In this section we are not going to enter into a detailed account of SOC phenomena; we illustrate them by two pedagogical examples that exemplify the main classes of SOC models. The first example is the Bak–Tang–Wiesenfeld (BTW) model of avalanches in sandpiles, introduced very schematically in Fig. 4.1(b). The driving mechanism is performed by adding sand grains at random positions, and its competition with the loss of particles through the boundaries of the sandpile drives the system to its self-organized critical behavior. The criterion for producing an avalanche is described in terms of the local slope of the sandpile (as is the case in a real one) rather than on the local height (as done, for the sake of simplicity, in Fig. 4.1(a)). Some quantitative aspects concerning this model are still not completely understood and are debated. The second example is the Bak–Sneppen model, where SOC emerges as a consequence of an extremal dynamic rule, which amounts to the extinction of the less fit species in an ecosystem. This model has been widely investigated and many of its quantitative aspects have been successfully determined by numerical and analytical methods.

4.3.1 The Bak–Tang–Wiesenfeld Model

The BTW model introduced by Per Bak, Chao Tang, and Kurt Wiesenfeld originally aimed at describing the phenomenon of the occurrence of avalanches in a sandpile. Pictorially, we can think about a sandpile contained in a box, where sand grains (particles) topple to lower heights and may eventually escape from the box boundaries. If the toppling process stops, we proceed by adding single sand grains at random positions until an avalanche forms.

The evolution rule amounts to a cellular automaton, whose bulk dynamics conserves the number of particles. For pedagogical reasons we describe the BTW rule in a $d = 1$ lattice made of N sites, since its extension to higher dimensions is straightforward. We define the integer height h_i ($i = 1, \dots, N$) of the sandpile at site i, i.e., the number of sand grains piled up at that position. The toppling process from site i to site $i+1$ occurs if the height difference $z_i = h_i - h_{i+1}$ exceeds a threshold value $Z > 0$. In this case a sand grain moves from site i to site $i+1$, thus yielding the evolution rules (see Fig. 4.1(b))

$$z_i \to z_i - 2 \tag{4.6}$$

$$z_{i\pm1} \to z_{i\pm1} + 1. \tag{4.7}$$

Notice that in this way the evolution of the system can reach an absorbing or stable state every time $z_i \leq Z \quad \forall i$, because the toppling process stops. Accordingly, the total number of stable states is Z^N. In order to reactivate the toppling process we add one sand grain at a randomly selected site i, so that

$$z_i \to z_i + 1 \tag{4.8}$$

$$z_{i-1} \to z_{i-1} - 1. \tag{4.9}$$

The boundary conditions are open on the right and closed on the left, because particles can leave the system from $i = N$ if $h_N > Z$, while they cannot roll to $i = 1$ from the left.

These boundary conditions can be formally implemented writing $z_0 = 0$ (i.e., $h_0 = h_1$) and $z_N = h_N$ (i.e., $h_{N+1} = 0$).

In one dimension the BTW model exhibits quite simple features. In fact, starting from an empty system and adding grains at random positions after a transient time the pile will eventually reach a state S_0, where $z_i = Z \quad \forall i$, which is the maximally unstable state among the Z^N absorbing states.[3] The addition of a grain at any position will produce the propagation of the toppling process from left to right, until the grain will leave the pile from the right boundary and the pile will be again in S_0. Said differently, in the $d = 1$ BTW model a local perturbation propagates through the entire lattice that always returns to its maximally unstable state. If we measure the size of avalanches as the number of sites that topple after the addition of a single grain, it is clear that this is a purely random variable, because this size is equivalent to the distance between the randomly selected position, where the sand grain is initially added, and the right boundary.

For $d > 1$ the BTW model exhibits completely different features, because the long-time robustness of S_0 is lost and the size n and duration τ of avalanches are found to be distributed according to power-law distributions,

$$P(n) \sim n^{-\theta}, \qquad P(\tau) \sim \tau^{-\eta}, \tag{4.10}$$

with $\theta \simeq 0.98$ in $d = 2$ and $\theta \simeq 1.35$ in $d = 3$, while $\eta \simeq 0.42$ in $d = 2$ and $\eta \simeq 0.90$ in $d = 3$.

We want to point out that one of the main difficulties associated with the study of this model is the control of finite-size effects, which are particularly disturbing when performing numerical simulations. In particular, the determination of the values of the exponent η in $d = 2, 3$ and even the robustness of this self-organized critical behavior is still the object of a long-standing debate. Here we want to mention that the BTW model has the merit of highlighting the existence of a new class of simple dynamical models, which exhibit interesting similarities with some natural phenomena. Moreover, despite their simplicity, this class of models seems to capture basic mechanisms that belong to a still widely unexplored realm of mathematical sciences.

4.3.2 The Bak–Sneppen Model

A power-law behavior like the Gutenberg and Richter one (see Eq. (4.5)), is also seen in biological phenomena. For instance, paleontological studies have revealed that the distribution of massive species extinctions as a function of the number of extinct species n obeys a power law of the type

$$P(n) \sim n^{-\delta}, \tag{4.11}$$

[3] The reader may think there is a contradiction in the wording. There are Z^N absorbing states and S_0 is named the maximally unstable one, because the adding of a single particle in any site leads to an avalanche, but at the end of the avalanche S_0 is restored. Instead, any other absorbing state may be stable with respect to the deposition process of one single particle, but once that dynamics is activated the system eventually reaches the state S_0.

where $\delta \approx 2$. It has been conjectured that this peculiar law could be explained as a correlation between massive extinction events and catastrophic events, like earthquakes that obey similar power-law statistics, which could have been at the origin of the extinction process. Anyway, we are still far from any clear evidence that could support such a conjecture. An alternative explanation could be traced back to the competition mechanism among species, which is the cornerstone of Darwin's evolutionary theory. On this basis, at the beginning of the 1990s Per Bak and Kim Sneppen proposed a very simple and seminal model of an ecosystem where N different species coexist. It can be considered the first example of the class of extremal models, because the evolution rule will be seen to involve the choice of the species with the lowest fitness.

In order to simplify the representation of the ecosystem, in the Bak–Sneppen (BS) model species are organized on a one-dimensional lattice made of N sites. The spatial organization is a crucial element of the model, because nearby species are assumed to have direct influence on each other, as in a prey–predator scenario or by reciprocal mutuality or parasitic relations. Anyway, the BS model does not specify the kind of interaction and each species is attributed a fitness value f, which can be viewed as a quantity measuring its adaptability to the ecosystem. In Darwin's language the fitness can be interpreted as the ability of each species to reproduce itself by taking advantage of the resources available in the ecosystem. The dynamics in this ecosystem is governed by a selection mechanism that takes the form of an extremal rule: the species i with lowest value of f_i is removed from the lattice site i and substituted there with a new species, whose fitness value is chosen at random from a uniform distribution that can be assumed, without prejudice of generality, to be in the interval $I = [0, 1]$. If that is all, the fitness has a clear drift toward higher values and the distribution of fitness is crushed upward, with each f_i asymptotically equal to one.

In order to account for an effective species interaction, the BS model assumes that this local extinction also modifies the fitness of the nearby species at site $i \pm 1$: again we attribute from the same uniform probability distribution two new values to $f_{i\pm1}$. This mechanism may induce an avalanche of extinctions of neighboring species, whose origin is less noticeable than that of the BTW model. In fact, in the present case, the random choice of fitness would seem to amount to the neutrality of such events, because the update of the fitness next to an extinct species may result in favorable or unfavorable events and in a completely uncorrelated way with the previous history.

For a better comprehension of the physical mechanism underlying avalanches in the BS model, we can focus on the resulting stationary state. Starting from a random initial condition, where the values of the f_i are extracted from the uniform probability distribution, numerical simulations show that after a sufficiently long transient evolution the dynamics eventually yields a state where the f_i are typically distributed above a threshold value $f_T = 2/3$. This threshold value changes with the dimension of the lattice and with the number of interacting species, although the overall scenario remains qualitatively the same.

Because of such steady distribution, two competing mechanisms appear to play a role in this oversimplified evolution model. Natural selection eliminates the less fit species, but the mutations induced in the nearby species oppose the evolutionary pressure, because nearby species with high fitness typically acquire a lower fitness as a consequence of the induced mutation. When the *lower* fitness is the *lowest* fitness, an avalanche starts. Let us discuss their statistics.

An avalanche is characterized by its duration τ and by the number of species n involved in the massive extinction event.[4] Computer simulations indicate that in both cases we obtain power-law distributions, namely,

$$P(\tau) \sim \tau^{-\eta} \tag{4.12}$$

and

$$P^*(n) \sim n^{-\theta}. \tag{4.13}$$

In particular, it has been found that $1 < \eta < 3/2$, thus yielding a divergent average duration of avalanches,[5]

$$\langle \tau \rangle = \int_1^\infty d\tau \tau P(\tau) \sim \tau^{2-\eta} \Big|_1^\infty = \infty. \tag{4.14}$$

The distributions of the duration τ and size n of avalanches are not independent, because it has been found that the size grows with the duration according to the power law

$$n(\tau) \sim \tau^{\mu}, \tag{4.15}$$

with $0 \leq \mu \leq 1$. Since the number of avalanches is a given quantity, independent of the variable used to characterize it, the two distributions have to fulfill the condition

$$P^*(n)dn = P(\tau)d\tau, \tag{4.16}$$

i.e.,

$$\frac{dn}{n^\theta} \sim \frac{d\tau}{\tau^\eta}. \tag{4.17}$$

Using Eq. (4.15) we find

$$\tau^{\mu-1}\tau^{\eta}d\tau \sim \tau^{\mu\theta}d\tau, \tag{4.18}$$

which provides a scaling relation between the exponents,

$$\theta = 1 + \frac{\eta - 1}{\mu}. \tag{4.19}$$

In $d = 1$ numerical estimates give $\eta \approx 1.073$ and $\mu \approx 0.42$, so that θ is slightly larger than 1 and both $\langle \tau \rangle$ and $\langle n \rangle$ diverge. These divergences (see also the note 5 above) simply indicate that there is a finite probability for a complete extinction of the ecosystem.

At this point we could open a long debate about the reliability of the BS model in providing a detailed description of realistic events. On the other hand, such a debate is of poor practical interest. In fact, the main message of the BS model is that a completely

[4] These quantities are not equivalent because the species occupying a given site can enter more than once in an avalanche.

[5] Note that a perfect separation of time scales is necessary in order to have power-law behaviors at any scale. In practice, this means that the duration τ of an avalanche must be much smaller than the time interval between two consecutive avalanches, which limits, in a real system, the maximal duration and size of an avalanche. This fact removes the unphysical divergences.

nontrivial scaling behavior is found to characterize a fairly simple evolutionary dynamics, which combines minimal ingredients such as randomness and fitness in an ecosystem.

4.4 The TASEP Model

The standard model introduced in the previous chapter is the simplest example of a system driven out of equilibrium. It is a lattice gas model where particles feel a force E favoring hops in the $+x$-direction, while opposing hops in the opposite, $-x$-direction. This force, along with boundary conditions, breaks detailed balance, so that steady states are not equilibrium states.

In the following we make things as simple as possible, by reducing the model to the bare minimum: in fact, in the limit $E \to \infty$ hops to the right are always accepted and hops to the left are always rejected. We also consider a one-dimensional system, so that the hops in directions perpendicular to x, not affected by the field, are absent. It is important to keep in mind that one-dimensional systems are very peculiar, because equilibrium phase transitions are not allowed for short-range interactions, as is the case here. However, we will show this is not true for nonequilibrium systems, so that our restriction to $d = 1$ makes the analysis simpler, but not too simple.

4.4.1 Periodic Boundary Conditions

The acronym for our model is TASEP, Totally ASymmetric Exclusion Process, whose meaning is straightforward: particles can move in only one direction (because the process is totally asymmetric) and each site can be occupied by one particle at the most (because of exclusion). We start with the simplest boundary conditions breaking detailed balance: periodic boundary conditions; see Fig. 4.2. The dynamical evolution works as follows. Starting from any configuration with N particles, we choose randomly a site k: if it is occupied ($n_k = 1$) and the site to the right is empty ($n_{k+1} = 0$), the particle moves from site k to site $k+1$. After some transient, dependent on the initial conditions, we can imagine the system sets in a nonequilibrium steady state, i.e., a state whose statistical properties do not depend on time. Such a steady state must satisfy Eq. (3.116) for any s. What are the properties of this state?

We can prove that for periodic boundary conditions the steady state is characterized by a nonequilibrium stationary probability $p^{\text{ne}}(s)$, which does not depend on s: all microscopic states have the same probability. This proof makes reference to Fig. 4.2, where a general configuration s is charaterized by the number m_s of blocks of particles. A block is a sequence of adjoining particles. According to Eqs. (3.116), $p^{\text{ne}}(s) =$ constant is a solution if

$$\sum_{s'} w_{s,s'} = \sum_{s'} w_{s',s}. \tag{4.20}$$

Since in the present model $w_{s,s'}$ is equal either to 0 (forbidden transition) or to 1 (allowed transition), the previous condition is equivalent to saying we can go from s toward a

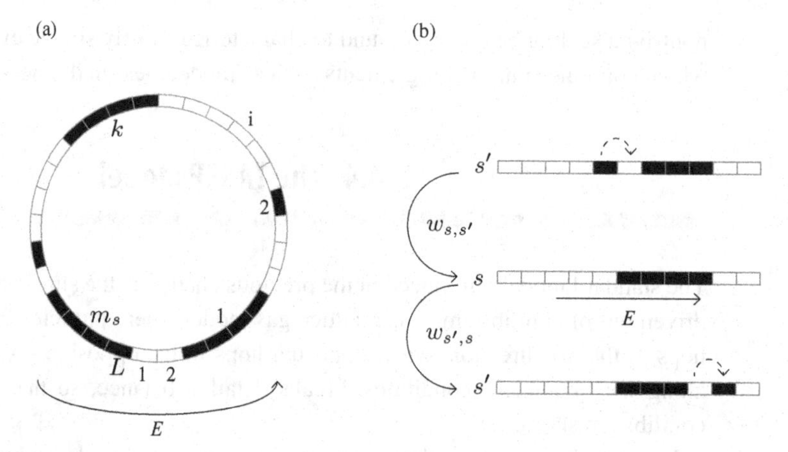

Fig. 4.2 The TASEP model with periodic boundary conditions. (a) Outer indexing refers to sites, $i = 1, 2, \ldots, L$. Inner indexing refers to blocks of particles, $k = 1, 2, \ldots, m_s$. (b) Detail of a single block of particles (center), with the configuration from which it originates (top) and the configuration to which it evolves (bottom).

number $N_{\text{out}}(s)$ of configurations s' that must be equal to the number $N_{\text{in}}(s)$ of configurations from which we can go to s. Just a word of caution–the number of arrival states is equal to the number of states of origin, but arrival states are different from states of origin. As a matter of fact, they *must* be different, because in TASEP $w_{s,s'} = 1$ necessary implies $w_{s',s} = 0$. This is another important difference with equilibrium dynamics, because if detailed balance is satisfied, at $T \neq 0$, $w_{s,s'} \neq 0$ implies $w_{s',s} \neq 0$.

Now we are going to prove that both $N_{\text{out}}(s)$ and $N_{\text{in}}(s)$ are equal to m_s, the number of blocks of particles characterizing the microscopic configuration s. In fact (see Fig. 4.2(b)), arrival states are obtained from s by taking the particle on the extreme right of any block and moving it to the right. Therefore, $N_{\text{out}}(s) = m_s$. Analogously, states of origin are obtained from s by taking the particle on the extreme left of any block and moving it to the left, so that $N_{\text{in}}(s) = m_s$.

The equiprobability of microscopic states implies the absence of correlations, e.g., between the occupation probabilities of site i and site j, so that mean-field theory is exact. Let us check this point in more detail. Two quantities play a major role in TASEP models, the probability p_i that site i is occupied and the current of particles, $J_{i,i+1}$, between site i and site $i+1$ (this current is different from zero if site i is occupied and site $i+1$ is empty):

$$p_i = \langle n_i \rangle \tag{4.21}$$

$$J_{i,i+1} = \langle n_i(1 - n_{i+1}) \rangle . \tag{4.22}$$

In a steady state, translational invariance implies that both quantities do not depend on i, so $p_i = p = N/L$, where N is the constant number of particles and L is the number of sites. As for the correlator $\langle n_i n_{i+1} \rangle$, in general it should be written as

$$\langle n_i n_{i+1} \rangle = \frac{\sum_s p^{\text{ne}}(s) n_i(s) n_{i+1}(s)}{\sum_s p^{\text{ne}}(s)} . \tag{4.23}$$

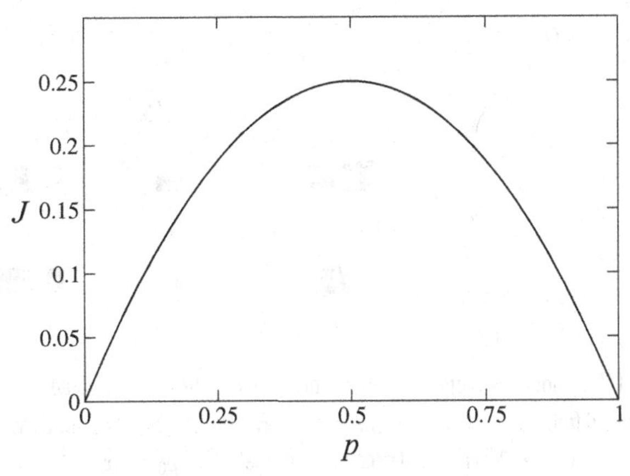

Fig. 4.3 The current $J(p) = p(1-p)$ of the TASEP model.

Because of the equiprobability of microscopic states, $p^{\text{ne}}(s)$ is constant and the above expression reduces to counting states,

$$\langle n_i n_{i+1} \rangle = \frac{\text{\# states where both } i \text{ and } i+1 \text{ are occupied}}{\text{total \# of states}} = \frac{N(N-1)}{L(L-1)}. \tag{4.24}$$

The only correlation surviving in our model is due to finiteness: once a site is occupied (with probability N/L), we are left with $N-1$ particles and $L-1$ sites. In the limit $N, L \gg 1$ we can forget these finite-size correlations and write $\langle n_i n_{i+1} \rangle = \langle n_1 \rangle \langle n_{i+1} \rangle$, so that $J_{i,i+1} = p(1-p)$. This current is plotted in Fig. 4.3, showing that the maximal current state is obtained for $p = \frac{1}{2}$, because this value maximizes the probability of finding a given site occupied and its right site empty.

Despite its apparent simplicity, in the following chapter we are going to show that the TASEP model with periodic boundary conditions is not trivial at all, when we study how the steady state is attained and how its statistical properties depend on L. However, the reason we introduce it here is that it is the necessary starting point to study more complicated boundary conditions, which is what we are going to do in the next section.

4.4.2 Open Boundary Conditions

Let us now consider open boundary conditions, illustrated in Fig. 4.4. We allow particles to be injected at rate α to the left (site $i = 1$) and to be removed at rate β from the right (site $i = L$). The system evolves as follows. We choose randomly an integer $k = 0, \ldots, L$. If $k = 0$ and $n_1 = 0$, we inject a new particle with probability α. If $0 < k < L$, $n_k = 1$, and $n_{k+1} = 0$, we move the particle from k to $k+1$. If $k = L$ and $n_L = 1$, we remove the particle with probability β. It is clear that the system does not satisfy translational invariance any longer and we do not expect mean-field approximation to be exact, as in the case of periodic boundary conditions. Anyway, along with direct simulations of the model, a mean-field treatment is the first approximation to be used, so we will start our discussion from that.

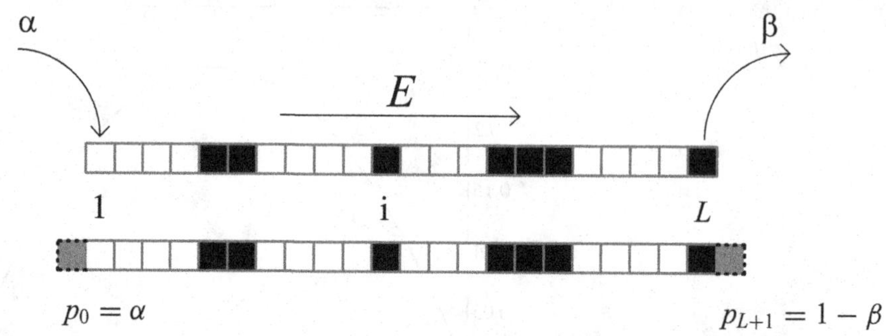

Fig. 4.4 The TASEP model with open boundary conditions. Particles are injected in site $i = 1$ (if empty) at rate α and are removed from site $i = L$ (if occupied) at rate β. These boundary conditions are equivalent to introducing fictitious sites $i = 0, L + 1$, with fixed occupation probabilities, $p_0 = \alpha$ and $p_{L+1} = 1 - \beta$.

A steady configuration is characterized by time-independent averages. Therefore, even if $p_i = \langle n_i \rangle$ is site dependent, the current $J_{i,i+1}$ cannot depend on site i in the steady state, because an unbalance between $J_{i-1,i}$ and $J_{i,i+1}$ would determine a temporal variation of p_i,

$$\begin{aligned} \frac{d\langle n_i \rangle}{dt} &= \langle n_{i-1}(1 - n_i) \rangle - \langle n_i(1 - n_{i+1}) \rangle \\ &= p_{i-1}(1 - p_i) - p_i(1 - p_{i+1}), \end{aligned} \tag{4.25}$$

where the last equality applies in a mean-field approximation. The conditions $J_{i,i+1} = J$ produce $L + 1$ coupled equations, along with an equal number of unknowns ($p_1, \ldots, p_L$ and J itself):

$$\alpha(1 - p_1) = J \tag{4.26}$$

$$p_i(1 - p_{i+1}) = J, \quad i = 1, \ldots, L - 1 \tag{4.27}$$

$$\beta p_L = J. \tag{4.28}$$

It is worth noting that boundary conditions (4.26, 4.28) can be rewritten in the form (4.27) if we assume fictitious sites $i = 0$ and $i = L + 1$, with $p_0 = \alpha$ and $p_{L+1} = (1 - \beta)$. Therefore, mean-field equations correspond to a set of recursive equations. However, while in a kinetic Monte Carlo simulation (see Appendix J) α and β are the natural input data and J is evaluated by computing the average of $n_i(1 - n_{i+1})$, when solving Eqs. (4.26)–(4.28) the line of reasoning is different. We should rather start with a test value for J and check if it is consistent: in fact, assuming some J, Eq. (4.26) provides p_1 and Eqs. (4.27) determine all p_i up to p_L. Finally, Eq. (4.28) allows us to check if the chosen value of J is correct. This way of proceeding is not practical, neither for a numerical solution of recursive equations nor for their qualitative analysis.

Recursive equation (4.27) is also called map in the language of dynamical systems, where its solution gives the discrete time i evolution of the quantity p_i. In our case, i is a site index, not a time index, and the evolution of Eq. (4.27) for increasing/decreasing i just corresponds to moving to the right/left of the system. Far from the edges, we can expect the density p_i is approximately constant: in the language of maps, that means we are in proximity of a fixed point, but now a few words on maps are in order.

A recursive map is a function

$$x_{n+1} = F(x_n), \tag{4.29}$$

and a fixed point x^* is a solution of the equation $x^* = F(x^*)$: if $x_0 = x^*$, $x_n = x^*\ \forall n$. If x_0 is close to x^*, we can determine its evolution via a linear analysis of Eq. (4.29): $x_n = x^* + \epsilon_n$, so $\epsilon_{n+1} = F'(x^*)\epsilon_n$, whose solution is $\epsilon_n = (F'(x^*))^n\epsilon_0$. If $|F'(x^*)| < 1$, $\epsilon_n \to 0$ exponentially and the fixed point is linearly stable. If $|F'(x^*)| > 1$, $|\epsilon_n| \to \infty$ exponentially and the fixed point is linearly unstable. In both cases, the sign of $F'(x^*)$ indicates whether ϵ_n has a constant or an oscillating sign. If $|F'(x^*)| = 1$, we must expand $F(x)$ to the next order. Let us do that explicitly for $F'(x^*) = 1$: $\epsilon_{n+1} = \epsilon_n + \frac{1}{2}F''(x^*)\epsilon_n^2$. The discrete equation cannot be iterated analytically, but we can pass to continuum and solve the resulting differential equation,

$$\frac{d\epsilon}{dn} = \frac{1}{2}F''(x^*)\epsilon^2, \tag{4.30}$$

getting

$$\epsilon(n) = \frac{1}{\dfrac{1}{\epsilon(0)} - \dfrac{1}{2}F''(x^*)n}. \tag{4.31}$$

If $\epsilon(0)$ has the same sign as $F''(x^*)$, x_n moves away from the fixed point (the unphysical change of sign of $\epsilon(n)$ at increasing n occurs when ϵ is so large that the above analysis is no longer valid). If $\epsilon(0)$ has the opposite sign of $F''(x^*)$, x_n approaches the fixed point, with $|\epsilon(n)|$ vanishing as $1/n$. All the qualitative features concerning the stability of fixed points can be found by an easy graphical analysis: see Fig. 4.5, where we have assumed a positive slope and a negative curvature. As for the map of our interest, see Fig. 4.6.

We are now ready to apply the above analysis to the map associated with the recursive equation of TASEP model, $p_i(1 - p_{i+1}) = J$, i.e.,

$$p_{i+1} = 1 - \frac{J}{p_i} \equiv F(p_i), \tag{4.32}$$

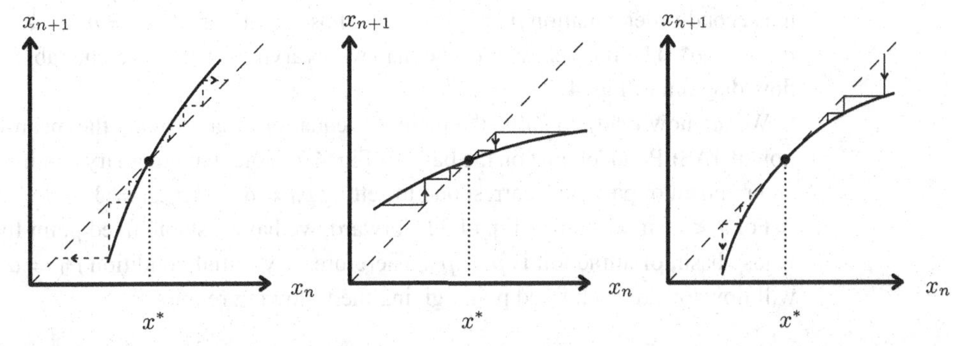

Fig. 4.5 Graphical analysis of the stability of a fixed point, in the three cases $F'(x^*) > 1$ (unstable, left), $F'(x^*) < 1$ (stable, center), and $F'(x^*) = 1$ (marginal, right). Solid lines are the function $F(x)$. Long dashed lines represent the diagonal $x_{n+1} = x_n$, so their intersection with the solid lines (solid circle) is the fixed point x^*. The evolution of a point is shown graphically by thin solid lines (stability) and thin dashed lines (instability).

Table 4.1 Number and stability of fixed points with varying J

$J < \frac{1}{4}$	Two fixed points	$p^*, 1-p^*$	$F'(p^*) > 1, F'(1-p^*) < 1$
$J = \frac{1}{4}$	One fixed point	$p^* = \frac{1}{2}$	$F'\left(\frac{1}{2}\right) = 1$
$J > \frac{1}{4}$	No fixed points		

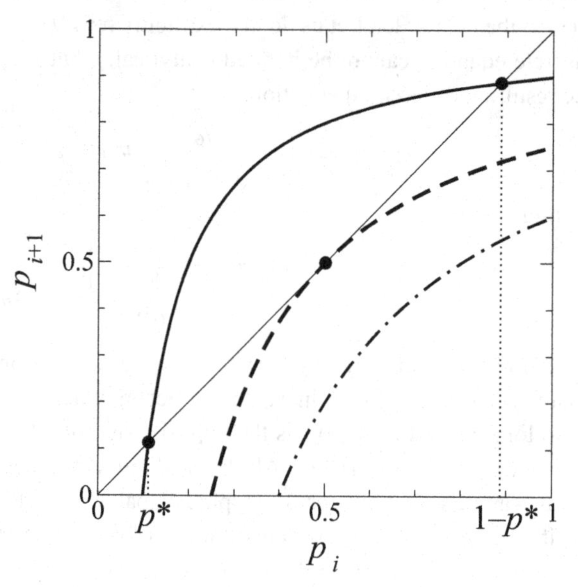

Fig. 4.6 The map associated with the TASEP model, Eq. (4.32), for $J < \frac{1}{4}$ (solid line), $J = \frac{1}{4}$ (dashed line), and $J > \frac{1}{4}$ (dot-dashed line).

which is plotted in Fig. 4.6 for different values of J. The fixed points are determined by the second-order equation $p_f(1 - p_f) = J$, whose solutions are $p_f = p^* \equiv \frac{1}{2} - \sqrt{\frac{1}{4} - J}$ and $p_f = 1 - p^*$. The full behavior of the map with varying J is reported in Table 4.1 and in the flow diagram of Fig. 4.7.

We are now ready to solve the recursive equations that provide the mean-field description of TASEP model, just on the basis of Fig. 4.7. Note that boundary conditions, injection and removal of particles, correspond to setting $p_0 = \alpha$ and $p_{L+1} = 1 - \beta$.

For $J < \frac{1}{4}$, if we evolve Eq. (4.32) forward, we have a stable fixed point for $p = 1 - p^*$ whose basin of attraction is $p > p^*$. Therefore, any initial condition $p_0 = \alpha$ in this basin will flow toward such fixed point, giving the following relations:

$$p_0 = \alpha > p^*, \qquad p_i \underset{i\nearrow}{\longrightarrow} (1 - p^*) = 1 - \beta, \qquad p^* < \frac{1}{2}. \tag{4.33}$$

Therefore, $p^* = \beta$ and the above solution applies in the region $\alpha > \beta$ and $\beta < \frac{1}{2}$. Except for a finite region close to the left edge, the density is equal to the stable fixed point,

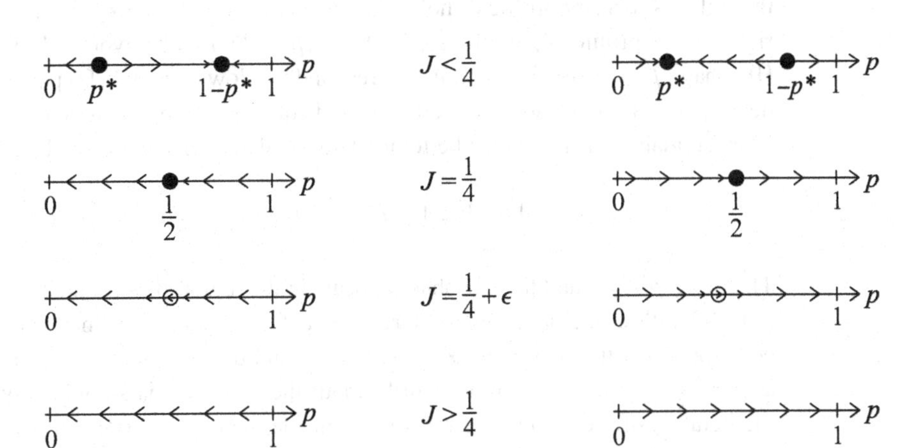

Fig. 4.7 Map flows for different values of J in the TASEP model. Arrows indicate the evolution of the map: the smaller the arrow, the smaller the quantity $|p_{i+1} - p_i|$. Solid circles are fixed points. The open circle appearing for $J = \frac{1}{4} + \epsilon$ signals the memory of the fixed point for $J = \frac{1}{4}$.

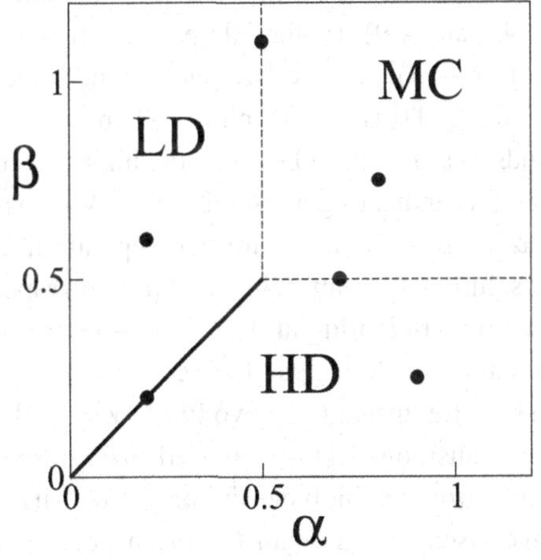

Fig. 4.8 Phase diagram of TASEP model. The six full dots locate the points (α, β) which have been used in Fig. 4.9 to plot the corresponding density profiles.

$p_{\text{f}} = 1-\beta$, and the current is equal to $J = \beta(1-\beta)$. Since the density is almost everywhere larger than $1/2$, this phase is called a *high density* phase (HD, see Fig. 4.8). In simple words, this solution appears when we are injecting more particles than we are removing ($\alpha > \beta$) and removal is not fully efficient ($\beta < \frac{1}{2}$). In these conditions, particle density gets constant toward the exit. As for the profile of particle density, p_i, it may be either increasing or decreasing, as easily understood from Fig. 4.7: it is increasing/decreasing if $p_0 = \alpha$ is smaller/larger than $p_{L+1} = 1 - \beta$.

The motion of particles driven to the right can be understood as the motion of holes (empty sites) driven to the left. Since the hole density is one minus the particle density, holes are injected on the right at rate β and removed on the left at rate α. This means that the right-left density profile of holes that we get for $(\alpha, \beta) = (\alpha^*, \beta^*)$ is equal to the left-right density profile of particles for $(\alpha, \beta) = (\beta^*, \alpha^*)$. In other words, the symmetric of the HD phase, $\beta > \alpha$ and $\alpha < \frac{1}{2}$, must correspond to a low density (LD) phase, where particle density is constant close to the left edge and equal to the low density fixed point, $p = p^*$. More formally, this phase can be found iterating the inverse map (see Fig. 4.7),

$$p_{L+1} = 1 - \beta < 1 - p^*, \qquad p_i \underset{i\searrow}{\longrightarrow} p^* = \alpha, \qquad p^* < \frac{1}{2}. \tag{4.34}$$

Therefore, $p^* = \alpha$ and the LD phase appears in the region $\beta > \alpha$ and $\alpha < \frac{1}{2}$, as expected.

If we follow the same lines of reasoning for $J = \frac{1}{4}$, we find that this value of J corresponds to the two lines, $\alpha > \frac{1}{2}, \beta = \frac{1}{2}$ and $\alpha = \frac{1}{2}, \beta > \frac{1}{2}$. The remaining phase $\alpha, \beta > \frac{1}{2}$ requires a few more words about the map. If J is significantly larger than $\frac{1}{4}$, the iteration of the map will always lead outside the physical domain $p = [0, 1]$ within a few steps. However, if $J - \frac{1}{4} \ll 1$, the map will stay for a long "time" close to $p = \frac{1}{2}$, a "time" that diverges when $(J - \frac{1}{4})$ vanishes. This phase is called maximal current (MC), because J reaches its greatest possible value, $J = \frac{1}{4}$, and it corresponds to $p_0 = \alpha > \frac{1}{2}$ and $p_{L+1} = 1 - \beta < \frac{1}{2}$, i.e., $\beta > \frac{1}{2}$.

We can now summarize the properties of the three different phases (see also the Table 4.2 and Figs. 4.8 and 4.9): (i) the HD phase, where the density is equal to $p_f = 1 - \beta > \frac{1}{2}$ and $J = \beta(1 - \beta)$; (ii) the LD phase, where the density is equal to $p_f = \alpha < \frac{1}{2}$ and $J = \alpha(1 - \alpha)$; and (iii) the MC phase, where $p = \frac{1}{2}$ and $J = \frac{1}{4}$. The separation line HD/LD corresponds to a transition between the high density fixed point and the low density fixed point, therefore indicating a discontinuity. What happens on that line?

When $\alpha = \beta < \frac{1}{2}$, there are two separate fixed points, at $p_f = \alpha$ and $p_f = 1 - \alpha$. Since the solution of mean-field equations corresponds to the evolution of the map starting at $p_0 = \alpha$ and terminating at $p_{L+1} = 1 - \alpha$, this trajectory simply connects the unstable fixed point at $p = \alpha$ to the stable fixed point at $p = 1 - \alpha$. The resulting density profile is shown in Fig. 4.9. Instead, the two lines ($\alpha = \frac{1}{2}, \beta > \frac{1}{2}$) and ($\alpha > \frac{1}{2}, \beta = \frac{1}{2}$) represent the continuous transitions LD/MC and HD/MC, respectively. In both cases the density goes continuously from the high density or low density fixed point toward $p = \frac{1}{2}$.

We have discussed in detail the mean-field approximation, because it gives a faithful representation of the phase diagram. In fact, the TASEP model has an exact solution that

Table 4.2 Different phases in the parameter space of TASEP model

Region	Name	Density p	Current J
$\alpha > \beta, \quad \beta < \frac{1}{2}$	HD	$1 - \beta$	$\beta(1 - \beta)$
$\alpha < \beta, \quad \alpha < \frac{1}{2}$	LD	α	$\alpha(1 - \alpha)$
$\alpha \geq \frac{1}{2}, \quad \beta \geq \frac{1}{2}$	MC	$\frac{1}{2}$	$\frac{1}{4}$

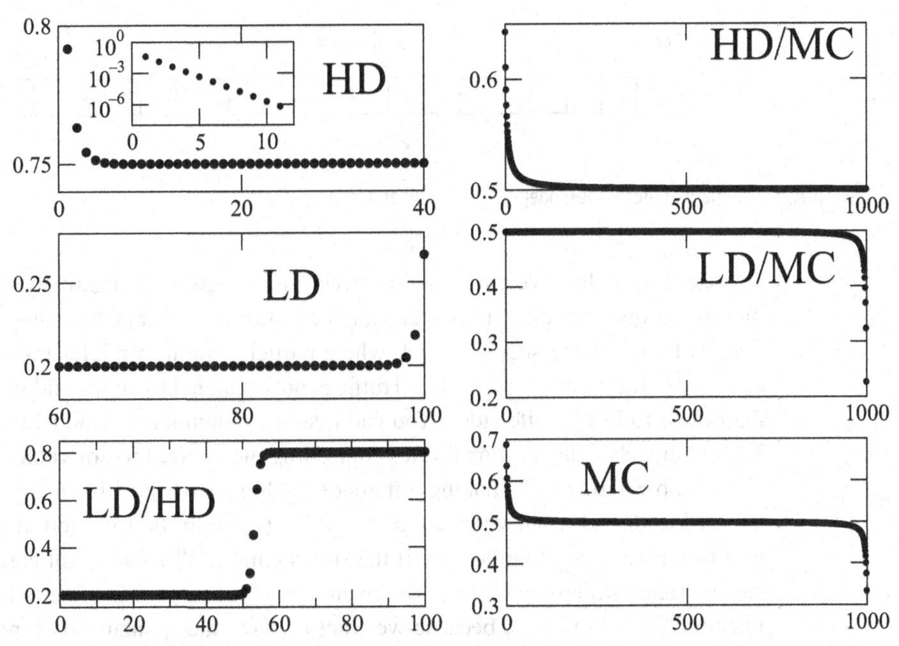

Fig. 4.9 Steady-state density profiles of TASEP model in mean-field approximation for the different phases (HD, LD, MC) and in the transition lines between them.

essentially confirms the above picture.[6] Finally, we discuss a potentially misleading point that also allows us to introduce the problem of spontaneous symmetry breaking, discussed in the next section. The line $\alpha = \beta < \frac{1}{2}$ separates the HD and LD phases. If we solve MF equations for a point on that line, we get a stationary, stable profile of the density p_i that connects the two fixed points and is plotted in Fig. 4.9. We stress the wording "stable profile," because it has been found via the MF time evolution of the density $p_i(t)$; see Eqs. (4.25). It is reasonable to expect that fluctuations (which are absent in MF) might be relevant in this case, allowing the domain wall between LD and HD phases to fluctuate and diffuse. This is exactly what happens: the wall diffuses on a time scale of order L^2, without one phase dominating over the other. Therefore, this process should not be confused with spontaneous symmetry breaking. An explicit example of symmetry breaking will be discussed in the next section.

4.5 Symmetry Breaking: The Bridge Model

We have discussed in detail the phase diagram of TASEP because such a model is the building block of the bridge model. In order to study symmetry breaking we need two

[6] The mean-field approximation fails, for example, to reproduce the exact steady profile in the MC phase, whose region of constant density p is approached with a power law equal to one, see Eq. (4.31), while the exact result gives a square root at denominator.

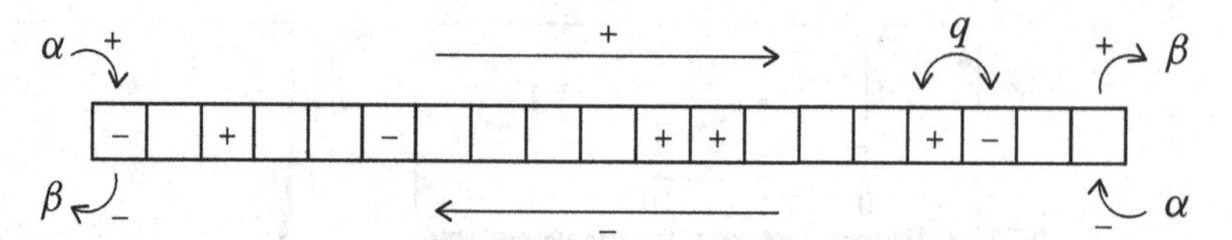

Fig. 4.10 Schematic of the bridge model; see Eqs. (4.35) and (4.36).

classes of particles. We can imagine having positively and negatively charged particles moving in opposite directions under the action of a field, but the name "bridge model" derives from a macroscopic model, where particles mimic vehicles traveling in opposite directions along a one-lane bridge. Traffic is not regulated by lights and vehicles may enter the bridge from opposite sides. You can imagine obtaining this model by combining two TASEP models with additional rules controlling the intersection of vehicles/particles.

In simple terms and making reference to Fig. 4.10, positive/negative particles are injected to the left/right with rate α and extracted from the right/left at rate β. Injection of a new particle is possible only if the site is empty. When different particles meet, they can exchange with probability q. We assume that the two classes of particles have the same injection/extraction rates, because we want to study the spontaneous symmetry breaking. As a matter of fact, if (for example) their injection rates were different, it would not be surprising that the resulting positive/negative currents and densities are also different. Instead, the problem of spontaneous symmetry breaking arises when kinetic rules for the two classes of particles are exactly the same.

More rigorously, the bridge model is defined as follows. Each site $i = 1, \dots, L$ is occupied by a positive particle $\oplus_i$, by a negative particle $\ominus_i$, or it is empty $\bigcirc_i$. We choose randomly an integer $k = 0, 1, \dots, L$. If $k \neq 0, L$, the only microscopic processes giving rise to evolution are

$$\begin{array}{lcll} \oplus_k \bigcirc_{k+1} & \to & \bigcirc_k \oplus_{k+1} & \text{with probability } 1 \\ \bigcirc_k \ominus_{k+1} & \to & \ominus_k \bigcirc_{k+1} & \text{with probability } 1 \\ \oplus_k \ominus_{k+1} & \to & \ominus_k \oplus_{k+1} & \text{with probability } q. \end{array} \tag{4.35}$$

All other processes involving sites $k, k+1$ have zero probability. As for boundary conditions, they enter when $k = 0$ or $k = L$. The only permitted processes are

$$\begin{aligned} &\text{if } k = 0, \quad \begin{cases} \ominus_1 \to \bigcirc_1 & \text{with probability } \beta \\ \bigcirc_1 \to \oplus_1 & \text{with probability } \alpha \end{cases} \\ &\text{if } k = L, \quad \begin{cases} \oplus_L \to \bigcirc_L & \text{with probability } \beta \\ \bigcirc_L \to \ominus_L & \text{with probability } \alpha. \end{cases} \end{aligned} \tag{4.36}$$

Despite appearances, evolution rules are simple. Overall (see Eqs. (4.35)), particles move as in TASEP with the additional condition that opposite particles can exchange their position with probability q. At boundaries (see Eqs. (4.36)), particles are injected with rate α and extracted with rate β. One caveat should be stressed: exchange of particles at boundaries is not allowed; therefore, a particle cannot get onto the bridge if the entry site is occupied by an opposite particle waiting to get off. We will come back to this point later.

In summary, we have two classes of particles traveling in opposite directions and having the same dynamical rules. Is it possible that a spontaneous symmetry breaking occurs, making currents and densities of positive and negative particles different? This is a natural question if we think of the regime $\beta \ll 1$ and $\alpha, q \approx 1$, because in this regime each flux of particles would tend to stabilize itself into the HD phase, which is manifestly impossible, because the system cannot sustain a density larger than $\frac{1}{2}$ for two different classes of particles at the same time. The solution might be an equal reduction of both densities. In fact, as we will see in MF a symmetric solution exists for any point of the phase diagram (α, β), and in the region $\beta \ll \alpha, q$ it is a symmetric LD phase. However, this phase may be unstable in much the same way an equilibrium symmetric state with no magnetization is unstable below T_C in a ferromagnetic material.

We will see that this instability leads one type of particles to dominate, therefore producing a symmetry breaking. Which class of particles dominates depends on the initial conditions and on fluctuations: the same state may lead to a dominance of positive particles or to a dominance of negative particles. The key point, however, is that a true phase transition occurs only in the thermodynamic limit, so it will be of great interest to study the dynamics with varying the size L.

4.5.1 Mean-Field Solution

In the TASEP model we only need the variable $n_i = 1, 0$, which indicates whether site i is occupied or not by a positive particle, while in the bridge model we also need the variable $\tau_i = 1, 0$, which gives the same information for a negative particle. The currents of positive/negative particles between two adjacent sites ($i = 1, \ldots, L-1$) are

$$
\begin{aligned}
(J_+)_{i,i+1} &= \langle n_i(1 - n_{i+1} - \tau_{i+1})\rangle + q\langle n_i\tau_{i+1}\rangle = \langle n_1(1 - n_{i+1} - (1-q)\tau_{i+1})\rangle \\
(J_-)_{i+1,i} &= \langle \tau_{i+1}(1 - \tau_i - n_i)\rangle + q\langle \tau_{i+1}n_i\rangle = \langle \tau_{i+1}(1 - \tau_i - (1-q)n_i)\rangle,
\end{aligned}
$$

where each current counts two distinct processes, the hopping toward an empty site and the exchange with an opposite particle.

These equations must be supplemented with boundary conditions, according to which

$$
\begin{array}{lll}
(J_+)_{01} = \alpha\langle 1 - n_1 - \tau_1\rangle, & (J_-)_{10} = \beta\langle \tau_1\rangle, & i = 1 \\
(J_+)_{L,L+1} = \beta\langle n_L\rangle, & (J_-)_{L+1,L} = \alpha\langle 1 - n_L - \tau_L\rangle, & i = L,
\end{array}
\tag{4.37}
$$

where the notations for the current have been used in order to make valid the following evolution equations for any $i = 1, \ldots, L$,

$$
\frac{d\langle n_i\rangle}{dt} = (J_+)_{i-1,i} - (J_+)_{i,i+1} \tag{4.38}
$$

$$
\frac{d\langle \tau_i\rangle}{dt} = (J_-)_{i+1,i} - (J_-)_{i,i-1}. \tag{4.39}
$$

We are primarily interested in the stationary state, which is characterized by site-independent currents. Within a mean-field approach, variables on different sites are independent and we can reduce to equations for the single site average values

$$
p_i = \langle n_i\rangle, \qquad m_i = \langle \tau_i\rangle. \tag{4.40}
$$

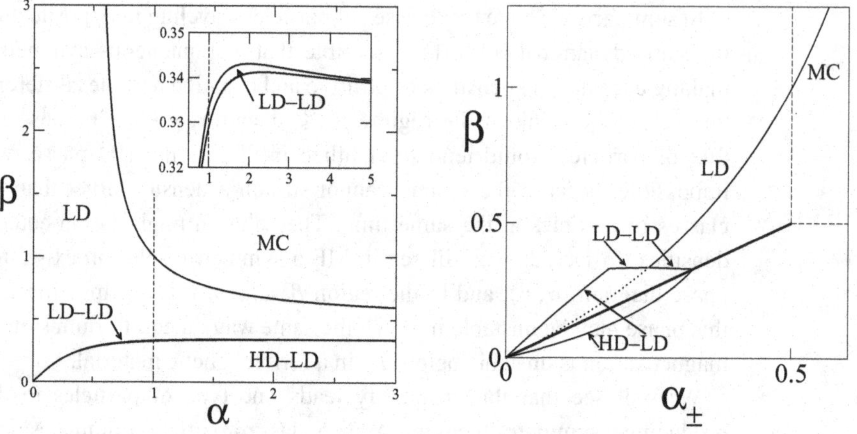

Fig. 4.11 Left: Mean-field phase diagram of the bridge model with a blow-up in the inset. Right: If we fix $\alpha = 1$ and let β increase from $\beta = 0$ to $\beta > 1$, we plot the corresponding $(\alpha_\pm, \beta)$ pairs as the thin straight lines. When the two curves join we pass from an asymmetric phase to a symmetric one. The dotted line is the unstable, symmetric LD phase. The thick straight line and the dashed lines separate the different regions of the TASEP phase diagram; see Fig. 4.8.

In terms of p_i, m_i, Eqs. (4.38) and (4.39) can be rewritten as

$$\begin{aligned} J_+ &= \alpha(1 - p_1 - m_1) = p_i(1 - p_{i+1} - (1-q)m_{i+1}) = \beta p_L, \\ J_- &= \alpha(1 - p_L - m_L) = m_{i+1}(1 - m_i - (1-q)p_i) = \beta m_1, \end{aligned} \qquad i = 1, \ldots, L. \tag{4.41}$$

The above equations show that the two fluxes of particles are coupled via boundary conditions and, if $q < 1$, because it is difficult to exchange their positions. If $q = 1$, their coupling in the bulk disappears because a particle is no longer able to discriminate between an opposite particle and an empty site. In this case the above equations simplify to

$$\begin{aligned} J_+ &= \alpha(1 - p_1 - m_1) = p_i(1 - p_{i+1}) = \beta p_L, \\ J_- &= \alpha(1 - p_L - m_L) = m_{i+1}(1 - m_i) = \beta m_1, \end{aligned} \qquad i = 1, \ldots, L. \tag{4.42}$$

We have already noticed how the model does not allow the exchange of particles at boundaries. We can now remark that allowing it would make the two fluxes completely independent for $q = 1$. In full general terms, the model might allow an exchange q_b at boundaries and an exchange q in the bulk. If $q_b = q = 1$, the model is trivially uncoupled and no phase transition may appear. The case we are going to study in a mean-field approximation corresponds to $q_b = 0, q = 1$: it allows an exact treatment and is not trivial.

We can now rephrase boundary conditions in a more tractable way. We start from boundary conditions in the present form

$$\begin{aligned} J_+ &= \alpha(1 - p_1 - m_1) = \beta p_L \\ J_- &= \alpha(1 - p_L - m_L) = \beta m_1, \end{aligned} \tag{4.43}$$

to realize that coupling occurs through the entry site only, not through the exit site. Therefore, our goal is to encapsulate the coupling in effective injection rates $\alpha_\pm$, which means we want to write

$$\begin{aligned} J_+ &= \alpha_+(1-p_1) \\ J_- &= \alpha_-(1-m_L). \end{aligned} \tag{4.44}$$

From Eqs. (4.43) we have $\frac{J_+}{\alpha} = (1-p_1) - \frac{J_-}{\beta}$ and $\frac{J_-}{\alpha} = (1-m_L) - \frac{J_+}{\beta}$. Using Eqs. (4.44) to express $(1-p_1)$ and $(1-m_L)$ in terms of J_+, α_+ and J_-, α_-, respectively, we can write

$$\alpha_+ = \frac{J_+}{\frac{J_+}{\alpha} + \frac{J_-}{\beta}} \quad \text{and} \quad \alpha_- = \frac{J_-}{\frac{J_-}{\alpha} + \frac{J_+}{\beta}}. \tag{4.45}$$

In conclusion, for $q = 1$ the bridge model can be reformulated in terms of two TASEP models with new injection rates $\alpha_\pm = g_\pm(\alpha, \beta, J_+, J_-)$. In the previous section we have solved the TASEP model, so we know the functions $J_\pm = f(\alpha_\pm, \beta)$; see Table 4.2. We can therefore solve the bridge model self-consistently, assuming that the bridge model is in a given phase and using the above equations to find whether such a solution exists for some values of α and β.

The explicit calculations are done in Appendix K for symmetric and asymmetric phases. A symmetric phase means that $\alpha_+ = \alpha_-$ and $J_+ = J_-$: the two types of particles are not distinguishable from their dynamics. In the appendix we find the LD and MC symmetric phases, while the HD symmetric phase cannot exist, as argued here above. An asymmetric phase means that $\alpha_+ \neq \alpha_-$ and $J_+ \neq J_-$: in the appendix we find two different types of asymmetric phases, the HD–LD and the LD–LD one. In Fig. 4.11 we plot the different phases of the bridge model in the mean-field approximation. If we increase β, keeping $\alpha > \frac{1}{2}$, fixed we first encounter the asymmetric HD–LD phase, then the asymmetric LD–LD phase, then the symmetric LD phase, and finally the symmetric MC phase. If $\alpha < \frac{1}{2}$, the above sequence of transitions stops with the symmetric LD phase.

The continuous or discontinuous character of the transitions can be determined making reference to the right panel of Fig. 4.11, while taking into account the phase diagram of the TASEP model; see Fig. 4.8. Using these figures, we can conclude that the first transition (HD–LD) → (LD–LD) is discontinuous, while the other ones are continuous.

4.5.2 Exact Solution for $\beta \ll 1$

The mean-field approximation has the advantage of providing a full phase diagram, but fluctuations can break the mean-field picture. In equilibrium systems this is especially true in low dimensions, so an alternative proof that nonequilibrium symmetry breaking does occur in the bridge model would be welcome. Such evidence will be given in this section, where we focus on the limiting case $\beta \ll \alpha, q$.

In fact, in this limit there are two main time scales: a fast time scale, during which particles travel in the bulk and new particles possibly enter in the bridge, and a slow time scale, during which particles leave the bridge. This separation of time scales is better understood if we imagine setting $\beta = 0$ and monitoring the resulting dynamics, then switching on β. If $\beta = 0$, particles cannot leave the bridge, therefore as soon as a positive particle has attained the site $i = L$ and a negative particle has attained the site $i = 1$, the following dynamics conserves matter, because such particles act as a stopper, preventing the entry of new ones. From now on, dynamics leads to a final, steady configuration where

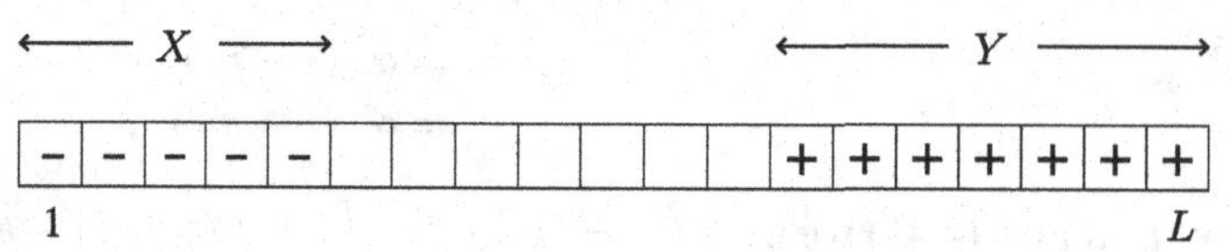

Fig. 4.12 Typical configuration of the bridge model for $\beta \to 0$.

all (X) negative particles are grouped to the left, all (Y) positive particles are grouped to the right, and $(L - X - Y)$ empty sites are in the middle; see Fig. 4.12.

Now let us switch on β, therefore allowing particles to leave the bridge, but keeping $\beta \ll \alpha, q$. Now, after a time $\tau_{ex} \approx \beta^{-1}$ one of the two particles acting as stopper may get out, therefore leaving an empty site in its place. This site can be taken up by the second particle of the same type in the queue for the exit or by a new particle of opposite type entering into the bridge. In the former case a new stopper again closes the exit. In the latter case the new particle will travel freely toward its queue on the opposite side of the bridge while the entrance has been closed again. In both cases, it is necessary to wait a time τ_{ex} because something new happens. The net result of having switched β has been to get the same type of configuration as before, with different lengths X, Y of the queues of positive and negative particles. In other words, the dynamics of the bridge model, if observed on a time $t > \tau_{ex}$, can be described in a simple, discrete two-dimensional phase space (X, Y), with suitable transition rates between a lattice point and its neighbors.

Before discussing dynamics we need to clarify which aspects are relevant to settle the existence of a spontaneous symmetry breaking in the bridge model. For this reason we focus on a system that is well known to have an ordered phase as a result of a spontaneous symmetry breaking: the two-dimensional Ising model. Let's consider such a model on a finite lattice of linear size L at temperature $T < T_c$. If $L \gg \xi(T)$, the correlation length, a typical snapshot of the system will provide a finite average magnetization m, very close to the equilibrium value $\overline{m}(T)$ for the infinite system. On a short time scale, fluctuations do not destroy order and as time goes by $m(t)$ fluctuates around $\overline{m}(T)$. However, since L is finite the system has a nonvanishing probability of overcoming the barrier between the present state and the symmetric state with $m = -\overline{m}(T)$. The key point is the time $\tau(L)$ required for inverting the magnetization: it increases exponentially with L, so that finite but large systems practically stay in one minimum forever.[7]

A second aspect of spontaneous symmetry breaking at equilibrium is worth discussing, the effect of a field. In fact, an arbitrarily small field destroys the phase transition. If we prepare the Ising model with negative magnetization and apply a small, positive field, the reversal of the magnetization occurs even in the thermodynamic limit, because it takes place via nucleation of the opposite phase. The nucleation process will be studied in detail in Section 6.6; here it suffices to say that at equilibrium the reversal time does not increase exponentially any longer with the size of the system if an arbitrarily small field is applied.[8]

[7] This picture does not apply to the one-dimensional Ising model. If $T = 0$ the system cannot increase its energy even momentarily, which forbids magnetization reversal. If $T > 0$, as soon as $L > \xi(T)$ the system is disordered and the average magnetization fluctuates around zero at any time.

[8] In the nonequilibrium context of the bridge model there is another reason to consider the effect of a field, i.e., an asymmetry between the two classes of particles. While in the Ising model there is a natural symmetry between up and down spins, deriving from the Hamiltonian describing the system, here there is no a priori reason to require that rules for positive and negative particles are *exactly* the same.

Based on the above discussion, we are led to consider two problems: (i) we prepare the system filled with negative particles and evaluate the time $\tau(L)$ necessary to obtain the configuration where the system is filled with positive particles; (ii) we do the same thing in the presence of a "field" favoring the HD phase of positive particles. Since the two problems can be treated in the same manner,[9] we discuss the probabilities for the transition $(X, Y) \to (X', Y')$ in the general case of a field, problem (ii). If the field is zero, we recover problem (i). To begin with, what is a field in the context of the bridge model? It is something that favors a phase with respect to the other. The + phase, e.g., may be favored by an injection rate a_+ larger than the injection rate a_- for negative particles (we use $a_\pm$ rather than $\alpha_\pm$ to avoid confusion with the previous section, where such notation had a completely different meaning).

Let us now describe the microscopic dynamics in the (X, Y)-plane. If we start with X negative particles and Y positive particles, after a typical time τ_{ex} one of the following four processes take place:

1. *The particle $\ominus_1$ leaves the bridge and the particle $\ominus_2$ takes its place.* Then all the remaining negative particles automatically move to the left. The new state is $(X-1, Y)$.
2. *The particle $\ominus_1$ leaves the bridge and a $\oplus$ particle is injected from the left.* This particle travels all along the bridge to join the other positive particles. The new state is $(X-1, Y+1)$.
3. *The particle $\oplus_L$ leaves the bridge and the particle $\oplus_{L-1}$ takes its place.* Then all other positive particles move to the right. The new state is $(X, Y-1)$.
4. *The particle $\oplus_L$ leaves the bridge and a $\ominus$ particle is injected from the right.* This particle travels to the left until it joins the other negative particles. The new state is $(X+1, Y-1)$.

In order to determine the probabilities of different processes, we must not confuse competing dynamical processes with processes that automatically follow a given occurrence. The four competing dynamical processes are highlighted in italics above. Since the rate of hopping (processes 1 and 3) is equal to 1 and the rate of injecting a new particle is a_+ for process 2 and a_- for process 4, the probabilities p_i of the four processes are $p_1 = c, p_2 = ca_+, p_3 = c, p_4 = ca_-$, with c a suitable normalization constant. The existence of a "field" producing the asymmetry between a_+ and a_- can be made evident writing $a_\pm = \alpha(1 \pm H)$. With this notation, the condition $\sum_{i+1}^4 p_i = 1$ implies $c = 1/(2+2\alpha)$ and the probabilities finally are as follows:

$$(X, Y) \to \begin{cases} (X-1, Y), & p_1 = \dfrac{1}{2(1+\alpha)} \\ (X-1, Y+1), & p_2 = \dfrac{\alpha(1+H)}{2(1+\alpha)} \\ (X, Y-1), & p_3 = \dfrac{1}{2(1+\alpha)} \\ (X+1, Y-1), & p_4 = \dfrac{\alpha(1-H)}{2(1+\alpha)}. \end{cases} \tag{4.46}$$

[9] This is another relevant difference with the equilibrium case, where the physics of magnetization reversal is different for $H = 0$ and $H \neq 0$.

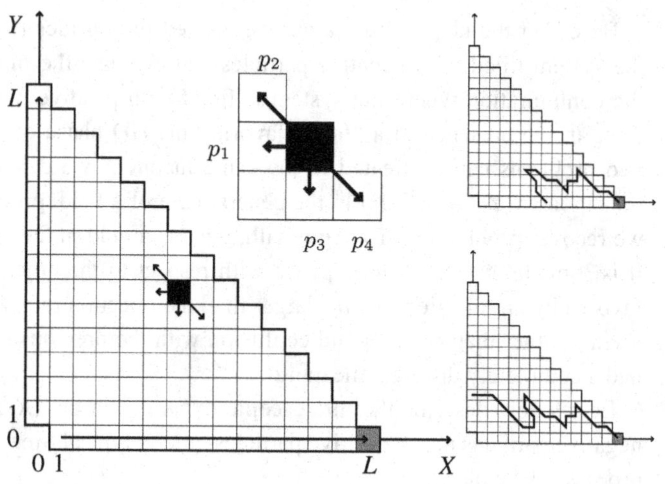

Fig. 4.13 Phase space of the bridge model for $\beta \ll 1$ and schematic of dynamics. The gray site of coordinates $(L, 0)$ is the initial configuration (bridge filled with negative particles) and the black site is a generic configuration whose allowed transitions are indicated by the four arrows. Each move has a probability p_i, given in Eq. (4.46). On the right we sketch two possible trajectories: in the top panel the representative point crosses the horizontal axis and the system fails to invert its population, while in the bottom panel the representative point crosses the vertical axis and the system has a population inversion, passing from a dominance of negative particles to a dominance of positive ones.

Possible moves are described in Fig. 4.13. Because of the constraints on allowed moves, the motion of the system is confined to a triangular region in the quadrant $X, Y \geq 0$ and it can cross its border only through the horizontal axis or through the vertical axis (not through the origin). In the former case ($Y = 0$) all $\oplus$ particles are out, no stopper exists on the right and the system fills with $\ominus$ particles; see the horizontal arrow crossing all sites $(X, 0)$. A symmetric process happens for the $X = 0$ axis. In conclusion, all configurations $(X, 0)$ immediately pass to $(L, 0)$ and all configurations $(0, Y)$ immediately pass to $(0, L)$.

The system is initially in a full negative phase and we wonder how the probability of passing to a full positive phase scales with the system size L. This means we start with the system in $(L, 0)$ (see the gray box in Fig. 4.13), and we wonder what the probability is of leaving the triangle crossing the Y-axis rather than the X-axis (these two possibilities are sketched in the same figure, on the right). The key point is that the system diffuses with a net drift in both directions, because

$$\langle \Delta X \rangle = \sum_{i=1}^{4} p_i (\Delta X)_i = -\frac{1 + 2\alpha H}{2(1 + \alpha)} \tag{4.47a}$$

$$\langle \Delta Y \rangle = \sum_{i=1}^{4} p_i (\Delta Y)_i = -\frac{1 - 2\alpha H}{2(1 + \alpha)}. \tag{4.47b}$$

Therefore the system has a horizontal drift in the negative direction and a vertical drift that is negative for $H < 1/(2\alpha)$. If we monitor the motion in the (X, Y) directions, the problem of passing from the negative to the positive phase can be rephrased as follows. The X variable starts from value L and the Y value starts from zero. If Y is moved from zero to one, what is the probability that X attains zero before Y comes back to zero? We can get a simple, semi-quantitative answer if we treat the motions in the X and Y directions as independent and we approximate the motion along X with its average over noise.

In a continuum picture, with variables $x(t), y(t)$, dynamics is written in terms of Langevin equations,

$$\begin{aligned} \dot{x}(t) &= -v_x + \eta_x(t), \quad \langle \eta_x(t) \rangle = 0, \quad \langle \eta_x(t)\eta_x(t') \rangle = \Gamma_x \delta(t-t') \\ \dot{y}(t) &= -v_y + \eta_y(t), \quad \langle \eta_y(t) \rangle = 0, \quad \langle \eta_y(t)\eta_y(t') \rangle = \Gamma_y \delta(t-t'), \end{aligned} \tag{4.48}$$

where

$$\begin{aligned} -v_x &= \frac{\langle \Delta x \rangle}{\Delta t} = \langle \Delta X \rangle, \qquad \Gamma_x = \langle (\Delta X)^2 \rangle \\ -v_y &= \frac{\langle \Delta y \rangle}{\Delta t} = \langle \Delta Y \rangle, \qquad \Gamma_y = \langle (\Delta Y)^2 \rangle. \end{aligned} \tag{4.49}$$

Variances are easily calculated as the average values above,

$$\langle (\Delta X)^2 \rangle = \langle (\Delta Y)^2 \rangle = \frac{1+2\alpha}{2(1+\alpha)}. \tag{4.50}$$

The motion along the x-direction starts at $x(0) = L$ and the average time to reach the origin is

$$t^* = \frac{L}{v_x}. \tag{4.51}$$

We wonder what the probability $S(t^*)$ is that a particle moving in the y-direction according to (4.48) and such that $y(0) = y_0 > 0$ has not crossed the origin $y = 0$ within time t^*. Using Eq. (D.75), we get

$$S(t^*) \simeq \frac{y_0}{\sqrt{\pi}} \frac{\sqrt{2\Gamma_y}}{v_y^2} \frac{\exp\left(\frac{y_0 v_y}{\Gamma_y}\right)}{(t^*)^{3/2}} \exp\left(-\frac{v_y^2}{2\Gamma_y} t^*\right). \tag{4.52}$$

For $H = 0$, we obtain

$$S(L) \simeq y_0 \left(\frac{2}{\pi}(1+2\alpha)\right)^{1/2} \frac{\exp\left(\frac{y_0}{1+2\alpha}\right)}{L^{3/2}} \exp\left(-\frac{1}{2(1+2\alpha)} L\right). \tag{4.53}$$

The average time to pass from one pure phase, e.g., the negative one, to another pure phase, the positive one, is

$$\tau_{-\to+}(L) = S(L)^{-1} = f(\alpha) L^{3/2} e^{g(\alpha)L} \tag{4.54}$$

and it grows exponentially with the size of the system. This exponential growth is the key result, proving that a phase transition indeed exists in the limit of vanishing β. The exponential growth arises because the motion along the y-direction has a negative drift v_y. Since

$$v_y = -\frac{1-2\alpha H}{2(1+\alpha)}, \tag{4.55}$$

the above scenario is preserved as long as $v_y < 0$, i.e., for

$$H < \frac{1}{2\alpha}. \tag{4.56}$$

Therefore, critical nonequilibrium systems display two important differences with respect to critical equilibrium systems: we may have a spontaneous symmetry breaking even in one dimension and the application of a small field does not destroy the transition.

4.6 Bibliographic Notes

A recent reference book about nonequilibrium phase transitions, including an extended illustration of models with many absorbing states, is M. Henkel, H. Hinrichsen, and S. Lübeck, *Non-equilibrium Phase Transitions, volume 1: Absorbing Phase Transitions* (Springer, 2008).

A reference book about driven diffusive systems and their critical properties is B. Schmittmann and R. K. P. Zia, *Statistical Mechanics of Driven Diffusive Systems*, Phase Transitions and Critical Phenomena, vol. 17 (Academic Press, 1995).

An overview about the modern evolution of Markov theory of stochastic processes toward interacting particle models such as spin models, the voter model, and contact processes is contained in T. M. Liggett, *Interacting Particle Systems* (Springer, 1985).

A book containing an overview about self-organized critical phenomena is P. Bak, *How Nature Works: The Science of Self-Organized Criticality* (Copernicus Press, 1996). Further instructive readings on this topic are the original papers by P. Bak, C. Tang, and K. Wiesenfeld, Self-organized criticality, *Physical Review A* **38** (1988) 364–374, and by P. Bak and K. Sneppen, Punctuated equilibrium and criticality in a simple model of evolution, *Physical Review Letters* **71** (1993) 4083–4086. Finally, the model to pass from absorbing phase transitions to SOC is A. Vespignani et al., Absorbing-state phase transitions in fixed-energy sand-piles, *Physical Review E*, **62** (2000) 4564–4582.

The exact solution of the TASEP model in one dimension can be found in the papers by G. Schütz and E. Domany, Phase transitions in an exactly soluble one-dimensional exclusion process, *Journal of Statistical Physics*, **72** (1993) 277–296, and by B. Derrida, S. A. Janowsky, J. L. Lebowitz, and E. R. Speer, Exact solution of the totally asymmetric simple exclusion process: shock profiles, *Journal of Statistical Physics*, **73** (1993) 813–842.

The exact solution of the ASEP model in one dimension can be found in the paper by B. Derrida, M. R. Evans, V. Hakim, and V. Pasquier, Exact solution of a 1D asymmetric exclusion model using a matrix formulation, *Journal of Physics A: Mathematical and General* **26** (1993) 1493–1517.

The original paper where the bridge model was introduced is C. Godréche, J. M. Luck, M. R. Evans, D. Mukamel, S. Sandow, and E. R. Speer, Spontaneous symmetry-breaking: exact results for a biased random walk model of an exclusion process, *Journal of Physics A: Mathematical and General* **28** (1995) 6039–6072.

5 Stochastic Dynamics of Surfaces and Interfaces

This book began with the study of the stochastic dynamics of a Brownian particle, then generalized to the Brownian motion of a general physical quantity. Next, we considered the stochastic dynamics of an ensemble of interacting particles. In the present chapter we address the stochastic motion of a line or a surface. More precisely, we intend to study the dynamics of a driven interface in the presence of noise (if it were not driven, it would be at equilibrium, at least asymptotically). It is interesting that noise plus drive can induce a delocalization of the interface, as evidenced by the standard deviation of the height fluctuations of the interface. This quantity is called roughness (W); it depends on the time t and on the spatial scale L, over which such fluctuations are evaluated and it may occur that $W(L, t)$ diverges in the limit $L, t \to \infty$. When this happens, we have a kinetic roughening phenomenon.

In addition to the process of delocalization, the main consequence of kinetic roughening is to determine a self-affine interface, i.e., an object that is statistically the same under a suitable anisotropic rescaling of space, time, and "direction of motion." This property means that fluctuations have universal features and the function $W(L, t)$ is a homogeneous function, characterized in the limit $L, t \to \infty$ by suitable critical exponents, which define universality classes, much in the same way as in equilibrium and nonequilibrium phase transitions. The two most important universality classes are called Edwards–Wilkinson (EW) and Kardar–Parisi–Zhang (KPZ).

The EW universality class can be described by a linear theory, which is somewhat surprising: it is not evident that a linear model could account for a sort of nonequilibrium critical phenomenon. Given the linearity, the model can be solved in any spatial dimension by Fourier transform, obtaining the exact form of $W(L, t)$. The KPZ universality class is nonlinear and comes out in very different physical contexts, including the diffusing, interacting particle systems studied in the previous chapters. There has been a renewed interest in the KPZ topic, motivated by important, exact analytical results in $d = 1$, by large-scale simulations in greater dimension, and (finally!) by clear experimental evidence of real systems belonging to the KPZ universality class. We will try to give a short account of all these aspects, but will keep our student readers in mind.

In spite of the considerable differences between EW and KPZ universality classes, they have a basic, common feature: they concern local processes and are described by local models. For this reason we have decided to provide a short introduction to nonlocal models, focusing on diffusion limited aggregation (DLA), which should be familiar to the reader, at least pictorially.

5.1 Introduction

A freshman student is likely to have a partial notion of what surfaces and interfaces are in science: a surface is a mathematical object, a two-dimensional manifold, while an interface might be what allows us to interact with an electronic device. A student in physics certainly meets these concepts when studying fluids (surface tension) and solid state (*p*–*n* junctions). In this chapter we use the two terms, surfaces and interfaces, interchangeably and we adopt a broad point of view, according to which an interface separates two regions with some different physical property. Restrictions come next: we are interested in the dynamics of these objects, when they are set in motion in the presence of noise. What type of noise will be discussed later, but noise is the key ingredient.

These few words should have clarified the title of the present chapter. Now let's give a freewheeling list of examples of interfaces and of the driving forces setting them in motion. A class of examples, which will be pedagogically useful, concerns deposition and growth processes, where a solid phase is in contact with either a vapour phase or a liquid phase. In the former case the solid grows, because particles attach to its surface and the flux of particles is the driving force. In the latter case, we can imagine a solid in contact with its melt, because a temperature gradient is applied. If the solid/melt interface is set in motion with respect to the gradient, the solid phase can grow and the liquid recedes. This method can be used for the production of monocrystalline silicon.

A domain wall in a ferromagnet, separating regions where the magnetization has different orientations, is another typical example of interface, whose motion can be easily tuned by applying a magnetic field (or an electric field in a ferroelectric material); see Fig. 5.1. A conceptually similar example is the interface between different turbulent states in a nematic liquid crystal undergoing a convection process. We will come back to this later.

There are other examples in different domains and at different scales: a flame or burning front, driven by a contact process and possibly by air currents; a wetting front in a porous medium, driven by capillarity, by gravity, or by imposing a pressure gradient to the flowing fluid; a liquid front on an inclined plane, driven by gravity; or the growing front of a bacterial colony in a Petri dish, driven by nutrients.

As for the noise, i.e., fluctuations, it may have different origins.

The most familiar one is certainly thermal noise: for instance, due to thermal fluctuations, the interface between magnetic domains in an Ising system at equilibrium cannot be a straight line.

A second type of fluctuations, which have a relevant role in the statistical description of a disordered system, is the so-called *quenched* disorder, which exemplifies the role of randomly distributed structural defects in a material. A simple illustration of quenched disorder is given in Fig. 5.1, where the motion of the domain wall is hindered by defects, whose main effect is to pin the wall. At equilibrium (which is not the case in the figure) we should first determine the partition function Z summing over the thermal degrees of freedom, i.e., the spins $\{s_i\}$, then we should average the free energy $F = -T \ln Z$ over the quenched degrees of freedom, i.e., the positions $\{\mathbf{r}_i\}$ of the defects. In the present

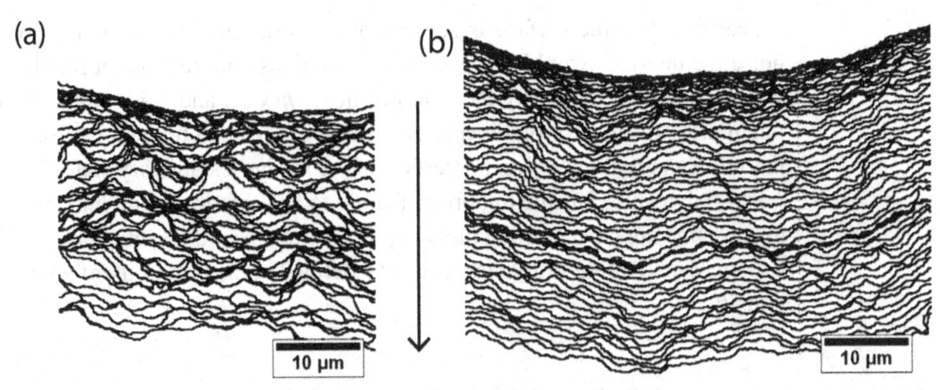

Fig. 5.1 Domain wall propagation in Pt/Co(0.65 nm)/Pt films. The Co ultrathin magnetic film is first magnetized in the down direction, then a magnetic field is applied in order to nucleate domains of positive magnetization. Once such a domain has been created, the domain wall is set in motion (from top to bottom) by subsequent applications of a field pulse $H_a = 77$ Oe (a) and $H_b = 260$ Oe (b). Pulse durations were between 2 and 8 s in (a) and between 20 and 100 μs in (b). The fields H_a and H_b are smaller than the depinning field, $H_{\text{dep}} \approx 750$ Oe, which plays the role of the static friction F_c when we pull a body in contact with a surface with a force F. In this case, if $F < F_c$, the (macroscopic) body remains at rest, but in the microscopic world of magnetic domains, the wall can move even if $H < H_{\text{dep}}$, provided $T > 0$, through a thermally activated dynamics. Courtesy of Peter Metaxas. P. J. Metaxas, Domain wall dynamics in ultrathin ferromagnetic film structures: disorder, coupling, and periodic pinning, PhD thesis, The University of Western Australia and Université Paris-Sud 11, 2009.

out-of-equilibrium situation it happens that if the driving force is too weak the interface is blocked, because the disorder acts as an ensemble of pinning centers and may prevent its motion (see the concept of depinning field, introduced in the figure legend).

Deposition and growth processes undergo kinetic fluctuations, whose nature is fairly similar to the fluctuating force acting on a Brownian particle, studied in Chapter 1. In practice, this noise is the result of more microscopic degrees of freedom. This is the main type of noise we focus on in the present chapter.

Finally, strange as it may seem, noise can actually be present even if it should be absent, because the dynamics is deterministic. This is the case for deterministic chaos, and in some cases it is possible to make a rigorous connection between a chaotic, deterministic evolution and a stochastic evolution due to kinetic noise. The most relevant example (and here perfectly appropriate) is given by the Kuramoto–Sivashinsky equation,[1] which exhibits a deterministic, spatiotemporal chaotic dynamics: this is found to be statistically equivalent to the stochastic dynamics produced by the Kardar–Parisi–Zhang equation, which is one of the main subjects of this chapter.

The statement about the statistical equivalence of a deterministic, chaotic equation and a stochastic one should be understood as a sign of universality and a suggestion toward a treatment of stochastic dynamics, which is as general as possible. But what are the main features of dynamics and the main physical quantities we are interested in? The central physical concept appearing in this chapter is kinetic roughening, a dynamic phenomenon

[1] In one spatial dimension, the Kuramoto–Sivashisnky equation has the form: $\partial_t h(x,t) = -\partial_{xx}h - \partial_{xxxx}h - h\partial_x h$.

appearing when the surface is set in motion in the presence of noise; in the absence of noise the interface would stay flat, but fluctuations tend to make it rough. An interface in d dimensions is defined by its local position,[2] $h(\mathbf{x}, t)$, and it is rough when mean square fluctuations of its position diverge on large space (L) and time (t) scales. Therefore, in the thermodynamic limit, a rough interface never attains a stationary state. However, for large L and t, dynamics has the important feature to be scale invariant, and this property allows us to define appropriate roughness exponents, much in the same way equilibrium critical exponents and universality classes can be defined in the proximity of a second-order phase transition.

5.2 Roughness: Definition, Scaling, and Exponents

Deposition models represent the simplest examples of a stochastically moving interface. They allow us to illustrate the definition of roughness and the role of noise. Furthermore, they can be easily simulated, offering a practical way to test scale invariance and universality and to evaluate roughness exponents.

We can imagine having a flat, d-dimensional substrate of size L^d and depositing particles in a sequential way, one particle after the other. First, we randomly choose a substrate site. Then, we assign some rule for attaching the incoming particle to the growing solid. This separation of the evolution rule in two steps should highlight the role of noise, because the first step is entirely stochastic. Let h_i be the discrete height of site i ($i = 1, \dots, V = L^d$). Periodic boundary conditions in all spatial directions are assumed. Once we have defined the average height of the surface, $\overline{h}$, and the average square height, $\overline{h^2}$,

$$\overline{h}(t) = \frac{1}{V}\sum_{i=1}^{V} h_i(t), \qquad \overline{h^2}(t) = \frac{1}{V}\sum_{i=1}^{V} h_i^2(t), \tag{5.1}$$

the roughness $W(L, t)$ is given by

$$W^2(L, t) = \left\langle \frac{1}{V}\sum_{i=1}^{V}\left(h_i(t) - \overline{h}(t)\right)^2 \right\rangle \tag{5.2}$$

$$= \left\langle \overline{\left(h_i(t) - \overline{h}(t)\right)^2} \right\rangle \tag{5.3}$$

$$= \left\langle \overline{h^2}(t) - \overline{h}^2(t) \right\rangle. \tag{5.4}$$

Two different types of average appear in the definition of $W(L, t)$: the spatial average, $\overline{(\cdots)}$, and the average over noise, $\langle \cdots \rangle$. While the meaning of the former is evident from Eqs. (5.1), the meaning of the latter requires a few more words. The evolution of

[2] This possibility should not be taken for granted, because real or simulated systems may be characterized by overhangs, therefore resulting in a multivalued function. In these cases the function $h(\mathbf{x}, t)$ should be understood as the local position of an effective interface.

the dynamics is stochastic: starting from the same initial conditions we obtain different functions $h_i(t)$. This is obvious in a simulation (provided we update the *seed* of the pseudorandom number generator[3]) and the average over noise just means averaging over different simulation runs. In analytical calculations, as shown in the following, we will average over a suitable function $\eta(\mathbf{x}, t)$, which represents noise. When trying to measure $W(L, t)$ in an experiment the meaning of average over noise is not so clear, because the researcher may well not have the possibility of repeating the experiment many times. However, in an experiment the interface may have a total linear size that may extend over a very large scale, so that we can evaluate $\left(\overline{h^2}(t) - \overline{h}^2(t)\right)$ for different regions of linear size L and average over them (self-averaging).[4]

For future use, it is useful to write the roughness for a continuum system as well,

$$\overline{h}(t) = \frac{1}{V}\int_V d\mathbf{x} h(\mathbf{x}, t), \qquad \overline{h^2}(t) = \frac{1}{V}\int_V d\mathbf{x} h^2(\mathbf{x}, t) \tag{5.5}$$

$$W^2(L, t) = \left\langle \frac{1}{V}\int_V d\mathbf{x} \left(h(\mathbf{x}, t) - \overline{h}(t)\right)^2 \right\rangle \tag{5.6}$$

$$= \left\langle \overline{\left(h(\mathbf{x}, t) - \overline{h}(t)\right)^2} \right\rangle \tag{5.7}$$

$$= \left\langle \overline{h^2}(t) - \overline{h}^2(t) \right\rangle, \tag{5.8}$$

where the system has a linear size L in each dimension and $V = L^d$.

We conclude this preliminary part stressing that "average over noise" and "space average" can be reversed, because they are independent of each other. If $A(\mathbf{x}, \eta)$ is a generic space-dependent and stochastic function, we have

$$\left\langle \overline{A(\mathbf{x}, \eta)} \right\rangle = \left\langle \frac{1}{V}\int_V d\mathbf{x} A(\mathbf{x}, \eta) \right\rangle \tag{5.9}$$

$$= \frac{1}{V}\int_V d\mathbf{x} \, \langle A(\mathbf{x}, \eta) \rangle \tag{5.10}$$

$$= \overline{\langle A(\mathbf{x}, \eta) \rangle}. \tag{5.11}$$

Now, to be more concrete, we make reference to a specific deposition model, a simple but not trivial model that allows us to highlight the general behavior of $W(L, t)$. It is called random deposition with relaxation (RDR) and is defined as follows. We choose randomly a site $i = 1, \ldots, V$ and deposit a particle on it. Then, if the particle can reduce its height moving to a neighboring site k, it does, with k (if necessary) chosen randomly. The algorithm is shown in Fig. 5.2 for a one-dimensional lattice. This model has been simulated and in Fig. 5.3 we plot the roughness as a function of time for different sizes L of the lattice. We remark on a few properties, which are common to different physical systems undergoing kinetic roughening. (i) If we fix L, W first increases in time following

[3] A sequence of pseudorandom numbers is produced deterministically, starting from an initial integer number, called seed. Different runs of the program with the same seed produce the same sequences.

[4] This self-averaging is effective only if different regions are essentially decoupled, which means that L is much larger than the in-plane correlation length, $L \gg \xi_\parallel$. See the continuation of this section for the definition of $\xi_\parallel$.

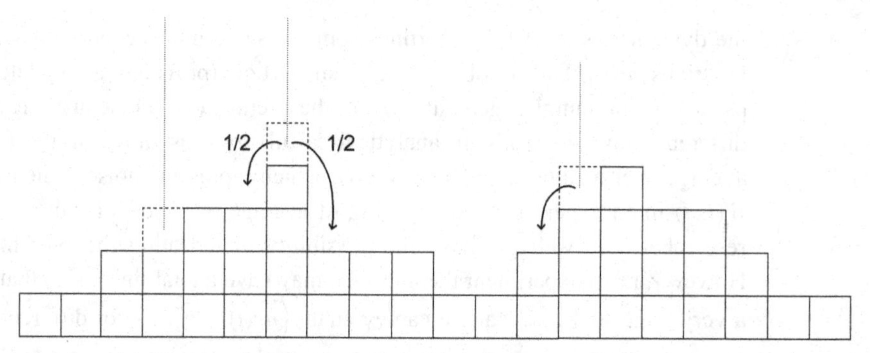

Fig. 5.2 The model of random deposition with relaxation in $d = 1$. A site is chosen randomly and a new (dashed) particle is deposited on it, then left and right neighbors are evaluated. If the new deposited particle can reduce its height by moving to a nearest-neighbor site, it does (center and right cases). If more than one move is allowed (center case), the move is chosen randomly.

a power law, then it saturates to $W_{\text{sat}}(L)$. (ii) If we use a larger L, the short time behavior of W is the same, but saturation occurs at a larger time. (iii) The saturation value depends on L, following a power law.

If t_c is the crossover time between the first regime (W increases in time) and the second regime (W saturates), we can summarize the results as

$$W(L,t) \approx \begin{cases} t^{\beta}, & t \ll t_c \\ L^{\alpha}, & t \gg t_c. \end{cases} \tag{5.12}$$

These expressions define the growth exponent β and the roughness exponent α. In the RDR model we have $\beta \simeq \frac{1}{4}$ and $\alpha \simeq \frac{1}{2}$. At the crossing time we must have $t_c^{\beta} \approx L^{\alpha}$, which implies

$$t_c \approx L^z, \qquad z = \frac{\alpha}{\beta}, \tag{5.13}$$

where z is called dynamic exponent. In the RDR model, according to numerical estimates, $z \simeq 2$.

Fig. 5.3 suggests we might rescale $W(L,t)$ by $W_{\text{sat}}(L)$ so as to obtain a common asymptotic value and to rescale t by $t_c(L)$ so as to have a common crossover time. These rescalings do not affect the power-law behaviors (5.12) and should lead to a collapse of different curves, which is what we observe in Fig. 5.4 if we plot $W(L,t)/L^{\alpha}$ versus t/L^z. Therefore, the roughness is expected to satisfy the scaling form, suggested by Fereydoon Family and Tamás Vicsek,

$$W(L,t) = L^{\alpha} w\left(\frac{t}{L^z}\right), \tag{5.14}$$

with

$$w(u) \approx \begin{cases} u^{\beta}, & u \ll 1 \\ \text{const.}, & u \gg 1. \end{cases} \tag{5.15}$$

We may wonder what the in-plane correlation length, $\xi_{\parallel}$ is, mentioned in note 4 above. If we plot W versus L for different times t, we obtain curves very much similar to those

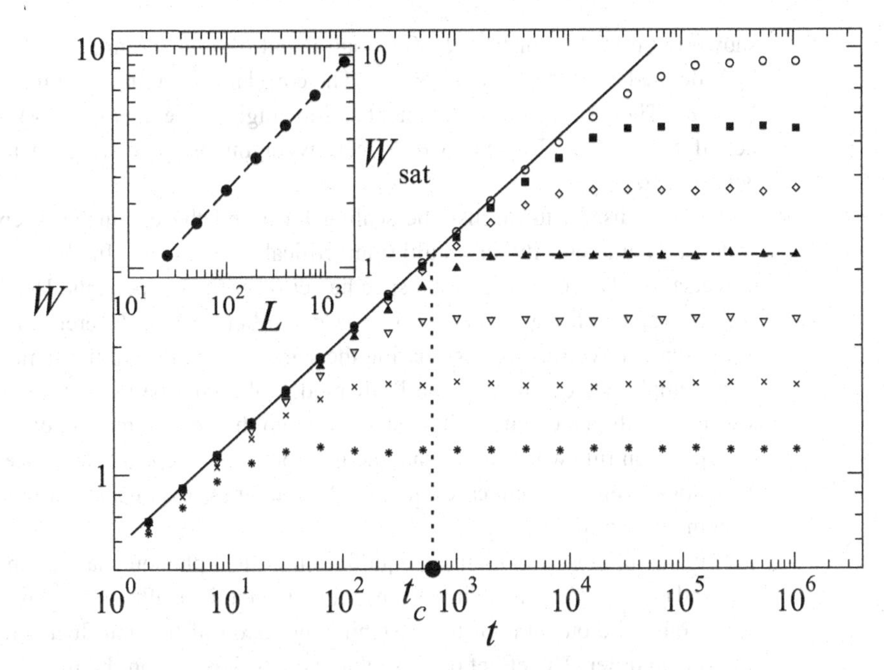

Fig. 5.3 Simulation of the RDR model, depicted in Fig. 5.2. The main graph shows roughness as a function of time for $L = 25$ (asterisks), 50, 100, 200, 400, 800, 1600 (open circles). The fit of the short time regime gives $\beta \simeq 0.247$. The crossing between the saturation value $W(L, t_\infty)$ and the short-time, power-law behavior defines the crossover time, $t_c(L)$. In the inset we plot $W(L, t_\infty)$ versus L, obtaining the roughness exponent $\alpha \simeq 0.502$.

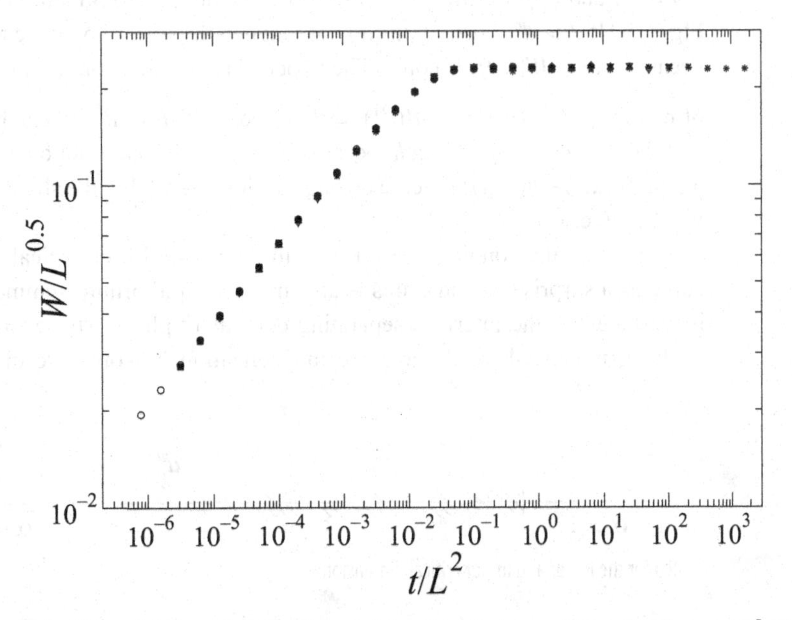

Fig. 5.4 Simulation of the RDR model. Dynamic scaling through data collapse is shown, plotting W/L^α versus t/L^z. We have used the values $\alpha = \frac{1}{2}$ and $\zeta = 2$.

shown in Fig. 5.3, with $W \approx L^\alpha$ for small L and $W \approx t^\beta$ for large L. The crossover length L_c is defined by $L_c^\alpha \approx t^\beta$, so $L_c \approx t^{1/z}$. The correlation length is just this crossover length, $\xi_\parallel = L_c$. Therefore, we have efficient self-averaging (see again note 4) if $L \gg \xi_\parallel \approx t^{1/z}$, i.e., if $t \ll t_c(L)$. In other words, self-averaging is effective to determine β, not to determine α.

It is now useful to discuss the scaling form and the exponents keeping in mind the concept of universality in equilibrium critical phenomena. In the equilibrium case, a universality class is defined by a specific set of critical exponents. Here, it is defined by specific values of, e.g., α and z ($\beta = \alpha/z$). In both cases, different models may belong to the same universality class, sharing the same exponents and the same scaling function. For example, we can modify the RDR model, allowing particles to be incorporated in a lower site within a certain finite distance δ from the deposition site, or we can modify the incorporation rule when more than one incorporation site is allowed (see Fig. 5.2, middle deposition event). In both cases, neither the roughness exponents nor $w(u)$ are affected by such modifications.[5]

Finally and most important, in equilibrium critical phenomena we can define the lower (d_c^ℓ) and upper (d_c^u) critical dimensions for a given universality class. An equilibrium phase transition is the outcome of the destabilizing action of thermal fluctuations, which break long-range order. The effect of these fluctuations depends on the physical dimension d of the system and it increases with decreasing d. If d is small enough ($d < d_c^\ell$), fluctuations are so important that an ordered phase may survive at $T = 0$ K only. Conversely, if d is large enough ($d > d_c^u$), fluctuations are negligible and the exponents can be determined within a mean-field theory.

Lower and upper critical dimensions can be defined for kinetic roughening as well (see Fig. 5.5). If $d > d_c^u$, fluctuations do not roughen the surface on large time and length scales, because $\lim_{L,t\to\infty} W(L,t) = \text{const}$. The upper critical dimension is signaled by the vanishing of α and β, i.e., $\alpha(d_c^u) = \beta(d_c^u) = 0$. Instead, if $d < d_c^\ell$, fluctuations are so important that the surface is *super-rough*, because $\alpha > 1$. In this case not only does the difference of height $(h(\mathbf{x},t) - h(\mathbf{x}',t))$ diverge on large scales $|\mathbf{x}-\mathbf{x}'|$, but the slope $|h(\mathbf{x},t) - h(\mathbf{x}',t)|/|\mathbf{x}-\mathbf{x}'|$ also diverges.

The use of the condition $\alpha(d_c^\ell) = 1$ to define the lower critical dimension should not come as a surprise, because this is also true for equilibrium, symmetry-breaking systems if we focus on the interface separating equivalent phases (for example, two domains of an Ising model where the average magnetization has opposite directions). In fact, the

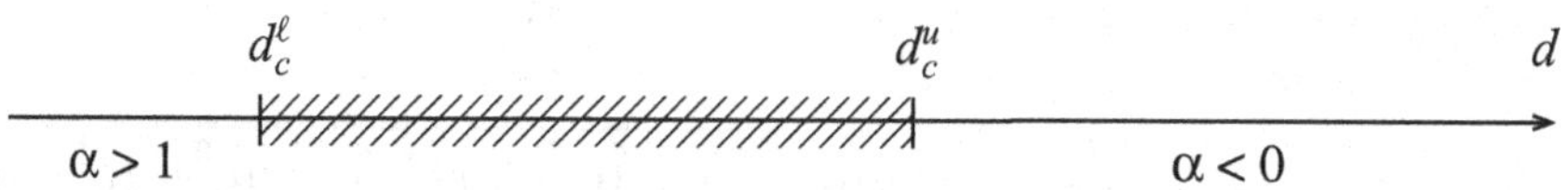

d_c^ℓ d_c^u d

$\alpha > 1$ $\alpha < 0$

Fig. 5.5 Sketch for the lower and upper critical dimensions.

[5] More precisely, $w(u)$ depends on δ or on the incorporation rule through irrelevant constants, as will be shown when we study its continuum description, the Edwards–Wilkinson equation.

interface separating two equivalent domains of size L has a certain roughness defined by the exponent α_E, where the subscript emphasizes that we are now at equilibrium. This means that the fluctuations of the domain wall in the direction perpendicular to it are of order L^{α_E}, but these fluctuations must be smaller than L; otherwise, a domain of size L cannot be defined, because the domain must have a size L in any direction. When $\alpha_E = 1$, phase coexistence breaks: we have attained the lower critical dimension.

We can finally comment about the choice of $W(L,t)$ to describe kinetic roughening phenomena. In fact, we could also use the correlation function,

$$G(\mathbf{r},t) = \left\langle \overline{(h(\mathbf{x}+\mathbf{r},t) - h(\mathbf{x},t))^2} \right\rangle, \tag{5.16}$$

where, as before, $(\overline{\cdots})$ means spatial average over $\mathbf{x}$ and $\langle \cdots \rangle$ means average over the noise. In the limit of diverging $\mathbf{r}$ we have

$$G(\infty,t) = 2W^2(\infty,t), \tag{5.17}$$

while for general (large) $\mathbf{r}, t$, we have

$$G(\mathbf{r},t) = r^{2\alpha} g\left(\frac{t}{r^z}\right), \tag{5.18}$$

with

$$g(u) \approx \begin{cases} u^{2\beta}, & u \ll 1 \\ \text{const} & u \gg 1. \end{cases} \tag{5.19}$$

It is interesting to observe (and, in Section 5.6 we will show) that the lower critical dimension has an additional signature in terms of $G(\mathbf{r},t)$, because the constant appearing in Eq. (5.19) diverges when $d < d_c^\ell$. Thus,

$$\begin{aligned} \lim_{r,t\to\infty} G(\mathbf{r},t) &= \infty \qquad \text{for} \quad d \le d_c^u \\ \lim_{t\to\infty} G(\mathbf{r},t) &= \infty \qquad \text{for} \quad d \le d_c^\ell. \end{aligned} \tag{5.20}$$

5.3 Self-Similarity and Self-Affinity

The reader should be familiar with self-similarity and scale invariance, because these concepts appear when studying equilibrium phase transitions. As discussed in Chapter 3, according to the scaling hypothesis the correlation length ξ of a critical system is the only characteristic length. Since $\xi \to \infty$ when $T \to T_c$, at criticality the system is scale invariant. In the proximity of the critical point, i.e., for vanishing reduced temperature[6] $t^* = (T - T_c)/T_c$ and field H, thermodynamic quantities are homogeneous functions of t^* and H. In the context of kinetic roughening, the critical point corresponds to large time and large L, the typical size over which roughness is evaluated, and roughness $W(L,t)$ in

[6] We use the symbol t^* instead of t to avoid confusion with time.

some sense plays the role of the free energy density, $f(H, t^*)$. Furthermore, as the limits $H, t^* \to 0$ do not commute, neither do the limits $t, L \to \infty$.

In order to determine the correct scaling of $\mathbf{x}$, t, and h to obtain self-similarity, let us consider the scaling form (5.14) for the roughness, which we rewrite here:

$$W(L, t) = L^\alpha w\left(\frac{t}{L^z}\right).$$

The question can be put in the following terms: if we rescale space by a factor[7] b, how should we rescale t and h in order to have a statistically similar profile? Since W depends on time through the function $w(t/L^z)$, if $\mathbf{x} \to b\mathbf{x}$, i.e., $L \to bL$, we need $t \to b^z t$ so that the ratio (t/L^z) remains unchanged. With space rescaling we also have $W \to b^\alpha W$. On the other hand, W is linear in the height h (see definition (5.8)), therefore it must be $h \to b^\alpha h$. In conclusion, the scaling relations are

$$\mathbf{x} \to b\mathbf{x} \tag{5.21a}$$

$$t \to b^z t \tag{5.21b}$$

$$h \to b^\alpha h. \tag{5.21c}$$

Let us comment on transformations (5.21). If we are in the stationary regime ($t \gg L^z$), time transformation (5.21b) has no effect and invariance under Eqs. (5.21a) and (5.21c) just means that a snapshot of the interface profile is statistically invariant under a suitable, anisotropic magnification (the horizontal axis is scaled by a factor b, the vertical axis by a factor b^α). If we are not confined to the stationary regime, a given snapshot, after above anisotropic rescaling, should be compared with a snapshot taken at a different time, and Eq. (5.21b) tells us what the appropriate time is.

The fact that the magnification of a snapshot of the profile is anisotropic, $\mathbf{x}$ and h scaling differently, means that the stationary profile is not self-similar but is self-affine. The simplest example of a random, self-affine object is the trajectory of a one-dimensional random walk. If we plot its position as a function of time, we obtain the nonanalytic curve drawn in Fig. 1.3. It is well known that during the time interval Δt the position of the particle changes by a quantity Δx, with $\langle \Delta x \rangle = 0$ and $\langle (\Delta x)^2 \rangle \sim \Delta t$. This means that rescaling the horizontal axis by a factor b must be accompanied by a rescaling of the vertical axis by a factor $b^{1/2}$, in order to get a statistically equivalent profile.

We have deliberately talked about rescaling of horizontal/vertical axes, rather than rescaling of time and position, because we will see that two different deposition models are characterized by the same value $\alpha = \frac{1}{2}$ in $d = 1$. This means that in the stationary regime, $t \gg t_c(L)$, the height profile $h(x, t)$ of the interface is self-affine under rescaling of the horizontal axis by a factor b and rescaling of the vertical axis by a factor $b^{1/2}$. In other words, if $\alpha = \frac{1}{2}$ the long-time height profile is statistically equivalent to the trajectory of a random walk.

[7] The factor b here plays the same role as Λ in Eq. (3.152).

5.4 Continuum Approach: Toward Langevin-Type Equations

5.4.1 Symmetries and Power Counting

Discrete models are successful for simulations, but continuum models are easier to be treated analytically (even if the resulting continuum equation may finally have to be solved numerically and by a discrete algorithm). Here we plan to write a sufficiently general equation for a moving interface, i.e., a partial differential equation for the interface position $h(\mathbf{x}, t)$,

$$\partial_t h(\mathbf{x}, t) = \mathcal{A}[h] + \eta(\mathbf{x}, t). \tag{5.22}$$

First, we use symmetries and invariances in order to get rid of many terms on the right-hand side, then scale transformation and power counting will provide us with the most relevant terms on large scales.

Eq. (5.22) is not the most general one, because higher-order time derivatives might appear on the left-hand side. However, such terms would be negligible with respect to $\partial_t h$, much in the same way that the term $\ddot{x}$ is asymptotically negligible with respect to $\dot{x}$ in the Langevin equation for a Brownian particle. As a matter of fact, Eq. (5.22) appears to be a generalized Langevin equation, and some known tools, like the Fokker–Planck equation, will be used (see Sections 1.6.1 and 1.6.2 for their introduction).

As for the noise, we limit ourselves here to additive, white, and uncorrelated noise, i.e., to a function $\eta(\mathbf{x}, t)$ having the properties

$$\begin{aligned} \langle \eta(\mathbf{x}, t) \rangle &= 0 \\ \langle \eta(\mathbf{x}, t)\eta(\mathbf{x}', t') \rangle &= \Gamma \delta(\mathbf{x} - \mathbf{x}')\, \delta(t - t'). \end{aligned} \tag{5.23}$$

The first, robust hypothesis is that the functional $\mathcal{A}$ does not depend explicitly on $\mathbf{x}$ and t. This means we assume invariance under time shift ($t \to t + t_0$) and spatial translation ($\mathbf{x} \to \mathbf{x} + \mathbf{x}_0$). A time-dependent incoming flux in a deposition process or a space-dependent wind in a flame front would break such invariances. A second, less robust hypothesis is the spatial invariance in the "growth" direction, i.e., under the transformation $h(\mathbf{x}, t) \to h(\mathbf{x}, t) + h_0$. This assumption implies that $\mathcal{A}[h]$ cannot depend explicitly on h, but only on its spatial derivatives. This condition is not satisfied, e.g., by deposition processes where incoming particles interact with the substrate. A typical example is the so-called heteroepitaxial growth, where a chemical species A is deposited on top of a different species B. The epitaxial constraint forces the growth process to occur under compression or tension conditions, inducing long-range, elastic effects. However, $\mathcal{A}$ does not depend on h in most of the examples cited in Section 5.1 and we will assume such independence also here. In conclusion, our starting point is

$$\partial_t h(\mathbf{x}, t) = \mathcal{A}[\partial_i h, \partial_{ij} h, \dots] + \eta(\mathbf{x}, t), \tag{5.24}$$

where $\partial_i h \equiv \partial h/\partial x_i$, with x_i denoting the ith component of $\mathbf{x}$. The noise term is now set aside, our focus being on $\mathcal{A}$, which can be expanded as follows, according to the order of the derivative and to its power,

$$\mathcal{A} = A_0 + \sum_{i=1}^{d} A_i \partial_i h + \sum_{i,j=1}^{d} A_{ij} \partial_i h \partial_j h + \sum_{i,j=1}^{d} B_{ij} \partial_{ij} h + \cdots . \tag{5.25}$$

Now we wonder if a scale transformation as (5.21) allows us to single out certain terms as the most relevant ones on large scales, i.e., $b \to \infty$. First of all, applying Eq. (5.21b) we find that a time derivative of order n, $\partial_t^n h$, gains a factor b^{-nz}, which explains why we kept only the lowest order time derivative, $n = 1$, on the left-hand side. Before applying Eqs. (5.21a) and (5.21c), we can easily get rid of the first two terms on the right-hand side of Eq. (5.25) with a suitable redefinition of h. The constant term, A_0, disappears if we define $h'(\mathbf{x}, t) = h(\mathbf{x}, t) - A_0 t$, which corresponds to "measuring" the height of the surface with respect to a reference moving at speed A_0. Similarly, the linear term $\sum_i A_i \partial_i h$, disappears under a Galilean transformation,

$$h'(\mathbf{x}, t) = h(\mathbf{x} - \mathbf{v}t, t), \tag{5.26}$$

which implies

$$\begin{aligned} \partial_t h' &= \partial_t h - \sum_i v_i \partial_i h, \\ \partial_i h' &= \partial_i h. \end{aligned} \tag{5.27}$$

Therefore, if we choose $v_i = A_i$, the second term on the right-hand side of Eq. (5.25) disappears and we are finally left with the expression

$$\mathcal{A} = \sum_{i,j=1}^{d} A_{ij} (\partial_i h)(\partial_j h) + \sum_{i,j=1}^{d} B_{ij} \partial_{ij} h + \sum_{i,j,k=1}^{d} A_{ijk} (\partial_i h)(\partial_j h)(\partial_k h) + \cdots . \tag{5.28}$$

It is useful to label each term of Eq. (5.28) with a pair of integers $\{p, q\}$, where p is the number of factors h appearing in the term (regardless of the derivatives) and q is the total number of spatial derivatives. Using this notation, Eq. (5.28) reads

$$\mathcal{A} = \{2, 2\} + \{1, 2\} + \{3, 3\} + \cdots \tag{5.29}$$

It is noteworthy that it must be $p \le q$, otherwise an explicit dependence on h would appear. Think, for example, of the term $\{2, 1\} = \sum_i \tilde{A}_i h(\partial_i h)$.

Under the transformations $\mathbf{x} \to b\mathbf{x}$ and $h \to b^\alpha h$, we obtain

$$\{p, q\} \to b^{p\alpha - q} \{p, q\}. \tag{5.30}$$

Since $p \le q$, for any fixed p and large scale ($b \to \infty$), we have

$$\{p,p\} \gg \{p,p+1\} \gg \{p,p+2\} \ldots, \tag{5.31}$$

with the term $\{p, p\}$ transforming according to

$$\{p,p\} \to b^{p(\alpha-1)} \{p,p\}. \tag{5.32}$$

If $\alpha > 1$, no specific term can be singled out and the expansion (5.25) is useless. If $\alpha < 1$, i.e., for $d > d_c^\ell$, Eqs. (5.31) and (5.32) give a clear outcome of our expansion: the most relevant term is the term $\{p,p\}$ with the smallest, nontrivial p. The first two terms on the right-hand side of (5.25) correspond to $\{0,0\}$ and $\{1,1\}$, respectively. These terms are trivial, because we got rid of them with a simple redefinition of h. Therefore, we expect the relevant term to be

$$\{2,2\} \equiv \sum_{i,j=1}^{d} A_{ij}(\partial_i h)(\partial_j h). \tag{5.33}$$

Since the matrix A_{ij} is real and can be symmetrized,[8] it can be diagonalized, corresponding to a rotation of axes x_i. After this rotation, we have

$$\{2,2\} \equiv \sum_{i=1}^{d} A_{ii}(\partial_i h)^2, \tag{5.34}$$

and the resulting equation is

$$\partial_t h(\mathbf{x},t) = \sum_{i=1}^{d} A_{ii}(\partial_i h)^2 + \eta(\mathbf{x},t), \tag{5.35}$$

which reads, in one and two dimensions,

$$\begin{aligned} \partial_t h(x,t) &= A_{11}(\partial_x h)^2 + \eta(x,t), & d=1 \\ \partial_t h(x,y,t) &= A_{11}(\partial_x h)^2 + A_{22}(\partial_y h)^2 + \eta(x,y,t), & d=2. \end{aligned} \tag{5.36}$$

Changing the sign of the interface profile ($h \to -h$) changes the sign of all coefficients A_{ij} on the right-hand side (a change of sign of the noise is irrelevant), and rescaling x and y ($x \to a_x x$, $y \to a_y y$) allows us to rescale separately the coefficients to make $|A_{11}| = |A_{22}|$, without changing their signs. These facts allow us to rewrite Eqs. (5.36) as

$$\begin{aligned} \partial_t h(x,t) &= |A_{11}|(\partial_x h)^2 + \eta(x,t), & d=1 \\ \partial_t h(x,y,t) &= |A_{11}|\left[(\partial_x h)^2 \pm (\partial_y h)^2\right] + \eta(x,y,t), & d=2. \end{aligned} \tag{5.37}$$

where the sign $+$ $(-)$ applies if A_{11} and A_{22} have the same (opposite) sign.

As already noted (and exploited), Eq. (5.35) does not satisfy the reflection symmetry $h \to -h$, while it satisfies the symmetry $\mathbf{x} \to -\mathbf{x}$ (in fact, it satisfies the stronger symmetry $x_i \to -x_i$ for each direction i, separately). In case the $h \to -h$ symmetry does hold (and that depends on the specific physical process under investigation), the term $\{2,2\}$ (see Eq. (5.33)) cannot survive, because p must be odd. If we also require the symmetry $\mathbf{x} \to -\mathbf{x}$, q must be even. Finally, the most relevant term would be

$$\{1,2\} \equiv \sum_{i,j=1}^{d} B_{ij}\partial_{ij} h, \tag{5.38}$$

[8] If not symmetric, it suffices to rewrite $\sum_{ij} A_{ij}\partial_i h \partial_j h = \sum_{ij} \frac{1}{2}(A_{ij}+A_{ji})\, \partial_i h \partial_j h$ and the new matrix $\frac{1}{2}(A_{ij}+A_{ji})$ is symmetric.

which simplifies to

$$\{1,2\} \equiv \sum_{i=1}^{d} B_{ii}\partial_{ii}h \tag{5.39}$$

after a rotation of the axes or if the stronger symmetry $x_i \to -x_i$ applies.

In the following, we consider rotational invariance in the d-dimensional space, so that the term $\{2,2\}$ is proportional to $(\nabla h)^2$ and the term $\{1,2\}$ is proportional to $\nabla^2 h$. In conclusion, we are led to consider the following two equations,

$$\partial_t h(\mathbf{x},t) = \nu\nabla^2 h + \eta(\mathbf{x},t) \qquad \text{EW} \tag{5.40}$$

$$\partial_t h(\mathbf{x},t) = \nu\nabla^2 h + \frac{\lambda}{2}(\nabla h)^2 + \eta(\mathbf{x},t). \quad \text{KPZ} \tag{5.41}$$

Eq. (5.40) is called Edwards–Wilkinson (EW) equation. It looks like the diffusion equation with a noise term, which explains why we require $\nu > 0$. It is a linear equation that satisfies the reflection symmetry $h \to -h$ and has the conserved form $\partial_t h(\mathbf{x},t) = -\nabla \cdot \mathbf{J} + \eta(\mathbf{x},t)$, which will be discussed in more detail in Section 5.6.

Eq. (5.41) is called Kardar–Parisi–Zhang (KPZ) equation and it does *not* satisfy the symmetry $h \to -h$. Furthermore, it does not have a conserved form. KPZ is nonlinear and the sign of λ is irrelevant, because changing the sign of h also changes the sign of λ. It is usual to assume $\lambda > 0$. In spite of its irrelevance at large scales, the linear term is present, because it regularizes the equation at small scales. This is obvious from Eq. (L.9) in the deterministic case, where the width of the Burgers front vanishes for $\nu = 0$, inducing angular points between parabolas; see Eq. (L.10). In the stochastic case, white noise induces fluctuations at arbitrarily small scales, which must be smoothed out by the addition of the linear, diffusive term. The KPZ equation will be studied in detail in Section 5.7.

5.4.2 Hydrodynamics

In addition to the previous, general derivation of continuum equations we propose here a hydrodynamic derivation of the KPZ equation from particle models, in one dimension. We will do that in two slightly different ways: first, by starting from an explicit ASymmetric Exclusion Process (ASEP), which also allows us to define a surface growth model; second, in more generality, using a phenomenological approach based on the continuity equation.

The mapping of ASEP into a surface growth model is illustrated in Fig. 5.6. In the upper part of the figure we sketch the topology and the dynamics of ASEP: (i) we have a one-dimensional lattice where sites can be occupied by a particle (gray sites) or they can be empty (white sites), and (ii) each particle can move to its right with probability c_+ and to the left with probability c_-, if the arrival site is empty. Now, the mapping works as follows: an occupied (empty) site i corresponds to a piece of surface of negative (positive) slope $m_i = -1$ ($m_i = +1$). As for dynamics, moving a particle to the right in the ASEP model corresponds to depositing a rhomboidal particle in a local minimum in the growth model and moving a particle to the left corresponds to evaporating a particle from a local maximum. In order to implement periodic boundary conditions in the growth model, we

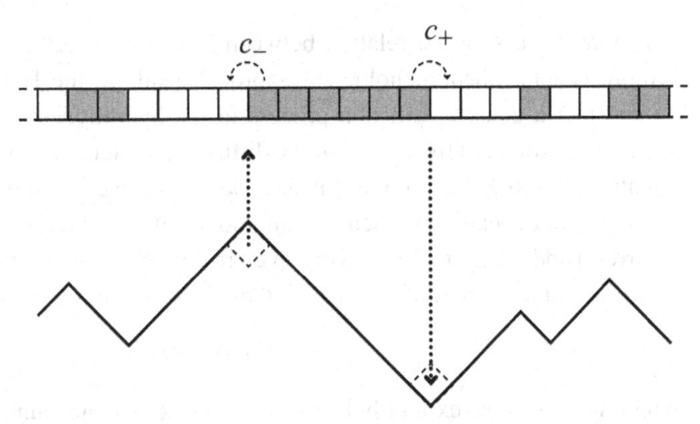

Fig. 5.6 Mapping of ASEP to the single-step model. A move of a particle to the right (left) corresponds to the deposition (evaporation) of a particle in a local minimum (from a local maximum). Driving the particle model out of equilibrium means creating an asymmetry between hops to the right and to the left: in the surface growth model, this asymmetry implies an unbalancing between deposition and evaporation.

require an equal number of occupied and empty sites in the ASEP model; otherwise, helical boundary conditions should be used. In any case the number of particles is a constant of motion and it determines the average slope of the surface. The resulting growth model is called single step model.

In the ASEP model, the probability that site i is occupied, p_i, depends on time because the net current $J_{i-1,i}$ between sites $i-1$ and i does not counterbalance the net current $J_{i,i+1}$. More precisely,

$$\tau_0 \frac{dp_i}{dt} = c_+[p_{i-1}(1-p_i) - p_i(1-p_{i+1})] + c_-[p_{i+1}(1-p_i) - p_i(1-p_{i-1})], \quad (5.42)$$

where the time constant τ_0 has been introduced to make the relation dimensionally correct. To pass to the continuum, we write $p_i(t) = p(x,t)$ and $p_{i\pm 1} = p(x,t) \pm a_0 p'(x,t) + \frac{1}{2}a_0^2 p''(x,t)$, where a_0 is the lattice constant and primes indicate spatial derivatives. If we write $c_\pm = \frac{1}{2} \pm \delta$, we obtain

$$\tau_0 \frac{\partial p}{\partial t} = -2\delta a_0 p' + \frac{1}{2}a_0^2 p'' + 4\delta a_0 p p'. \quad (5.43)$$

We can now pass from the variable density, p, to the variable slope, $m = 1 - 2p$, arriving at the so-called Burgers equation,

$$\partial_t m(x,t) = \nu m_{xx} + \lambda m m_x, \quad (5.44)$$

where[9] $\nu = a_0^2/(2\tau_0)$ and $\lambda = -(2\delta a_0)/\tau_0$.

[9] The continuum limit must be handled with care, because if we take $a_0, \tau_0 \to 0$, it is manifest that the nonlinear term dominates over the linear one. This is a well-known fact when we consider anisotropic diffusion: in fact, along with $a_0 \to 0$ we should also assume a vanishing drift, $\delta \to 0$. The physical reason is that the combination of a finite drift with a vanishing lattice constant a_0 would lead to a divergent driving force in the continuum limit.

Before discussing the relation between Burgers and KPZ equations, let us prove how a more general phenomenological approach leads to the Burgers equation as well. Let us consider a generic diffusion process in the hydrodynamic limit, i.e., for large spatial scales. Regardless of the details of the diffusion, matter conservation implies the continuity equation, $\partial_t \rho + \partial_x J = 0$. On a mesoscopic scale we assume that J is composed of two parts: a term depending on density inhomogeneities (which is the simplest "force" driving a current) and a term that is at work even if ρ is spatially constant. The latter term depends on the asymmetric part of the diffusion process. Summing up, we can write

$$J = J_0(\rho) - D\partial_x \rho + \xi, \tag{5.45}$$

where we have now explicitly introduced a noise term accounting for the stochastic nature of the diffusion process. We can therefore write

$$\frac{\partial \rho}{\partial t} + J_0{}'(\rho)\partial_x \rho - D\partial_{xx}\rho + \partial_x \xi = 0. \tag{5.46}$$

If we now assume that ρ has small deviations from some values ρ_0, $\rho = \rho_0 + u(x,t)$, and expand with respect to u, we have

$$\partial_t u + \left(J_0{}'(\rho_0) + J_0{}''(\rho_0)u\right)\partial_x u - D\partial_{xx} u + \partial_x \xi = 0. \tag{5.47}$$

If we define the function $m(x,t)$, corresponding to the profile $u(x,t)$ traveling at speed $J_0{}'(\rho_0)$, $u(x,t) = m(x - J_0{}'(\rho_0)t, t)$, we obtain

$$\partial_t m = \nu \partial_{xx} m + \lambda m \partial_x m - \partial_x \xi, \tag{5.48}$$

where we have defined $\lambda = -J_0{}''(\rho_0)$ and replaced D with ν. The above equation is the stochastic Burgers equation, to be compared with Eq. (5.44).

The mapping of ASEP to the single-step model, whose hydrodynamic description is provided by the stochastic Burgers equation, indicates that m is a slope rather than a height h. Therefore, we write $m(x,t) = \partial_x h(x,t)$ and integrate both sides of Eq. (5.48). Up to an irrelevant constant and to the irrelevant sign of ξ, we obtain

$$\partial_t h(x,t) = \nu h_{xx} + \frac{\lambda}{2} h_x^2 + \xi(x,t), \tag{5.49}$$

which is recognized as the KPZ equation. Its deterministic version, corresponding to $\xi \equiv 0$ and equivalent to the Burgers equation (5.44), is studied in Appendix L.

5.5 The Random Deposition Model

Let us now discuss specific models, starting with the random deposition (RD) model, which could not be simpler: we choose randomly a site and add a particle, then we choose a new site, and so on. Since there is no postdeposition process and the choice of the deposition site is fully random, there is manifestly no coupling between different sites, which means that the size and even the spatial dimension of the system are irrelevant.

For the same reason, with the absence of coupling, a spatially continous model would produce singularities, as we are going to argue below. In a spatially discrete model the evolution equation of each site i is simply

$$\partial_t h_i(t) = \eta_i(t), \tag{5.50}$$

with $\langle \eta_i(t) \rangle = 0$ and $\langle \eta_i(t)\eta_j(t') \rangle = \Gamma \delta_{ij}\delta(t - t')$. A constant term A_0 on the right-hand side has been eliminated by redefining $h_i(t)$ as the local height with respect to the average height.

The reader can easily recognize Eq. (5.50) as the evolution equation of a free, random walk. Therefore, the random deposition model simply corresponds to a collection of independent random walkers. We can therefore immediately conclude that the roughness, i.e., the mean square fluctuations of the interface, grows in time as the square root of time, $W \sim \sqrt{t}$. So, the growth exponent is $\beta = \frac{1}{2}$, while there is no saturation of the roughness at large t: because of the absence of coupling between different sites, the system never attains a stationary state and the roughness exponent α is not defined in this model.

Finally, we can evaluate explicitly the roughness and the correlation function, which allows us to highlight the singularity that would appear in a continuum approach. We need the following quantities,

$$\overline{h}(t) = \int_0^t dt' \overline{\eta_i(t)} \tag{5.51}$$

$$\overline{h^2}(t) = \int_0^t dt' \int_0^t dt'' \overline{\eta_i(t')\eta_i(t'')} \tag{5.52}$$

$$\overline{h_{i+r}(t)h_i(t)} = \int_0^t dt' \int_0^t dt'' \overline{\eta_{i+r}(t')\eta_i(t'')}. \tag{5.53}$$

Before performing the average over the noise we remark that $\overline{\eta_i(t)} = 0$, because with the spatial average we are averaging over an infinite set of independent realizations of noise. Finally, since the "average over noise" and the "space average" can be reversed, we obtain

$$W^2(L,t) = \left\langle \overline{h^2}(t) - \overline{h}^2 \right\rangle = \int_0^t dt' \int_0^t dt'' \Gamma \delta_{ii} \delta(t' - t'') = \Gamma t \tag{5.54}$$

$$G(r,t) = 2\left\langle \overline{h^2}(t) - \overline{h_{i+r}(t)h_i(t)} \right\rangle \tag{5.55}$$

$$= 2\Gamma(\delta_{ii} - \delta_{i+r,i})t = 2\Gamma(1 - \delta_{r,0})t. \tag{5.56}$$

In a continuum approach the Kronecker delta is replaced by a Dirac delta and the quantity δ_{ii} would be replaced by $\delta(0)$. This is the reason why the RD model is not consistent within a continuum picture.

5.6 The Edwards–Wilkinson Equation

The simplest, nontrivial continuum equation is the EW equation,

$$\partial_t h(\mathbf{x}, t) = \nu \nabla^2 h + \eta(\mathbf{x}, t), \tag{5.57}$$

which satisfies the symmetry $h \to -h$ and is rotationally invariant.

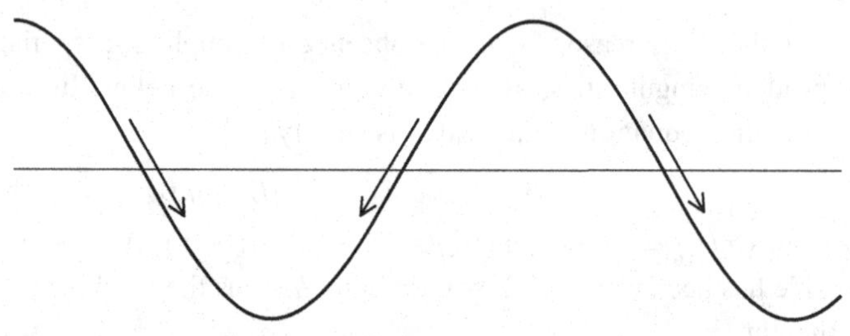

Fig. 5.7 Pictorial image of the current **J** of the EW model.

Its deterministic part is conserved,[10] because the equation has the form

$$\partial_t h = -\nabla \cdot \mathbf{J} + \eta, \tag{5.58}$$

with $\mathbf{J} = -\nu \nabla h$. This implies that the average height of the interface is constant,

$$\frac{d\bar{h}}{dt} = \frac{1}{V}\int_V d\mathbf{x}\eta(\mathbf{x}, t) = 0, \tag{5.59}$$

because the integral of the divergence vanishes for periodic boundary conditions.

The deterministic part looks like the diffusion equation and its form means there is a net current of matter from high regions (large h) to low regions (small h), as exemplified in Fig. 5.7. This current produces a smoothing of the surface and competes with noise, which tends to roughen the interface, as clearly shown by the random deposition model. In a deposition process it seems natural to relate **J** to the relaxation mechanism studied in Fig. 5.2: after deposition a particle moves to a neighboring site if it can reduce its height.

If this reasoning is correct, we expect that the RDR model belongs to the EW universality class, therefore sharing the same exponents α, β: this is actually what happens. The hypothesis that the RDR model is correctly described by the EW equation can be understood if we give h_i a different meaning: not the local height of site i, but the number of particles in site i. Then, the RDR model corresponds to a diffusion-like process in the presence of a uniform supply of particles from some external reservoir. Within this interpretation, the parameter ν is exactly the diffusion coefficient and it is also clear why the flux does not break the up/down symmetry.

Before determining the roughness exponents and the scaling function for the EW equation, we discuss a further property of this equation: it is derivable from a potential. In fact, the equation has the form (see Appendix L.1)

$$\partial_t h = -\frac{\delta \mathcal{F}}{\delta h} + \eta \tag{5.60}$$

with

$$\mathcal{F}[h] = \frac{\nu}{2}\int_V d\mathbf{x}(\nabla h)^2. \tag{5.61}$$

[10] The noise itself might conserve matter. This is not the case in deposition processes, where noise represents fluctuations of an external source.

This form implies that the dynamics evolves in order to minimize the potential, because

$$\frac{d\mathcal{F}}{dt} = \int_V d\mathbf{x} \frac{\delta \mathcal{F}}{\delta h}\frac{\partial h}{\partial t} \tag{5.62}$$

$$= -\int_V d\mathbf{x} \left(\frac{\delta \mathcal{F}}{\delta h}\right)^2 + \int_V d\mathbf{x}\frac{\delta \mathcal{F}}{\delta h}\eta(\mathbf{x},t), \tag{5.63}$$

so that[11]

$$\left\langle \frac{d\mathcal{F}}{dt} \right\rangle \leq 0. \tag{5.64}$$

The meaning of $\mathcal{F}$ is simple, because it represents the total area $\mathcal{S}$ of the surface $h(\mathbf{x})$, in the small slope approximation,

$$\mathcal{S} = \int_V d\mathbf{x}\sqrt{1+(\nabla h)^2} = V + \frac{1}{2}\int_V d\mathbf{x}(\nabla h)^2 + O\left((\nabla h)^4\right). \tag{5.65}$$

Therefore, neglecting terms that are irrelevant for power counting, we can write

$$\mathcal{F} = \nu \mathcal{S}, \tag{5.66}$$

and the minimization of $\mathcal{F}$ acquires the trivial meaning of minimizing the total surface area.

The fact that dynamics is driven by the potential (5.61) has a simple consequence for the properties of the stationary state (which is attained at $t \gg t_c \simeq L^z$): it is statistically equivalent to an equilibrium state described by the Hamiltonian $\mathcal{H} = \mathcal{F}$.

5.6.1 Dimensional Analysis

Let us rewrite the EW equation and the noise correlation function,

$$\partial_t h(\mathbf{x},t) = \nu \nabla^2 h + \eta(\mathbf{x},t) \tag{5.67}$$

$$\langle \eta(\mathbf{x},t)\eta(\mathbf{x}',t')\rangle = \Gamma \delta(\mathbf{x}-\mathbf{x}')\delta(t-t'). \tag{5.68}$$

Two parameters enter in these expressions, the "diffusion coefficient" ν and the strength of noise $\sqrt{\Gamma}$, appearing in the noise correlation function. Indicating with $[A]$ the dimension of the physical quantity A, we can start writing $\mathtt{L} = [\mathbf{x}]$, $\mathtt{T} = [t]$, and $\mathtt{H} = [h]$. The first two relations are trivial: we just state that $\mathbf{x}$ is a spatial variable and t is a time variable. Instead, the third one may appear arbitrarily redundant: why should we introduce a *new* dimension $\mathtt{H}$ for a height variable? Shouldn't we simply write $\mathtt{H} = \mathtt{L}$? In fact, no: the EW equation describes a variety of different physical problems, not only deposition processes, and there is no reason to assume that h represents a length. There is also another motivation to keep $\mathtt{H}$ and $\mathtt{L}$ distinct, even in deposition processes, where h can be measured in centimeters as is x: the EW equation describes the dynamics of a self-affine interface, not a self-similar one; so, the space parallel to the substrate behaves differently from the space orthogonal to it.

[11] The second term in Eq. (5.63) may be positive and occasionally dominate the deterministic, negative term, but it averages to zero.

We can now impose that the different terms appearing in Eqs. (5.67) and (5.68) have the same dimensions. We can write, respectively,

$$\frac{\mathtt{H}}{\mathtt{T}} = \frac{[\nu]\,\mathtt{H}}{\mathtt{L}^2} = [\eta] \tag{5.69a}$$

$$[\eta]^2 = \frac{[\Gamma]}{\mathtt{L}^d\mathtt{T}}, \tag{5.69b}$$

which give

$$[\nu] = \frac{\mathtt{L}^2}{\mathtt{T}} \tag{5.70a}$$

$$[\Gamma] = \frac{\mathtt{L}^d\mathtt{H}^2}{\mathtt{T}}. \tag{5.70b}$$

We should now use the Family–Vicsek scaling for $W(L, t)$ including an explicit dependence on Γ and ν,

$$W(L,t) = \Gamma^{a_1}\nu^{b_1}L^\alpha w\left(\Gamma^{a_2}\nu^{b_2}\frac{t}{L^z}\right), \tag{5.71}$$

and hopefully determine all the exponents by imposing that left-hand side and right-hand side have the same dimension. This amounts to saying that the argument of the function w is adimensional and the factor multiplying w has the dimension $[W] = \mathtt{H}$,

$$[\Gamma]^{a_1}\,[\nu]^{b_1}\mathtt{L}^\alpha = \mathtt{H}, \tag{5.72a}$$

$$[\Gamma]^{a_2}\,[\nu]^{b_2}\mathtt{T} = \mathtt{L}^z. \tag{5.72b}$$

Using the above results for $[\nu]$ and $[\Gamma]$, we can finally write

$$\left(\frac{\mathtt{L}^d\mathtt{H}^2}{\mathtt{T}}\right)^{a_1}\left(\frac{\mathtt{L}^2}{\mathtt{T}}\right)^{b_1}\mathtt{L}^\alpha = \mathtt{H}, \tag{5.73a}$$

$$\left(\frac{\mathtt{L}^d\mathtt{H}^2}{\mathtt{T}}\right)^{a_2}\left(\frac{\mathtt{L}^2}{\mathtt{T}}\right)^{b_2}\mathtt{T} = \mathtt{L}^z. \tag{5.73b}$$

We can see that the consistency conditions on $\mathtt{H}$ impose $a_1 = \frac{1}{2}$ and $a_2 = 0$. The skeptical reader will be reassured by an alternative derivation of such results: in fact, $a_1 = \frac{1}{2}$ and $a_2 = 0$ mean that $W(L, t)$ is proportional to $\sqrt{\Gamma}$; see Eq. (5.71). The linearity between W and the strength of the noise simply comes from the linear character of the EW equation. Once we have determined a_1 and a_2, Eq. (5.73b) gives $b_2 = 1$ and $z = 2$, while Eq. (5.73a) gives $b_1 = -\frac{1}{2}$ and $\alpha = \frac{2-d}{2}$. Finally, using the scaling relation $z = \alpha/\beta$ we obtain β. In conclusion, a simple dimensional analysis has allowed us to find the roughness exponents

$$\alpha = \frac{2-d}{2}, \quad \beta = \frac{2-d}{4}, \quad z = 2. \qquad \text{(EW equation)} \tag{5.74}$$

From the conditions $\alpha(d_c^\ell) = 1$ and $\alpha(d_c^u) = \beta(d_c^u) = 0$, we find that the lower and upper critical dimensions are $d_c^\ell = 0$ and $d_c^u = 2$. For $d = 1$ we obtain $\alpha = \frac{1}{2}$ and $\beta = \frac{1}{4}$, which agree with the numerical values found for the RDR model, thus attesting that it belongs to the EW universality class.

We have also been able to write the scaling function in the universal form

$$W_{\text{EW}}(L,t) = c_1 \sqrt{\frac{\Gamma L^{2-d}}{\nu}} w_{\text{EW}}\left(c_2 \frac{\nu t}{L^2}\right), \tag{5.75}$$

where the numerical factors c_1, c_2 are the only non-universal quantities. In other words, different models belonging to the EW universality class must have the above scaling function and possibly differ only in the values of $c_{1,2}$. The explicit expression of the function w will be derived in the next section.

We conclude this part by considering the effect of scale transformations (5.21) on the EW equation. In order to do that, we need to know how the noise scales, which can be deduced from its correlation function

$$\langle \eta(\mathbf{x},t)\eta(\mathbf{x}',t')\rangle = \Gamma\delta(\mathbf{x}-\mathbf{x}')\,\delta(t-t'). \tag{5.76}$$

If $\eta \to b^\chi \eta$, using Eqs. (5.21) the condition that left-hand side and right-hand side scale in the same way implies

$$b^{2\chi} = \frac{1}{b^d}\frac{1}{b^z}, \tag{5.77}$$

so

$$\chi = -\frac{d+z}{2}. \tag{5.78}$$

We can finally determine how the different terms of the EW equation scale with b,

$$b^{\alpha-z}\partial_t h(\mathbf{x},t) = \nu b^{\alpha-2}\nabla^2 h + b^{-\frac{d+z}{2}}\eta(\mathbf{x},t). \tag{5.79}$$

It is straightforward to check that all terms scale in the same way, if the correct expressions (5.74) for the roughness exponents are used. Therefore, we might have used Eq. (5.79) the other way and determine α and z by *imposing* the condition that all terms must scale in the same way.

5.6.2 The Scaling Functions

The conserved character of the EW equation implies $\overline{h(\mathbf{x},t)} \equiv 0$, and the roughness has a simpler expression,

$$W^2(L,t) = \left\langle \overline{h^2(\mathbf{x},t)} \right\rangle = \overline{\langle h^2(\mathbf{x},t)\rangle}, \tag{5.80}$$

where we have used the possibility of exchanging the two types of average (see Section 5.2). In fact, if the system is translationally invariant, once we average $h^2(\mathbf{x},t)$ over the noise realizations, the spatial dependence disappears, so we simply have

$$W^2(L,t) = \langle h^2(\mathbf{x},t)\rangle. \tag{5.81}$$

The linear EW equation can be solved in Fourier space, so we write

$$h(\mathbf{x},t) = \frac{1}{(2\pi)^d}\int d\mathbf{q} e^{i\mathbf{q}\cdot\mathbf{x}} h(\mathbf{q},t), \tag{5.82}$$

and, from Eq. (5.81),

$$W^2_{\mathrm{EW}}(L,t) = \frac{1}{(2\pi)^{2d}}\int d\mathbf{q}\int d\mathbf{q}' e^{i(\mathbf{q}+\mathbf{q}')\cdot\mathbf{x}} \langle h(\mathbf{q},t) h(\mathbf{q}',t)\rangle . \tag{5.83}$$

Replacing the Fourier expansion in the EW equation, we find that each Fourier component $h(\mathbf{q},t)$ satisfies the equation

$$\partial_t h(\mathbf{q},t) = -\nu q^2 h + \eta(\mathbf{q},t), \tag{5.84}$$

where $q^2 = \mathbf{q}\cdot\mathbf{q}$. Its solution is[12]

$$h(\mathbf{q},t) = h(\mathbf{q},0)e^{-\omega_{\mathbf{q}}t} + \int_0^t dt' \eta(\mathbf{q},t')e^{-\omega_{\mathbf{q}}(t-t')}, \tag{5.85}$$

where we have used the simplified notation $\omega_{\mathbf{q}} = \nu q^2$. The first term on the right-hand side vanishes for an initially flat surface ($h(\mathbf{q},0) \equiv 0$), or it vanishes anyway after a transient, because of the exponential factor. Therefore, for long times the second term on the right-hand side always dominates, which also proves the linearity between h and the noise. So, we can write

$$h(\mathbf{q},t) = \int_0^t dt' \eta(\mathbf{q},t')e^{-\omega_{\mathbf{q}}(t-t')} \tag{5.86}$$

and the correlator appearing in Eq. (5.83) takes the form

$$\langle h(\mathbf{q},t) h(\mathbf{q}',t)\rangle = \int_0^t dt' e^{-\omega_{\mathbf{q}}(t-t')} \int_0^t dt'' e^{-\omega_{\mathbf{q}'}(t-t'')} \langle \eta(\mathbf{q},t')\eta(\mathbf{q}',t'')\rangle. \tag{5.87}$$

Therefore, we need to evaluate the correlator of noise in the Fourier space. Since $\eta(\mathbf{q},t) = \int_V d\mathbf{x} e^{-i\mathbf{q}\cdot\mathbf{x}}\eta(\mathbf{x},t)$, we find

$$\begin{aligned}\langle \eta(\mathbf{q},t')\eta(\mathbf{q}',t'')\rangle &= \int d\mathbf{x}\int d\mathbf{x}' e^{-i\mathbf{q}\cdot\mathbf{x}} e^{-i\mathbf{q}'\cdot\mathbf{x}'} \langle \eta(\mathbf{x},t')\eta(\mathbf{x}',t'')\rangle \\ &= \Gamma\delta(t'-t'')\int d\mathbf{x} e^{-i\mathbf{q}\cdot\mathbf{x}} e^{-i\mathbf{q}'\cdot\mathbf{x}} \\ &= (2\pi)^d \Gamma \delta(t'-t'')\delta(\mathbf{q}+\mathbf{q}'),\end{aligned} \tag{5.88}$$

which can be replaced in (5.87), obtaining

$$\begin{aligned}\langle h(\mathbf{q},t)h(\mathbf{q}',t)\rangle &= (2\pi)^d\Gamma \int_o^t dt' e^{-\omega_{\mathbf{q}}(t-t')} e^{-\omega_{-\mathbf{q}}(t-t')}\delta(\mathbf{q}+\mathbf{q}') \\ &= (2\pi)^d\Gamma \left(\frac{1-e^{-2\omega_{\mathbf{q}}t}}{2\omega_{\mathbf{q}}}\right)\delta(\mathbf{q}+\mathbf{q}').\end{aligned} \tag{5.89}$$

[12] The solution of the differential equation $\dot{g}(t) = -\omega g + f(t)$, found by the method of variation of parameters, is $g(t) = g(0)e^{-\omega t} + \int_0^t dt' f(t')e^{-\omega(t-t')}$.

Finally, this result can be inserted in Eq. (5.83), which yields

$$W^2_{\mathrm{EW}}(L,t) = \frac{\Gamma}{(2\pi)^d}\int_{V_{\mathbf{q}}} d\mathbf{q}\frac{1-e^{-2\omega_{\mathbf{q}}t}}{2\omega_{\mathbf{q}}}, \tag{5.90}$$

where the domain of integration is $V_{\mathbf{q}} \equiv \{\frac{\pi}{L} \leq |q_i| \leq \frac{\pi}{a_0}\}$, a_0 being a cutoff for avoiding integration over arbitrarily small wavelengths, and it is justified by the ultimate discrete nature of matter. As expected, the right-hand side of Eq. (5.90) does not depend on $\mathbf{x}$ even if no spatial average has been performed: averaging over noise restores translational invariance.

Result (5.90) has a much larger applicability than expected. In fact, it is valid for any equation of the form

$$\partial_t h(\mathbf{x},t) = \mathcal{L}[h] + \eta(\mathbf{x},t) \tag{5.91}$$

where $\mathcal{L}[h]$ is a generic[13] linear operator. Fourier-transforming the equation,

$$\partial_t h(\mathbf{q},t) = -\omega_{\mathbf{q}} h(\mathbf{q},t) + \eta(\mathbf{q},t), \tag{5.92}$$

we obtain the function $\omega_{\mathbf{q}}$, which encapsulates the properties of $\mathcal{L}$ and appears in Eq. (5.90). Let us consider a possible generalization of the EW equation, called the (stochastic) Herring–Mullins equation,

$$\partial_t h(\mathbf{x},t) = \nu\nabla^2 h - K(\nabla^2)^2 h + \eta(\mathbf{x},t). \tag{5.93}$$

The minus sign in front of the quartic term is necessary to have a positive $\omega_{\mathbf{q}} = \nu q^2 + Kq^4$, so that the deterministic component of each Fourier mode $h(\mathbf{q},t)$ decays in time as $\exp(-\omega_{\mathbf{q}}t)$ while the stochastic component fluctuates; see Eq. (5.85). From a physical point of view, the quadratic and quartic terms represent different atomistic processes. The quadratic term, as shown in Fig. 5.7, comes from a surface current proportional to the slope, while the quartic term, as discussed in more detail in Section 6.4.2, comes from a chemical potential related to the curvature of the surface. According to power counting, the higher-order term $(\nabla^2)^2 h$ should be negligible with respect to the EW linearity $\nabla^2 h$, and this is reflected in the smallness of the q^4 term with respect to the q^2 term for large spatial scales, i.e., for $q \to 0$. However, the microscopic processes determining the EW linearity might be absent, or, if present, ν might be small. In the latter case we have a crossover from a regime dominated by the quartic term (for $q \gg \sqrt{\nu/K}$) to the EW regime (for $q \ll \sqrt{\nu/K}$). Here we will limit ourselves to considering a single, dominant linear term with spatial derivatives of order δ, so that $\omega_{\mathbf{q}} = cq^\delta$.

If the system has to be rough, W must diverge with increasing L and t. This means we require (see Eq. (5.90)) that,

$$W^2(\infty,\infty) = \lim_{L\to\infty}\frac{\Gamma}{(2\pi)^d}\int_{V_{\mathbf{q}}}\frac{d\mathbf{q}}{2\omega_{\mathbf{q}}} = \infty, \tag{5.94}$$

[13] In fact, in our derivation we have used $\omega_{\mathbf{q}} = \omega_{-\mathbf{q}}$, which means the operator $\mathcal{L}$ should be invariant under parity, $\mathbf{x} \to -\mathbf{x}$.

and such divergence comes out from small-**q** integration. A simple evaluation of the small-**q** contribution allows us to determine the upper critical dimension. In fact,

$$W^2(\infty,\infty) \approx \int_0 \frac{d\mathbf{q}}{\omega_\mathbf{q}} \approx \int_0 dq \frac{q^{d-1}}{q^\delta}, \tag{5.95}$$

which diverges if $d < \delta$. So, $d_c^u = \delta$, extending the result $d_c^u = 2$, based on dimensional arguments we found for EW. In the following we assume $d < \delta$, in which case the integral has a divergence for small **q**, while it converges for large **q**. This allows us to replace the upper limit of the integral, of order π/a_0, with infinity. Let's now define $V_\mathbf{q}^\infty \equiv \{|q_i| \geq \frac{\pi}{L}\}$ and write

$$W^2(L,t) = \frac{\Gamma}{(2\pi)^d} \int_{V_\mathbf{q}^\infty} d\mathbf{q} \frac{1 - e^{-2\omega_\mathbf{q} t}}{2\omega_\mathbf{q}}, \qquad d < \delta. \tag{5.96}$$

With a simple change of variable, $\mathbf{s} = L\mathbf{q}$ we can prove that W satisfies the Family–Vicsek scaling form:

$$\begin{aligned} W^2(L,t) &= \frac{\Gamma}{(2\pi)^d} \int_{|q_i|>\pi/L} d\mathbf{q} \frac{1 - e^{-2cq^\delta t}}{2cq^\delta} \\ &= \frac{\Gamma L^{\delta-d}}{2c} \frac{1}{(2\pi)^d} \int_{|s_i|>\pi} d\mathbf{s} \frac{1 - e^{-2\frac{ct}{L^\delta} s^\delta}}{s^\delta} \\ &\equiv \left[\sqrt{\frac{\Gamma L^{\delta-d}}{2c}} w\left(\frac{ct}{L^\delta}\right) \right]^2, \end{aligned} \tag{5.97}$$

with[14]

$$w^2(u) = \frac{1}{(2\pi)^d} \int_{|s_i|>\pi} d\mathbf{s} \frac{1 - e^{-2us^\delta}}{s^\delta}. \tag{5.98}$$

Eq. (5.97) suggests that $\alpha = (\delta - d)/2$ and $z = \delta$. However, in order to complete the proof, we have to show that $w(u) \approx u^\beta$ for $u \ll 1$ and $w(u)$ goes to a constant for large u; see Eq. (5.15). The limit $u \gg 1$ is trivial, because $w(\infty)$ is finite,

$$w^2(\infty) = \frac{1}{(2\pi)^d} \int_{|s_i|>\pi} \frac{d\mathbf{s}}{s^\delta} < \infty, \tag{5.99}$$

for $d < \delta$. The opposite limit $u \ll 1$ requires a new change of variable, $\mathbf{y} = u^{1/\delta}\mathbf{s}$, so

$$w^2(u) = \frac{1}{(2\pi)^d} u^{\frac{\delta-d}{\delta}} \int_{|y_i|>\pi u^{1/\delta}} d\mathbf{y} \frac{1 - e^{-2y^\delta}}{y^\delta}. \tag{5.100}$$

In the limit $u \ll 1$ the integral goes to a constant and

$$w(u) \approx u^{\frac{\delta-d}{2\delta}}, \tag{5.101}$$

therefore satisfying the relation $w(u) \approx u^\beta$, with $\beta = \alpha/z$.

[14] We consistently denote the modulus of **s** by s.

The above calculations have allowed us to find the roughness exponents and the scaling function for a general linear theory whose dominant term has the form $(\nabla^2)^{\delta/2}h$. The result for the exponents is[15]

$$\alpha = \frac{\delta - d}{2}, \qquad \beta = \frac{\delta - d}{2\delta}, \qquad z = \delta. \tag{5.102}$$

The lower and upper critical dimensions are $d_c^\ell = \delta - 2$ and $d_c^u = \delta$. Therefore, passing from the EW equation ($\delta = 2$) to the quartic, stochastic Herring–Mullins equation ($\delta = 4$) shifts the physically relevant interval (d_c^ℓ, d_c^u) from $(0, 2)$ to $(2, 4)$, while the scaling function $w(u)$ depends on δ according to Eq. (5.98).

We conclude this part by evaluating the scaling form for the correlation function of the interface height,

$$G(\mathbf{r}, t) = \left\langle \overline{(h(\mathbf{x}, t) - h(\mathbf{x} + \mathbf{r}, t))^2} \right\rangle \tag{5.103}$$

where, as we have done for computing the roughness, we can exchange the two averages and remove the spatial average, because of translational invariance. Using Eq. (5.82), we can write

$$h(\mathbf{x}, t) - h(\mathbf{x} + \mathbf{r}, t) = \frac{1}{(2\pi)^d} \int d\mathbf{q} e^{i\mathbf{q}\cdot\mathbf{x}} \left(1 - e^{i\mathbf{q}\cdot\mathbf{r}}\right) h(\mathbf{q}, t), \tag{5.104}$$

so

$$G(\mathbf{r}, t) = \frac{1}{(2\pi)^{2d}} \int d\mathbf{q} e^{i\mathbf{q}\cdot\mathbf{x}} \left(1 - e^{i\mathbf{q}\cdot\mathbf{r}}\right) \int d\mathbf{q}' e^{i\mathbf{q}'\cdot\mathbf{x}} \left(1 - e^{i\mathbf{q}'\cdot\mathbf{r}}\right) \langle h(\mathbf{q}, t) h(\mathbf{q}', t) \rangle. \tag{5.105}$$

The correlation function appearing in the previous equation has been evaluated in (5.89), therefore yielding the result

$$G(\mathbf{r}, t) = \frac{\Gamma}{(2\pi)^d} \int d\mathbf{q} \left(1 - e^{i\mathbf{q}\cdot\mathbf{r}}\right)\left(1 - e^{-i\mathbf{q}\cdot\mathbf{r}}\right) \left(\frac{1 - e^{-2\omega_\mathbf{q} t}}{2\omega_\mathbf{q}}\right), \tag{5.106}$$

Finally, since

$$\left(1 - e^{i\mathbf{q}\cdot\mathbf{r}}\right)\left(1 - e^{-i\mathbf{q}\cdot\mathbf{r}}\right) = 4 \sin^2\left(\frac{\mathbf{q}\cdot\mathbf{r}}{2}\right), \tag{5.107}$$

we get

$$G(\mathbf{r}, t) = \frac{2\Gamma}{(2\pi)^d} \int d\mathbf{q} \sin^2\left(\frac{\mathbf{q}\cdot\mathbf{r}}{2}\right) \frac{\left(1 - e^{-2\omega_\mathbf{q} t}\right)}{\omega_\mathbf{q}}. \tag{5.108}$$

The $\mathbf{q}$ integration in the previous expression is not bounded from below, because the dependence on $\mathbf{r}$ is now explicit and the spatial integration in the definition of G (see Eq. (5.103)) is extended to the entire physical space, which is not restricted to a sample of linear size L. As before, we will assume the general form $\omega_\mathbf{q} = cq^\delta$.

The upper critical dimension can be recovered from (5.108), determining when such integral diverges at large r and t. In this limit, $\sin^2\left(\frac{\mathbf{q}\cdot\mathbf{r}}{2}\right) \to \frac{1}{2}$ and

$$G(\infty, \infty) = \frac{\Gamma}{(2\pi)^d} \int \frac{d\mathbf{q}}{cq^\delta}. \tag{5.109}$$

[15] The exponents could also be found by generalizing the dimensional analysis of Section 5.6.1, but the explicit expression of the scaling function can be found only with the present analysis.

The integral diverges in $\mathbf{q} = 0$ if $d < \delta$, confirming that $d_c^u = \delta$. With this constraint on the dimension, the upper bound of the integral, of order $1/a_0$, can be extended to infinity. With the change of variable $\mathbf{s} = r\mathbf{q}$ we obtain the scaling form

$$G(r,t) = \frac{2\Gamma r^{\delta-d}}{c} g\left(\frac{ct}{r^\delta}\right) \tag{5.110}$$

$$g(u) = \frac{1}{(2\pi)^d} \int d\mathbf{s} \sin^2\left(\frac{\mathbf{s}\cdot\hat{\mathbf{r}}}{2}\right) \frac{\left(1 - e^{-2us^\delta}\right)}{s^\delta}, \tag{5.111}$$

where the direction $\hat{\mathbf{r}} = \mathbf{r}/r$ is irrelevant.

The above scaling form for the correlation function should be compared with the scaling function for $W^2(L,t)$; see Eqs. (5.97) and (5.98). The main point we are making here is about the limit $t = \infty$ for finite r. Unlike the roughness, which is always finite for finite L, the correlation function may diverge for finite r. We must take the limit

$$\lim_{u\to\infty} g(u) = \frac{1}{(2\pi)^d} \int d\mathbf{s} \sin^2\left(\frac{\mathbf{s}\cdot\hat{\mathbf{r}}}{2}\right) \frac{1}{s^\delta} \tag{5.112}$$

and focus on a possible divergence in $\mathbf{s} = 0$. Evaluating the contribution of the vanishing $\mathbf{s}$ region, we can expand the sine for small argument and obtain

$$g(\infty) = \frac{1}{4(2\pi)^d} \int_{S_{d-1}} d\Omega \cos^2\theta \int_0 \frac{ds}{s^{\delta-d-1}} + \text{finite contribution}, \tag{5.113}$$

where S_d is the d-dimensional sphere and $d\Omega$ is the solid angle. It is straightforward to conclude that $g(\infty) = \infty$ for $d < \delta - 2$, i.e., $d < d_c^\ell$. It is noteworthy that such divergence requires an infinite sample: if we evaluate $\langle (h(\mathbf{x},t) - h(\mathbf{x}+\mathbf{r},t))^2 \rangle$ for a finite sample of size $L > r$, the divergence disappears.

We can sum up the difference between the scaling forms of $W^2(L,t)$ and $G(r,t)$ as follows. For $d < d_c^u$ the interface is rough and both scaling forms have the same asymptotic behaviors. In particular, for finite L, r and infinite t,

$$W^2(L,\infty) = \frac{\Gamma}{2c} w^2(\infty) L^{\delta-d} \tag{5.114a}$$

$$G(r,\infty) = \frac{2\Gamma}{c} g(\infty) r^{\delta-d}. \tag{5.114b}$$

While $w(\infty)$ is always finite, $g(\infty)$ diverges for $d < d_c^\ell$. As we already noticed, this divergence can be healed by taking a finite sample, i.e., adding the length scale L. If we do that, for $d < d_c^\ell = \delta - 2$ we can write a more complicated scaling function involving r, t, and L. This is the so-called anomalous scaling.

5.7 The Kardar–Parisi–Zhang Equation

In Section 5.4.1 we have shown that the KPZ nonlinearity, $(\nabla h)^2$, is the most relevant term at large scales, which means that a generic model should belong to the KPZ universality class. The RDR model introduced in Section 5.2 does not belong to it for two reasons: (i) it has the symmetry $h \to -h$, and (ii) it has the conserved form (5.58). On the contrary, the KPZ equation,

$$\partial_t h(\mathbf{x}, t) = \nu \nabla^2 h + \frac{\lambda}{2}(\nabla h)^2 + \eta(\mathbf{x}, t), \tag{5.115}$$

does not satisfy the up/down symmetry and it does not conserve the volume, because $\overline{h}(t)$ is not conserved:

$$\overline{\partial_t h} = \frac{\lambda}{2}\overline{(\nabla h)^2}. \tag{5.116}$$

This absence of conservation is a strong indication of which processes may or may not belong to the KPZ universality class. In the RDR model we have a deposition process accompanied by a relaxation: matter is conserved because all deposited particles are incorporated in the aggregate and the attachment/relaxation processes prevent the formation of voids and overhangs. In fact, the conservation of $\overline{h}(t)$ indicates that the volume V_{agg} is conserved, because $V_{agg}(t) = V \cdot \overline{h}(t)$, where V is the area of the substrate.

In addition to the above two features (symmetry breaking and nonconservation), the KPZ equation is generally associated with a third feature, lateral growth, which is sketched and compared with the EW-type of growth in Fig. 5.8. The EW case (a) corresponds to attachment via deposition and the growth direction is $\hat{z}$, regardless of the slope of the interface. This picture is appropriate for processes where deposition and attachment have a specified direction. The KPZ case (b) corresponds to a surface that grows orthogonally to itself. This picture applies when growth can occur in any direction. We have, respectively,

$$\begin{aligned} &(\text{case a}) \qquad \partial_t h = v_0 \\ &(\text{case b}) \qquad \partial_t h = v_0\sqrt{1 + (\nabla h)^2} \approx v_0 + \frac{v_0}{2}(\nabla h)^2 . \end{aligned} \tag{5.117}$$

The equation for case (b), with the small slope expansion justified by power counting, tells us that the KPZ nonlinearity is the natural outcome (and signature) of a dynamical process where *lateral growth* occurs. Eq. (5.117) also suggests that λ, the coefficient of

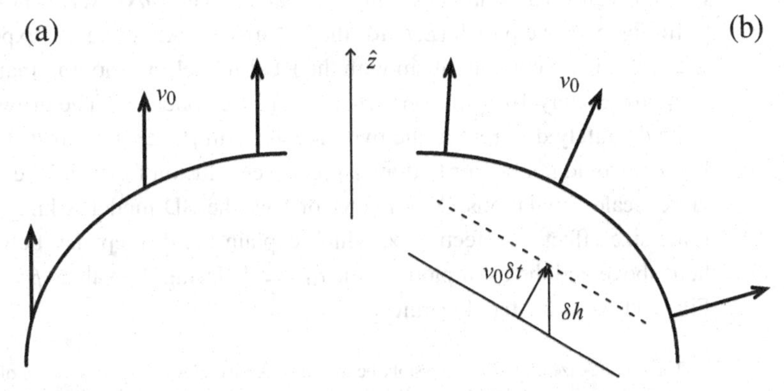

Fig. 5.8 Comparison of EW- and KPZ-type growth. In the absence of relaxation processes, EW-type growth (a) is characterized by a constant, vertical velocity, while KPZ-type growth (b) is characterized by a constant, lateral growth, i.e., perpendicular to the local profile of the interface. We also show how a lateral velocity v_0 determines the growth rate, $\partial_t h = v_0\sqrt{1 + (\nabla h)^2} \simeq v_0 + (v_0/2)(\nabla h)^2$.

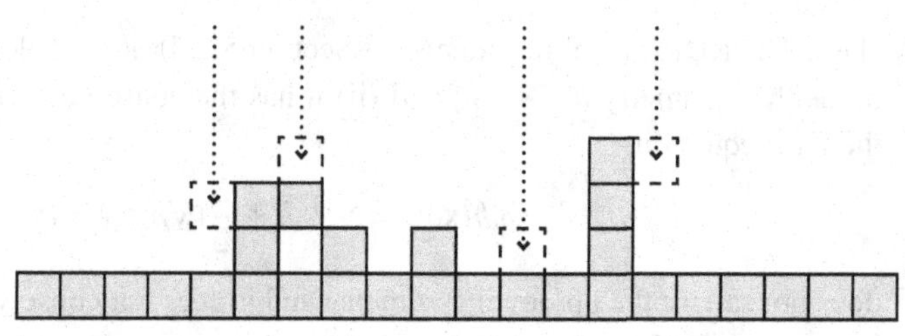

Fig. 5.9 Ballistic deposition model. Gray bricks are deposited particles at time t. A void (white brick) is clearly visible. Dashed bricks represent different, possible deposition events at time $t+1$. In two cases an overhang forms.

KPZ nonlinearity, is related to the interface velocity. In fact, for a surface of negligible curvature and average slope $\mathbf{m} = \overline{\nabla h}$,

$$\overline{(\nabla h)^2} \simeq (\overline{\nabla h})^2 = m^2. \tag{5.118}$$

Therefore, Eq. (5.116) tells us that the average interface velocity is proportional to the square of the slope,

$$\overline{\partial_t h} \simeq \frac{\lambda}{2} m^2, \tag{5.119}$$

and the constant of proportionality is just $\lambda/2$. This relation allows us to test if a given deposition model has an orientation-dependent average velocity by tuning the orientation of the substrate.

A simple deposition model where $\int_V d\mathbf{x} h(\mathbf{x}, t)$ is not conserved and lateral growth occurs is the ballistic deposition (BD) model; see Fig. 5.9. Here, once a deposition site $\mathbf{x}$ has been chosen randomly, the particle sticks to the closest, incorporated site found by the particle during its ballistic trajectory. The evolution rule of this model is easily implemented. If $\mathbf{x}$ is the deposition site,

$$h(\mathbf{x}) \to \max_{\text{nn}}\{h(\mathbf{x}) + 1, h(\mathbf{x} + \boldsymbol{\delta}_{\text{nn}})\}, \tag{5.120}$$

where $\boldsymbol{\delta}_{\text{nn}}$ links a site to its nearest neighbors. In this model, matter is conserved but volume is not, because voids and overhangs appear, so that $\int_V d\mathbf{x} h(\mathbf{x}, t)$ is not conserved.

In Fig. 5.10 we plot $W(L, t)$ for the BD model, extracting the exponents $\beta \simeq 0.32$ and $\alpha \simeq 0.45$. If we compare them with the RDR model, β is approximately 30% larger and α is approximately 10% smaller: what should we conclude? The growth exponents β seem to be definitely different in the two models, with $\beta_{\text{BD}} \simeq \frac{1}{3}$ against $\beta_{\text{RDR}} \simeq \frac{1}{4}$. On the other hand, it would not be fair to draw a precise conclusion for α and we should rather perform larger scale simulations. As a matter of fact, the BD model is known to suffer important finite-size effects[16] affecting α, which explain the discrepancy between the value found here above and the asymptotic value $\alpha_{\text{BD}} = \frac{1}{2}$. Using the values $\beta_{\text{BD}} = \frac{1}{3}$ and $\alpha_{\text{BD}} = \frac{1}{2}$, in Fig. 5.11 we show the dynamic scaling.

[16] These finite-size effects can possibly be traced back to a relevant *intrinsic width* of the BD model, i.e., to the residual roughness, when L is small and the scaling term $L^\alpha w(t/L^z)$ virtually vanishes. This intrinsic roughness is expected to be important for the BD model, since the difference in height of neighboring sites, $(\Delta h)_{\text{nn}}$, can be quite large. For the same reason, the RSOS model (introduced in Section 5.7.4) is expected to have a small intrinsic width, because $(\Delta h)_{\text{nn}} \leq 1$. In fact, the RSOS model does not suffer equally important finite-size effects. On the contrary, it is used for large-scale simulations.

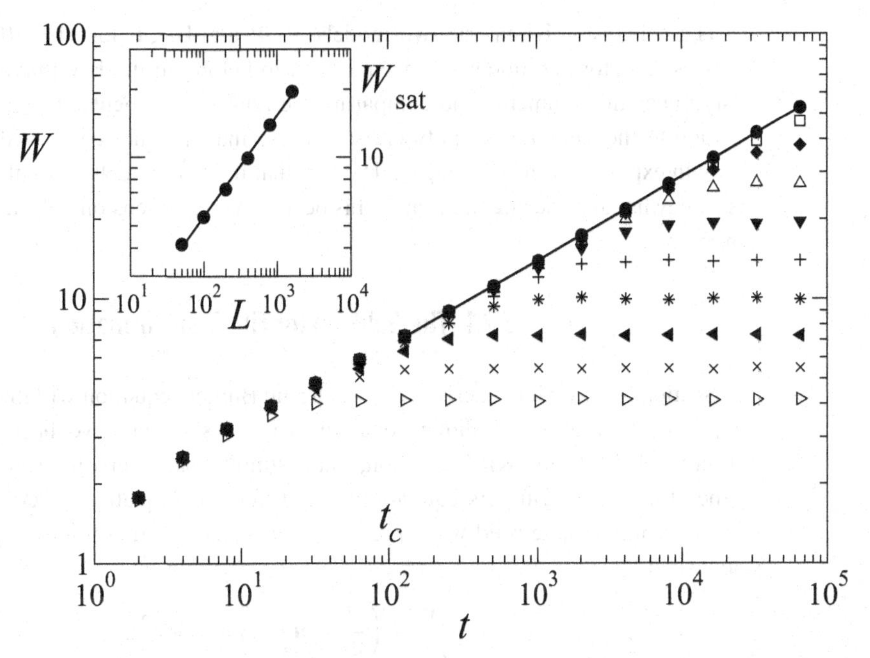

Fig. 5.10 Simulation of the BD model. Roughness as a function of time for $L = 50$ (bottom curve), 100, 200, 400, 800, 1600, 3200, 6400, 12,800, and $L = 51{,}200$ (top curve). The fit of the short time regime gives $\beta \simeq 0.32$. In the inset we plot $W(L, t_\infty)$ versus L, providing the roughness exponent $\alpha \simeq 0.45$.

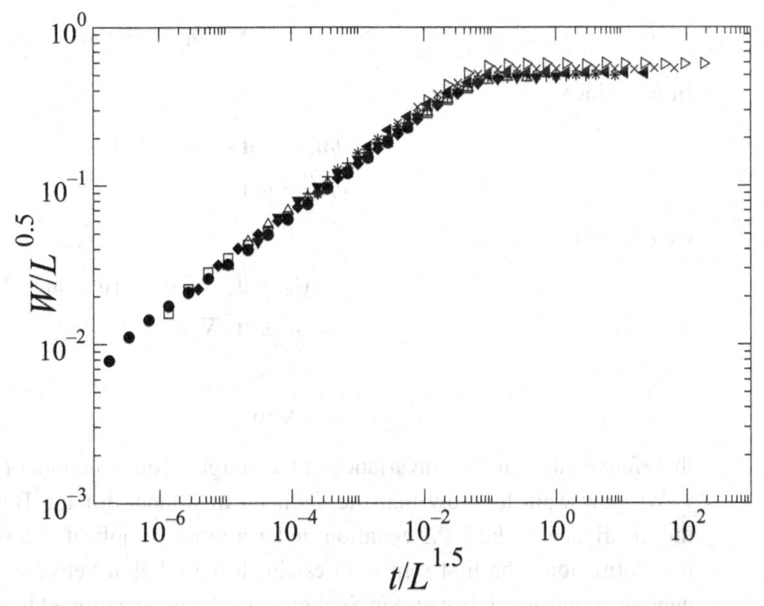

Fig. 5.11 Simulation of the BD model. We show dynamic scaling through data collapse plotting $W/\sqrt{L}$ versus $t/L^{1.5}$.

The conclusion is that the two models, RDR and BD, belong to different universality classes. The former model has been argued to belong to the EW universality class, both using analytic arguments and comparing the relative exponents. The latter model cannot belong to the same universality class, because matter is not conserved, and, in fact, the growth exponent differs from β_{EW}. The fact that the two models share the same value of α is somewhat of an accident, because this occurs in one dimension only, as will be discussed in Section 5.7.2.

5.7.1 The Galilean (or Tilt) Transformation

In Section 5.4.2 we have derived the stochastic Burgers equation within the hydrodynamic approximation of a one-dimensional diffusive system and we have shown that it is strictly related to the KPZ equation, via a simple space derivative of the latter one. In general terms, the Burgers equation is the differential equation describing the dynamics of an incompressible fluid when pressure effects and external forces, e.g., gravity, can be neglected,[17]

$$\frac{d\mathbf{u}}{dt} \equiv \left(\frac{\partial}{\partial t} + \mathbf{u}\cdot\nabla\right)\mathbf{u} = \nu\nabla^2\mathbf{u}. \tag{5.121}$$

The operator d/dt is the material time derivative, typical of the mechanics of continuous media, while the linear term on the right-hand side accounts for viscous friction. The presence of the material derivative ensures the invariance under the Galilean transformation

$$\mathbf{u}'(\mathbf{x},t) = \mathbf{u}(\mathbf{x}-\mathbf{u}_0 t, t) + \mathbf{u}_0. \tag{5.122}$$

In fact, since

$$\partial_t\mathbf{u}' = \partial_t\mathbf{u} - (\mathbf{u}_0\cdot\nabla)\mathbf{u} \tag{5.123a}$$

$$\partial_i\mathbf{u}' = \partial_i\mathbf{u}, \tag{5.123b}$$

we can write

$$\begin{aligned}(\partial_t + \mathbf{u}'\cdot\nabla)\mathbf{u}' &= (\partial_t - \mathbf{u}_0\cdot\nabla)\mathbf{u} + \big((\mathbf{u}+\mathbf{u}_0)\cdot\nabla\big)\mathbf{u}\\ &= (\partial_t + \mathbf{u}\cdot\nabla)\mathbf{u}\\ &= \nu\nabla^2\mathbf{u}\\ &= \nu\nabla^2\mathbf{u}',\end{aligned} \tag{5.124}$$

therefore confirming the invariance of the Burgers equation under (5.122).

We now want to show that the Galilean invariance for the Burgers equation implies the invariance of the KPZ equation under a more complicated transformation, called tilt transformation. The first step is to establish the relation between the two equations in a generic dimension d, because in Section 5.4.2 we were confined to $d = 1$. In order to make

[17] The reader can start from the Navier–Stokes equation (P.1) and remove the pressure term and the gravity term.

the relation more transparent, let us define $\mathbf{m} = \nabla h$ and take the (partial) derivative ∂_i of both sides of KPZ equation,

$$\partial_i(\partial_t h) = \partial_i\left(\nu\nabla^2 h + \frac{\lambda}{2}\sum_j(\partial_j h)(\partial_j h) + \eta\right) \tag{5.125a}$$

$$\partial_t m_i = \nu\nabla^2 m_i + \lambda\sum_j m_j\partial_j m_i + \partial_i\eta \tag{5.125b}$$

$$\partial_t \mathbf{m} = \nu\nabla^2\mathbf{m} + \lambda(\mathbf{m}\cdot\nabla)\mathbf{m} + \nabla\eta, \tag{5.125c}$$

and finally,

$$(\partial_t - \lambda(\mathbf{m}\cdot\nabla))\,\mathbf{m} = \nu\nabla^2\mathbf{m} + \nabla\eta. \tag{5.126}$$

The above equation is manifestly related to the Burgers equation. In fact, if we define $\mathbf{u} = -\lambda\mathbf{m} \equiv -\lambda\nabla h$, we obtain

$$(\partial_t + (\mathbf{u}\cdot\nabla))\,\mathbf{u} = \nu\nabla^2\mathbf{u} - \lambda\nabla\eta\,. \tag{5.127}$$

In conclusion, if h satisfies the KPZ equation, $\mathbf{u} = -\lambda\nabla h$ satisfies the Burgers equation with conserved noise. What does Galilean invariance for the Burgers equation imply for KPZ? Let us rewrite (5.122) in terms of h as follows,

$$-\lambda\nabla h'(\mathbf{x},t) = -\lambda\nabla h(\mathbf{x}-\mathbf{u_0}t,t) + \mathbf{u_0} \tag{5.128}$$

or, after integration over the space variable $\mathbf{x}$, as

$$h'(\mathbf{x},t) = h(\mathbf{x}-\mathbf{u_0}t,t) - \frac{1}{\lambda}\mathbf{u_0}\cdot\mathbf{x} + a(t)\,, \tag{5.129}$$

where the integration constant a may depend on time. The function $a(t)$ can be determined by replacing the above expression in the KPZ equation, but a shorter approach can be used, imposing that the relation (5.116),

$$\partial_t\overline{h} = \frac{\lambda}{2}\overline{(\nabla h)^2}, \tag{5.130}$$

is invariant under the tilt transformation (5.129). From Eq. (5.129) we derive

$$\begin{aligned}\partial_t\overline{h'} &= \partial_t\overline{h} - \mathbf{u_0}\cdot\overline{\nabla h} + \dot{a}\\ &= \frac{\lambda}{2}\overline{(\nabla h)^2} - \mathbf{u_0}\cdot\overline{\nabla h} + \dot{a}\\ &= \frac{\lambda}{2}\overline{(\nabla h')^2} - \frac{u_0^2}{2\lambda} + \dot{a},\end{aligned} \tag{5.131}$$

where we have also used that $\mathbf{u_0}\cdot\nabla h' = \mathbf{u_0}\cdot\nabla h - u_0^2/\lambda$. In conclusion, from the invariance of the relation giving the average interface velocity we derive it must be $\dot{a}(t) = u_0^2/(2\lambda)$. Finally, the KPZ equation is invariant under the transformation[18]

$$h'(\mathbf{x},t) = h(\mathbf{x}-\mathbf{u_0}t,t) - \frac{1}{\lambda}\mathbf{u_0}\cdot\mathbf{x} + \frac{u_0^2}{2\lambda}t. \tag{5.132}$$

[18] A direct proof of this statement is greatly simplified for an infinitesimal transformation, $u_0 \to 0$, because quadratic terms in u_0 can be neglected.

We should not forget noise, which transforms according to

$$\eta'(\mathbf{x}, t) = \eta(\mathbf{x} - \mathbf{u_0}t, t). \tag{5.133}$$

If η is δ-correlated in space and time, η' has the same properties, but if η has finite time correlations, η' has different spectral properties. Therefore, the tilt transformation is actually valid only for δ-correlated noise, which is what we are interested in here.

If we have devoted so much space to prove the invariance of the KPZ equation with respect to the tilt transformation it is because it has a consequence of primary importance: the possibility of deriving a relation between roughness exponents, which is exact in any dimension. In fact, since the tilt transformation explicitly depends on λ and the invariance must be preserved at all length scales, this combination of facts implies that λ cannot be renormalized by a change of scale. To be more precise, we can extend Eq. (5.79) to the KPZ nonlinearity, obtaining

$$b^{\alpha-z}\partial_t h(\mathbf{x}, t) = \nu b^{\alpha-2}\nabla^2 h + \frac{\lambda}{2}b^{2\alpha-2}(\nabla h)^2 + b^{-\frac{d+z}{2}}\eta(\mathbf{x}, t), \tag{5.134}$$

which can also be rewritten as

$$\partial_t h(\mathbf{x}, t) = \nu b^{z-2}\nabla^2 h + \frac{\lambda}{2}b^{\alpha+z-2}(\nabla h)^2 + b^{\frac{z-2\alpha-d}{2}}\eta(\mathbf{x}, t). \tag{5.135}$$

If $\lambda = 0$ the theory is linear, the coefficient ν and the noise are not affected by the scale b, and in fact exponents can be determined by such a condition. If $\lambda \neq 0$ the theory is nonlinear and in general we expect that parameters ν, λ, and η are all affected by the scale and the renormalization group is the correct tool to investigate how they are (see Appendix M for a short account). However, the invariance under tilt transformation tells us that λ does not renormalize, which requires that $b^{\alpha+z-2} = 1$, i.e.,

$$\alpha + z = 2. \tag{5.136}$$

In conclusion, this relation implies that only one exponent is actually undetermined in the KPZ model, the other being fixed by Eq. (5.136). Then next section shows that in $d = 1$ it is possible to determine α, therefore solving the problem, while for generic d the problem is still open.

5.7.2 Exact Exponents in $d = 1$

In one dimension it is possible to determine α using the Fokker–Planck formalism. The generalized Langevin equation

$$\partial_t h(\mathbf{x}, t) = \mathcal{N}[h] + \eta(\mathbf{x}, t) \tag{5.137a}$$

$$\langle \eta(\mathbf{x}, t) \rangle = 0 \tag{5.137b}$$

$$\langle \eta(\mathbf{x}, t)\eta(\mathbf{x}', t') \rangle = \Gamma\delta(\mathbf{x} - \mathbf{x}')\delta(t - t') \tag{5.137c}$$

is associated with the Fokker–Planck equation,

$$\frac{\partial P}{\partial t} = \int d\mathbf{x}\frac{\delta}{\delta h}\left[-\mathcal{N}P + \frac{\Gamma}{2}\frac{\delta P}{\delta h}\right], \tag{5.138}$$

where $P(h,t)$ is the probability that a given profile $h(\mathbf{x})$ occurs at time t. In principle, the knowledge of $P(h,t)$ allows us to determine any average over noise,

$$\langle F(h)\rangle = \frac{\int \mathcal{D}hF(h)P(h,t)}{\int \mathcal{D}hP(h,t)}, \tag{5.139}$$

therefore, in particular, $F(h) = \overline{h^2(\mathbf{x},t)}$ and $F(h) = \left(\overline{h(\mathbf{x},t)}\right)^2$, which are the two building blocks to evaluate the roughness. However, finding the general solution of (5.138) is not possible. A simpler task is to find its time-independent solution, which describes the stationary state, attained asymptotically by a finite system. Such a solution would allow us to determine the roughness exponent α. This is actually possible for the EW equation, because, as shown below, its derivability from a Lyapunov functional automatically provides the stationary solution of the Fokker–Planck equation. The important point we want to discuss here is that, in $d = 1$ (and only in $d = 1$!), KPZ has the same stationary solution as EW. Therefore, it must be

$$\alpha_{\text{KPZ}} = \alpha_{\text{EW}} = \frac{1}{2}, \qquad d = 1. \tag{5.140}$$

The rest of this section is devoted to proving that in $d = 1$ the Fokker–Planck equations for EW and KPZ have the same stationary solution.

A stationary solution $P_s(h)$ of (5.138) is surely found if we are able to solve the equation

$$-\mathcal{N}P_s + \frac{\Gamma}{2}\frac{\delta P_s}{\delta h} = 0 \tag{5.141}$$

or

$$\frac{\delta P_s}{\delta h} = \frac{2}{\Gamma}\mathcal{N}P_s. \tag{5.142}$$

For a problem with one degree of freedom and order parameter $h(t)$, $\mathcal{N}$ is a function of h and so is P, and the functional derivative is replaced by the usual derivative. Every function $\mathcal{N}(h)$ would be integrable,

$$\mathcal{U}(h) = -\int dh\mathcal{N}(h), \tag{5.143}$$

and the solution of (5.142) would simply be $P_s(h) = e^{-2\mathcal{U}(h)/\Gamma}$. For our generalized Langevin equation, the "integrability" of $\mathcal{N}[h]$ is equivalent to saying that (5.137a) is derivable from a Lyapunov functional $\mathcal{F}$. If this happens,

$$\mathcal{N}[h] = -\frac{\delta\mathcal{F}}{\delta h} \tag{5.144}$$

and

$$P_s = e^{-\frac{2\mathcal{F}}{\Gamma}} \tag{5.145}$$

is the stationary solution of the Fokker–Planck equation.

For the EW equation,

$$\partial_t h = \mathcal{N}_{\text{EW}} + \eta\,, \tag{5.146}$$

and we know that $\mathcal{N}_{\text{EW}} = -\delta \mathcal{F}_{\text{EW}}/\delta h$, where

$$\mathcal{F}_{\text{EW}} = \frac{\nu}{2} \int d\mathbf{x} (\nabla h)^2, \tag{5.147}$$

so that we also know the stationary solution

$$P_s^{\text{EW}} = e^{-\frac{2\mathcal{F}_{\text{EW}}}{\Gamma}}. \tag{5.148}$$

The KPZ equation,

$$\partial_t h = \mathcal{N}_{\text{KPZ}} + \eta, \tag{5.149}$$

admits the same Fokker–Planck stationary solution if (see the right-hand side of Eq. (5.138))

$$\int d\mathbf{x} \frac{\delta}{\delta h(\mathbf{x})} \left[-\mathcal{N}_{\text{KPZ}} P_s^{\text{EW}} + \frac{\Gamma}{2} \frac{\delta P_s^{\text{EW}}}{\delta h(\mathbf{x})} \right] = 0. \tag{5.150}$$

Since $\mathcal{N}_{\text{KPZ}} = \mathcal{N}_{\text{EW}} + (\lambda/2)(\nabla h)^2$, the above equation corresponds to

$$-\frac{\lambda}{2} \int d\mathbf{x} \frac{\delta}{\delta h(\mathbf{x})} \left[(\nabla h(\mathbf{x}))^2 P_s^{\text{EW}} \right] + \int d\mathbf{x} \frac{\delta}{\delta h(\mathbf{x})} \left[-\mathcal{N}_{\text{EW}} P_s^{\text{EW}} + \frac{\Gamma}{2} \frac{\delta P_s^{\text{EW}}}{\delta h(\mathbf{x})} \right] = 0. \tag{5.151}$$

The second integral vanishes automatically, because P_s^{EW} is the stationary solution of the Fokker–Planck equation for EW. Finally, P_s^{EW} is also stationary solution of the Fokker–Planck equation for KPZ if (and only if)

$$\int d\mathbf{x} \frac{\delta}{\delta h(\mathbf{x})} \left[(\nabla h(\mathbf{x}))^2 P_s^{\text{EW}} \right] = 0. \tag{5.152}$$

Taking the functional derivative, we obtain[19]

$$\begin{aligned} \int d\mathbf{x} \frac{\delta}{\delta h(\mathbf{x})} \left[(\nabla h(\mathbf{x}))^2 P_s^{\text{EW}} \right] &= -2P_s^{\text{EW}} \delta(0) \int d\mathbf{x} \nabla^2 h(\mathbf{x}) + \frac{2\nu}{\Gamma} P_s^{\text{EW}} \int d\mathbf{x} (\nabla h(\mathbf{x}))^2 \nabla^2 h(\mathbf{x}) \\ &= \frac{2\nu}{\Gamma} P_s^{\text{EW}} \int d\mathbf{x} (\nabla h(\mathbf{x}))^2 \nabla^2 h(\mathbf{x}), \end{aligned} \tag{5.153}$$

where the integral $\int d\mathbf{x} \nabla^2 h(\mathbf{x})$ vanishes because of periodic boundary conditions and the origin of the term $\delta(0)$ is discussed in Appendix L.1. Let us now evaluate the second integral in $d = 1$ and $d = 2$:

$$\int d\mathbf{x} (\nabla h)^2 \nabla^2 h = \begin{cases} \int dx h_x^2 h_{xx} = \int dx \frac{1}{3} (h_x^3)_x = 0, & d = 1 \\ \int d\mathbf{x} (h_x^2 + h_y^2)(h_{xx} + h_{yy}) \neq 0, & d = 2. \end{cases} \tag{5.154}$$

Therefore, in $d = 1$, the KPZ and the EW equations share the same long-time statistics of stationary fluctuations. This property implies that $\alpha_{\text{KPZ}} = \alpha_{\text{EW}} = \frac{1}{2}$. Since $\alpha_{\text{KPZ}} + z_{\text{KPZ}} = 2$, we can derive $z_{\text{KPZ}} = \frac{3}{2}$ and consequently $\beta_{\text{KPZ}} = \frac{1}{3}$.

[19] The treatment of the first term of the right-hand side, proportional to $\delta(0)$, can be made more rigorous within a discrete notation in the Fourier space, but the second term (see Eq. (5.154)) is more manageable in the continuum, real space.

In conclusion, in $d = 1$, $\alpha_{KPZ} = \alpha_{EW}$ and $\beta_{KPZ} > \beta_{EW}$. This means that in one dimension a KPZ surface roughens faster than an EW surface, but the final, stationary value scales with L in the same way. In higher dimensions both KPZ exponents are larger than EW exponents (which actually vanish for $d \geq 2$). The fact that a KPZ surface is generally rougher than an EW surface can be traced to the weaker smoothing mechanism of its deterministic part. The EW smoothing process is the typical diffusional smoothing, which induces an exponentially fast relaxation, while the parabolas appearing in Fig. L.1 flatten following a power law.[20] It is therefore reasonable to speculate that in KPZ, noise should lead to a rougher surface than in EW.

5.7.3 Beyond the Exponents

The determination of the roughness exponents is an important step, but it does not allow the solution and the comprehension of a given model. This is especially true for the KPZ universality class, which is expected to represent the generic behavior of a fluctuating interface (and more; see below and the next section). The exponents α, β represent average properties: how does the standard deviation of the height fluctuations depend on the time t and on the size L of the region under observation? A complete study should also provide the distribution of fluctuations, $\rho(\chi)$, where $\chi = (h(\mathbf{x}, t) - \overline{h})/W(L, t)$. The factor W in the denominator allows to have $\overline{\langle \chi^2 \rangle} = 1$.

Brownian motion is a useful example of what we mean. Its properties are not limited to stating that the average distance Δx traveled by a Brownian particle in time t is of order $\sqrt{t}$; we also know the probability distribution of Δx, which follows a Gaussian distribution. In fact, we can even say where its universal properties come from: from the central limit theorem (see Appendix A), according to which the sum of *independent* variables follows a normal distribution. We expect the same to hold for the Edwards–Wilkinson equation. In fact, if we combine Eqs. (5.82) and (5.86) we obtain (for $h(\mathbf{x}, 0) \equiv 0$)

$$h(\mathbf{x}, t) = \frac{1}{(2\pi)^d} \int d\mathbf{q} e^{i\mathbf{q}\cdot\mathbf{x}} \int_0^t dt' \eta(\mathbf{q}, t') e^{-\omega_{\mathbf{q}}(t-t')}, \tag{5.155}$$

which shows that the height, at given point $\mathbf{x}$ and time t, is a linear combination of independent random variables $\eta(\mathbf{q}, t')$. Therefore, the random variable $h(\mathbf{x}, t)$ follows a Gaussian distribution with zero mean and standard deviation σ_h^{EW} equal to the roughness,

$$\sigma_h^{EW} = \sqrt{\langle h^2(\mathbf{x}, t) \rangle} \equiv W(L, t). \tag{5.156}$$

If we now come back to the (one-dimensional) KPZ equation, we immediately acknowledge that $h(x, t)$ is no longer a linear function of noise and there is no reason to expect the random variable $h(x, t)$ to follow a Gaussian distribution. Nonetheless, if the system is finite the distribution of h must be Gaussian in the stationary regime at large t, because we have shown that EW and KPZ fluctuations are described by the same stationary, Fokker–Planck solution; see Eq. (5.145). The open question is: what is the distribution of h in the thermodynamic limit (or for $t \ll L^z$, if L is finite)? This problem has given new life to

[20] Only parabolas of negative curvature flatten; we refer the reader to Appendix L for a detailed discussion.

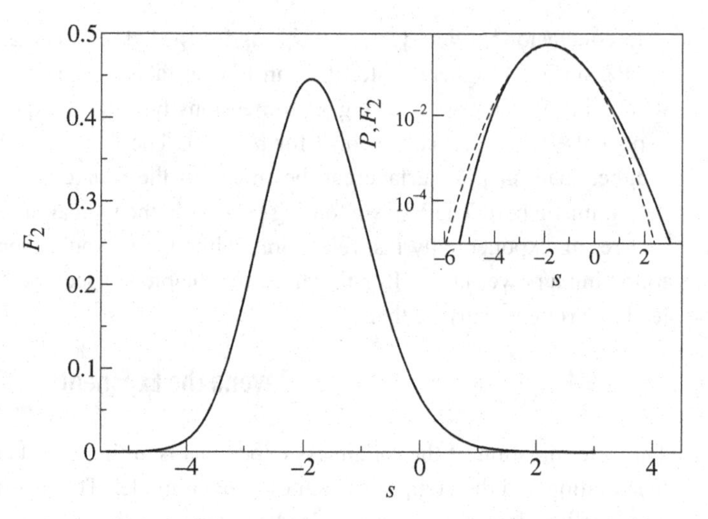

Fig. 5.12 The Gaussian unitary ensemble Tracy–Widom distribution, $F_2(s)$. In the inset we plot it on a lin-log scale in order to highlight the difference between the TW distribution (solid line) with the Gaussian distribution (dashed line), plotted for the same average value $\bar{s} = -1.771$ and the same standard deviation, $\sigma = 0.9017$.

"KPZ and all that" for a couple of reasons. First, because such a distribution, called Tracy–Widom (TW) distribution, can be found in completely different domains. Second, because the concept of universality has been somewhat updated.

Whereas uncorrelated variables produce a Gaussian distribution, as in Eq. (5.155), correlated variables produce a distribution that cannot be determined a priori: there is no equivalent to the central limit theorem for correlated variables. It is therefore fairly surprising that problems as diverse as a fluctuating interface, the largest eigenvalue of a random matrix, and the longest increasing sequence of shuffled integers give rise to the same TW distribution, which is plotted and compared with the Gaussian curve in Fig. 5.12. Further comments are given in Appendix A on the central limit theorem.

It is worth discussing the random matrix problem, because it highlights more easily the existence of distinct distributions, and therefore the existence of different KPZ universality classes. A random matrix A is simply a matrix with random elements A_{ij}, chosen according to some weight. For example, we might have independent elements following the same Gaussian distribution, with zero mean and fixed variance $\langle |A_{ij}|^2 \rangle$. If the matrix is Hermitian, all eigenvalues are real and we can ask what their distribution is, a question that has physically relevant consequences, because it is related, e.g., to the distribution of the energy levels of atomic nuclei.

In particular, we may ask what the distribution of the *largest* eigenvalue is. The answer is the TW distribution and it depends on the real or complex character of the matrix: in the former case A is symmetric and we obtain the Gaussian orthogonal ensemble (GOE) TW distribution, while in the latter case we have the Gaussian unitary ensemble (GUE) TW distribution. Strange as it may seem, these two distributions reproduce the height fluctuations of a flat and a curved KPZ interface, respectively. These results have been found analytically and there are experimental results, discussed in Section 5.8, that provide a scaling function with the one-dimensional KPZ exponents and also reproduce the two TW distributions.

5.7.4 Results for $d > 1$

We might think of applying the dimensional analysis of Section 5.6.1 to the KPZ equation. It is not necessary to do it explicitly to realize that such an analysis cannot provide the roughness exponents. In fact, the KPZ equation contains an additional (nonlinear) term, whose prefactor has been indicated with λ. Therefore, Eq. (5.71) should include on the right-hand side the additional factors λ^{c_1} and λ^{c_2}, giving a total of seven unknown exponents (a_i, b_i, c_i with $i = 1, 2$ and the roughness exponent α, z being determined by Eq. (5.136)). However, we still have six conditions, because each one of Eqs. (5.73) implies three conditions, one for each dimension $\mathtt{L}, \mathtt{T}$, and $\mathtt{H}$. Conclusion: we have insufficient conditions to determine the exponents.

The power counting argument can be useful instead. Following scale transformations (5.21), $\mathbf{x} \to b\mathbf{x}$ and $h \to b^\alpha h$, the EW linearity $\nabla^2 h$ acquires the term $b_{\text{EW}} = b^{\alpha-2}$, while the KPZ nonlinearity gains the factor $b_{\text{KPZ}} = b^{2\alpha-2}$. When the surface is rough, $\alpha > 0$ and b_{KPZ} dominates. However, we have found that $\alpha_{\text{EW}} = (2-d)/2$, so that it is negative for $d > 2$. We might conclude that the KPZ nonlinearity is irrelevant for $d > 2$. This statement is questionable, because we used the EW result, $\alpha = (2-d)/2$, to evaluate the scale factors $b_{\text{EW}}, b_{\text{KPZ}}$, which might be inconsistent. As a matter of fact, this procedure has a perturbative spirit and it can be studied more rigorously within a perturbative renormalization group (pRG) approach, which confirms that a sufficiently small KPZ nonlinearity is irrelevant for $d > 2$. The detailed calculations will not be done; we limit our discussion here to the results, while the procedure is sketched in Appendix M.

The pRG provides the following picture for the KPZ equation. For $d \leq 2$ the nonlinearity is relevant and the surface is rough, while for $d > 2$ two phases exist, depending on the value of the effective coupling constant $g^2 = \lambda^2\Gamma/(2\nu^3)$: for $g < g^*$ the nonlinearity is irrelevant because g renormalizes to zero, the KPZ equation reduces to the EW equation, and the surface is smooth; for $g > g^*$ we have a strong coupling phase. The analysis presented in Appendix M, because of its perturbative nature, says nothing about this phase, nor about the dependence of g^* on d, i.e., about the upper critical dimension.

A variety of approaches has been used to gain information on KPZ behavior for $d > 1$ ($d = 1$ being the only case where a clear analytical picture is available): simulation of suitable deposition models, direct numerical integration of the KPZ equation, nonperturbative RG and real space RG, mapping to the directed polymer problem. All these tools agree about the phase diagram for $d < 4$: they confirm the pRG picture and a strong coupling, rough phase. The behavior of the KPZ equation for $d \geq 4$ is still controversial, however, and the main bone of contention is the value of d_c^u: while field-theoretic approaches seem to suggest[21] $d_c^u = 4$, numerics (direct integration or simulation of deposition models) support $d_c^u > 4$ and the real-space RG method gives $d_c^u = \infty$.

Computation simplicity has given a special role to the simulation of the so-called restricted solid on solid model (RSOS). In this model the random deposition is not accompanied by a relaxation (as in the RDR model) and the new particle is incorporated on top of the particles deposited in the same site, in contrast to the BD model. However, it differs from the RD model because of a constraint on the height difference between

[21] Or at least that something physically relevant occurs at $d_c = 4$.

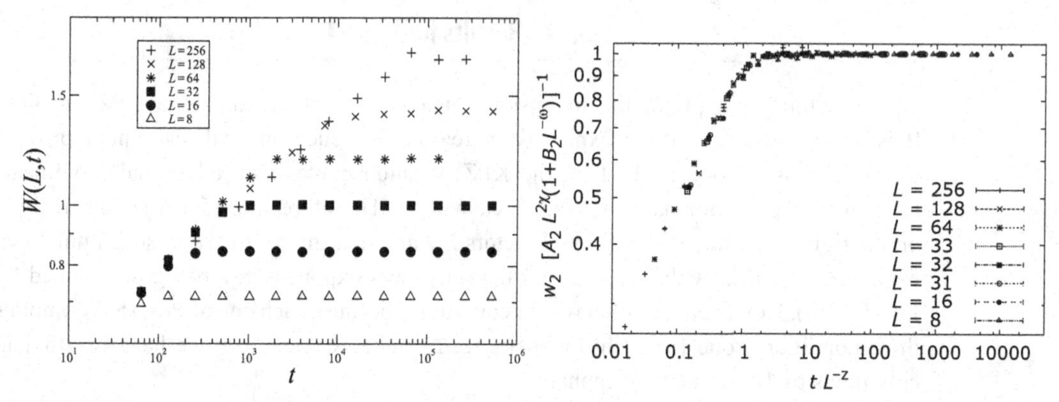

Fig. 5.13 Left: $W(L,t)$ for the RSOS model in $d=4$. The asymptotic value increases with the size L, clearly showing that $\alpha > 0$ and that $d=4$ is not the upper critical dimension. Data courtesy of Andrea Pagnani. Right: The data collapse of the square roughness $W^2(L,t)$. It is necessary to introduce the sub-leading term, as explained in the original paper, A. Pagnani and G. Parisi, Multisurface coding simulations of the restricted solid-on-solid model in four dimensions, *Physical Review E*, **87** (2013) 010102.

neighboring sites, which must be preserved during growth: $|h(\mathbf{x}+\boldsymbol{\delta}_{\rm nn}) - h(\mathbf{x})| \leq 1 \quad \forall \boldsymbol{\delta}_{\rm nn}$, where $\boldsymbol{\delta}_{\rm nn}$ connects each site to its nearest neighbors. If the deposition of a particle in the chosen site breaks this constraint, the particle is not deposited and a new lattice site is chosen. Because of this sticking rule the attachment rate depends on the local environment, and the resulting growth process does not conserve matter. The model should therefore belong to the KPZ universality class for power-counting arguments, but also because the sticking rule clearly suggests a slope-dependent "evaporation" process, with a deposited particle that is more likely to be rejected if the surface is tilted (therefore giving rise to a negative nonlinearity, $\lambda < 0$).

Fig. 5.13 plots $W(L,t)$ of the RSOS model in $d=4$, giving a firm support to the claim that $d_c^u > 4$ in this model. The right panel also shows data collapse using $\alpha \simeq 0.25$ and $z = 2-\alpha$. More precise values for $\alpha(d)$ are given in Table 5.1. These values also suggest that a simple, rational conjecture made by Jin Min Kim and John Kosterlitz, $\alpha = \frac{2}{d+3}$, is not correct.

5.8 Experimental Results

5.8.1 KPZ $d=1$

Convincing experimental evidence of KPZ behavior for a one-dimensional interface has been found in the study of the electroconvection of nematic liquid crystals. Here, the convection is induced by an ac electric field applied to a thin container of liquid crystal, and turbulence sets in. Two different, spatiotemporal chaotic states may appear, one is named DSM1 and is metastable, the other is named DSM2 and is stable. The stable state can be

Table 5.1 Numerical values of the roughness exponents for the RSOS model

d^a	α^b	β^c	z^d
1	$\frac{1}{2}$	$\frac{1}{3}$	$\frac{3}{2}$
2	0.3869(4)	0.2398	1.6131
3	0.3135(15)	0.186	1.6865
4	0.2537(8)	0.1453	1.7463

[a]For $d = 1$ we report the exact values.

[b]The numbers within parentheses are the standard deviation. Numerical data from A. Pagnani and G. Parisi, Numerical estimate of the Kardar–Parisi–Zhang universality class in (2+1) dimensions, *Physical Review E*, **92** (2015) 010101, for $d = 2$; from E. Marinari, A. Pagnani, and G. Parisi, Critical exponents of the KPZ equation via multi-surface coding numerical simulations, *Journal of Physics A: Mathematical and General*, **33** (2000) 8181–8192, for $d = 3$; from A. Pagnani and G. Parisi, Multisurface coding simulations of the restricted solid-on-solid model in four dimensions, *Physical Review E*, **87** (2013) 010102, for $d = 4$.

[c]Determined by the relation $\beta = \alpha/z$.

[d]Determined by the relation $z = 2 - \alpha$.

nucleated from the metastable one by inducing a defect with a laser pulse. Once a DSM2 cluster is created, it grows and the interface DSM1/DSM2 gets rough. The laser pulse can be shot so as to produce a circular geometry or a flat geometry, which are visualized in the two snapshots on top of Fig. 5.14, the dark region corresponding to DSM2.

The Family–Vicsek scaling is shown in Fig. 5.15, both for the roughness (a, c) and the correlation function (b, d). The exponents are the same, independently of the circular (a, b) or flat (c, d) geometry. The geometry is relevant instead for the distribution of height fluctuations, as shown in Fig. 5.14. In both cases the curves follow Tracy–Widom distributions: the circular geometry is related to the Gaussian unitary ensemble and the flat geometry is related to the Gaussian orthogonal ensemble.

5.8.2 KPZ $d = 2$

A deposition process on a real, two-dimensional substrate seems to be the best candidate to look for KPZ behavior, but this has remained elusive for a long time in spite of the expected universal character of this equation. In addition, it is not possible to determine the universality class of a given roughening process by simply looking at approximate roughness exponents; checking some distribution or correlation function is essential. Fig. 5.16 plots height fluctuations and roughness distribution for the growth process of vapor-deposited organic thin films. The extracted roughness exponents, $\alpha = 0.45 \pm 0.04$ and $\beta = 0.28 \pm 0.05$, are in marginal agreement with numerical KPZ values (see Table 5.1), but the two distributions shown in the figure are more convincing. The left panel shows the distribution of normalized height fluctuations in the regime $t \ll t_c = L^z$, L being the sample

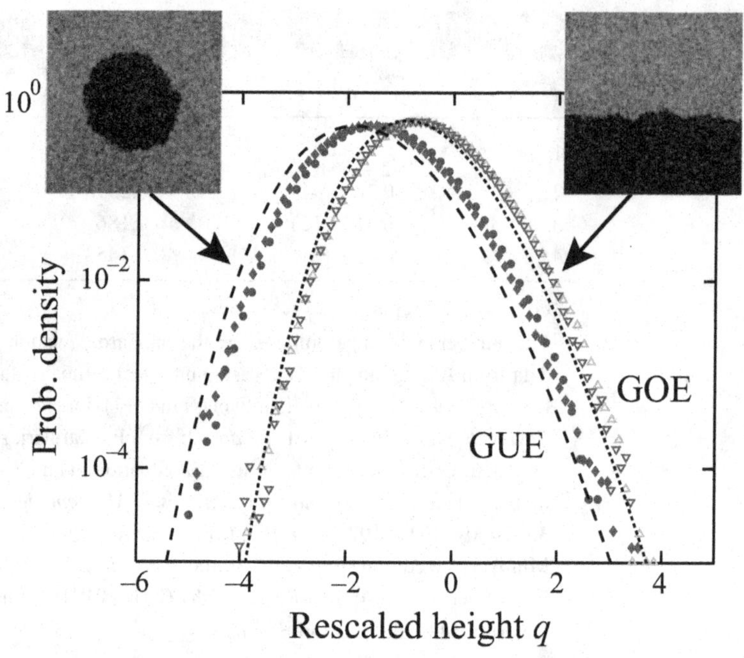

Fig. 5.14 Histogram of the rescaled local height $q \equiv (h - v_\infty t)/(\Gamma t)^{1/3}$ for the circular (solid symbols) and flat (open symbols) interfaces. The dashed and dotted curves show the GUE and GOE TW distributions, respectively. The snapshots show the connection between the geometry of the interface and the pertinent ensemble to evaluate the TW distribution. From K. A. Takeuchi and M. Sano, Evidence for geometry-dependent universal fluctuations of the Kardar–Parisi–Zhang interfaces in liquid-crystal turbulence, *Journal of Statistical Physics*, **147** (2012) 853–890.

size, and it is the equivalent of Fig. 5.14. The right panel refers to the regime $t \gg t_c = L^z$, i.e., $L \ll \xi_\parallel$, where in this case L is the sampling scale. In this regime fluctuations are stationary and the distribution of the normalized fluctuations of the roughness does not depend on L. Experimental results are compared with simulation and analytical results.

5.9 Nonlocal Models

The deposition models and the equations studied in this chapter all have a common feature: they are local models. For a discrete model this means that the evolution rule of a given site depends on the height of the site itself and the height of neighboring sites, within a finite distance. For example, in the RDR model we may choose a site and the deposited particle will be incorporated in the lowest height site within a certain distance. In a continuum description this means that the equation has a finite number of relevant terms. For example, both EW and KPZ have the form $\partial_t h = \mathcal{N}[h] + \eta$, where $\mathcal{N}[h]$ is a single (linear) term for EW and is the sum of two terms for KPZ. We can add other terms, but they are irrelevant in the sense of the dynamical renormalization group.

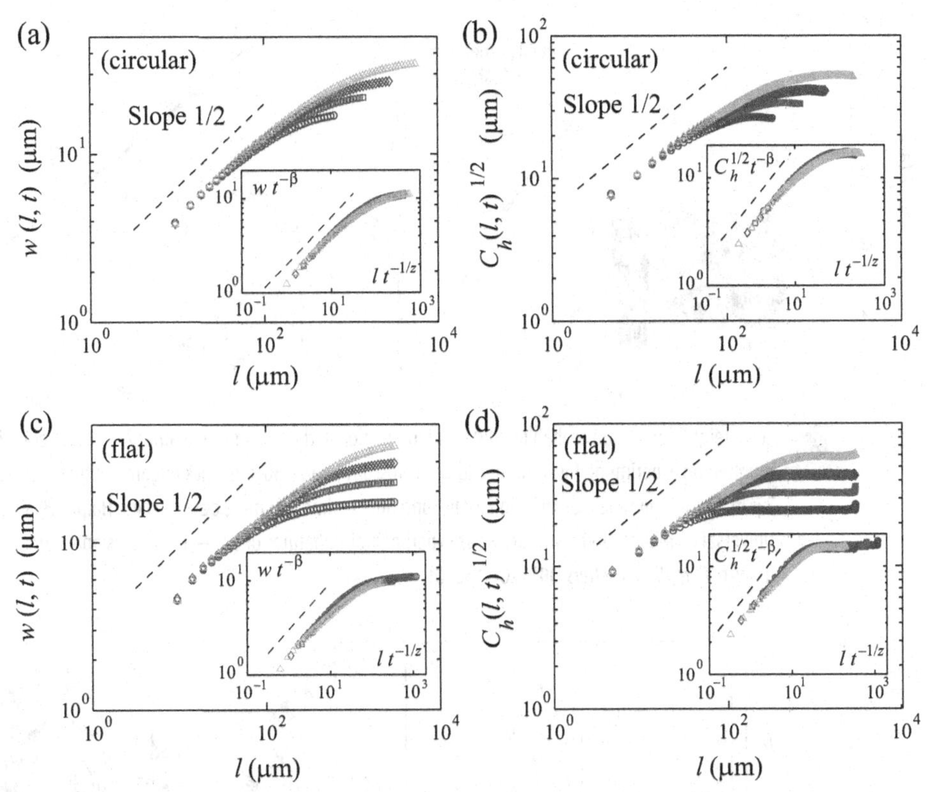

Fig. 5.15 The interface width $w(l,t)$ (a, c) and the square root of the correlation function $C_h(l,t)^{1/2}$ (b, d) are shown for different times t for the circular (a, b) and the flat (c, d) interfaces. Time varies from $t = 2$ s to $t = 30$ s for the circular interface and from $t = 4$ s to $t = 60$ s for the flat interface, time increasing from bottom to top. The insets show the same data with the rescaled axes, with the KPZ exponents $\beta = \frac{1}{3}$ and $z = \frac{3}{2}$. The dashed lines are guides for the eyes indicating the slope for the KPZ exponent $\alpha = \frac{1}{2}$. From K. A. Takeuchi and M. Sano, Evidence for geometry-dependent universal fluctuations of the Kardar–Parisi–Zhang interfaces in liquid-crystal turbulence, *Journal of Statistical Physics*, **147** (2012) 853–890.

However, there are *nonlocal* processes where the growth velocity at a given point depends on the profile of the entire system. A simple way to obtain one such process is to modify the growth rule of the BD model, where particles are deposited ballistically along the vertical direction and stick to the aggregate as soon as the particle occupies a site close to the aggregate itself. There are two main modifications we can do that both lead to a nonlocal model: either change the flux orientation of incoming particles or change from ballistic to diffusional deposition. The former case is illustrated in the left panel of Fig. 5.17 with the so-called grass model, where each blade grows in proportion to the light it receives (in a particle model this is equivalent to saying that incoming particles arrive uniformly from all possible directions). Here nonlocality is due to shadowing effects: a high part of the profile may capture a larger fraction of an oblique flux than a low part of the profile, much in the same way a tall building shades a short one if the sun is out of the zenith. The latter case corresponds to particles that arrive not ballistically but diffusionally,

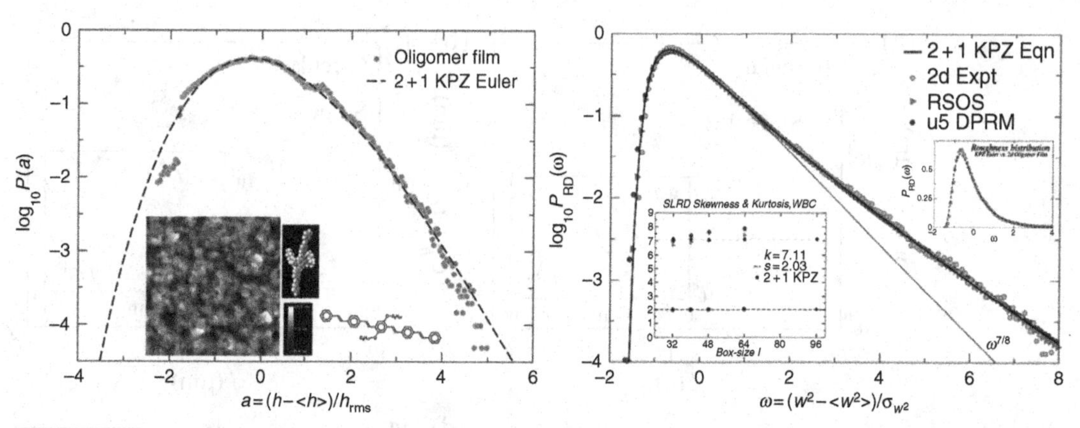

Fig. 5.16 Left: Height fluctuation probability distribution function in the short time regime: 2 + 1 KPZ equation versus kinetic roughening experiment. Inset: Atomic force microscopy image, 50-nm-thick oligomer film; lateral scan direction 2 μm. Right: Roughness distribution in the long time regime. For more details, see T. Halpin-Healy and G. Palasantzas, Universal correlators and distributions as experimental signatures of (2 + 1)-dimensional Kardar–Parisi–Zhang growth, *Europhysics Letters*, **105** (2014) 50001.

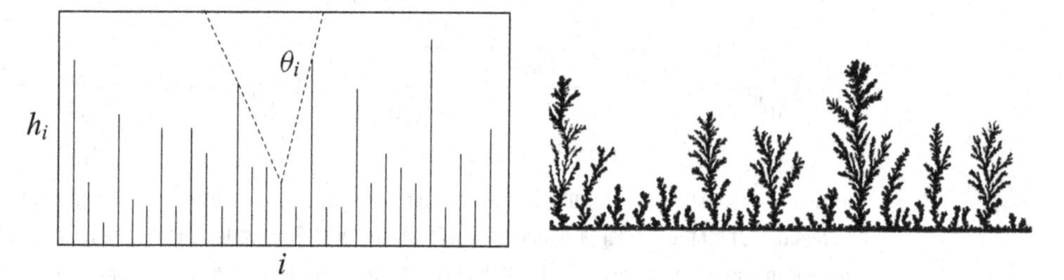

Fig. 5.17 Left: The grass model for a one-dimensional lattice: each blade of grass has a height h_i and grows as a function of the light it receives, which is proportional to the angle θ_i. Right: A forest of zinc metal trees obtained by electrodeposition and representative of the diffusional growth for a one-dimensional substrate. A similar morphology can be obtained with a simulation where particles are released, one by one, at a large distance above the substrate. Each particle performs a standard random walk until it sticks to the aggregate or to the substrate. At that point the particle is incorporated and a new particle is released. Reprinted from M. Matsushita, Y. Hayakawa, and Y. Sawada, Fractal structure and cluster statistics of zinc-metal trees deposited on a line electrode, *Physical Review A*, **32** (1985) 3814–3816.

as illustrated in the right panel of the same figure[22]: as in the BD model, particles stop and are incorporated in the growing aggregate as soon as they occupy a nearest-neighbor site of the substrate or of the aggregate. Here nonlocality is a natural outcome of diffusion, and in the rest of this section we will focus on this case.

The resulting morphology of diffusional growth is fairly different from those discussed previously in this chapter. Two striking differences are immediately visible: diffusional

[22] Here we are limited to considering the case where the total dimension of space (where diffusion takes place) is equal to $d + 1$, but we might have a total space of dimension $d + n$. For example, the case $d = 1$, $n = 2$, would correspond to three-dimensional diffusion and attachment to a one-dimensional wire.

growth produces a fractal, rather than a compact morphology, and the growing aggregate is an ensemble of separated (but strongly interdependent) tree-shaped clusters.[23] The different morphology implies that we also need different quantities to characterize the aggregate as the fluctuations of the local height are no longer appropriate. The main relevant quantity is now the cluster-size distribution function, $n_s(N)$: if V is the total number of substrate sites and we have deposited N particles per substrate site, we have $\mathcal{N}_s(N)$ clusters of size s, whence the function

$$n_s(N) = \frac{\mathcal{N}_s(N)}{V}. \tag{5.157}$$

In practice, N plays the role of time. In fact, in a simulation N is exactly the number of Monte Carlo steps.

The above quantity is found to exhibit the scaling form

$$n_s(N) = \frac{1}{s^\tau} f\left(\frac{s}{s^*(N)}\right), \tag{5.158}$$

where $s^* \sim N^\sigma$ is the mean cluster size and $f(x)$ is a function that is a constant for small arguments ($x \ll 1$) and it vanishes exponentially fast for large arguments ($x \gg 1$), because it is extremely unlikely to have clusters much larger than s^*. In practice we may think that $n_s(N)$ is a power-law function with an upper cut-off that scales with s^*. The normalization condition imposes a relation between σ and τ, because

$$N = \sum_s s n_s(N) = \sum_s s \frac{1}{s^\tau} f\left(\frac{s}{s^*(N)}\right) \simeq \int^{s^*} ds \frac{1}{s^{\tau-1}} \simeq (s^*)^{2-\tau}, \tag{5.159}$$

so $s^* \simeq N^{1/(2-\tau)}$ and $\sigma = (2-\tau)^{-1}$. This relation implies $\tau < 2$ because s^* must increase with N.

If a cluster of size s has a typical linear size (therefore also a typical height) $h(s) \simeq s^\theta$, the average height of the whole system, which is a function of N, will be given by

$$\bar{h}(N) = \frac{1}{N} \sum_s s h(s) n_s(N) \simeq \frac{1}{N} \int^{s^*} ds \frac{s s^\theta}{s^\tau} \simeq N^{\frac{\theta}{2-\tau}}. \tag{5.160}$$

Therefore, we have introduced two exponents, θ and τ, to characterize the statistics of the morphology resulting from diffusional growth. However, up to now we have not yet used the fractal character of such morphology. We do it now, finding that both exponents can be expressed in terms of the fractal dimension D of the aggregate and of the Euclidean dimension d of the substrate.

Since each cluster is a fractal of dimension D, the quantity of matter contained in a volume of linear size $h(s)$ scales according to $s \sim h^D$, so $\theta = 1/D$. The whole aggregate has a fractal dimension D as well, so the condition that in a volume of linear size $\bar{h}(N)$ there must be N particles per substrate site is

$$\frac{\bar{h}^D}{\bar{h}^d} \simeq N, \tag{5.161}$$

[23] In this context a cluster can be defined as a collection of "deposited" particles connected to the same substrate site through nearest neighbors.

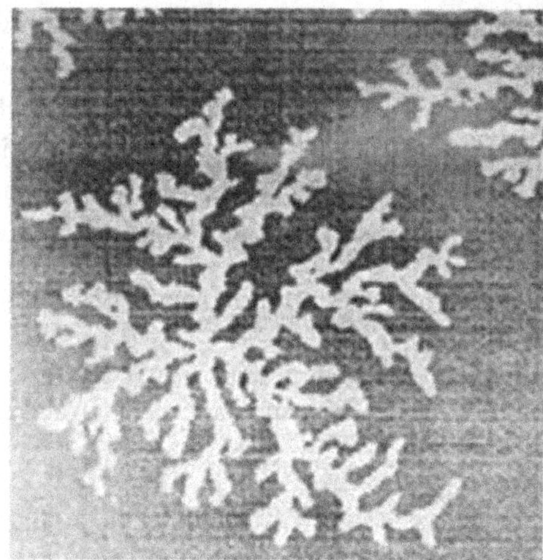
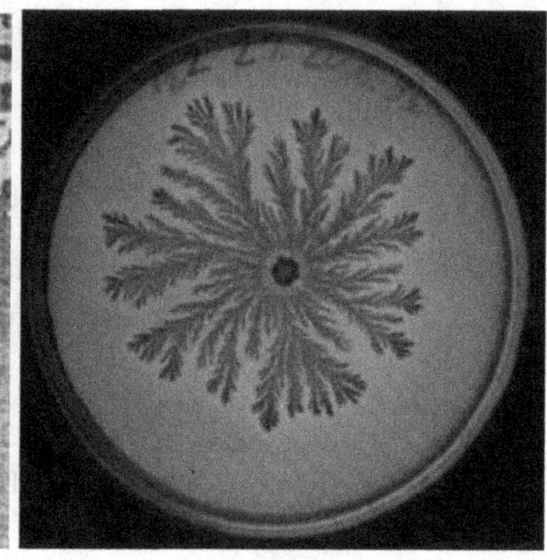

Fig. 5.18 Left: Gold atoms deposited on top of a rubidium substrate form DLA aggregates. Reprinted from R. Q. Hwang et al., Fractal growth of two-dimensional islands: Au on Ru(0001), *Physical Review Letters*, **67** (1991) 3279–3282. Right: Similar morphology in a bacterial colony growing in a Petri dish. Reprinted from E. Ben-Jacob et al., Generic modelling of cooperative growth patterns in bacterial colonies, *Nature*, **368** (1994) 46–49.

which means $\bar{h}(N) \simeq N^{1/(D-d)}$. The relations $\theta = 1/D$ and $\theta/(2-\tau) = 1/(D-d)$ imply

$$\tau = 1 + \frac{d}{D}. \tag{5.162}$$

In order to focus on the essential features of the "diffusional" morphology it is easier to consider a simpler geometry, where the aggregate grows from a single seed, rather than a whole substrate. In fact, this is the geometry originally used by Thomas A. Witten and Leonard M. Sander in 1981 to define the celebrated diffusion-limited aggregation (DLA) model. Here the aggregate is originally composed of a single seed placed at the origin of a regular lattice and new particles are released homogeneously, far from the aggregate. Once released a particle diffuses until it sticks to the aggregate (or is lost into outer space). Rather than plotting a simulated DLA geometry (whose branches look very similar to the trees shown in Fig. 5.17 on the right) we refer the reader to the experimental morphologies shown in Fig. 5.18. The panel on the left is a good experimental realization of the DLA model: gold atoms are deposited on a rubidium substrate and diffuse at its surface, and when two gold atoms meet they stop diffusing and form the seed for a growing layer, which will collect other diffusing Au atoms.[24] The reason why such an intricate structure emerges is apparently simple: once a fjord-like recess is formed, it will be harder to be filled by diffusing particles.

The right panel of Fig. 5.18 displays a similar morphology, for a similar reason: a growing process related to a diffusive field. The experiment of bacterial colony growth

[24] The experiment is not a perfect realization of DLA for two main reasons: first, Au atoms may be deposited close to the growing cluster (if not on the top of it); second, once Au atoms stick the aggregate they may have a residual diffusion motion along the cluster edge. Both effects contribute to smooth the morphology of the growing cluster, making it "less fractal."

consists in preparing an agar-based growth medium in a Petri dish at whose center a droplet of bacteria is inoculated. Then, the presence of nutrients allows the bacterial colony to grow and multiply. With an appropriate tuning of parameters, first of all the food concentration, it is possible to obtain the morphology shown in the figure. Without pretending to oversimplify a process that is more complex than DLA, it is nevertheless correct to say that both experiments of Fig. 5.18 display a branching instability of diffusional character. Diffusion is what makes a straight profile unstable and is also what makes the problem nonlocal. The instability mechanism is simple: if a fluctuation creates a small amplitude wave-like profile, the bumps collect more diffusing material and growth is faster, therefore reinforcing the wavy structure of the profile (positive feedback). This mechanism would not be active if the quantity attaching to the front moved ballistically rather than diffusionally.[25] Nonlocality comes from the screening effect that the growing front exerts on itself.

Let us now give a more formal description of a diffusional growth process, because it allows us to understand its generality, making reference to the DLA model. In a continuum picture the concentration $c(\mathbf{x}, t)$ of diffusing particles satisfies the equation

$$\frac{\partial c}{\partial t} = D\nabla^2 c, \tag{5.163}$$

with the boundary condition, $c(\mathbf{x}_F, t) = 0$ at the points $\mathbf{x}_F$ of the growing front, denoting that a sticking particle is definitely incorporated in the aggregate. The local growth velocity of the front, v, is perpendicular to the front itself and is equal to the current of atoms attaching to it,

$$v = D\frac{\partial c}{\partial n}, \tag{5.164}$$

where ∂_n means the derivative along the outward normal. In DLA and also in some experiments the growth velocity v is weak enough and we can replace the diffusion equation (5.163) with the Laplace equation, $\nabla^2 c(\mathbf{x}, t) = 0$. This approximation is justified if the typical time τ_f to relax the field $c(\mathbf{x}, t)$ to the stationary solution, corresponding to fixed boundary conditions, is much smaller than the time τ_b required to move the boundary. In DLA, τ_f is the time interval between two consecutive releases of a particle. In order to move the boundary, on average, by a lattice constant we need a number of releases equal to the number of boundary sites, N_b, which increases and diverges in time. Therefore, $\tau_f/\tau_b \approx 1/N_b \to 0$. In conclusion, the approximation becomes asymptotically exact.

It is therefore clear why DLA and similar models are examples of "Laplacian growth," whose continuum formulation reduces to studying an electrostatic problem with a growth velocity proportional to the local electric field. This electrostatic equivalence explains why electrodeposition patterns may give rise to DLA-like morphologies. In fact, we might go even further and find similar morphologies as the result of the injection of an inviscid fluid in a viscous one. The resulting interface is unstable and viscous fingering forms. However, these phenomena fall in the domain of pattern formation, studied in Chapter 7, rather than in the domain of kinetic roughening, because they originate from a deterministic instability.

[25] In that case, depending on the orientation of the flux, we may have shadowing effects, which are what characterize the grass model.

5.10 Bibliographic Notes

A general introduction to the topics discussed in this chapter is A.-L. Barabási and H. E. Stanley, *Fractal Concepts in Surface Growth* (Cambridge University Press, 1995). Here the reader can find references to experiments and many simulations.

The book by Paul Meakin, *Fractals, Scaling and Growth Far from Equilibrium* (Cambridge University Press, 1998), has a special emphasis to models and an extensive list of references.

The book by A. Pimpinelli and J. Villain, *Physics of Crystal Growth* (Cambridge University Press, 1998), is not formal and contains many remarks and physical considerations. However, it is oriented to growth processes, rather than to generic processes of kinetic roughening.

A detailed discussion of EW and KPZ equations can be found in the review paper by J. Krug, Origins of scale invariance in growth processes, *Advances in Physics*, **46** (1997) 139–282, which is also a good source for the topic of scale invariance in general.

The details of the perturbative RG approach to KPZ can be found in the books by Barabasi and Stanley and in Pimpinelli and Villain.

Recent results and useful physical discussions about KPZ in $d = 1$ can be found in K. A. Takeuchi, M. Sano, T. Sasamoto, and H. Spohn, Growing interfaces uncover universal fluctuations behind scale invariance, *Scientific Reports*, **1** (2011) 34. A review paper on KPZ that connects this chapter to the previous one is T. Kriecherbauer and J. Krug, A pedestrian's view on interacting particle systems, KPZ universality and random matrices, *Journal of Physics A: Mathematical and Theoretical*, **43** (2010) 1–41. A short history of KPZ in the context of kinetic roughening is T. Halpin-Healy and K. A. Takeuchi, A KPZ cocktail–shaken, not stirred . . . , *Journal of Statistical Physics*, **160** (2015) 794–814.

A simple account of DLA is in L. M. Sander, Diffusion-limited aggregation: a kinetic critical phenomenon?, *Contemporary Physics*, **41** (2000) 203–218.

6 Phase-Ordering Kinetics

This chapter addresses the following, basic question: how does a system with equivalent ground states relax to equilibrium from a disordered state? Two types of processes are involved in the ensuing dynamics: coarsening and nucleation. Coarsening, the main focus of this chapter, is a collective and delocalized process, which is characterized by the increase in time of the typical size of ordered regions. Nucleation, instead, is a localized and thermally activated process, which requires an energy barrier to be overcome.

Rather than introducing general concepts and equations, we prefer to begin addressing a well-defined issue, which is a paradigm of the general problem: what is the ordering dynamics of an Ising model if its temperature is rapidly quenched from $T > T_c$ to $T < T_c$? We use this approach for two reasons. First, the reader can focus on the process of phase-ordering using tools already in hand: the Ising model, the Metropolis algorithm, the Ginzburg–Landau free energy, the diffusion equation, and so on. Second, the reader can easily gain insight into notable differences with equilibrium dynamics: the irrelevance of the physical dimension of the system (with some caveats) and the relevance of conservation laws.

After introductory considerations we go on to analyze the coarsening process of an Ising system in $d = 1$ (Section 6.2) and in $d > 1$ (Section 6.3). When studying nonconserved dynamics we use a spin language, and the basic, microscopic process is the spin-flip. Instead, when studying conserved dynamics we will use a particle language (as for the lattice gas), and the basic, microscopic process is the hopping of a particle to a neighboring lattice site. In the spin language this process is called spin-exchange. The most remarkable result is that the coarsening exponent n, which describes how the linear size L of ordered regions increases in time, $L(t) \sim t^n$, differs between conserved and nonconserved dynamics (the former being slower than the latter), while it does not depend on the physical dimension of the space where phase-ordering occurs. This is even more striking when comparing the different descriptions of the phase-ordering process in $d = 1$ (a stochastic description) and in $d > 1$ (a deterministic description).

In Section 6.4 we present more general concepts and tools, which allow us, once again, to distinguish between conserved and nonconserved dynamics: the Langevin equation description, the correlation function, the structure factor, the domain size distribution, and the differences between critical and off-critical quenches. The continuum description is used in Section 6.5 to study nonscalar systems. Finally, in the last section we present the nucleation process. Due to the introductory and pedagogical character of this book we discuss this topic on the basis of relatively simple and qualitative arguments, without entering into the details of a reliable quantitative approach.

6.1 Introduction

Phase transitions are among the most relevant equilibrium phenomena in a many-body system. The para-ferromagnetic transition is an archetypal phase transition. In the absence of a magnetic field, it is a continuous, second-order transition, where the average magnetization (the order parameter) vanishes for $T > T_c$ and it is different from zero in the ordered phase, $T < T_c$. In the low-temperature phase, several equivalent phases exist, depending on the symmetry of the order parameter. Here and in most of this chapter we focus on the Ising symmetry, with a scalar magnetization, $M(T, H)$, that depends, in the general case, on temperature and magnetic field. See Chapter 3 for details of the Ising model and of its equilibrium properties.

In Fig. 6.1 we plot the magnetization curves for an ultrathin magnetic film Fe/W, a good experimental realization of a two-dimensional Ising model. Because of the symmetry $M(T, -H) = -M(T, H)$, we limit ourselves to $M > 0, H > 0$. The thick line, called curve of spontaneous magnetization, is the coexistence curve, because if $H = 0$ and $T < T_c$, the equilibrium state corresponds to two distinct values of the magnetization, $|M(T, 0)|$ and $-|M(T, 0)|$. The coexistence curve separates two different regions of the phase diagram:

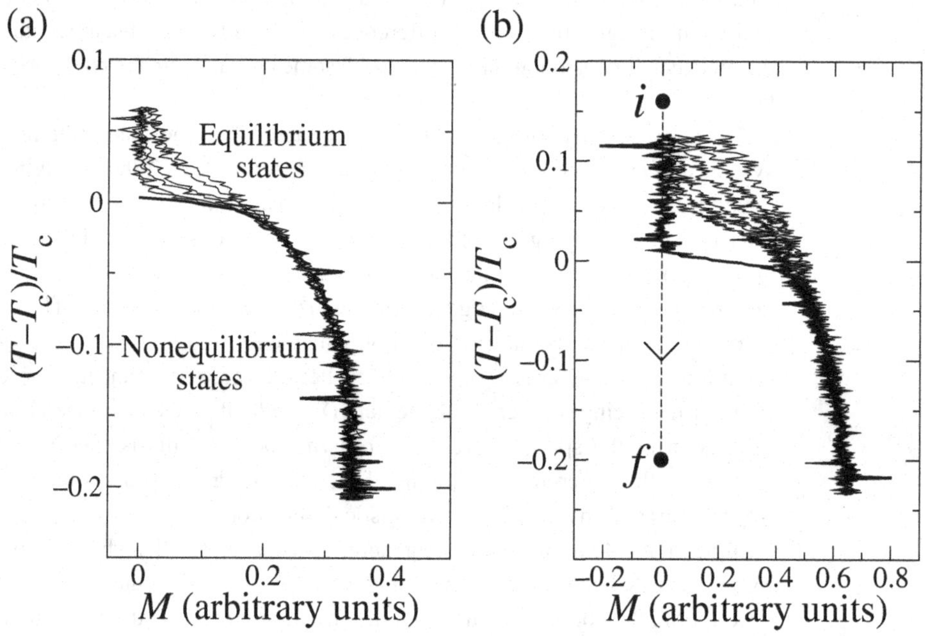

Fig. 6.1 Magnetization of equilibrium of a Fe/W(110) magnetic film (1.8 ± 0.1 atomic layers) as a function of temperature T and field H for two different samples. (a) Sample 1. $H = 0, 0.05, 0.1, 0.5, 1, 3, 5$ Oe. (b) Sample 2. $H = 0, 5, 10, 20, 50, 100, 200$ Oe. Thick lines correspond to $H = 0$, thin lines depart increasingly more with increasing H. The dashed line in (b) represents a variation of the temperature from $T_i > T_c$ to $T_f < T_c$. Raw experimental data courtesy of Danilo Pescia. C. H. Back et al., Experimental confirmation of universality for a phase transition in two dimensions, *Nature*, **378** (1995) 597–600.

the states for $T < T_c$ and $|M| < M(T, H = 0^+)$ are nonequilibrium states, which cannot be attained with a reversible transformation. Instead, it is possible to tune temperature and field in order to attain, at equilibrium, all other states.[1]

It is possible to attain the inner, nonequilibrium region with a sudden change of control parameters, a procedure called quenching. Indeed, the goal of this chapter is to study the dynamics after a similar quenching. Let us first try to understand what happens if we prepare the system in the paramagnetic phase ($T_i > T_c$, $H = 0$), then reduce the temperature, either adiabatically or suddenly, to a temperature in the ferromagnetic phase, $T_f < T_c$; see the dashed line in Fig. 6.1(b).

There is a major difference between an adiabatic change of T and a quenching: crossing the critical point adiabatically allows the system to pass from a disordered homogeneous state (the paramagnetic phase) to an ordered homogeneous state (the broken symmetry, ferromagnetic phase), because the energy barrier between the two equivalent ferromagnetic states is vanishingly small at the critical point and the system has time to relax to the equilibrium state in each moment. The choice between the two ordered states is a random choice made by fluctuations, which are symmetric in the absence of stray fields.

Instead, after quenching the disordered system cannot choose between the two equivalent, ferromagnetic ground states and a fluctuation cannot allow an infinite barrier to be overcome. Nonetheless, the system will tend to decrease its free energy by creating ordered regions (magnetic domains) of increasing size $L(t)$. This process, called coarsening, is the *core business* of phase-ordering kinetics, if we perform a symmetric quenching, i.e., in the absence of an asymmetry between the two ground states. An example of coarsening process is visible in Fig. 6.2, obtained by kinetic Monte Carlo (KMC) simulations; see Appendix J.

Phase-ordering is not limited to magnetic systems, of course, because it occurs whenever there is phase coexistence. A binary alloy, which undergoes an order–disorder transition, is a nice and useful additional example. In brief, an *AB* solid alloy can be described by a lattice model, where each site is occupied by an *A* or *B* atom and a pair of nearest-neighbor atoms has an energy ϵ_{AA}, ϵ_{BB}, or ϵ_{AB}, depending on the type of atoms. This model is equivalent to an Ising model, as proved in Appendix H. If $\epsilon_{AA} + \epsilon_{BB} < 2\epsilon_{AB}$, the system exhibits a demixing transition with decreasing temperature, equivalent to the ferromagnetic transition. The total number of *A* and *B* atoms, N_A and N_B, is constant, which means, in the magnetic language, that the total magnetization is constant.

In Fig. 6.3 we show some well-known snapshots of the phase separation process of a Fe_xAl_{1-x} alloy, for different concentrations x (as visible from the white to black regions ratio). The middle column and the two side columns have a clearly different morphology. The middle sequence of images from top to bottom is qualitatively similar to the coarsening process of a magnet, described here above and visualized in Fig. 6.2. The two side columns correspond to qualitatively different initial conditions, where the two phases are unbalanced and dynamics is conserved (i.e., such unbalancing must be globally preserved). This case is called off-critical quench and will be studied in Section 6.4.5.

[1] With the obvious constraint that M is smaller than the maximum physically acceptable value, corresponding to the value attained at the lowest T.

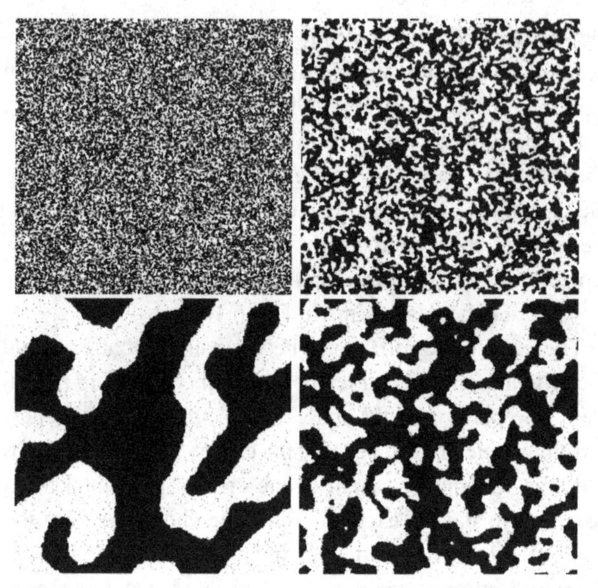

Fig. 6.2 KMC simulations of a 1000×1000 Ising model with nonconserved, spin-flip dynamics, after quenching from $T_i = \infty$ to $T_f = 0.661T_c$. Clockwise, from top left: Snapshots of the system at times $t = 10, 10^2, 10^3, 10^4$. Simulation data are a courtesy of Federico Corberi.

Most of this chapter will deal with the two model systems we have just mentioned, a uniaxial ferromagnet and a binary alloy, and with a lattice gas, where we have a transition between a disordered, gas phase and a condensed phase. The global, qualitative features of the dynamics of all these models are similar, but the quantitative aspects are different, because in the case of the ferromagnet the order parameter is not conserved, while it is conserved in the other cases, binary alloy and lattice gas. This conservation law will be seen to be relevant for the ordering process.

Even if these models are the focus of this chapter, we should not forget there are many systems of interest that require a more complicated modeling. The simplest generalization, which will be studied in Section 6.5, is a nonscalar system with $O(n)$ symmetry, i.e., $n > 1$. A magnetic system described by the Heisenberg Hamiltonian is an example for the case $n = 3$. A more complex example is represented by a binary liquid mixture. In this case the order parameter, i.e., the density, must be coupled to the fluid velocity, which satisfies the Navier–Stokes equation. Convective motion allows a faster transport of the order parameter, which asymptotically speeds up the growth of $L(t)$. However, at short times convective effects can be neglected and the results discussed in this chapter can be applied. This is specially true for high-viscosity fluids, where convection is delayed.

The main feature of the coarsening process is self-similarity, which has been introduced in Chapter 3. Self-similarity and the derived property of dynamic scaling will be seen to hold at large times, when $L(t)$ is much greater than any other scale. Which other scales? To begin with, the lattice constant of a discrete system (which is the smallest scale of the problem). Another scale is the width of a domain wall, the region between two neighboring

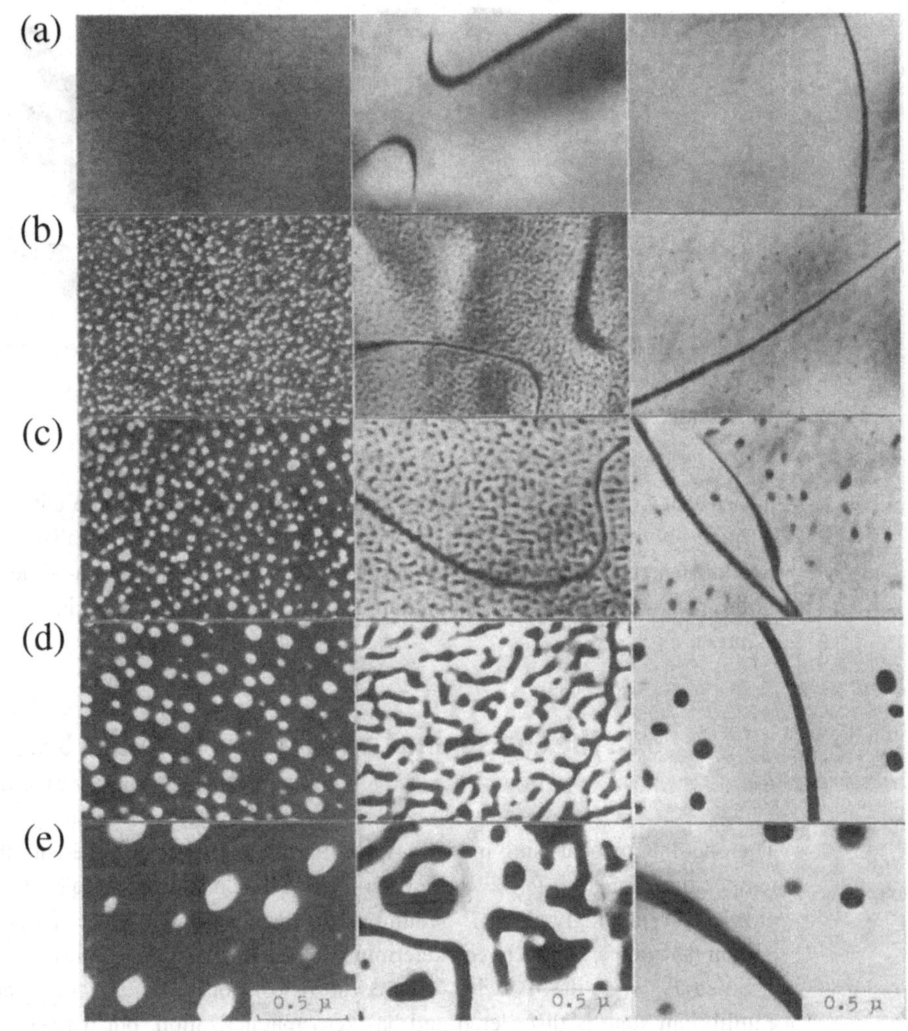

Fig. 6.3 Electron microscopy study of the phase separation in a Fe–Al alloy. The temperature is quenched from $T_i = 630°C$ to $T_f = 570°C$ (left and center images) and to $T_f = 568°C$ (right images). The fraction of Al is 23% (left), 24.7% (center), and 24.9% (right). The center panels correspond to a near-critical quench, the case discussed in these pages. The left and right panels correspond to off-critical quenches, discussed in Section 6.4.5: (a) as quenched; (b) annealed for 15 min in left and right columns and for 10 min in center column; (c) 100 min; (d) 1,000 min; (e) 10,000 min. From K. Oki, H. Sagane, and T. Eguchi, Separation and domain structure of $\alpha + B_2$ phase in Fe–Al alloys, *Journal de Physique Colloques*, **38** (1977) C7-414–C7-417.

domains. Its size depends on temperature, but not on time. Finally, the domain wall itself might be rough, which introduces a new scale depending on time, but this scale is smaller than $L(t)$, because the roughness of an interface of linear size L cannot be larger than L^α, which is smaller than L, because, as discussed in the previous chapter, the roughness exponent is $\alpha < 1$.

In simple terms, self-similarity means that two pictures taken at different times, t_1 and t_2, look statistically the same if we rescale them so as $L(t_1) = L(t_2)$, as we do in Fig. 6.4. This

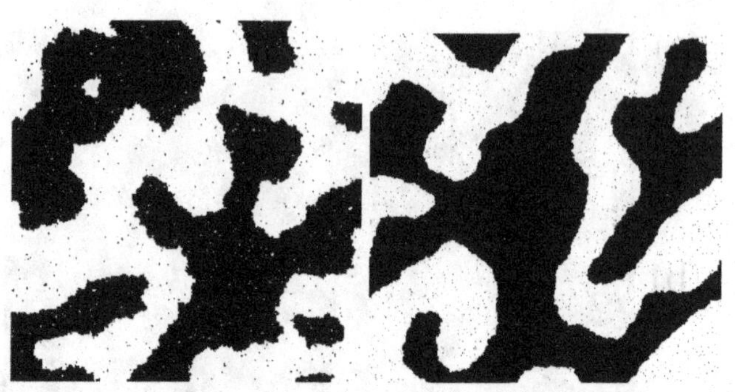

Fig. 6.4 A simple visual example of self-similarity. We compare the configurations of the nonconserved model at $t = 10^3$ and $t = 10^4$ (see Fig. 6.2) by enlarging a piece of the $t = 10^3$ configuration by a factor $L(t = 10^4)/L(t = 10^3)$. The two images are statistically indistinguishable.

means that $L(t)$ is the only relevant scale in the problem and time enters in a statistical average only through $L(t)$. We will discuss more widely and formalize self-similarity in Section 6.4, but the existence of a single relevant length scale at time t will be used in the next sections to study quantitatively the coarsening process in a model with Ising symmetry.

6.2 The Coarsening Law in $d = 1$ Ising-like Systems

The one-dimensional Ising model has peculiar features, because the equilibrium ordering temperature vanishes, $T_c(d = 1) = 0$. In spite of this, it is possible to study phase-ordering at zero or vanishingly small T, gaining useful insights on the effect of a conservation law and on the comparison between deterministic and stochastic dynamics.

Since $T_c = 0$, the model is ordered at $T = 0$ only. At any finite temperature, the equilibrium state is disordered and has zero magnetization, but it is composed of large domains of up and down spins, whose typical size, equal to the correlation length ξ, diverges exponentially when $T \to 0$,

$$\xi(T) \simeq \exp(2J/T). \tag{6.1}$$

This means that the coarsening process in a one-dimensional Ising model cannot last forever, because it will stop when $L(t) \approx \xi(T)$. Zero temperature coarsening might be perpetual because $\xi(T = 0) = \infty$, but $T = 0$ dynamics is restricted to processes that conserve or lower the energy, and, as we will see, this may forbid coarsening at all.

In Appendix H we show how the Ising model may be used for apparently different physical systems. We recall here two of them, the magnet and the lattice gas. For a magnet, the variable $s_i = \pm 1$ is the local spin and the order parameter $m = \frac{1}{N}\sum_i s_i$ is not conserved, meaning that m is allowed to change over time. The ground state of the magnet corresponds to a fully magnetized state, with all spins up ($s_i = +1\ \forall i$) or down ($s_i = -1\ \forall i$). For a lattice gas it is worth using the variable $n_i = (s_i + 1)/2 = 0, 1$, which denotes the presence

($n_i = 1$) or the absence ($n_i = 0$) of a particle in a given site i. The order parameter $\rho = \frac{1}{N}\sum_i n_i$ is conserved because the total number of particles is a constant of motion. The ground state corresponds to a fully condensed phase, with particles joined together to form one single, compact aggregate whose actual shape depends on boundary conditions.

Consequently, when studying the dynamics of the two systems (the magnet and the lattice gas) we have to implement different rules. The simplest possible microscopic rules, which conserve or do not conserve the magnetization, respectively, are exchange (Kawasaki) dynamics and spin-flip (Glauber) dynamics. In spin language, exchange dynamics amounts to exchanging a pair of neighboring spins; spin-flip dynamics amounts to reversing a single spin. In both cases, the elementary process is accepted or refused according to some probability, which must satisfy the detailed balance condition in order to guarantee the convergence toward equilibrium. If we refer to the popular Metropolis algorithm (see Section 1.5.5), the microscopic move is accepted with probability 1 if it lowers or keeps the energy constant, while it is accepted with probability $\exp(-\Delta E/T)$ if the final energy $E_f = E_i + \Delta E$ is higher than the initial energy E_i. Let's now discuss in more detail the conserved and nonconserved dynamics.

6.2.1 The Nonconserved Case: Spin-Flip (Glauber) Dynamics

The dynamical picture is made easier if we represent a given spin configuration using domain walls (DWs), i.e., fictitious points located between spins of opposite orientation (see Fig. 6.5). There are three types of elementary processes for DWs: (a) The hopping of a DW to its right or to its left. This corresponds to flipping a spin i (the thick spin in Fig. 6.5) between two spins of opposite orientations ($s_{i+1} = -s_{i-1}$). This process does not change the energy, $\Delta E = 0$. (b) The annihilation of two neighboring DWs. This corresponds to flipping a spin i located between two spins, both antiparallel to it ($s_{i+1} = s_{i-1} = -s_i$). This process decreases the energy, $\Delta E < 0$. (c) The creation of two neighboring DWs. This corresponds to flipping a spin i that is parallel to its neighbors ($s_{i-1} = s_i = s_{i+1}$). This process increases the energy, $\Delta E > 0$.

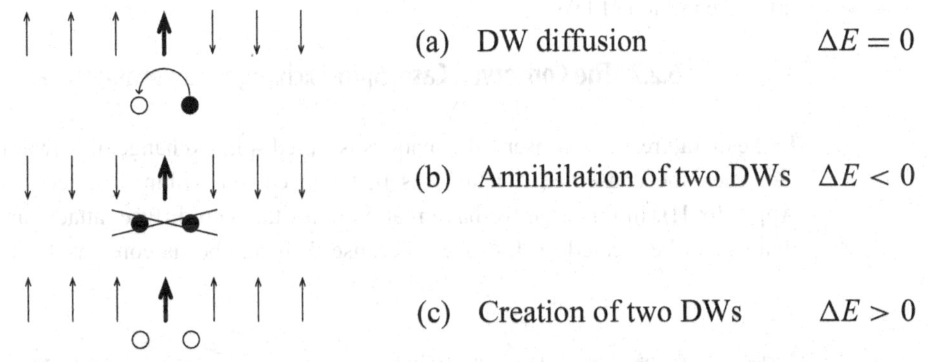

Fig. 6.5 Nonconserved, spin-flip (Glauber) dynamics. We sketch the possible transitions using spin language and domain wall (DW) language. The thick spin is reversed and circles represent the DWs before (solid circles) and after (open circles) the flipping process.

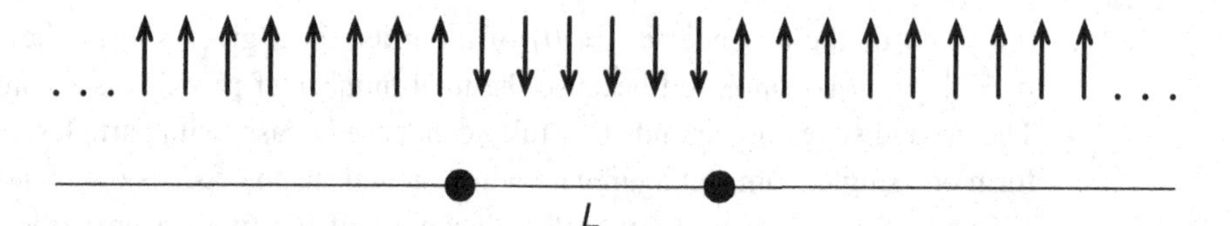

Fig. 6.6 A domain of down spins of size L is represented through two pseudo-particles, located at DW positions.

At $T = 0$, process (c) is forbidden, but processes (a) and (b) are sufficient to drive the system toward the ordered state. Therefore, perpetual coarsening occurs at $T = 0$ with nonconserved dynamics. We will see that this is not the case with conserved order parameter. Zero temperature dynamics in terms of DWs is easy: these (pseudo)particles diffuse; when two of them occupy the same site they annihilate, leading to a decreasing number of DWs, i.e., to an increasing of their average distance $L(t)$, which is nothing but the average size of ordered regions (magnetic domains). At $T > 0$, process (c) would also allow the creation of new DWs: this is the mechanism by which a finite correlation length $\xi(T)$ eventually arises, since large magnetic domains are broken into smaller parts by this process.

The key point to evaluating $L(t)$ is that self-similarity assures the existence of one single, physically relevant length scale, whose time dependence can be easily determined studying the dynamics of a suitably chosen configuration. The simplest one, for nonconserved dynamics, is one domain of (say) negative spins immersed in a sea of positive ones; see Fig. 6.6. In the DW language, we have two particles at distance L that can diffuse, and we wonder, what is the time t required to ensure they meet and annihilate, i.e., such that $L(t) = 0$? Since the quantity L itself performs a random walk, the time necessary to pass from $L(0) = L$ to $L(t) = 0$ scales as L^2, $t \sim L^2$. Assuming that there is just one length scale at time t, the coarsening law can simply be found by reversing the above relation, giving

$$L(t) \approx t^n \qquad \left(n = \frac{1}{2}, \quad \text{nonconserved dynamics} \right). \tag{6.2}$$

In conclusion, the (nonconserved) coarsening exponent $n = \frac{1}{2}$ simply comes from the diffusive motion of DWs.[2]

6.2.2 The Conserved Case: Spin-Exchange (Kawasaki) Dynamics

In the literature, local conserved dynamics is called spin-exchange or Kawasaki dynamics. However, the conserved dynamics is better discussed within a lattice gas model (see Appendix H). In this case we have real particles that can diffuse, attach, and detach and that cannot be created or destroyed, because their number is conserved. The hopping of

[2] The reader familiar with random walks can object that the chosen configuration, one single minority domain in an infinite system, is *too* simple. In fact, if a random walker is in $x = L$ at $t = 0$, the mean time to cross the origin is actually infinite, while the median time scales as L^2, as shown in Appendix D.5. In order to have a finite mean time we should consider a finite sample of size $2L$ with periodic boundary conditions.

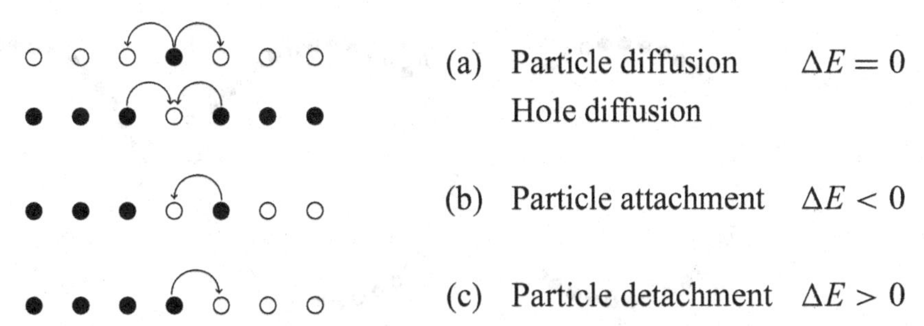

Fig. 6.7 Conserved, spin-exchange (Kawasaki) dynamics. We plot the possible transitions using the lattice gas picture. Solid circles are particles (up spins), open circles are holes (down spins). Arrows depict possible moves of a single particle.

a particle in an empty site corresponds, in spin language, to exchange an up spin with a neighbouring down spin, from which comes the name spin-exchange dynamics.

Different types of processes (see Fig. 6.7) can be listed according to the variation $\Delta\bar{z}$ in the number of nearest neighbors of the hopping particle, i.e., to the change of the energy. Since hopping is restricted to nearest neighbors, a hopping particle cannot be enclosed by two particles, neither before nor after hopping. So, the possible processes are: (a) particle/hole diffusion ($\Delta\bar{z} = 0$). If $\bar{z} = 0$ we speak of particle diffusion, if $\bar{z} = 1$ we speak of hole diffusion. This process does not cost energy; $\Delta E = 0$. (b) Particle attachment ($\Delta\bar{z} = 1$): the particle gains one neighbor after hopping, lowering the energy; $\Delta E < 0$. (c) Particle detachment ($\Delta\bar{z} = -1$): the particle loses one neighbor after hopping, raising the energy; $\Delta E > 0$.

Therefore, there are similarities and differences between lattice gas dynamics and domain wall dynamics. Similarities refer to single particle dynamics, which is free diffusion in both cases. Differences refer to the "coupling" between particles: in the lattice gas we have real particles with an attractive nearest-neighbor interaction; in the domain wall case we have fictitious particles, which annihilate when they meet and which are created in pairs via a thermally activated process.[3]

Let us now review the lattice gas dynamics, making reference to Fig. 6.7. At $T = 0$, only diffusion, process (a), and attachment, process (b), are permitted, but they are not sufficient to provide a meaningful dynamics. In fact, zero temperature dynamics stops as soon as all particles are attached to other particles and all holes are attached to other holes; in other words, when domains have a minimal length of two. Therefore, if we want to study coarsening in this conserved model we have to switch temperature on, being aware that phase separation will not last forever: it will stop at the equilibrium correlation length.

Because of conservation, we cannot consider the problem equivalent to that studied in the nonconserved case: the disappearance of a single domain of particles in a sea of holes. We need at least two domains of particles that can exchange matter. The simplest configuration giving rise to coarsening is plotted in Fig. 6.8, where two particle domains of size L are separated by two hole domains of the same size, with periodic boundary conditions. It is

[3] There is also another difference: the lattice gas model is valid in any dimension d, while domain walls can be represented as fictitious particles only in $d = 1$.

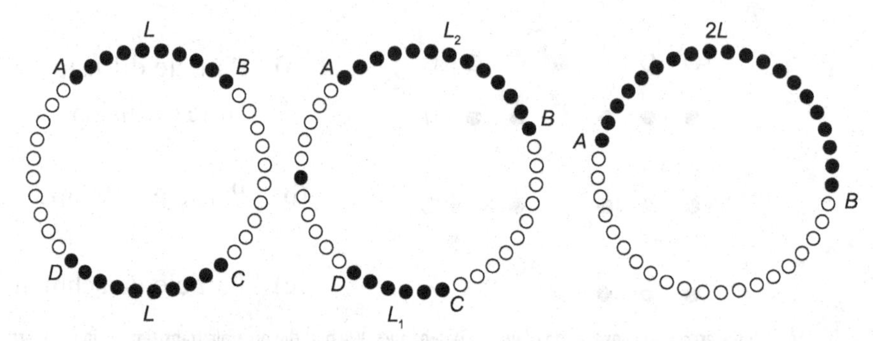

Fig. 6.8 Left: Two domains of particles of initial size L are separated by a distance L, with periodic boundary conditions. Center: Particles can detach at domain edges A, B, C, D, diffuse and either attach to the other domain or reattach to the same domain. Right: After a time $t(L)$ one domain has evaporated and all particles are gathered in a domain of size $2L$.

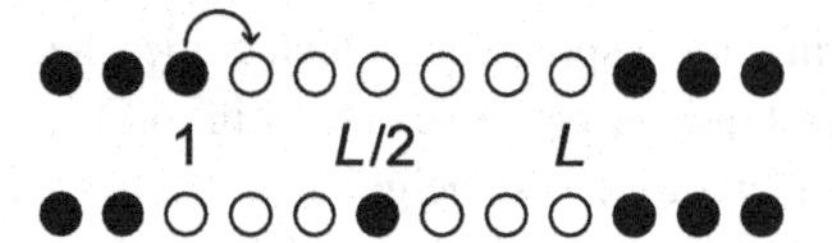

Fig. 6.9 Two domains are separated by holes. Domains can exchange matter if a particle detaches, e.g., by the left domain and attains the right domain (site L) before coming back to the original site 1. If the particle is in site $L/2$, the probabilities to attain the starting domain or the arrival domain are the same.

apparent that the system cannot evolve at $T = 0$, because dynamics requires the detachment of particles from domain edges, a process that costs energy. The probability of such process, $\exp(-2J/T)$, sets a time scale $\tau_0 \approx \exp(2J/T)$, which is extremely long at low temperature. For a fixed size L of the domains, we can choose T such that τ_0 is much longer than any other, nonthermally activated process, as particle diffusion (which scales as L^2) or particle attachment (which is of order 1). Once a particle has detached by one of the four edges (A, B, C, D) there are two possible outcomes: either the particle attaches (with probability p) to the next domain or it goes back (with probability $1-p$) to the starting site.

Dynamics can be summarized as follows. Every time τ_0 a particle detaches from an edge, with probability p it attaches to the other domain. Therefore, every time $\Delta t = \tau_0/p$, $L_1 \to L_1 - 1$ and $L_2 \to L_2 + 1$ (if the particle passes from the first to the second domain) or $L_1 \to L_1 + 1$ and $L_2 \to L_2 - 1$ (if the particle passes from the second to the first domain). Therefore, the length of each block of particles performs a random walk, and after a time $t(L) \approx L^2 \Delta t$ one of the two intervals has disappeared, leading to coarsening. We still have to calculate Δt, i.e., the probability $p(L)$ that a detached particle attaches to the neighboring interval, rather than going back to the same interval. This is equivalent to saying that the detached particle attains a distance L before coming back to the starting site.

The probability $p(L)$ can be easily evaluated as follows (see Fig. 6.9). Suppose that particle in site 1 detaches from the left domain (top). In order to join the other domain, namely, to reach site L, it must first reach site $L/2$ (bottom). Once the particle has attained the distance $L/2$ from the starting point, which occurs with probability $p(L/2)$, for symmetry reasons it has the same probability of attaining the starting domain or the other domain. Therefore,

$$p(L) = \frac{p(L/2)}{2}, \tag{6.3}$$

whose solution is $p(L) = c/L$. Since $p(2) = \frac{1}{2}$, $p(L) = 1/L$. In conclusion, $\Delta t = \tau_0/p = \tau_0 L$ and $t(L) \approx L^2 \Delta t \approx \tau_0 L^3$, with τ_0 depending on temperature, not on L. This relation leads to

$$L(t) \approx t^n \qquad \left(n = \tfrac{1}{3}, \quad \text{conserved dynamics}\right). \tag{6.4}$$

Therefore, the conservation law slows down coarsening and n passes from $\frac{1}{2}$ to $\frac{1}{3}$. This result shows the utmost importance of conservation laws in phase-ordering dynamics.

6.3 The Coarsening Law in $d > 1$ Ising-like Systems

If we try to extend the arguments of the previous section to more than one dimension, we realize that they cannot be applied, because domains now have complicated shapes (see Fig. 6.2), and the large variety of possible local environments of spins make it difficult to study the basic processes of spin-flip or spin-exchange. In addition to this, passing to higher dimension also introduces a different physics. This is somewhat obvious, considering the equilibrium properties of the Ising model, which has a vanishing critical temperature for $d = 1$ and a finite one for $d > 1$.

A simple remark helps us to focus on a relevant difference between one and larger dimensions. Nonconserved (Glauber) dynamics of the one-dimensional Ising model can be studied at $T = 0$ and leads to the coarsening exponent $n = 1/2$. More precisely, every initial condition leads to coarsening and in a finite system of size L_0 the ground state (all spins up or down) is always attained in a time of order L_0^2. If we pass to a d-dimensional model, we can easily observe that configurations varying along one direction only, i.e., sequences of domains infinitely straight in the remaining $(d-1)$ directions, are metastable structures, because a spin at the interface between two domains now has $(2d-1)$ parallel neighbors and just one antiparallel neighbor, so that its reversal costs energy. Therefore, these structures[4] are frozen at $T = 0$: if the system is trapped in one of these configurations, zero T dynamics stops.

We might think that these metastable configurations are essentially irrelevant for the dynamics, but this is not the case: numerical simulations show that the system is trapped with finite probability in $d = 2$ and with certainty in $d \geq 3$. However, this trapping is a finite-size effect, because the trapping time increases with the size of the system, so the coarsening process can be studied even if the final state is not a completely phase-separated state.

Actually, there is a deeper reason to be interested in the trapping of these metastable structures. Shortly after a quenching from $T = \infty$ the morphology of interconnected domains resembles the cluster morphology of critical percolation, and the spanning,

[4] In $d > 2$ there are also other, more complicated metastable structures.

percolation clusters are the precursors of the metastable, stripe-like states appearing at late times. We will see later that the parallel with percolation allows us to gain insight into the time dependence of the domain size distribution.

In the following two sections we consider models with Ising symmetry in $d > 1$, with nonconserved and conserved dynamics, respectively. As anticipated, we are not able to apply the simple lines of reasoning used in Section 6.2 to find out the coarsening exponent n. For each case our analysis first presents some Monte Carlo simulations and a numerical derivation of n; then, we offer a continuum approach to the coarsening process, using the scaling ansatz. In spite of the obvious differences with one-dimensional discrete models, our approach to determine $L(t)$ analytically is conceptually the same: starting with a simple configuration with one or two domains of size L, we determine the time $t(L)$ to get the equilibrium state. The coarsening law will be found by reversing the function $t(L)$.

6.3.1 The Nonconserved Case

Following the adage "A picture is worth a thousand words," in Fig. 6.2 we plot some snapshots at different times for the relaxation process of an Ising model with nonconserved dynamics. They refer to a quenching from $T_i = \infty$ to $T_f = 0.661T_c$. A quantitative analysis of such images via the correlation function will be made in Section 6.4. Here we want to focus on the coarsening process, which is clearly visible and is made quantitative in Fig. 6.10. The numerical data suggest a coarsening exponent $n = 1/2$, therefore the same as in $d = 1$.

This result seems to suggest that the coarsening exponent might not depend on the dimension d, which is surprising if we have in mind equilibrium critical phenomena, where critical exponents do depend on d. In fact, this statement (n does not depend on d) will be confirmed but also revised by the continuum approach we are going to present.

The first step is to write a continuum expression for the free energy of the Ising model, the so-called Ginzburg–Landau free energy. Since this functional is commonly used in the

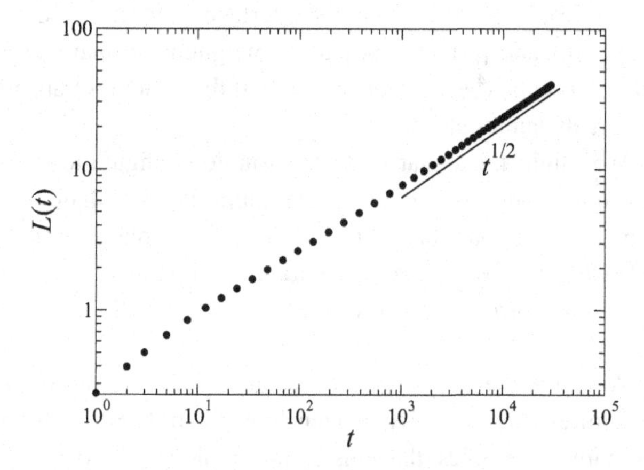

Fig. 6.10 KMC simulation of a 2000×2000 Ising model with nonconserved, spin-flip dynamics, after quenching from $T_i = \infty$ to $T_f = 0.661T_c$. The length scale L is determined by the width at half height of the correlation function $C(r, t)$, $C(L, t) = C(0, t)/2$. Simulation data courtesy of Federico Corberi.

study of phase transitions and the reader is likely to be familiar with it (see also Section 3.2.3), we give its derivation in Appendix I; see Eq. (I.19). The final result, valid for $T < T_c$, is

$$\mathcal{F} = f_0 \int d\mathbf{r} \left[\frac{1}{2}(\nabla m)^2 - \frac{1}{2}m^2 + \frac{1}{4}m^4 \right], \tag{6.5}$$

where $m(\mathbf{r}, t)$, the order parameter, is the magnetization density at point $\mathbf{r}$ and time t, normalized so that it varies in the range $(-1, +1)$.

The next, nonrigorous step is to write a phenomenological equation for the evolution of $m(\mathbf{r}, t)$. The goal is to have an equation with two main features: (i) the order parameter is not conserved, and (ii) dynamics proceeds so as to minimize $\mathcal{F}$. The simplest equation is a generalization of the Newton equation for a particle subject to friction, $m_0\ddot{x} = -\gamma^{-1}\dot{x} - U'(x)$. In the overdamped regime the second time derivative is negligible with respect to the first time derivative, yielding $\dot{x} = -\gamma U'(x)$. According to this equation, dynamics minimizes $U(x)$ because $dU/dt = U'(x)\dot{x} = -\gamma(U'(x))^2 < 0$.

We can generalize the equation $\dot{x} = -\gamma U'(x)$, passing from $x(t)$ to $m(\mathbf{r}, t)$, as

$$\frac{\partial m}{\partial t} = -\gamma \frac{\delta \mathcal{F}}{\delta m} = \gamma f_0 \left(\nabla^2 m + m - m^3 \right), \tag{6.6}$$

where the symbol $\delta [...] / \delta m$ means a functional derivative; see Appendix L.1.

The proof that dynamics minimizes $\mathcal{F}$ is just as easy, because

$$\frac{d\mathcal{F}}{dt} \equiv \int d\mathbf{r} \frac{\delta \mathcal{F}}{\delta m} \frac{\partial m}{\partial t} = -\gamma \int d\mathbf{r} \left(\frac{\delta \mathcal{F}}{\delta m} \right)^2 \leq 0. \tag{6.7}$$

Therefore, if $\mathcal{F}[m(\mathbf{r}, t)]$ is evaluated along a dynamical trajectory, it is a decreasing function of time (except when $m(\mathbf{r}, t)$ does not depend on time, in which case $\mathcal{F}$ also is constant in time).

According to Eq. (6.6) it is also trivial to remark that the order parameter is not conserved,

$$\frac{d}{dt}\overline{m}(t) \neq 0, \tag{6.8}$$

where $\overline{m}(t)$ is the spatial average of the order parameter.

Now we come back to Eq. (6.6), and rescale time, $t \to t/(\gamma f_0)$, so as to get an equation that is parameter free,

$$\frac{\partial m}{\partial t} = \nabla^2 m + m - m^3 \equiv \nabla^2 m - U'(m), \qquad U(m) = -\frac{m^2}{2} + \frac{m^4}{4}. \tag{6.9}$$

This equation is called time-dependent Ginzburg-Landau (TDGL) equation, and the general goal would be to solve it with an initial condition corresponding to a completely disordered state ($T_i = \infty$). We will not try to face such a general problem here. We rather focus on a simpler one, fully analogous to the one considered in Section 6.2.1: determining the time scale $t(L)$ necessary for a domain of negative magnetization in a sea of positive magnetization to disappear.

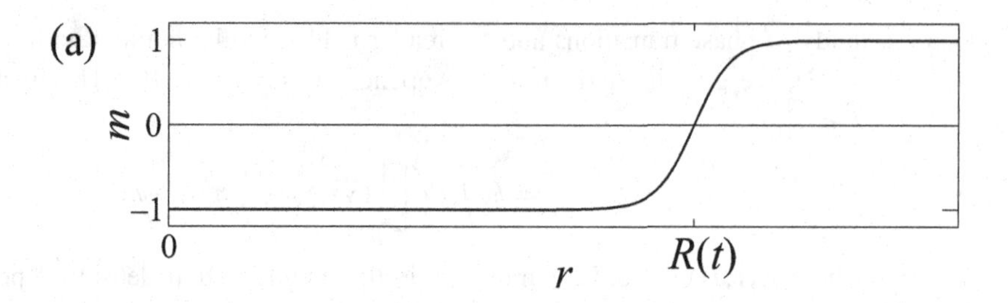

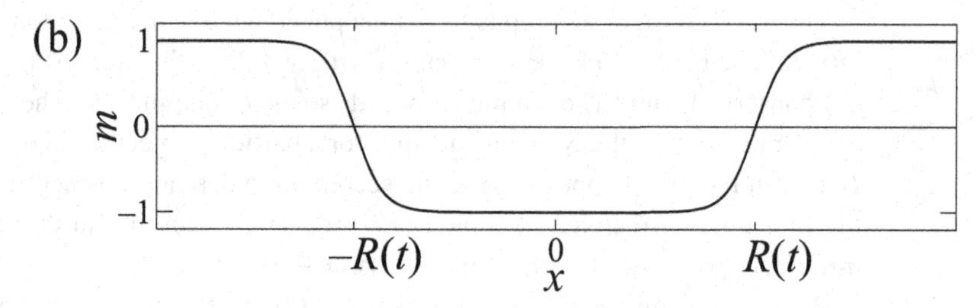

Fig. 6.11 Profiles of the order parameter for a single negative domain in a positive sea. (a) In $d > 1$ the domain is spherical and the profile of $m(r)$ is approximated by a kink centered in $r = R(t)$. (b) In $d = 1$ the domain is a segment and can be obtained by superposing two kinks: a negative kink centered in $x = -R(t)$ and a positive kink centered in $x = +R(t)$. Note that in $d > 1$, $r > 0$ because it is the radius in spherical coordinates, while in $d = 1$ x is the Cartesian coordinate, so it may be negative.

Let us consider the problem of a single spherical domain of radius R. For spherical symmetry, $m(\mathbf{r}, t) = m(r, t)$, and Eq. (6.9) is rewritten as

$$\partial_t m(r,t) = \partial_{rr} m + \frac{d-1}{r}\partial_r m - U'(m). \tag{6.10}$$

The potential $U(m)$ has minima in $m = \pm 1$, so a domain of radius $R(t)$ corresponds to a profile $m(r,t)$, which varies with continuity between $m = -1$ for $r \ll R$ to $m = +1$ for $r \gg R$; see Fig. 6.11. More precisely, what is the extension of the region around $r = R$ where m is significantly different from ± 1? We can answer rigorously in $d = 1$ and assume the answer is qualitatively correct for any d.

In $d = 1$, the TDGL equation is written as

$$\partial_t m(x,t) = \partial_{xx} m + m - m^3, \tag{6.11}$$

which has the exact stationary solution

$$m(x) = \tanh\left(\frac{x - R}{\sqrt{2}}\right). \tag{6.12}$$

The domain wall has a finite size, the profile $m(x)$ converging exponentially toward ± 1. Let us now come back to Eq. (6.10), with that in mind. We assume that while the domain shrinks, i.e., $R(t)$ decreases, the value of $m(r,t)$ just depends on the distance between r and the center of the domain wall,

$$m(r,t) = m(r - R(t)), \tag{6.13}$$

so that $\partial_t m = -\dot{R}(t)\partial_r m$, and Eq. (6.10) can be rewritten as

$$-\dot{R}m' = m'' + \frac{d-1}{r}m' - U'(m), \tag{6.14}$$

where the easier notation $m' = \partial_r m(r-R)$ has been used.

An ordinary differential equation for $R(t)$ can be obtained by multiplying both terms of Eq. (6.14) by m' and integrating between zero and infinity:

$$-\dot{R}\int_0^\infty dr(m')^2 = \int_0^\infty drm'\left[m'' + \frac{d-1}{r}m' - U'(m)\right] \tag{6.15}$$

$$= \left[\frac{1}{2}(m')^2 - U(m)\right]_{r=0}^{r=\infty} + (d-1)\int_0^\infty dr\frac{(m')^2}{r}. \tag{6.16}$$

The integral proportional to $(d-1)$ can be simplified, because $(m')^2$ is nonvanishing only in a small, finite region around $r = R$, so it can be evaluated by replacing r with R in the denominator. Defining $I = \int_0^\infty dr(m')^2$, we obtain

$$-\dot{R}I = -\frac{1}{2}(m'(r=0))^2 + U(m(r=0)) - U(m(r=\infty)) + \frac{d-1}{R}I, \tag{6.17}$$

where we have used that $m'(+\infty) = 0$.

Since the domain wall profile converges exponentially to the minima of U (see Eq. (6.12)), both $m'(r=0)$ and $[U(m(r=0)) - U(m(r=\infty))]$ are exponentially small with respect to $1/R$, so they are negligible when $d > 1$. In this case,

$$\dot{R} = -\frac{d-1}{R}, \tag{6.18}$$

whose solution is

$$R^2(t) = R^2(0) - 2(d-1)t. \tag{6.19}$$

If $R(0) = L$, the time to make the domain disappear is $t(L) = L^2/(2(d-1))$. This result, $t(L) \approx L^2$, is therefore in agreement with the coarsening exponent $n = 1/2$ found numerically in Fig. 6.10 for $d = 2$ and also found in Section 6.2 for the one-dimensional Ising model with nonconserved dynamics.

The last step is to solve Eq. (6.17) for $d = 1$, which gives an unexpected outcome. For $d = 1$ the last term on the right-hand side of Eq. (6.17) vanishes and the exponentially small remaining terms must be taken into account. In $d = 1$, the variable r is replaced by the spatial coordinate x and the profile $m(x,t)$ of a domain located between $x = -R$ and $x = R$ is symmetric so $m'(x=0)$ is exactly zero, leading to the result

$$\dot{R} = -\frac{1}{I}\left[U(m(x=0)) - U(m(x=\infty))\right]. \tag{6.20}$$

In order to get a negative domain between $x = -R$ and $x = R$ we have to suitably superpose a kink centered in $x_0 = R$ and a kink centered in $x_0 = -R$ (see Fig. 6.11(b)),

$$m(x) = \tanh\left(\frac{x-R}{\sqrt{2}}\right) - \tanh\left(\frac{x+R}{\sqrt{2}}\right) + 1, \tag{6.21}$$

where the constant term 1 is required to get $m(x) \to 1$ for $x \to \pm\infty$. Using the previous equation, for large R we obtain

$$m(0) = 1 - 2\tanh(R/\sqrt{2}) \approx -1 + 4\exp(-\sqrt{2}R) \tag{6.22}$$

and

$$U(m(x=0)) - U(m(x=\infty)) = 16e^{-2\sqrt{2}R}, \tag{6.23}$$

so that, using Eq. (6.20), we obtain

$$-I\dot{R} = 16e^{-2\sqrt{2}R}. \tag{6.24}$$

The equation is easily integrated by separation of variables, giving

$$\frac{1}{2\sqrt{2}}\left[e^{2\sqrt{2}R(t)} - e^{2\sqrt{2}R(0)}\right] = -\frac{16}{I}t. \tag{6.25}$$

Therefore, if $R(0) = L/2$ the domain disappears in a time $t(L)$,

$$t(L) = \frac{I}{32\sqrt{2}}\left(e^{\sqrt{2}L} - 1\right) \simeq \frac{I}{32\sqrt{2}}e^{\sqrt{2}L}. \tag{6.26}$$

In conclusion, the time to close a domain of size L increases exponentially with L and the coarsening is logarithmically slow,

$$L(t) \approx \ln t. \tag{6.27}$$

This result raises two questions: why does the continuum theory give such a slow coarsening in $d = 1$? And why does this result disagree with what we found in Section 6.2, where we got a power law coarsening? Eq. (6.17) for $R(t)$ shows that $\dot{R}$ is made up of two quantities, one of order $1/R$ and vanishing in $d = 1$, the other of order $\exp(-R)$. Both quantities give a negative contribution to the speed of the wall. The quantity $1/R$ is nothing but the curvature of the domain wall, which is clearly absent in $d = 1$, where walls reduce to points. Therefore, the TDGL equation in $d > 1$ represents a system whose evolution is driven by the curvature of domain walls. Instead, the exponentially small quantity represents an effective interaction between walls, which is the only term driving coarsening in $d = 1$, where curvature does not play any role. Since kinks have an exponential profile, their interaction is exponentially small and coarsening is logarithmically slow.

These considerations do not explain why Eq. (6.27) does not agree with the result we found for the one-dimensional Ising model. The reason lies in the fact that Glauber dynamics is stochastic, while Eq. (6.14) is deterministic. In order to take into account noise in our continuum model, the TDGL equation should be supplemented with a stochastic term, $\eta(\mathbf{r}, t)$, making Eq. (6.9) a generalized Langevin equation. Inserting the noise term, in $d = 1$ Eq. (6.14) is written as

$$-\dot{R}m' = m'' - U'(m) + \eta(x, t). \tag{6.28}$$

As before, we multiply by $m'(x - R)$ and integrate. Eq. (6.24) is modified accordingly, giving

$$-I\dot{R} = 16e^{-2\sqrt{2}R} + \int_0^\infty dx\, m'(x - R(t))\eta(x, t), \tag{6.29}$$

which can be rewritten as

$$\dot{R} = -\frac{16}{I} e^{-2\sqrt{2}R} + \bar{\eta}(t), \tag{6.30}$$

where

$$\bar{\eta}(t) = -\frac{1}{I} \int_0^\infty dx\, m'(x - R(t))\eta(x, t). \tag{6.31}$$

If η has zero mean and is δ-correlated in space and time,

$$\langle \eta(x, t) \rangle = 0, \qquad \langle \eta(x, t)\eta(x', t') \rangle = \Gamma\delta(x - x')\delta(t - t'), \tag{6.32}$$

the spectral properties of $\bar{\eta}(t)$ are

$$\begin{aligned}
\langle \bar{\eta}(t) \rangle &= -\frac{1}{I} \int_0^\infty dx m' \langle \eta(x, t) \rangle = 0 \\
\langle \bar{\eta}(t)\bar{\eta}(t') \rangle &= \frac{1}{I^2} \int_0^\infty dx \int_0^\infty dx' m'(x - R(t)) m'(x' - R(t')) \langle \eta(x, t)\eta(x', t') \rangle \\
&= \frac{1}{I^2} \left[\Gamma \int_0^\infty dx (m')^2 \right] \delta(t - t') \\
&= \frac{\Gamma}{I} \delta(t - t').
\end{aligned}$$

Therefore, $\bar{\eta}(t)$ is an effective Gaussian and uncorrelated noise.

In summary, the introduction of thermal noise in the one-dimensional TDGL equation for a single domain has the effect of adding a noise term to the effective equation for the position of the DW, which now (see Eq. (6.30)) looks like the equation for a random walker in an exponentially small potential. It is reasonable to assume that the exponentially small deterministic drift is negligible, therefore giving a time $t(L) \approx L^2$ because a domain of size L closes. Therefore, the effect of noise in $d = 1$ is to accelerate coarsening, which passes from logarithmic to power law with an exponent $1/2$, the same found in the one-dimensional Ising model with Glauber dynamics. If we add noise to the continuum model in $d > 1$, a rough generalization of Eq. (6.30) would simply lead to replacing the deterministic term, proportional to $e^{-2\sqrt{2}R}$, with the right-hand side of Eq. (6.18),

$$\dot{R} = -\frac{d-1}{R} + \bar{\eta}(t). \tag{6.33}$$

Since both terms on the right-hand side give, separately, the coarsening exponent $n = \frac{1}{2}$, we can assume that their combination does not modify it.

6.3.2 The Conserved Case

We begin this section with some snapshots of a two-dimensional Ising model, obtained with a kinetic Monte Carlo simulation of conserved dynamics, after a quench into the ordered phase; see Fig. 6.12. The numerical analysis of the growth of $L(t)$ (see Fig. 6.13) suggests $n = 1/3$, therefore prompting us to conclude that the coarsening exponent does not depend on d even in the conserved case.

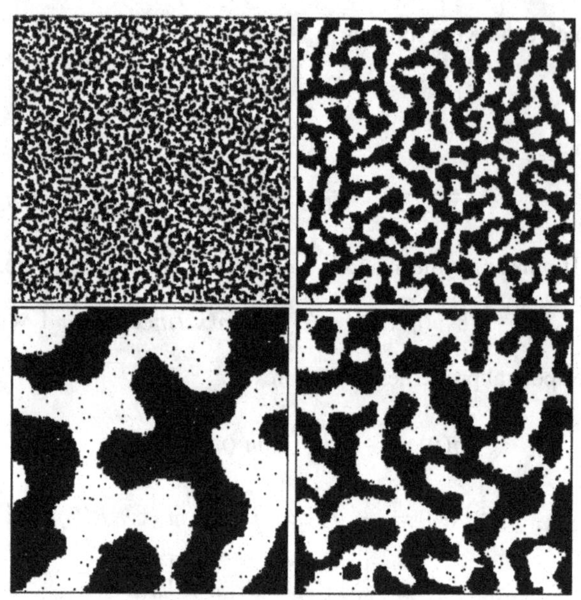

Fig. 6.12 KMC simulations of a 256×256 Ising model with conserved, spin-exchange dynamics, after a quenching from $T_i = \infty$ to $T_f = 0.661T_c$. Clockwise, from top left: Snapshots of the system at times $t = 10^2, 10^4, 10^5, 10^6$. Simulation data courtesy of Federico Corberi.

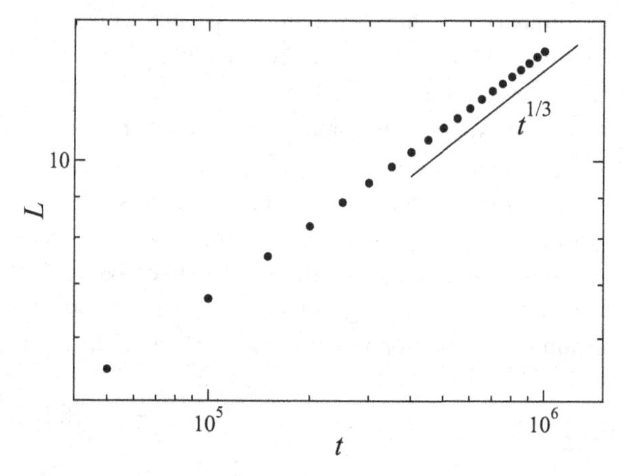

Fig. 6.13 KMC simulation of a 512×512 Ising model with conserved, spin-exchange dynamics, after quenching from $T_i = \infty$ to $T_f = 0.661T_c$. The length scale L is determined by the width at half height of the correlation function $C(r,t)$, $C(L,t) = C(0,t)/2$. Simulation data courtesy of Federico Corberi.

While a quantitative comparison of the domain morphology with the nonconserved case is deferred to the next section, we want first to offer an analytical derivation of the coarsening exponent. The continuum analysis of the nonconserved case was based on the study of the TDGL equation, whose derivation required two steps: a continuum approximation of the free energy of an Ising model and the writing of a simple dissipative

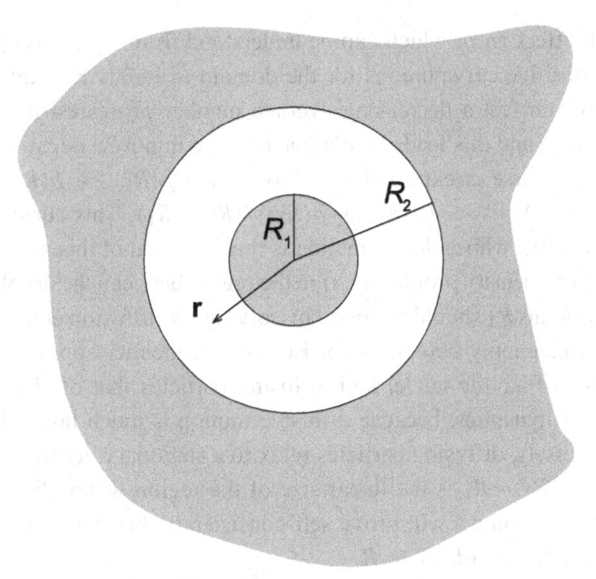

Fig. 6.14 Two-sphere geometry to study the conserved coarsening in a continuous model, in dimension $d > 1$.

equation, which is a generalized form of the equation $\dot{x} = -\gamma U'(x)$, valid for a single particle. We now consider dynamics that conserves the order parameter, so it must be

$$\frac{d}{dt}\int d\mathbf{r}\, m(\mathbf{r},t) = 0. \tag{6.34}$$

A new partial differential equation satisfying such a constraint will be introduced further on. However, its analysis is more elaborate and here we prefer to focus on diffusion, the physical process leading to coarsening in the conserved system, trying to perform a calculation similar in spirit to what we did for the one-dimensional Ising model. In that case we studied two domains that exchanged particles; see Fig. 6.8. We argued that extending such calculation to higher dimensions would have been difficult, so we are now going to perform such extension within a continuum picture. However, as in the nonconserved case, we have to consider the consequence passing from a stochastic to a deterministic model. As for $d = 1$, in the conserved case we need two domains that can exchange matter, finding the appropriate geometry to make the problem solvable. We suggest a spherical geometry where the external domain is infinite and is called a reservoir; see Fig. 6.14.

The idea is to model the process of particle detachment from the small domain of radius R_1, the particle diffusion in the empty region $R_1 < r < R_2$, and the particles' final attachment to the reservoir, of inner radius R_2. The main question is: why is there a net flux of particles between domain and reservoir? The answer is qualitatively simple and we discuss it here. It is less trivial to make it quantitative, because we must refer to the so-called Gibbs–Thomson effect, discussed in Appendix N.

Particles diffuse from high density to low density regions, and the density ρ of diffusing particles close to a domain depends on the curvature of the domain itself (Gibbs–Thomson effect). If a domain wall is straight, the equilibrium density of particles close to it has some value $\rho = \rho_0$, depending on temperature. If we now bend the domain, we can expect an effect on ρ, which can be understood from a microscopic point of view. If we impose a positive curvature, as for the domain of radius R_1, particles at the surface of the inner domain feel a decreasing average number of nearest neighbors, therefore it is easier to detach and this leads to a larger ρ. If we impose a negative curvature, as for the reservoir, the reverse is expected. For this reason $\rho(R_1) > \rho(R_2)$, leading to a current of order $J \approx -\nabla\rho \approx -(\rho(R_2) - \rho(R_1))/(R_2 - R_1)$. This current shrinks the domain in a time $t(R_1(0))$, whose determination is the final goal of this section.

The density profile $\rho(r,t)$ in the region between the droplet of radius R_1 and the reservoir of radius R_2 should be found by solving the diffusion equation, $\partial_t\rho - D\nabla^2\rho = 0$. However, we can apply here a sort of Born–Oppenheimer approximation, where domains play the role of atomic nuclei and diffusing particles that of electrons. In fact, we can decouple their dynamics, because diffusive motion is much faster than domain wall motion. More precisely, diffusing particles relax to a stationary density profile in a time $\tau_0 \approx R^2$, where $R = R_2 - R_1$ is the linear size of the region where diffusion takes place. At the end of this section we will prove self-consistently that during time τ_0 the boundaries move by a quantity of order $1 \ll R$.

In the approximation discussed above, we can solve the stationary diffusion equation, i.e., the Laplace equation, and determine the steady profile $\rho(r, R_1, R_2)$. From that we can derive the current of particles that attach to/detach from boundaries and, at the end, their time evolution $R_{1,2}(t)$. Using the obvious spherical symmetry of the problem, the Laplace equation, $\nabla^2\rho = 0$, reads

$$\rho'' + \frac{d-1}{r}\rho' = 0, \tag{6.35}$$

where r is the radius in spherical coordinates. Boundary conditions, qualitatively anticipated here above, are derived in Appendix N. The result is

$$\begin{aligned} \rho(R_1) &= \rho_0 + \frac{c}{R_1} \\ \rho(R_2) &= \rho_0 - \frac{c}{R_2}, \end{aligned} \tag{6.36}$$

with $c > 0$.

Once we know $\rho(r)$, we can determine the current in the vicinity of the inner domain edge,

$$j = -D\rho'(r)|_{R_1} \tag{6.37}$$

and finally get the evolution equation for R_1 via the relation $dV/dt = -jS$, where V and S are, respectively, the volume and the surface of the domain of radius R_1. Since

$$\frac{dV}{dt} = \frac{dV}{dR_1}\frac{dR_1}{dt} = S\dot{R}_1, \tag{6.38}$$

we find $-jS = S\dot{R}_1$, i.e.,

$$\frac{dR_1}{dt} = D\rho'(R_1). \tag{6.39}$$

Eq. (6.39), supplemented by the initial condition $R_1(0) = L$, allows us to determine the time $t(L)$ required to close the inner domain. As a matter of fact, we don't really need to determine the exact expression of $t(L)$; it is enough to obtain the functional dependence of t on L. Since we expect a power-law dependence, our task is equivalent to determining how $t(L)$ rescales when $R_1(0)$ rescales: if $t(L) \simeq L^{1/n}$, where n is the sought-after coarsening exponent, then

$$t(R_1(0) = L) = t(R_1(0) = 1)L^{1/n}. \tag{6.40}$$

The determination of the prefactor $t(R_1(0) = 1)$ requires us to solve the full problem, but the determination of $1/n$ can be done with a simple rescaling of variables. If $r = Lx$ and $R_{1,2} = Lx_{1,2}$, the boundary conditions (6.36) suggest the rescaling of the density[5]

$$\rho(r, t; L) = \rho_0 + \frac{1}{L}p(x, t). \tag{6.41}$$

Using the new variables p and x, the problem is rewritten as

$$\begin{aligned} p_{xx} + \frac{d-1}{x}p_x &= 0 \\ p(x_1) &= +\frac{c}{x_1} \\ p(x_2) &= -\frac{c}{x_2}. \end{aligned} \tag{6.42}$$

The central point is that Eqs. (6.42) do not depend on L anymore. Therefore, Eq. (6.39) can be rewritten with the new variables as

$$L\frac{dx_1}{dt} = \frac{D}{L^2}p_x(x_1). \tag{6.43}$$

The last step is to rescale time so as to get rid of L in this equation as well. It is straightforward to show that the correct rescaling is $t = L^3\tau$, from which we get

$$\frac{dx_1}{d\tau} = Dp_x(x_1). \tag{6.44}$$

Eqs. (6.42) and (6.44) allow us to determine $t(R_1(0) = 1)$, while the above scaling tell us that when $R_1(0) \to LR_1(0)$, $t(R_1(0)) \to L^3 t(R_1(0))$. In conclusion,

$$t(L) = t(1)L^3, \tag{6.45}$$

which means that the closure time of a domain of initial radius L is L^3 times larger than the closure time of a domain of initial radius equal to one, which is just a pure number. Therefore, inverting above relation, the coarsening exponent is found to be $n = \frac{1}{3}$, regardless of the dimension d.

Since a time of order L^3 is required to close an inner domain, whose size is of order L, on a time scale of order L^2 the radius of the inner domain varies by a quantity of order one. This is the result we have used at the beginning of this section to justify the approximation to replace the diffusion equation with the Laplace equation.

[5] Here (and only here) we make explicit that the density profile of the gas phase also depends on the initial condition, $R_1(t = 0) = L$.

In one dimension, the curvature effect on the particle density in proximity to a cluster is not effective; the above treatment would give a constant density profile in the region $R_1 < r < R_2$ and no net current between the cluster and the reservoir. As a matter of fact, the treatment in Section 6.2 revealed that coarsening in the one-dimensional Ising model is the result of the random exchange of particles between otherwise equivalent clusters. In order to take into account fluctuations, we would need to write a Langevin equation, much in the same way we did for the conserved case. We will do that in the next section, but without solving the equation, because the conservation constraint makes calculations more difficult.

The specific problem we have studied in this section, the exchange of particles between a central cluster and the reservoir surrounding it, was justified by its spherical symmetry and by the scaling hypothesis. In general, we should have a distribution of clusters that exchange particles, with a net current directed from clusters smaller than average toward cluster larger than average. This process, called Ostwald ripening, leads to the disappearance of the smallest clusters and to the increase in time of the average size $\overline{R}(t)$ of clusters, according to the same law we have just found, $\overline{R}(t) \simeq t^{1/3}$. It is also possible to find the limit of the self-similar distribution of cluster sizes, but we do not discuss it here.

6.4 Beyond the Coarsening Laws

6.4.1 Quenching and Phase-Ordering

We started this chapter with Fig. 6.1, plotting the equilibrium magnetization curves $M(T, H)$ for a real two-dimensional magnetic system and arguing how the region inside the curve $M(T, 0)$ was composed of nonequilibrium states, which are accessible only with a sudden quench of some control parameter and for a finite time, because such states are either unstable or metastable.

If the state is unstable, the destabilization is a global process, triggered by delocalized fluctuations of small amplitude: they originate many inhomogeneities, which are the precursors of domains. This is the case we have considered until now. Instead, if the initial, disordered state is metastable, it must be destabilized by localized fluctuations of finite amplitude, a process that is called nucleation and that is thermally activated. In this case there is the formation of a few big domains throughout the system.

To be more precise, we consider a system with order parameter[6] $\phi(\mathbf{x})$ and described by the free energy density,

$$\overline{f} = \frac{1}{V}\int_V d\mathbf{x}\left[\frac{1}{2}(\nabla\phi)^2 + V(\phi)\right]. \tag{6.46}$$

[6] In order to emphasize the generality of the approach, we use the symbol ϕ rather than a symbol related to a specific model or system.

Immediately before a quench the system is homogeneous, i.e., $\phi(\mathbf{x}) \equiv \phi_0$. We now perturb this state, $\phi(\mathbf{x}) = \phi_0 + \delta\phi(\mathbf{x})$, and evaluate how the post-quench free energy is affected by this perturbation,

$$\bar{f} = \frac{1}{V}\int d\mathbf{x}\left[\frac{1}{2}(\nabla\delta\phi)^2 + \mathcal{V}(\phi_0) + \mathcal{V}'(\phi_0)\delta\phi(\mathbf{x}) + \frac{1}{2}\mathcal{V}''(\phi_0)(\delta\phi(\mathbf{x}))^2\right] \tag{6.47}$$

$$= \mathcal{V}(\phi_0) + \frac{1}{2}\overline{(\nabla\delta\phi)^2} + \mathcal{V}'(\phi_0)\overline{\delta\phi} + \frac{1}{2}\mathcal{V}''(\phi_0)\overline{(\delta\phi)^2}, \tag{6.48}$$

where $\overline{\cdots}$ is a short-hand notation for the spatial average.

The term coming from the energy cost of spatial inhomogeneities, $\overline{(\nabla\delta\phi)^2}$, can be neglected for delocalized perturbations of large wavelength λ, because it is inversely proportional to λ^2. In conclusion, the variation of free energy, because of the perturbation, is

$$\Delta\bar{f} = \mathcal{V}'(\phi_0)\overline{\delta\phi} + \frac{1}{2}\mathcal{V}''(\phi_0)\overline{(\delta\phi)^2} + \mathcal{O}\left(\frac{1}{\lambda^2}\right). \tag{6.49}$$

For conserved dynamics, any perturbation must satisfy the condition $\overline{\delta\phi} \equiv 0$, which implies that the dominant term is the quadratic one, so a perturbation decreases the energy if $\mathcal{V}''(\phi_0) < 0$ and it increases the energy if $\mathcal{V}''(\phi_0) > 0$. Therefore, the system is linearly unstable in proximity to a maximum of $\mathcal{V}(\phi)$ (dashed line in Fig. 6.15) and it is metastable

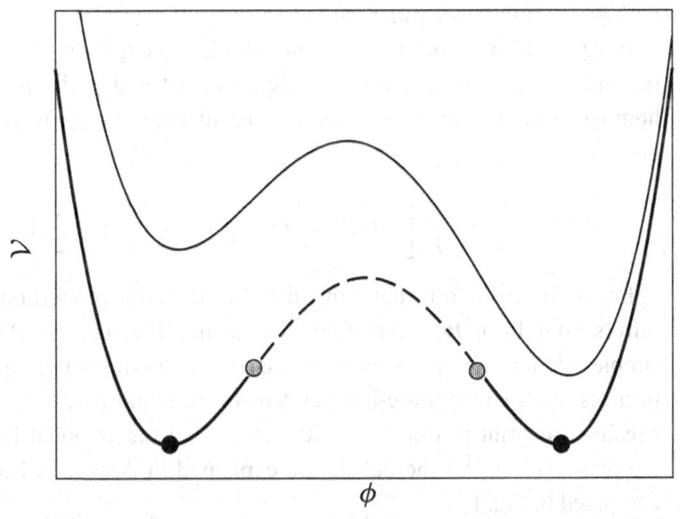

Fig. 6.15 A symmetric double-well potential (below) and an asymmetric potential (above). For the symmetric one we have indicated with a thick line the stable regions, $\mathcal{V}''(\phi) > 0$, and with the dashed line the unstable region, $\mathcal{V}''(\phi) < 0$. Stable and unstable regions are separated by gray circles, locating the points of change of curvature. For the conserved case, metastability arises after a quench in the stable regions (except for the black circles, indicating the ground states). The physically relevant values of ϕ for the phase-ordering process are in the range between the black circles, corresponding to the equilibrium, coexisting phases (see Figs. 6.16 and 6.18 for more details). For the nonconserved case, metastability may arise in the case of an asymmetric potential; see the upper curve. In practice, we might perform a quench in the presence of a magnetic field, then reverse the field.

in proximity to a minimum of $\mathcal{V}(\phi)$ (thick line in Fig. 6.15). The condition $\mathcal{V}''(\phi_0) = 0$ (gray circles), when plotted in the plane T, ϕ, defines the separatrix between the two regimes and is called *spinodal line*. The symmetric quench, corresponding to the maximum $\phi_0 = 0$, clearly leads to an unstable state.

For nonconserved dynamics, there is no constraint on $\overline{\delta\phi}$, and the linear term vanishes only at the extrema of the potential (black circles in the figure), $\mathcal{V}'(\phi_0) = 0$. If $\mathcal{V}''(\phi_0) > 0$, the system is in a local minimum; if it is an absolute minimum the system stays there forever, but if it is not an absolute minimum we have metastability, which may arise by making the potential asymmetric (see the thin line in the figure). For example, we might first apply a magnetic field (bringing the system into the absolute minimum), then reverse the field. What happens next is discussed in Section 6.6. The case $\mathcal{V}'(\phi_0) \neq 0$ for nonconserved dynamics will be discussed later, in the context of Langevin formalism.

Here we want to derive the mean-field coexistence curve and the so-called spinodal line. The coexistence curve (see Section 6.1), locates the equivalent minima of $\mathcal{V}(\phi)$, therefore it is defined by the condition $\mathcal{V}'(\phi) = 0$. The spinodal line, instead, locates the changes of curvature of the potential, therefore it is defined by the condition $\mathcal{V}''(\phi) = 0$. We will derive both curves for two different systems: a system described by the Ginzburg–Landau free energy with Ising symmetry (see Appendix I) and a fluid described by the van der Waals equation (see Appendix G). It is worth remembering that the Ising model can describe both a nonconserved and a conserved order parameter: it is the dynamics that discriminate between the two. However, in order to be specific, we will assume a magnetic language, with an order parameter $m(\mathbf{x}, t)$.

In Appendix I we derive the Helmholtz free energy density for a magnet, starting from the microscopic Hamiltonian of a ferromagnet and performing a coarse graining. The homogeneous part of the free energy, up to prefactor f_0, which can be taken as energy unit, is

$$f(m) = -\frac{m^2}{2} + \frac{T}{T_c}\left[-\ln 2 + \frac{1}{2}(1+m)\ln(1+m) + \frac{1}{2}(1-m)\ln(1-m)\right]. \quad (6.50)$$

Eq. (6.5), apart from the nonhomogeneous term, proportional to the square gradient, differs from this in two respects: (i) it is a simplified version of (6.50), because it uses the simplest double-well potential retaining the same symmetry properties, and (ii) it is written in a rescaled, temperature-independent form, while here we maintain the T-depedence, because we want to draw the coexistence and the spinodal lines, which are defined in the plane (T, m). All the details are explained in Appendix I and the two potentials are compared in Fig. I.1.

From Eq. (6.50) we can derive the equation of state by the thermodynamic relation $H = \partial f/\partial m$, where H is the magnetic field,

$$H = -m + \frac{T}{T_c}\left[\frac{1}{2}\ln(1+m) - \frac{1}{2}\ln(1-m)\right], \quad (6.51)$$

which can be rewritten in the more familiar form,

$$m = \tanh\left(\beta T_c (H+m)\right). \quad (6.52)$$

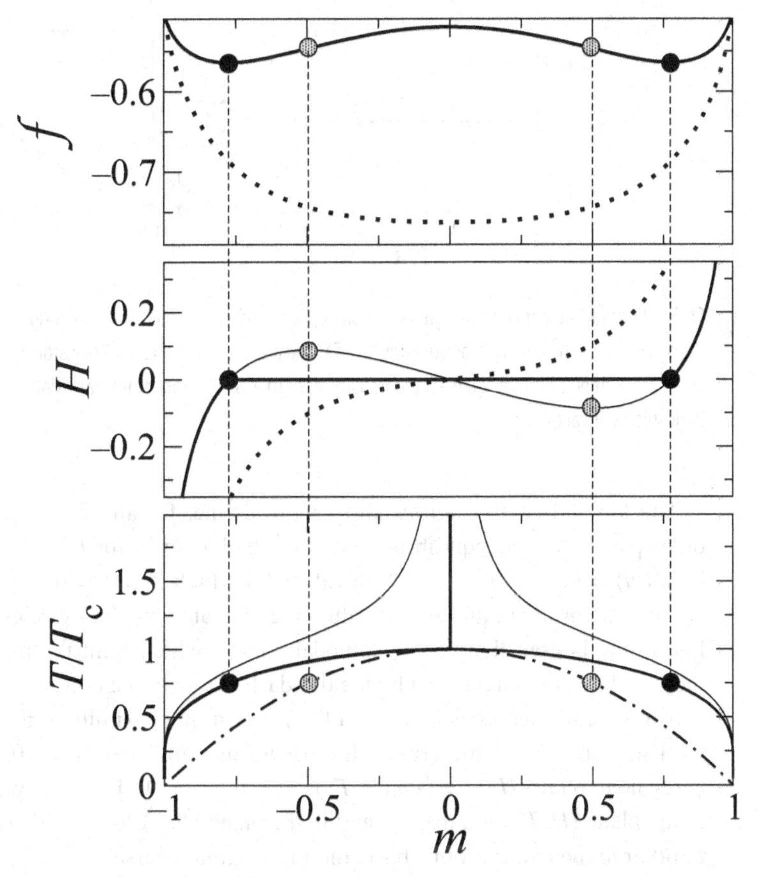

Fig. 6.16 Top: Free energy density of a magnet, Eq. (6.50), for $T < T_c$ (solid line) and for $T > T_c$ (dotted line). Center: The equation of state, Eq. (6.52), for $T < T_c$ (solid line) and for $T > T_c$ (dotted line). The thick, horizontal line corresponds to the Maxwell construction, which is fairly trivial in the case of a magnet, because of the symmetry $m(T, -H) = -m(T, H)$. The thin line corresponds to metastable states (segments between black and gray circles) or to unstable states (segment between the two gray circles). In both panels, black circles indicate the free energy minima, which coexist for $H = 0$ and $T < T_c$; gray circles indicate the change of curvature of f and the stability limit of metastable states when dynamics is conserved. Bottom: For each temperature T, we plot the values of the magnetization corresponding to coexistence (black circles) and metastability limit (gray circles), obtaining the coexistence curve (thick, solid line) and the spinodal line (dot-dashed line). The thin solid lines give the equilibrium curves $m(T, H)$ for a positive and a negative field H.

In Fig. 6.16 we plot the free energy (top) and the equation of state (center) for $T > T_c$ (dotted lines) and for $T < T_c$ (solid lines). For $H = 0$ and $T < T_c$, the two minima of the free energy are equivalent and we have phase coexistence. Passing from $H = 0^-$ to $H = 0^+$, the order parameter m displays a discontinuity (first-order transition) unless $T = T_c$ (second-order transition). The central panel shows the typical Maxwell construction, which is fairly trivial in this case, because of the symmetry of the problem, according to which $m(-H, T) = -m(H, T)$. Therefore, we have phase coexistence for $H = 0$, regardless of $T < T_c$.

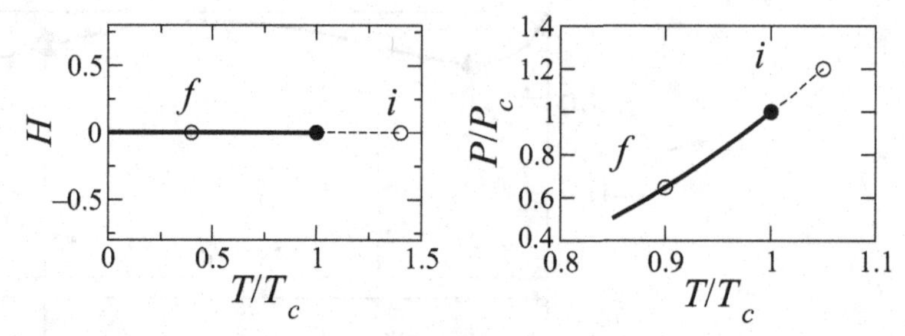

Fig. 6.17 Critical lines for the para-ferromagnetic transition (left) and for the gas–fluid transition (right). In the latter case, the thick line defines the critical isochore, $P_{iso}(T)$. The dashed line indicates a transformation between state i and state f. At equilibrium (quasi-static transformation) it is a second-order transition; at nonequilibrium it indicates a critical quench.

The top and central figures allow us to define, for any $T < T_c$, special values of the order parameter: the equilibrium values, which coexist for $H = 0$ and which are defined by $f'(m) = 0, f''(m) > 0$, and the values for which $f''(m) = 0$, which define the spinodal curve. The former (equilibrium) values are indicated by black circles and by the thick solid line in the bottom figure. The spinodal values, which limit the metastability region, are indicated by gray circles and by the dot-dashed line in the bottom panel.

Similar considerations apply to a fluid system in proximity to the gas–liquid transition, with the caveat that the critical line is not as simple as $H = 0$, because a symmetry equivalent to $m(-H, T) = -m(H, T)$ is now missing. In Fig. 6.17 we plot the critical lines in the plane (H, T) for a magnet and in the plane (P, T) for a fluid. These curves should be familiar to the reader from a basic thermodynamic course.

In order to draw the panels corresponding to Fig. 6.16 for a gas–fluid transition, it is simpler to start from the equation of state of a nonideal gas (see Section 3.2.1 and Appendix G), rather than from its free energy, which is poorly known. So, the starting point is the van der Waals equation,

$$P = \frac{T}{v - b} - \frac{a}{v^2}. \tag{6.53}$$

It is worth recalling that in the above equation, b represents the excluded volume effects (which reduce the volume per particle v, accessible to other particles) and a represents the attractive force between molecules, which reduce the pressure.

The function $P(v)$, much in the same way as the function $H(m)$, is a monotonic function for high T and displays the typical van der Waals loop for low T; see the central panel in Fig. 6.18. As shown in Appendix G, the free parameter form of the equation of state and the Gibbs free energy are

$$P^* = \frac{8}{3}\frac{T^*}{v^* - \frac{1}{3}} - \frac{3}{(v^*)^2} \tag{6.54}$$

$$G = P^* v^* - \frac{8}{3}T^* \ln\left(v^* - \frac{1}{3}\right) - \frac{3}{v^*}. \tag{6.55}$$

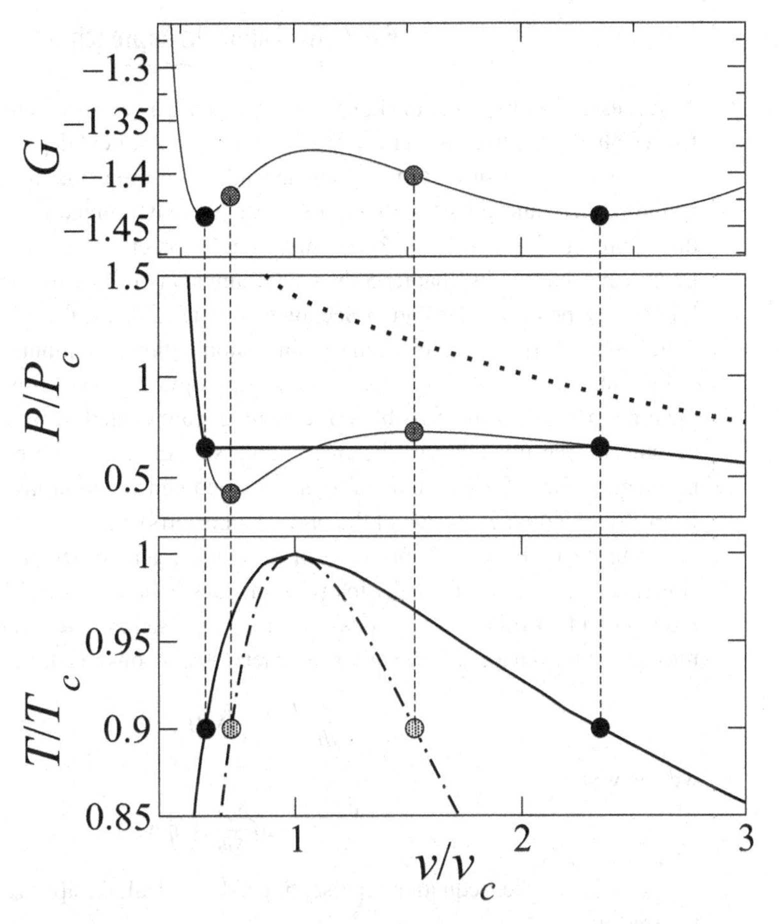

Fig. 6.18 This set of three panels is the equivalent of Fig. 6.16 for a van der Waals fluid. Free energy (top), equation of state (center), and coexistence/spinodal curves (bottom) for a fluid described by the van der Waals equation (6.54). Black circles indicate the free energy minima and define the coexistence curves (solid line, bottom panel). Gray circles indicate the change of curvature of G and the stability limit of metastable states; they define the spinodal line (dot-dashed line, bottom panel). In the central panel, the thick, solid line refers to $T < T_c$, the dotted line to $T > T_c$.

In the top panel of Fig. 6.18 we plot G for the point $(T^*, P^*) = (0.9, 0.647)$ of the critical line. In the middle panel we plot the equation of state (6.54) for $T^* = 0.9$ (solid line) and for $T^* > 1$ (dotted line). In the bottom panel we plot the coexistence and metastability curves.

The thick, solid lines in the bottom panels of Figs. 6.16 and 6.18 are the equilibrium coexistence lines. The region inside such curves corresponds to nonequilibrium states, which can be attained by a quench and which decay toward equilibrium via a phase-ordering process including spinodal decomposition or nucleation. The region outside such curves corresponds to equilibrium states.

6.4.2 The Langevin Approach

As discussed in Chapter 2, the Langevin approach is a generalization of the description of a Brownian particle to a problem where the unknown is a field depending on space and time, rather than a set of discrete variables depending on time (the three-dimensional position of the particle and possibly its momentum). We have already seen several examples of this approach: in Section 3.6.3 we studied the contact process with a phenomenological Langevin equation; in Chapter 5 the continuum study of kinetic roughening was based on Langevin-type equations; and in Section 6.3 we used this approach to face the problem of nonconserved coarsening in arbitrary dimension within a continuum picture.

In many cases it is possible to write an appropriate free energy $\mathcal{F}$, which depends on the order parameter and possibly on remaining slow variables, which cannot be taken into account by the noise term. The free energy should reflect the basic symmetries of the problem, while information about dynamics and conservation laws enters when we pass from $\mathcal{F}$ to the time evolution of the order parameter(s) and of the slow variables.

Let us focus on the simple case of a single, scalar order parameter $\phi(\mathbf{x}, t)$ without slow fields and with a fully dissipative dynamics, which is equivalent to the overdamped motion of a Brownian particle: this is the case representing the physical systems discussed throughout this chapter. If the order parameter is not conserved, i.e.,

$$\frac{d}{dt}\int d\mathbf{x}\phi(\mathbf{x}, t) \neq 0, \tag{6.56}$$

we can write

$$\frac{\partial \phi}{\partial t} = -M\frac{\delta \mathcal{F}}{\delta \phi} + \eta(\mathbf{x}, t), \tag{6.57}$$

which is the simplest equation representing a dynamical, dissipative process that decreases the free energy. Here,

$$\mathcal{F} = \int d\mathbf{x} f(\phi, \nabla\phi, \nabla^2\phi, \dots), \tag{6.58}$$

so the functional derivate means (see Appendix L.1)

$$\frac{\delta \mathcal{F}}{\delta \phi} = \frac{\partial f}{\partial \phi} - \sum_i \partial_i \frac{\partial f}{\partial(\partial_i \phi)} + \sum_{ij} \partial_{ij} \frac{\partial^2 f}{\partial(\partial_{ij}\phi)} + \cdots, \tag{6.59}$$

where M is the mobility and η is the noise, taking into account the fast degrees of freedom. As usual, the noise has zero average and its correlation function has the form

$$\langle \eta(\mathbf{x}, t)\eta(\mathbf{x}', t)\rangle = \Gamma\delta(\mathbf{x} - \mathbf{x}')\delta(t - t'). \tag{6.60}$$

Since the physical system is evolving toward equilibrium, the mobility M and the strength of the noise Γ must be related by the fluctuation–dissipation condition, similar to what we have seen in Chapter 1 for the Brownian particle,[7]

[7] In Section 1.4.1 (see Eqs. (1.42) and (1.48)), we have $\dot{v} = -\gamma v + \eta$ and $\Gamma = 2\gamma T/m$. However, if we want to write the Langevin equation in the form (6.57) we must introduce the energy $E = mv^2/2$, so $\dot{v} = -(\gamma/m)E'(v) + \eta$. Introducing the quantity $M = \gamma/m$, we have the fluctuation–dissipation relation, $\Gamma = 2MT$. It is also instructive to check that Eq. (6.61) is dimensionally correct.

$$\Gamma = 2MT. \tag{6.61}$$

Dynamics decreases the free energy, because we have

$$\frac{d\mathcal{F}}{dt} = \int d\mathbf{x} \frac{\delta \mathcal{F}}{\delta \phi} \frac{\partial \phi}{\partial t} = -\int d\mathbf{x} \left(\frac{\delta \mathcal{F}}{\delta \phi} \right)^2 + \int d\mathbf{x} \frac{\delta \mathcal{F}}{\delta \phi} \eta(\mathbf{x}, t). \tag{6.62}$$

The meaning of $d\mathcal{F}/dt$ is that we evaluate the free energy on the trajectory of the system, i.e., on a specific function $\phi(\mathbf{x}, t)$, the solution of Eq. (6.57). The deterministic part on the right-hand side of Eq. (6.62) leads to a decreasing $\mathcal{F}$. The noise term may occasionally lead to its momentary increase, but on average the noise term vanishes and

$$\left\langle \frac{d\mathcal{F}}{dt} \right\rangle = -\int d\mathbf{x} \left(\frac{\delta \mathcal{F}}{\delta \phi} \right)^2 \le 0. \tag{6.63}$$

If the order parameter is conserved, we must start by enforcing its conservation via a continuity equation,

$$\frac{\partial \phi(\mathbf{x}, t)}{\partial t} = -\nabla \cdot \mathbf{j}, \tag{6.64}$$

where the current $\mathbf{j}$ transports the order parameter ϕ. The current, in turn, is proportional to the gradient of the chemical potential, which is the thermodynamic force,

$$\mathbf{j} = -M\nabla\mu + \boldsymbol{\eta}(\mathbf{x}, t), \tag{6.65}$$

with the noise satisfying the usual relations,

$$\langle \eta_i(\mathbf{x}, t)\eta_j(\mathbf{x}', t')\rangle = \delta_{ij}\delta(\mathbf{x} - \mathbf{x}')\delta(t - t'). \tag{6.66}$$

A couple of remarks are in order here. First, the mobility M may depend on ϕ, but we neglect this possibility. Second, inserting the noise term in Eq. (6.65) rather than in Eq. (6.64) means that also noise is conserved, which is certainly correct if the only source of noise is thermal fluctuations in the transport process of the order parameter. Finally, the chemical potential μ can be expressed as the functional derivative of the free energy,

$$\mu = \frac{\delta \mathcal{F}}{\delta \phi}. \tag{6.67}$$

Summing up, we find

$$\frac{\partial \phi(\mathbf{x}, t)}{\partial t} = M\nabla^2 \frac{\delta \mathcal{F}}{\delta \phi} + \zeta(\mathbf{x}, t), \tag{6.68}$$

where $\zeta(\mathbf{x}, t) = -\nabla \cdot \boldsymbol{\eta}(\mathbf{x}, t)$ and its correlation function is

$$\begin{aligned}
\langle \zeta(\mathbf{x}, t)\zeta(\mathbf{x}', t)\rangle &= \langle \nabla \cdot \boldsymbol{\eta}(\mathbf{x}, t)\nabla \cdot \boldsymbol{\eta}(\mathbf{x}', t')\rangle && (6.69)\\
&= -\partial_i\partial_j\langle \eta_i(\mathbf{x}, t)\eta_j(\mathbf{x}', t')\rangle && (6.70)\\
&= -\Gamma\delta_{ij}\partial_i\partial_j\delta(\mathbf{x} - \mathbf{x}')\delta(t - t') && (6.71)\\
&= -\Gamma\nabla^2\delta(\mathbf{x} - \mathbf{x}')\delta(t - t'). && (6.72)
\end{aligned}$$

Again, the fluctuation–dissipation relation requires that $\Gamma = 2MT$.

The order parameter is conserved, because of the continuity equation, and dynamics decreases the free energy:

$$\left\langle \frac{d\mathcal{F}}{dt} \right\rangle = \left\langle \int d\mathbf{x} \left(\frac{\delta \mathcal{F}}{\delta \phi} \right) \frac{\partial \phi}{\partial t} \right\rangle \tag{6.73}$$

$$= M \int d\mathbf{x} \left(\frac{\delta \mathcal{F}}{\delta \phi} \right) \nabla^2 \left(\frac{\delta \mathcal{F}}{\delta \phi} \right) \tag{6.74}$$

$$= -M \int d\mathbf{x} \left(\nabla \frac{\delta \mathcal{F}}{\delta \phi} \right)^2 \leq 0. \tag{6.75}$$

In the above considerations we have not made explicit the free energy in order to be as general as possible in the derivation of Eqs. (6.57) and (6.68). If we focus on the Ginzburg–Landau free energy, which can be written, for $T < T_c$,

$$\mathcal{F} = \mathcal{F}_{\rm GL} \equiv \int d\mathbf{x} \left[\frac{1}{2} (\nabla \phi)^2 - \frac{\phi^2}{2} + \frac{\phi^4}{4} \right], \tag{6.76}$$

Eqs. (6.57) and (6.68) read, respectively,

$$\frac{\partial \phi}{\partial t} = M(\nabla^2 \phi + \phi - \phi^3) + \eta \qquad \text{model A (TDGL)} \tag{6.77}$$

$$\frac{\partial \phi}{\partial t} = -M\nabla^2(\nabla^2 \phi + \phi - \phi^3) + \zeta \qquad \text{model B (CH)} \tag{6.78}$$

The nonconserved Eq. (6.77) is called[8] model A of the dynamics and its deterministic version ($\eta \equiv 0$) is called time-dependent Ginzburg–Landau (TDGL) equation. These equations have already been encountered in Section 6.3.1. The conserved Eq. (6.78) is called model B of the dynamics and its deterministic version ($\zeta \equiv 0$) is called Cahn–Hilliard (CH) equation.

6.4.3 Correlation Function and Structure Factor

A relevant function that allows a more quantitative analysis of phase-ordering is the one time correlation function,

$$C(r,t) = \langle \phi(\mathbf{x},t)\phi(\mathbf{x}+\mathbf{r},t) \rangle - \langle \phi(\mathbf{x},t) \rangle^2, \tag{6.79}$$

which depends on $r = |\mathbf{r}|$ only, because of translational and rotational invariances in space. This function, which varies from $C(0,t) = \langle \phi^2(\mathbf{x},t) \rangle - \langle \phi(\mathbf{x},t) \rangle^2$ to $C(\infty,t) = 0$, may provide different information on different scales. For small distances, $r \ll L(t)$, the presence of domains has practically no effect and $C(r,t)$ probes equilibrium, thermal fluctuations, which are stationary and are known to decay over a distance ξ, the equilibrium correlation length. In the opposite limit, $r \gg L(t)$, $C(r,t)$ probes the domain structure, i.e., the nonequilibrium nature of the physical system. It may therefore be useful to think of $C(r,t)$ as the sum of an equilibrium contribution and a nonequilibrium one. However, if we

[8] According to the classification of Pierre Hohenberg and Bertrand Halperin.

are not close to T_c, ξ is of order of the lattice constant and the equilibrium contribution to the correlation function can be neglected. This is what we are assuming here below, where scaling considerations will be applied to $C(r,t)$. If the equilibrium part was not negligible, it should be subtracted from $C(r,t)$ before applying scaling.

The above remarks point to the need to be careful with the value T_f of the temperature after a quenching from the disordered phase. Nonetheless, for large time and large distances the equilibrium part is surely irrelevant and this suggests that the exact values of T_f should be irrelevant for the coarsening law and the asymptotic form of the correlation function. This statement, as well as a similar statement about the irrelevance of the initial temperature T_i, is corroborated by studies of the dynamical renormalization group applied to continuum models. The case $d = 1$, where $T_c = 0$, deserves special attention, of course, as discussed in Section 6.3.

We can now use scaling to further simplify the expression of $C(r,t)$, thanks to self-similarity: two snapshots of the system at different times $t, t' = bt$ are statistically equivalent if we rescale space appropriately, $\mathbf{x}' = b^n\mathbf{x}$. Therefore,[9]

$$C(r', t') = C(b^n r, bt) = C(r,t). \tag{6.80}$$

This relation must be valid for any b, in particular for $b = 1/t$, which implies

$$C\left(\frac{r}{t^n}, 1\right) = C(r,t). \tag{6.81}$$

In conclusion, if we define the length $L(t) = t^n$, we have

$$C(r,t) = f\left(\frac{r}{L(t)}\right), \tag{6.82}$$

a result that is also valid for logarithmic coarsening.

In Figs. 6.19 and 6.20 we plot the correlation function at different times for the simulations of the two-dimensional Ising model with Glauber and Kawasaki dynamics, respectively. The insets show the data collapse when $C(r,t)$ is plotted as $C(r/L(t))$, where the length $L(t)$ has been evaluated through the condition[10]

$$C(L,t) = \frac{C(0,t)}{2}. \tag{6.83}$$

The correlation functions of the nonconserved and conserved models have a distinct difference: the former, Fig. 6.19, is monotonously decreasing, while the latter, Fig. 6.20, is oscillating. These different behaviors are related to the different distributions of domain size, $n(A)$, where A is the area of the domain. This connection is discussed in the next section.

[9] In the most general case, scaling would imply a rescaling of the function itself, $C(r', t') = b^z C(b^n r, bt)$. This is the case, e.g., for the roughness function studied in the previous chapter (see Eq. (5.14)), or for the scaling form of the Gibbs free energy in equilibrium critical phenomena. In general terms, such rescaling is related to the fractal morphology of the structure under study, but in the present context domains have a fractal structure only if the temperature after quenching is the critical temperature, $T_f = T_c$. We are not considering this case.

[10] If scaling were perfect, the choice of the factor 2 in the denominator would be as good as any other. In practice, finite size and finite statistics effects spoil the simulation data for $C(r,t)$ at large r. It is common practice to determine $L(t)$ through Eq. (6.83) or, if $C(r,t)$ oscillates, through its first zero, $C(L,t) = 0$.

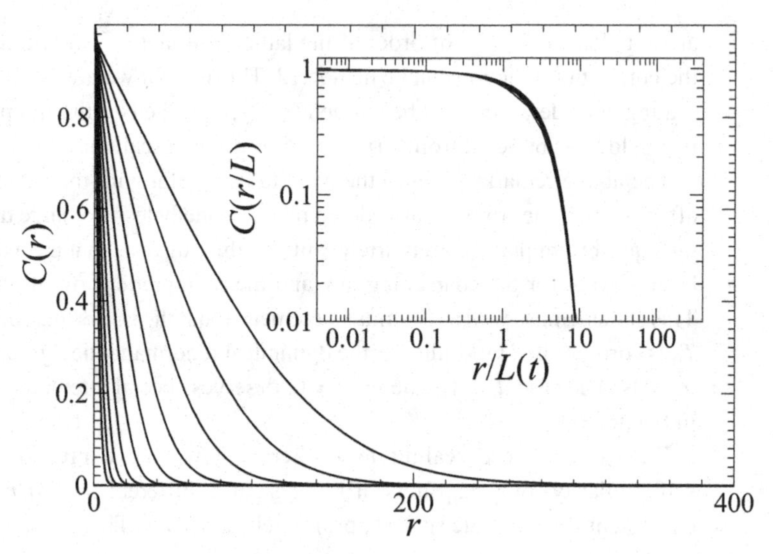

Fig. 6.19 The correlation function $C(r,t)$ for the two-dimensional Ising model with nonconserved (Glauber) dynamics, after quenching from $T_i = \infty$ to $T_f = 0.661T_c$, for different times $10 \leq t \leq 14,000$ (with increasing time the curve moves to the right). In the inset, C is plotted as a function of the reduced variable $r/L(t)$ to illustrate scaling. We use a log–log plot to make the graph clearer. Simulation data courtesy of Federico Corberi.

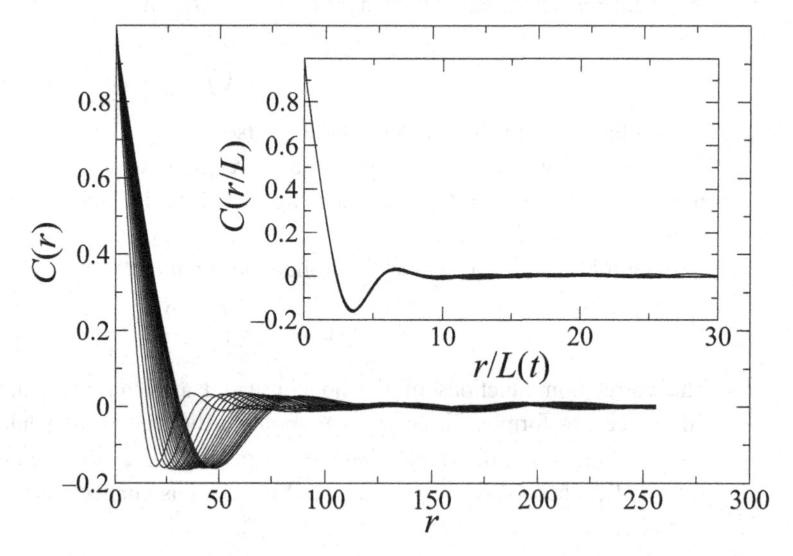

Fig. 6.20 The correlation function $C(r,t)$ for the two-dimensional Ising model with conserved (Kawasaki) dynamics, after quenching from $T_i = \infty$ to $T_f = 0.661T_c$, for different times $5 \times 10^4 \leq t \leq 10^6$ (with increasing time the curve moves to the right). In the inset, C is plotted as a function of the reduced variable $r/L(t)$ to illustrate scaling. Simulation data courtesy of Federico Corberi.

It is now useful to rewrite Eq. (6.82) in the **q**-space, because the structure factor $S(\mathbf{q}, t)$, defined as the Fourier transformation of the correlation function, is of interest for scattering experiments, being proportional to the scattering intensity. In **q**-space the dynamic scaling has the form

$$\begin{aligned} S(\mathbf{q}, t) &= \int d\mathbf{r} e^{-i\mathbf{q}\cdot\mathbf{r}} C(\mathbf{r}, t) = \int d\mathbf{r} e^{-i\mathbf{q}\cdot\mathbf{r}} f\left(\frac{r}{L(t)}\right) \\ &= L^d \int d\mathbf{s} e^{-iL\mathbf{q}\cdot\mathbf{s}} f(s) \\ &= L^d \tilde{f}(qL), \end{aligned} \tag{6.84}$$

where $\tilde{f}(q)$ is the d-dimensional Fourier transform of $f(s)$. We can also relate the structure factor to the correlation function of the order parameter in the **q**-space. If

$$\phi(\mathbf{q}, t) = \int d\mathbf{x} e^{-i\mathbf{q}\cdot\mathbf{x}} \phi(\mathbf{x}, t), \tag{6.85}$$

because of translational invariance we get

$$\begin{aligned} \langle \phi(\mathbf{q}, t)\phi(\mathbf{q}', t)\rangle &= \int d\mathbf{x} e^{-i\mathbf{q}\cdot\mathbf{x}} \int d\mathbf{r} e^{-i\mathbf{q}'\cdot(\mathbf{x}+\mathbf{r})} \langle \phi(\mathbf{x}, t)\phi(\mathbf{x}+\mathbf{r}, t)\rangle \\ &= \int d\mathbf{x} e^{-i(\mathbf{q}+\mathbf{q}')\cdot\mathbf{x}} \int d\mathbf{r} e^{-i\mathbf{q}'\cdot\mathbf{r}} (C(r, t) + \langle \phi(\mathbf{x}, t)\rangle^2) \\ &= (2\pi)^d \delta(\mathbf{q}+\mathbf{q}') \left[S(\mathbf{q}, t) + (2\pi)^d \delta(\mathbf{q}) \langle \phi(\mathbf{x}, t)\rangle^2 \right]. \end{aligned} \tag{6.86}$$

In Fig. 6.21 we offer experimental evidence of the dynamic scaling of the structure factor for the phase separation process of a Mn–Cu binary alloy. The quantity Q_{max} is proportional to $1/L(t)$, so the data collapse shown is equivalent to Eq. (6.84). We stress that dynamic scaling is a long-time, asymptotic property, as attested by the bottom panel and discussed in the figure legend.

6.4.4 Domain Size Distribution

The average size of domains, $L(t)$, is a relevant physical quantity, but the full distribution of domain sizes, $n_d(A, t)$, certainly contains much more information. In Fig. 6.22 we reproduce the size distribution of domains in the two-dimensional Ising model, both in the nonconserved (left) and in the conserved (right) case. The main difference between the two distributions is the presence of a maximum at finite size in the conserved case, while in the nonconserved case the function $n_d(A, t)$ decreases monotonically. In other words, the domain pattern is more regular for conserved dynamics. Why is it so?

The explanation must lie in the conservation law, which imposes a constraint on the size of neighboring domains of the same type: since they exchange particles, the shortening of one domain implies the lengthening of the other. In addition, the conservation law surely has an effect on the size distribution at short times, as shown by the linear stability analysis reported in Section 6.4.5 and discussed in detail in the next chapter: the comparison of Eqs. (6.92) and (6.98) indicates that the latter (conserved case) has a peak for finite q, i.e., for finite size.

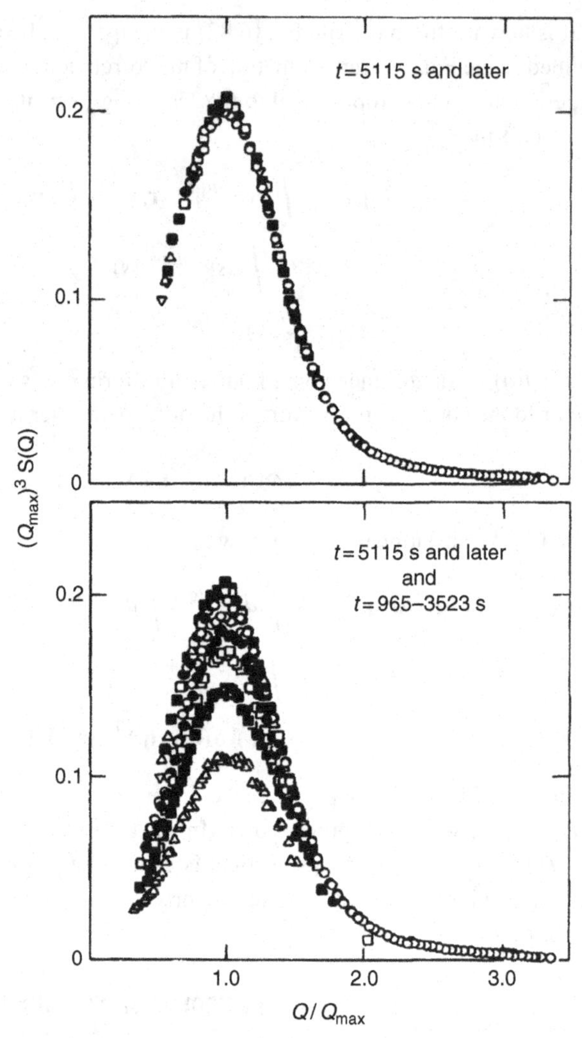

Fig. 6.21 Dynamic scaling of the coarsening process in a binary alloy ($Mn_{0.67}Cu_{0.33}$). The structure factor is proportional to the intensity of the neutron scattering experiment. The top panel shows clear dynamic scaling at late times (5115 s$< t <$ 8429 s). The bottom panel shows results at five different earlier times. There is no collapse and the departure from the scaling expression grows with decreasing time. Reprinted from B. D. Gaulin, S. Spooner, and Y. Morii, Kinetics of phase separation in $Mn_{0.67}Cu_{0.33}$, *Physical Review Letters*, **59** (1987) 668–671.

It is also interesting to note a possible connection between $n_d(A, t)$ and $C(r, t)$, exemplified in $d = 1$ by two limiting cases. In the first case we consider a perfectly periodic configuration: positive domains of length L alternate with negative domains of the same length. The distribution $n_d(\ell)$ is a simple Dirac delta, and the correlation function is an oscillating triangle wave. We can expect that destroying a bit of the periodic domain configuration will preserve the oscillating character of the correlation function, but it will make $C(r, t)$ vanish at large distance.

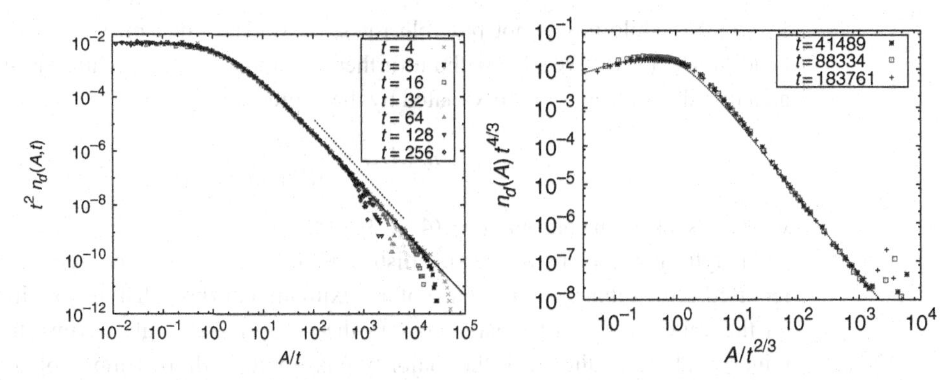

Fig. 6.22 The size distribution of domains in the two-dimensional Ising model, in the nonconserved (left) and in the conserved (right) case. Panel on left reprinted from J. J. Arenzon et al., Exact results for curvature-driven coarsening in two dimensions, *Physical Review Letters*, **98** (2007) 145701. Panel on right reprinted from A. Sicilia et al., Geometry of phase separation, *Physical Review E*, **80** (2009) 031121.

The second, limiting case is opposed to the first one, because it corresponds to a fully disordered state: each spin is equally likely to be positive or negative. Now a domain has the probability $n_d(\ell) = 1/2^\ell$ (a decreasing function) to have length ℓ, and the correlation function is trivially equal to $C(r) = 1$ if $r = 0$ and $C(r) = 0$ if $r \neq 0$; i.e., it does not oscillate. We guess that the different shapes of the correlation function for the nonconserved (Fig. 6.19) and for the conserved (Fig. 6.20) models are related to the different behaviors of $n_d(A, t)$ (see Fig. 6.22): a peaked distribution $n_d(A, t)$ in the conserved case produces an oscillating $C(r, t)$.

More rigorous considerations on domain size can be made for the nonconserved model, using the Allen–Cahn equation (see Appendix O). According to Eq. (O.7), the local velocity of a domain wall is proportional to its curvature, $v = -\frac{\lambda}{2\pi}K$. We have introduced a constant, which is not present in Eq. (O.7), because the latter has been derived from the parameter-free TDGL equation, while the constant is necessary if we want to compare with numerics or experiments. It is now simple to derive the time variation of the area A surrounded by a domain wall, because

$$\frac{dA}{dt} = \oint v dl = -\frac{\lambda}{2\pi}\oint K dl = -\lambda, \tag{6.87}$$

where we have used the Gauss–Bonnet theorem to evaluate the line integral of the curvature. Therefore, the area A decreases linearly in time and the distribution of areas is a function of $(A + \lambda t)$, because

$$n_a(A, t) = n_a(A + \lambda t, 0). \tag{6.88}$$

It is important to stress that the area A is defined as the region surrounded by any closed domain wall, which in turn may contain other domain walls. Therefore, the distribution n_a is not equal to the distribution of domains, $n_d(A, t)$, which only counts areas of equal order parameter. However, n_d and n_a are clearly related and there is numerical evidence they are almost identical. The reason to focus on n_a is that we can derive analytically

Eq. (6.88), while this is not possible for $n_d(A,t)$. The final, missing piece is the initial condition, $n_a(A,0)$. We state without further detail that for a quenching from $T_i \to \infty$, the area distribution can be approximated by the expression $n_a(A,0) = c/A^2$, so that

$$n_a(A,t) = \frac{c}{(A+\lambda t)^2}, \qquad (6.89)$$

where c is a known constant, $c = (4\pi\sqrt{3})^{-1} \simeq 0.046$.

Analytical treatments of the size distribution of domains for conserved dynamics (not reported here) support the presence of a maximum and show that the distribution depends on the value of the order parameter: in the case of a strongly asymmetric quench the minority phase is diluted in the majority phase with a droplet morphology and the size distribution has a cut-off. Asymmetric quenches are considered in the next section.

6.4.5 Off-Critical Quenching

Until now we have considered critical quenching, i.e., phase-ordering processes where $\langle\phi(\mathbf{x},t=0)\rangle = 0$. In fact, this corresponds to the simplest case, the symmetric one. However, we can think of cases where $\langle\phi(\mathbf{x},t=0)\rangle = \phi_0$. The first reaction might be to conclude that the phase-ordering process changes completely, because an initial value $\phi_0 \neq 0$ breaks the symmetry between the two equilibrium states and the system might evolve toward one of the two, according to the sign of ϕ_0, and coarsening disappears. However, we know that in the conserved case this process cannot occur, because the spatial average of $\langle\phi(\mathbf{x},t)\rangle$ does not depend on time.

A simple insight into the problem of off-critical quenching can be gained using a continuum approach, neglecting noise. For the nonconserved case we already know that the relevant equation is the TDGL:

$$\partial_t\phi(\mathbf{x},t) = \nabla^2\phi + \phi - \phi^3. \qquad (6.90)$$

When we perform the critical quenching, dynamics starts from a state where the average value of the order parameter is zero apart from small fluctuations. Therefore, the cubic term in the TDGL equation can be neglected, yielding the linear approximation

$$\partial_t\phi = \nabla^2\phi + \phi, \qquad (6.91)$$

which can be solved in Fourier space. We can limit ourselves here to a single Fourier component, assuming that $\phi(\mathbf{x},t) = \epsilon\exp(\sigma t + i\mathbf{q}\cdot\mathbf{x})$, where the small parameter ϵ justifies the irrelevance of the cubic term. Replacing it in Eq. (6.91), we find the dispersion curve

$$\sigma = 1 - \mathbf{q}^2. \qquad (6.92)$$

The positive character of σ for small $\mathbf{q}$ means that fluctuations of large wavelength are linearly unstable and their amplitude increases exponentially in time. This linear regime is the early time stage of the dynamics. Coarsening sets in when the amplitude is of order one and the nonlinear cubic term comes into play.

Let's now see what happens if $\langle\phi(\mathbf{x},t)\rangle = \phi_0 \neq 0$. While $\phi_0 = 0$ is a stationary solution of Eq. (6.90), a state with $\phi_0 \neq 0$ is not, except if it corresponds to an equilibrium state.

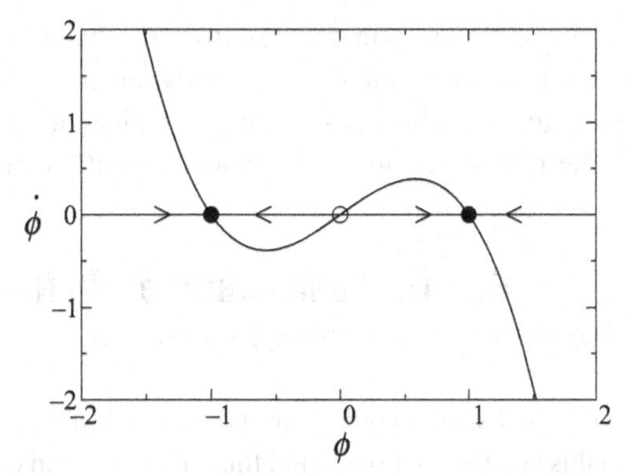

Fig. 6.23 Plot of $\dot{\phi} = \phi - \phi^3$. Solid and open circles are stable and unstable fixed points, respectively.

In fact, the TDGL equation has the uniform, spatially independent solution $\phi(t)$, which satisfies

$$\dot{\phi}(t) = \phi - \phi^3, \tag{6.93}$$

whose resulting dynamics is sketched in Fig. 6.23. The equilibrium states $m = \pm 1$ are fixed points and attract any uniform state ϕ_0 with positive and negative magnetization, respectively. So, the dynamics of the TDGL equation drives a constant profile toward the uniform equilibrium states $\phi = \pm 1$, according to the sign of ϕ_0.

If the order parameter is conserved, this scenario is forbidden. In this case we should rather refer to the CH equation,

$$\partial_t \phi(\mathbf{x}, t) = -\nabla^2(\nabla^2 \phi + \phi - \phi^3). \tag{6.94}$$

The first remark is that any state with constant magnetization, $\phi(\mathbf{x}) \equiv \phi_0$, is a stationary solution of the CH equation, because of the conservation of the order parameter. This is in striking contrast to the nonconserved case, where only the uniform states corresponding to extrema of the free energy $\mathcal{F}$ are steady states. If we study the stability of the state ϕ_0 at short times, we can write

$$\phi(\mathbf{x}, t) = \phi_0 + \epsilon(\mathbf{x}, t) \tag{6.95}$$

and linearize Eq. (6.94) with respect to ϵ, getting

$$\frac{\partial \epsilon}{\partial t} = \nabla^2(\nabla^2 \epsilon + \epsilon - 3\phi_0^2 \epsilon + \phi_0 - \phi_0^3) \tag{6.96}$$

$$= (1 - 3\phi_0^2)\nabla^2 \epsilon + (\nabla^2)^2 \epsilon. \tag{6.97}$$

Again, the stability spectrum is determined considering the evolution of a Fourier mode, $\epsilon(\mathbf{x}, t) = \epsilon_0 \exp(\sigma t + i\mathbf{q} \cdot \mathbf{x})$. Replacing such a functional form in the above equation, we find

$$\sigma(\mathbf{q}) = (1 - 3\phi_0^2)q^2 - q^4. \tag{6.98}$$

Therefore, the uniform solution with average order parameter equal to ϕ_0 is unstable (stable) if $(1 - 3\phi_0^2)$ is positive (negative). The condition $1 - 3\phi_0^2 = 0$, which separates the linearly stable uniform profiles from the linearly unstable ones, corresponds to the definition of the spinodal curve, already given in Section 6.4.1. We remark that the interpenetrating morphology of domains, typical of a symmetric quench, changes in favor of a droplet-like morphology if the quench is sufficiently asymmetric.

6.5 The Coarsening Law in Nonscalar Systems

So far we have discussed only Ising-type models, i.e., models with a scalar order parameter, while in this section we go beyond this limitation and consider $O(n)$ models. These models are characterized by an n-component, vector order parameter $\boldsymbol{\phi}$, which can freely rotate in an n-dimensional space and which lives in a d-dimensional space. We adopt a continuum description, making use of the Ginzburg–Landau free energy, suitably generalized to be applied to nonscalar models. In practice, since dynamics is purely dissipative, the continuum description of the ordering process for $O(n)$ models is a generalization of the TDGL (6.9) and CH (6.94) equations.

The starting point is the Ginzburg–Landau free energy for the vectorial order parameter, $\boldsymbol{\phi}(\mathbf{x}, t)$,

$$\mathcal{F}[\boldsymbol{\phi}] = \int \left[\frac{1}{2}(\nabla\boldsymbol{\phi})^2 + V(\boldsymbol{\phi})\right] d\mathbf{x} \equiv \int f d\mathbf{x}, \tag{6.99}$$

where

$$(\nabla\boldsymbol{\phi})^2 = \sum_{i=1}^{d}\sum_{j=1}^{n}(\partial_i\phi_j)^2 \tag{6.100}$$

with $\partial_i \equiv \partial/\partial x_i$ and $V(\boldsymbol{\phi}) = V_0(\boldsymbol{\phi}^2 - 1)^2$ is the usual, rotationally invariant, "Mexican hat" potential[11]; see Fig. 6.24.

As for the scalar models, dynamics is obtained by the equation[12] $\partial_t\boldsymbol{\phi} = -\delta\mathcal{F}/\delta\boldsymbol{\phi}$ for a nonconserved order parameter and by the equation $\partial_t\boldsymbol{\phi} = \nabla^2\delta\mathcal{F}/\delta\boldsymbol{\phi}$ for a conserved one. Here we focus on the nonconserved case, whose coarsening exponent can be derived without too much difficulty. A few words on the conserved models will be given at the end. The notation of vectorial functional derivative, $\delta\mathcal{F}/\delta\boldsymbol{\phi}$, means a vector, whose components are

$$\frac{\delta\mathcal{F}}{\delta\phi_i} = \frac{\partial f}{\partial\phi_i} - \sum_j \partial_j \frac{\partial f}{\partial(\partial_j\phi_i)}. \tag{6.101}$$

With this expression in mind, we find

$$\frac{\delta\mathcal{F}}{\delta\phi_i} = -\left(\nabla^2\phi_i - \frac{\partial V}{\partial\phi_i}\right), \tag{6.102}$$

[11] We write it in this form so that the free energy of the ground state ($\boldsymbol{\phi}^2 = 1$, $\nabla\boldsymbol{\phi} = 0$) vanishes.
[12] The mobility M (see Section 6.4.2) has been absorbed in the time t, to make these equations parameter-free.

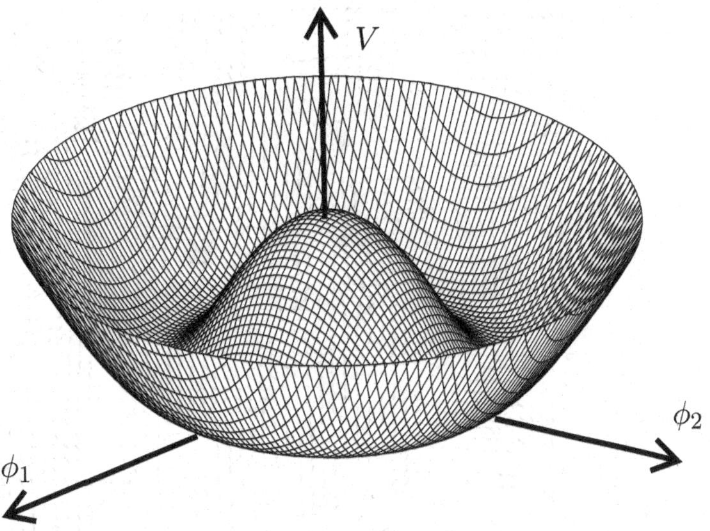

Fig. 6.24 Plot of the "Mexican hat" potential for $n = 2$, $V(\boldsymbol{\phi}) = V_0(\boldsymbol{\phi}^2 - 1)^2$.

so the equation for the $O(n)$ nonconserved model is

$$\frac{\partial \boldsymbol{\phi}}{\partial t} = \nabla^2 \boldsymbol{\phi} - \frac{\partial V}{\partial \boldsymbol{\phi}}. \tag{6.103}$$

The corresponding conserved model is obtained applying $-\nabla^2$ to the right-hand side.

Now that we have written the equations defining phase-ordering dynamics of $O(n)$ models, what are the relevant features making it different from that of the scalar model? In Fig. 6.25 we show a snapshot of the two-dimensional *XY*-model (i.e., $d = n = 2$) after a quench from the disordered to the ordered phase. This figure reveals ordered regions with $(\boldsymbol{\phi})^2 \simeq 1$, but with different orientations, separated by zones where the order parameter connects different minima of $V(\boldsymbol{\phi})$. We might think that the rotational symmetry should allow passage from one minimum to the other with continuity, but this is not the case if topological defects are stable, which is the case in which we are interested.

In Fig. 6.26 we plot a sketch of topological defects for $n = 1$ (the domain wall, or kink, which should be familiar), for $n = 2$ (the vortex), and for $n = 3$ (the monopole). It is worth stressing that each defect is strictly linked to a given value of n: the vortex cannot exist for $n = 1$ (which is fairly obvious), but if spins were allowed to have a third component ($n \geq 3$), they could rotate continuously and become parallel, so as to make the defect disappear. The same would happen with a kink for $n > 1$. As for the space dimension d, we will confine ourselves to the case $d \geq n$, because we require that the order parameter vanishes at the core of the defect. If d is strictly larger than n, the defect is translationally invariant in the remaining $(d-n)$ dimensions: see the case of the vortex in three dimensions in Fig. 6.26 or think of a domain wall in two or three dimensions. The evaluation of the energy and of the dynamical properties, e.g., the friction, of a defect are the required steps to determine the coarsening laws. We are going to accomplish this task.

Fig. 6.25 Snapshot of the ordering dynamics of the *XY* model. The squares and the triangles are the core regions of vortices and anti-vortices, respectively, where the magnitude of the order parameter is close to zero. For graphical reasons, not all the lattice sites are shown. From H. Qian and G. F. Mazenko, Vortex dynamics in a coarsening two-dimensional XY model, *Physical Review E*, **68** (2003) 021109.

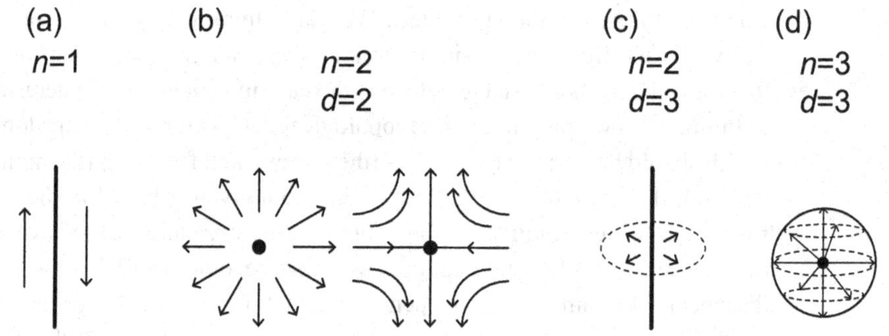

Fig. 6.26 Types of topological defects in the $O(n)$ model: (a) domain wall ($n = 1$), (b) vortex and antivortex ($n = 2$), (c) vortex ($n = 2$) in $d = 3$, (d) monopole or "hedgehog" ($n = 3$).

From a formal point of view, a defect is a time-independent, radially symmetric function,

$$\boldsymbol{\phi}_{\text{def}}(\mathbf{r}) = \hat{\mathbf{r}} f(r), \tag{6.104}$$

with $r = |\mathbf{r}|$ and $\hat{\mathbf{r}} = \mathbf{r}/r$, which is the solution of Eq. (6.103),

$$\nabla^2 \boldsymbol{\phi}_{\text{def}}(\mathbf{r}) - \frac{\partial V}{\partial \boldsymbol{\phi}_{\text{def}}(\mathbf{r})} = 0, \tag{6.105}$$

with the boundary conditions $f(0) = 0$ and $f(r = \infty) = 1$. These conditions correspond to saying that the order parameter $\boldsymbol{\phi}$ must vanish in the center of the defect and must tend to a minimum of the potential ($|\boldsymbol{\phi}| = 1$) for large r.

Using Eq. (6.104), we can write

$$\partial_i \left(\hat{r}_j f(r)\right) = \frac{\delta_{ij}}{r} f + \frac{r_i r_j}{r} \left(\frac{f}{r}\right)', \tag{6.106}$$

$$\partial_{ii} \left(\hat{r}_j f(r)\right) = \delta_{ij} \left(\frac{f}{r}\right)' \frac{r_i}{r} + \frac{r_j}{r} \left(\frac{f}{r}\right)' + \frac{r_i \delta_{ij}}{r} \left(\frac{f}{r}\right)' + \frac{r_i^2 r_j}{r} \left(\frac{1}{r} \left(\frac{f}{r}\right)'\right)' \tag{6.107}$$

where the prime means the derivative with respect to r. From Eq. (6.107), summing over i we find the expression

$$\nabla^2 \boldsymbol{\phi}_{\text{def}}(\mathbf{r}) = \hat{\mathbf{r}} \left[f''(r) + \frac{n-1}{r} f'(r) - \frac{n-1}{r^2} f(r) \right], \tag{6.108}$$

while it is easier to write

$$\frac{\partial V}{\partial \boldsymbol{\phi}_{\text{def}}(\mathbf{r})} = \hat{\mathbf{r}} V'(f). \tag{6.109}$$

Combining the previous two equations, we conclude that the profile $f(r)$ must satisfy the ordinary differential equation,

$$f''(r) + \frac{n-1}{r} f'(r) - \frac{n-1}{r^2} f(r) - V'(f) = 0. \tag{6.110}$$

The equation determining the kink shape for $n = 1$ is $f''(r) - V'(f) = 0$, whose solution (see Eq. (6.12)) is equal to the hyperbolic tangent: it tends exponentially to the limiting value $f = 1$ for $r \to \infty$. This means that $1 - f(r)$ is exponentially small for large r and the defect is localized, while this is not true for $n > 1$, as we are going to prove. If $f(r) = 1 - \epsilon(r)$, at the lowest order in ϵ we find

$$-\epsilon''(r) - \frac{n-1}{r} \epsilon'(r) - \frac{n-1}{r^2} + V''(1)\epsilon(r) = 0, \tag{6.111}$$

where we have used that $V'(1) = 0$. At the leading order, we obtain

$$\epsilon(r) = \frac{n-1}{V''(1)} \frac{1}{r^2}, \tag{6.112}$$

proving that the profile $f(r)$ attains the asymptotic value with a power law of exponent two.

The next step is to determine the energy of the defect, because dynamics is driven by the minimization of the energy of the system, which is concentrated in the defects. Since an isolated, single defect is translationally invariant in the $(d - n)$ directions orthogonal to the n-dimensional plane of the defect itself, let us determine its energy per unit length in such orthogonal directions (or, equivalently, let us assume that $d = n$). First of all (see Eq. (6.99)), we should evaluate $(\nabla\boldsymbol{\phi}_{\mathrm{def}})^2$. Taking the square of Eq. (6.106), we find

$$(\nabla\boldsymbol{\phi}_{\mathrm{def}})^2 = \sum_{ij}(\partial_i\phi_j)^2 = \frac{n-1}{r^2}f^2(r) + |\nabla f|^2, \tag{6.113}$$

so that, integrating Eq. (6.99) over angular variables, we obtain

$$E^n_{\mathrm{def}} = S_n \int r^{n-1}\left[\frac{n-1}{2r^2}f^2(r) + \frac{1}{2}|\nabla f|^2 + V(f)\right] dr, \tag{6.114}$$

where S_n is the surface of an n-dimensional sphere of unit radius.

Using the asymptotic form, $f(r) = 1 - \epsilon(r)$, with $\epsilon \approx r^{-2}$, we understand that the dominant term in square brackets here above is the first one, f^2/r^2, which decays as $1/r^2$ and makes the integral divergent for $n > 1$. So, what is the correct length scale we should use to evaluate the energy of the defect? Because of the scaling hypothesis, there is just one length scale, the size $L(t)$ of ordered regions, which means

$$E^n_{\mathrm{def}} = \begin{cases} \text{const}, & n = 1 \\ \ln L, & n = 2 \\ L^{n-2}, & n > 2, \end{cases} \tag{6.115}$$

where the superscript n indicates it is the defect energy in the n-dimensional plane of the defect. If $d > n$, the defect is translationally invariant in the remaining $(d - n)$ spatial directions and the total energy of the defect is obtained multiplying E^n_{def} by its linear size in such orthogonal directions elevated to the power $(d - n)$. Invoking scaling again, such linear size must be L and we finally obtain

$$E_{\mathrm{def}} = E^n_{\mathrm{def}} \times L^{d-n} = \begin{cases} L^{d-1}, & n = 1 \\ L^{d-2}\ln L, & n = 2 \\ L^{d-2}, & n > 2, \end{cases} \tag{6.116}$$

where we always assume that $d \geq n$.

Dynamics is driven by the fact that E_{def} depends on L and the driving force is $-dE_{\mathrm{def}}/dL$. Since dynamics is dissipative, we may think that the closing speed of a domain, $v = dL/dt$, and the driving force F are related by

$$F(L) = -\eta(L)\frac{dL}{dt}, \tag{6.117}$$

where we have indicated that the friction η (yet to be determined) may itself be dependent on the size L of the domain. A warning is in order here: the force $F(L)$ appearing in Eq. (6.117) is the force per unit length/surface of the $(d - n)$ directions, i.e.,

$$F(L) = \frac{1}{L^{d-n}}\left(-\frac{dE_{\mathrm{def}}}{dL}\right) = \begin{cases} 0, & d = n = 1 \\ L^{-1}, & d > n = 1 \\ L^{-1}, & d = n = 2 \\ L^{-1}\ln L, & d > n = 2 \\ L^{n-3}, & d \geq n > 2. \end{cases} \tag{6.118}$$

Before evaluating $\eta(L)$, we may focus on the case $n = 1$. For $d = 1$, $F(L) = 0$ and we obtain $dL/dt = 0$, which implies no coarsening. In fact, we know that for $n = d = 1$ it is necessary to take into account the exact profile of the kink, which has an exponential tail, while we have approximated the kink to a region of finite size, neglecting such exponentially small corrections. If they are taken into account, as we did in Section 6.3.1, we would get a logarithmically slow coarsening, $L(t) \sim \ln t$, rather than no coarsening at all. For $d > 1$, if the friction is assumed to be constant, we find $dL/dt = -1/L$, which gives $L(t) \sim t^{1/2}$, i.e., the result we already found in Section 6.3.1. This suggests that the friction should be constant for $n = 1$. As we are now going to show, this is no longer true for $n > 1$.

The friction $\eta(L)$ is now evaluated with regard to the scheme where a defect moves with constant velocity v_0 in a given direction (e.g., x_1), and we ask how its energy varies with time. As the force $F(L)$ appearing in Eq. (6.117) is the force per unit length/surface perpendicular to the n-dimensional space, the same "geometry" should be used to evaluate the friction $\eta(L)$. For this reason, the integrals below are restricted to such a space, $\mathbf{r} = (x_1, \ldots, x_n)$. We can now evaluate the desired quantity,

$$\frac{dE_{\mathrm{def}}}{dt} = \int d\mathbf{r} \left.\frac{\delta \mathcal{F}}{\delta \boldsymbol{\phi}}\right|_{\mathrm{def}} \cdot \frac{\partial \boldsymbol{\phi}_{\mathrm{def}}}{\partial t} \tag{6.119}$$

$$= -\int d\mathbf{r} \left(\frac{\partial \boldsymbol{\phi}_{\mathrm{def}}}{\partial t}\right)^2 \tag{6.120}$$

$$= -v_0^2 \int d\mathbf{r} \left(\frac{\partial \boldsymbol{\phi}_{\mathrm{def}}}{\partial x_1}\right)^2 \tag{6.121}$$

$$\equiv -\eta(L) v_0^2, \tag{6.122}$$

where we have assumed that the defect moves rigidly, $\boldsymbol{\phi}_{\mathrm{def}}(\mathbf{r}, t) = \boldsymbol{\phi}_{\mathrm{def}}(x_1 - v_0 t, x_2, \ldots)$. The result $d\mathcal{F}_{\mathrm{def}}/dt = -\eta(L)v_0^2$, which defines the friction $\eta(L)$, should not be a surprise, because it is the standard expression for the dissipated power by a particle moving at velocity v_0 and subject to the friction force $-\eta v_0$.

The last step is therefore to evaluate $\eta(L)$ for different values of n. Since

$$\eta(L) \equiv \int d\mathbf{r} \left(\frac{\partial \boldsymbol{\phi}_{\mathrm{def}}}{\partial x_1}\right)^2 = \frac{1}{n} \int d\mathbf{r} (\nabla \boldsymbol{\phi}_{\mathrm{def}})^2, \tag{6.123}$$

using Eq. (6.114) we observe that $\eta(L)$ has the same L-dependence as E^n_{def}, given by Eq. (6.115). Finally, using Eqs. (6.117) and (6.118), we obtain

$$\frac{dL}{dt} = -\frac{1}{\eta(L)} F(L) = \begin{cases} 0, & d = n = 1 \\ -L^{-1}, & d > n = 1 \\ -(L \ln L)^{-1}, & d = n = 2 \\ -L^{-1}, & d > n = 2 \\ -L^{-1}, & d \geq n > 2. \end{cases} \tag{6.124}$$

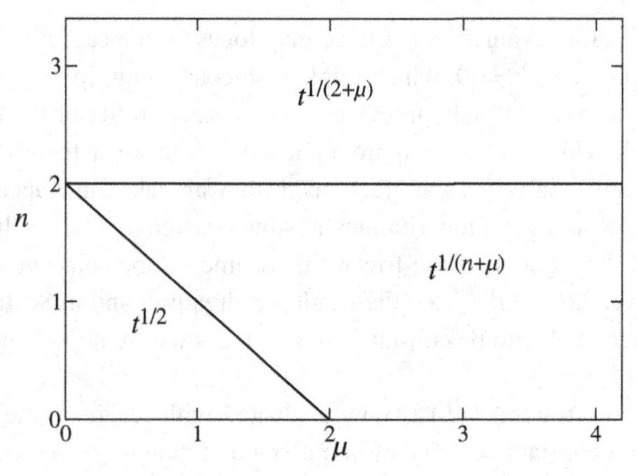

Fig. 6.27 Coarsening laws for purely dissipative dynamics, as a function of the dimension n of the order parameter and of the conservation law, represented by μ; see Eq. (6.127). Along the solid lines there are logarithmic corrections, which are related to scaling violations.

Therefore, apart from the (already commented on) special case $n = d = 1$ and a logarithmic correction for $n = d = 2$, we always have $dL/dt = -1/L$, whence $L(t) \sim t^{1/2}$. This result can be understood as follows. If we rewrite Eq. (6.124) as

$$\frac{dL}{dt} = -\frac{1}{\eta(L)}\frac{1}{L^{d-n}}\frac{d}{dL}(L^{d-n}E^n_{\text{def}}), \tag{6.125}$$

the fact that $\eta(L)$ and $E^n_{\text{def}}(L)$ scale alike implies $dL/dt \sim -1/L$, where the right-hand term $1/L$ comes from the derivative, because the involved quantities are powers of L. We can summarize the coarsening laws for the nonconserved case as

$$L(t) \sim \begin{cases} \ln t \quad \text{versus} \quad t^{1/2}, & n = d = 1 \\ \left(\dfrac{t}{\ln t}\right)^{1/2}, & n = d = 2 \\ t^{1/2}, & \text{otherwise.} \end{cases} \tag{6.126}$$

For $n = d = 1$ we have taken into account the correct results discussed in Sections 6.2.1 and 6.3.1, distinguishing between deterministic and noisy coarsening. As for the logarithmic correction for $n = d = 2$, it is due to a violation of the scaling, reminiscent of the ordered Kosterlitz–Thouless phase occurring in two dimensions in the XY- model.

For completeness, in Fig. 6.27 we graphically report the results, due to Alan Bray, for the coarsening laws for nonconserved and conserved models. The conservation law is represented by the parameter μ, which appears in the general equation[13]

$$\frac{\partial \phi}{\partial t} = (-\nabla^2)^{\mu/2}\left[\nabla^2\phi - \frac{\partial V}{\partial \phi}\right]. \tag{6.127}$$

[13] If μ is not an even integer, we have a fractional partial differential equation, which we have already encountered in Section 1.7.1.

The (nonconserved) Ginzburg–Landau equation and the (conserved) Cahn–Hilliard equation correspond to $\mu = 0$ and $\mu = 2$, respectively. Larger values of μ correspond to higher-order conservation rules, which we do not discuss here. However, there are two major remarks to be made with respect to the results shown in Fig. 6.27. The first, qualitative observation is that as soon as $\mu \neq 0$, the coarsening law depends on n. The second, more quantitative observation is that, except for the triangle covering the region of small μ and n, the coarsening law has the form $L(t) \sim t^{1/(p+\mu)}$, where p is some integer. Such μ-dependence simply corresponds to replacing the operator $(-\nabla^2)^{\mu/2}$ with the dimensional factor $1/L^{\mu}$, which would simply rescale t. Nonetheless, this simple dimensional analysis does not work for $n = 1, \mu = 0$; this is the reason why for Ising-like systems, when passing from the nonconserved ($\mu = 0$) to the conserved ($\mu = 2$) case, we pass from $t^{1/2}$ to $t^{1/3}$ and not to $t^{1/4}$.

6.6 The Classical Nucleation Theory

We have seen how a strongly asymmetric quench, beyond the spinodal decomposition line, leads to a state that is metastable, rather than linearly unstable. Therefore, the ensuing dynamics is different from what we have discussed until now, because an infinitesimal perturbation is no longer effective to destabilize the uniform state. What we need is a localized perturbation of finite amplitude and a thermally activated process, which requires an energy barrier to be overcome. This process is called *nucleation*, because it goes through the formation of nuclei of the new phase.

We can provide a qualitative picture of the nucleation process by making reference to a system described by a free energy, which has two equivalent minima if parameters are suitably tuned. For instance, in a ferromagnet such a condition corresponds to $T < T_c$ and $H = 0$, while in a fluid it corresponds to the critical isochore $P_{iso}(T)$; see Fig. 6.17. If $H \neq 0$, the two minima are not equivalent, and at equilibrium the system is magnetized in the direction of the field. Analogously, if $T < T_c$ and $P < P_{iso}(T)$ ($P > P_{iso}(T)$), the fluid is in the gas (liquid) phase.

Let us now suppose the system is in phase 1 (negative magnetization or gas phase). If we reverse the magnetic field or increase the pressure, the equilibrium state is characterized by a different phase 2 (positive magnetization or liquid phase), while phase 1 is now metastable. How does the transition between the metastable phase 1 and the stable phase 2 occur? Since phase 1 is metastable, such a transition cannot occur via a perturbation of vanishing amplitude; we need a perturbation of finite amplitude, which must necessarily be localized in space, otherwise its energetic cost would be infinite in a macroscopic sample.

The mechanism is easily exemplified by the Ising model, where, according to the above picture, we should imagine having a negatively magnetized state in the presence of a positive field. If we reverse an ensemble of spins we gain in terms of bulk free energy (reversed spins are now aligned with the field), but we lose in terms of surface free energy, because interface spins are antiparallel. Overall, we lose energy for a small cluster of reversed spins and we gain energy if the cluster is large enough. Let us now explain in more detail.

If $\delta g = g_1 - g_2 > 0$ is the difference in the Gibbs free energy density between the two phases and σ is the surface tension, the energy of a cluster of phase 2 and radius r within a sea of phase 1 is

$$\Delta G(r) = -\frac{4}{3}\pi r^3 \delta g + 4\pi r^2 \sigma . \tag{6.128}$$

This function initially grows with r, it has a maximum for $r = r^* = 2\sigma/\delta g$ and it is negative for $r > \frac{3}{2}r^*$. The size r^* defines the *critical* nucleus, because for $r > r^*$ the growth of the nucleus implies a lowering of its energy, so that it can grow spontaneously.

The quantity

$$\Delta G(r^*) = \frac{16\pi}{3} \frac{\sigma^3}{(\delta g)^2} \tag{6.129}$$

represents the energy barrier to be overcome to pass from phase 1 to phase 2, and the probability p to form a critical droplet is given by the Arrhenius formula (1.247),

$$p \sim \exp(-\Delta G(r^*)/T). \tag{6.130}$$

If we make reference to an Ising model with ferromagnetic exchange coupling J in the presence of a field H, the loss (per unit surface) in interface free energy is proportional to the former, $\sigma \sim J$, while the gain (per unit volume) in bulk free energy is proportional to the latter, $\delta g \sim H$. Therefore, the radius of critical nucleus is $r^* \sim (J/H)$, the energy barrier is $\Delta G(r^*) \sim J^3/H^2$, and $p \sim \exp(-aJ^3/H^2)$, where a is a numerical prefactor. In the next section we describe the formation of nuclei of different sizes using a kinetic description, with the goal to obtaining an approximate expression for the nucleation rate, i.e., the number of critical nuclei formed per unit time and volume in the metastable phase. This rate is found to be proportional to p.

If we return to the case of a strong asymmetric quench, between the spinodal and the coexistence curves, we realize that the nucleation process now has some peculiar features, because the dynamics is conserved. Using the binary mixture language (appropriate for the experimental results shown in Fig. 6.3), the initial, metastable state is a disordered, homogeneous state made up, e.g., of 80% of atoms A and 20% of atoms B. This ratio is conserved and dynamics occurs through diffusion: the formation of a nucleus of the minority phase is accompanied by a local increase of the majority phase in its surrounding. Regarding the difference of the free energy densities, $\delta g = g_1 - g_2$, in the present context g_1 should be understood as the free energy of the homogeneous mixture 80/20 and g_2 as the free energy of the nucleated new phase, which is 100% of atoms B. Similar remarks can be made for the interface energy σ between the two phases.

6.6.1 The Becker–Döring Theory

In this section we pass from purely thermodynamic considerations to a kinetic description, studying the formation of nuclei of the condensed, liquid phase. For simplicity and by tradition we use a language where "molecules" can attach or detach from clusters.

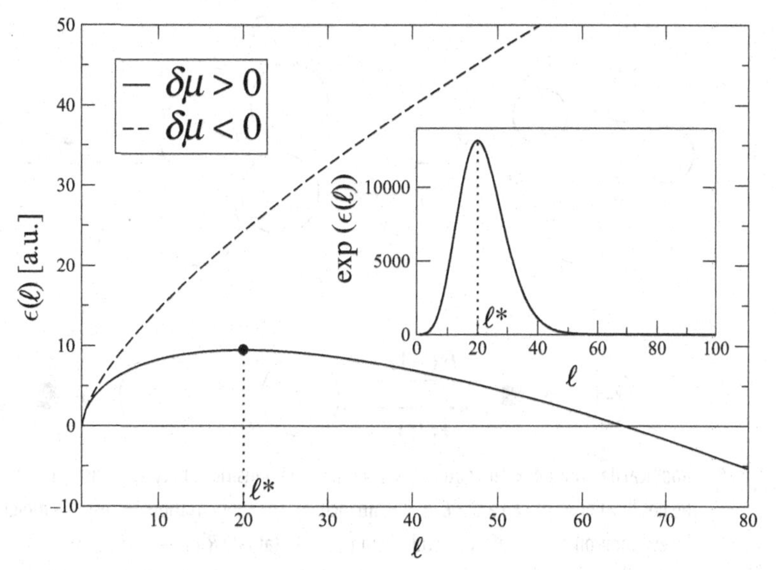

Fig. 6.28 Schematic plot of $\epsilon(\ell)$, Eq. (6.131). If $\delta\mu > 0$, it has a maximum for $\ell = \ell^*$. Inset: The function $\exp(\epsilon(\ell))$. Since $\epsilon(\ell)$ is in arbitrary units, the function is also representative of the inverse of the Boltzmann factor, $\exp(\epsilon(\ell)/T)$.

Therefore, rather than using a continuous variable for the radius of the cluster, it is more useful to consider the number ℓ of condensed molecules. Eq. (6.128) is now replaced by the energy of a cluster of size ℓ, given by[14]

$$\epsilon_\ell = -\delta\mu(\ell - 1) + \gamma(\ell - 1)^{2/3}, \tag{6.131}$$

where we have written $(\ell - 1)$ rather than ℓ in the right-hand side, because we are interested in the excess energy with respect to the gas phase, so it should be $\epsilon_1 = 0$. For $\delta\mu > 0$, ϵ_ℓ has a maximum for

$$\ell^* - 1 = \left(\frac{2\gamma}{3\,\delta\mu}\right)^3; \tag{6.132}$$

see Fig. 6.28.

The number of clusters composed of ℓ particles (from now on, ℓ-clusters) is indicated by $n_\ell(t)$; it can increase because of the condensation process of an atom on an $(\ell - 1)$-cluster and because of the evaporation of an atom from an $(\ell + 1)$-cluster, while condensation/evaporation on/from the ℓ-cluster itself leads to a decrease of n_ℓ. The processes of condensation and evaporation are described by certain kinetic coefficients, respectively, R_ℓ and R'_ℓ, which may depend on the size of the cluster, and the net balance between an ℓ- and an $(\ell + 1)$-cluster defines the current J_ℓ. All processes are graphically depicted in Fig. 6.29 and formally summarized in the following equations:

$$\frac{\partial n_\ell(t)}{\partial t} = J_{\ell-1}(t) - J_\ell(t) \tag{6.133}$$

$$J_\ell(t) = R_\ell n_\ell(t) - R'_{\ell+1} n_{\ell+1}(t). \tag{6.134}$$

[14] If v_0 is the volume per particle/spin, $\delta\mu = v_0\delta g$ and $\gamma = (36\pi)^{1/3} v_0^{2/3}\sigma$.

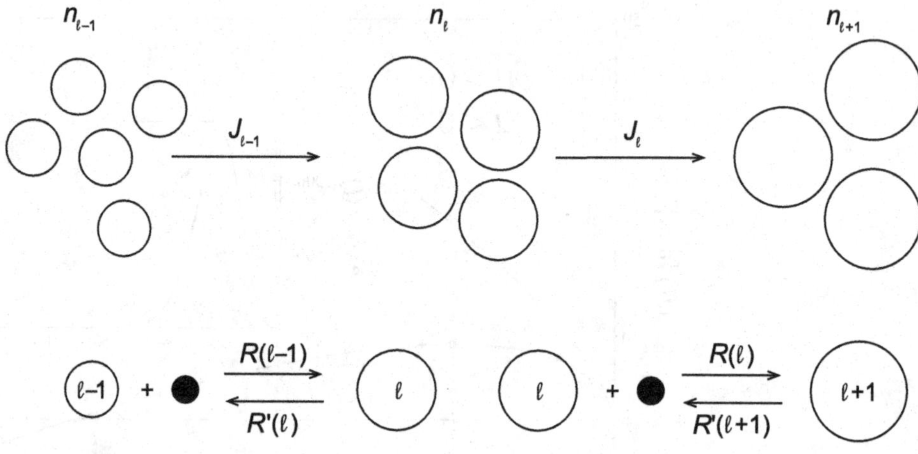

Fig. 6.29 Graphical description of rate equations governing the transitions between cluster (ℓ) and clusters $(\ell \pm 1)$: $J(\ell)$ is the net flux between (ℓ) and $(\ell+1)$ and includes the condensation process of a molecule on (ℓ) (at rate $R(\ell)$) and the evaporation process of a molecule from $(\ell+1)$ (at rate $R'(\ell+1)$).

Equations such as those above are called rate equations and are applicable to a wide variety of physical processes, whose details are contained in the explicit expressions of the kinetic coefficients R, R'. We are not going to make a specific ansatz on them; rather, we assume their values are the same as in equilibrium. This assumption allows us to find a relation between R and R', because at equilibrium a detailed balance holds, i.e., $J_\ell = 0$ for any ℓ. This condition is written as

$$R_\ell \langle n_\ell \rangle - R'_{\ell+1} \langle n_{\ell+1} \rangle = 0, \tag{6.135}$$

where $\langle n_\ell \rangle$ is the equilibrium value determined by the relation

$$\langle n_\ell \rangle = \langle n_1 \rangle e^{-\epsilon(\ell)/T}. \tag{6.136}$$

Therefore, kinetic coefficients must satisfy the relation

$$\frac{R_\ell}{R'_{\ell+1}} = \exp\left(\frac{\epsilon_{\ell+1} - \epsilon_\ell}{T}\right), \tag{6.137}$$

with ϵ_ℓ given in Eq. (6.131).

Then next step is to assume that, after a transient, the time evolution of the densities n_ℓ (see Eqs. (6.133) and (6.134)) yields a nonequilibrium steady state, therefore characterized by an ℓ-independent current, $J_{\ell-1} = J_\ell = I$, and by a steady population of clusters, $\overline{n}_\ell$. It is important to distinguish between the equilibrium steady state, defined by $J_\ell = 0$ and characterized by the population $\langle n_\ell \rangle$ given by the Boltzmann distribution (6.136), and the nonequilibrium steady state, defined by $J_\ell = I$ and characterized by the population $\overline{n}_\ell$. Finding the explicit expression of $\overline{n}_\ell$ is now our task.

Using Eqs. (6.134) and (6.137), we get a set of coupled, discrete equations

$$R_\ell \left[\overline{n}_\ell - \overline{n}_{\ell+1} \exp\left(\frac{\epsilon_{\ell+1} - \epsilon_\ell}{T}\right)\right] = I. \tag{6.138}$$

If we pass to a continuum variable ℓ, we can use standard methods for solving ordinary differential equations, so we expand at first order,

$$\overline{n}_{\ell+1} = \overline{n}_\ell + \frac{d\overline{n}}{d\ell}, \qquad \epsilon_{\ell+1} = \epsilon_\ell + \frac{d\epsilon}{d\ell}, \tag{6.139}$$

and we finally get

$$\frac{d\overline{n}(\ell)}{d\ell} = -\left(\frac{1}{T}\frac{d\epsilon}{d\ell}\right)\overline{n}(\ell) - \frac{I}{R(\ell)}, \tag{6.140}$$

where we have assumed that both $d\overline{n}/d\ell$ and $d\epsilon/d\ell$ are small quantities. Let us note that, in the above equation, $\epsilon(\ell)$ is the free energy of a cluster composed of ℓ atoms and its explicit expression is given in Eq. (6.131). The quantity $R(\ell)$ is the rate of the transition $(\ell) \longrightarrow (\ell+1)$ through adsorption of a molecule; see Fig. 6.29.

Eq. (6.140) has the form

$$\frac{d\overline{n}}{d\ell} = A'(\ell)\overline{n}(\ell) + B(\ell), \tag{6.141}$$

with $A(\ell) = -\epsilon(\ell)/T$ and $B(\ell) = -I/R(\ell)$. The general solution of the homogeneous equation is $\overline{n}(\ell) = ce^{A(\ell)}$ and a particular solution of the nonhomogeneous equation is found via variation of constants. Combining the two, we have

$$\overline{n}(\ell) = ce^{A(\ell)} - \int_\ell^\infty dx B(x) e^{(A(\ell)-A(x))}. \tag{6.142}$$

Since $A(\ell) \to +\infty$ for diverging ℓ, the condition $\overline{n}(\infty) = 0$ requires us to impose $c = 0$, while the finiteness of the integral requires that $B(x)$ vanishes for large x, i.e., that $R(x)$ diverges in the same limit. We will come back to this condition later. Finally, we find

$$\overline{n}(\ell) = I \int_\ell^\infty \frac{dx}{R(x)} \exp\left(\frac{\epsilon(x) - \epsilon(\ell)}{T}\right). \tag{6.143}$$

The above expression for the steady state distribution is exact, but the integral cannot be calculated exactly. It will be evaluated in the limits $\ell \ll \ell^*$ and $\ell \gg \ell^*$ assuming that $R(\ell)$ has a smooth behavior, while the function $\exp(\epsilon(x)/T)$ is strongly peaked in $x = \ell^*$ and rapidly vanishes for $x > \ell^*$; see Fig. 6.28. If $\ell < \ell^*$, the point $x = \ell^*$ is in the integration domain and we can evaluate the integral with the saddle point method. Therefore, we expand $\epsilon(x)$ around $x = \ell^*$,

$$\epsilon(x) = \epsilon(\ell^*) + \frac{1}{2}\epsilon''(\ell^*)(x - \ell^*)^2, \tag{6.144}$$

we evaluate $R(x)$ in $x = \ell^*$, and we change the integration limits to $\pm\infty$, getting a Gaussian integral,

$$\overline{n}(\ell) \simeq \frac{I}{R(\ell^*)} e^{(\epsilon(\ell^*)-\epsilon(\ell))/T} \int_{-\infty}^{\infty} dx \exp\left(\frac{\epsilon''(\ell^*)}{2T}(x-\ell^*)^2\right)$$

$$\simeq \boxed{\frac{I}{R(\ell^*)}\sqrt{\frac{2\pi T}{-\epsilon''(\ell^*)}} \exp\left[\frac{1}{T}\left(\epsilon(\ell^*) - \epsilon(\ell)\right)\right], \quad \ell < \ell^*} \tag{6.145}$$

If $\ell \gg \ell^*$, the dominant contribution to the integral (6.143) comes from the region x close to the lower limit, so we can expand $\epsilon(x)$ to the linear term,

$$\epsilon(x) = \epsilon(\ell) + \epsilon'(\ell)(x - \ell), \tag{6.146}$$

and obtain

$$\begin{aligned} \bar{n}(\ell) &\simeq \frac{I}{R(\ell)} \int_\ell^\infty dx \exp\left(-\frac{1}{T}|\epsilon'(\ell)|(x-\ell)\right) \\ &\simeq \boxed{\frac{I}{R(\ell)} \frac{T}{|\epsilon'(\ell)|}, \quad \ell \gg \ell^*} \end{aligned} \tag{6.147}$$

It is now interesting to compare the equilibrium distribution $\langle n_\ell \rangle$ (see Eq. (6.136)) with the newfound nonequilibrium stationary distribution. For large ℓ, $\langle n_\ell \rangle$ has a divergence (see Eq. (6.131)), which has been healed by the nonequilibrium treatment (see Eq. (6.147)). We also observe that since $|\epsilon'(\ell)|$ is a constant for large ℓ, $\bar{n}(\ell)$ vanishes in the same limit only if $R(\ell)$ diverges. In Eq. (6.151) we provide the explicit expression of $R(\ell)$, valid for gas–liquid transition, which clearly shows such divergence. As for the opposite limit of small ℓ, Eq. (6.145) shows that $\bar{n}(\ell)$ and $\langle n_\ell \rangle$ have the same ℓ-dependence, both being proportional to the Boltzmann factor,

$$\bar{n}(\ell) \approx \langle n_\ell \rangle \approx \exp\left[-\frac{\epsilon(\ell)}{T}\right], \quad \ell < \ell^*. \tag{6.148}$$

This result is certainly not accidental, because in the limit $\ell^* \gg 1$ (a limit that is tacitly assumed when we pass from discrete to continuum) the nucleation barrier $\epsilon(\ell^*)$ is high and the nucleation rate I is low. If we recall that for $I = 0$ the equilibrium steady state distribution must be recovered, for $\ell \ll \ell^*$ the two quantities $\bar{n}(\ell)$ and $\langle n_\ell \rangle$ must coincide. Therefore, in this limit $\bar{n}(\ell)$ and $\langle n_\ell \rangle$ must be equal,

$$\frac{I}{R(\ell^*)}\sqrt{\frac{2\pi T}{-\epsilon''(\ell^*)}} \exp\left[\frac{1}{T}(\epsilon(\ell^*) - \epsilon(\ell))\right] = \langle n_1 \rangle \exp\left[-\frac{\epsilon(\ell)}{T}\right]. \tag{6.149}$$

From this equation we obtain the nucleation rate,

$$I = \langle n_1 \rangle \sqrt{\frac{-\epsilon''(\ell^*)}{2\pi T}} R(\ell^*) \exp\left[-\frac{\epsilon(\ell^*)}{T}\right], \tag{6.150}$$

with $\epsilon(\ell^*) = \frac{4}{27}\frac{\gamma^3}{(\delta\mu)^2}$ and $\epsilon''(\ell^*) = -\frac{9}{8}\frac{(\delta\mu)^4}{\gamma^3}$. The quantity $\langle n_1 \rangle \exp\left[-\frac{\epsilon(\ell^*)}{T}\right]$ is the average number of nuclei attaining the top of the energy barrier by thermal fluctuations; not all of them continue to grow, but only the fraction given by the square root, $\sqrt{-\epsilon''(\ell^*)/(2\pi T)}$, called Zel'dovich factor.

In the case of the gas–liquid transition, the expression for $R(\ell)$ can be found within the kinetic theory, assuming that it is given by the number of gas atoms colliding with the ℓ-cluster per unit time, i.e.,

$$R(\ell) \simeq \frac{1}{6}\langle n_1 \rangle v (4\pi r^2), \tag{6.151}$$

where $v \simeq \sqrt{3T/m}$ is the thermal velocity[15] and r is the radius of the ℓ-cluster.

The most notable feature of the nucleation rate, Eq. (6.150), is its extreme sensitivity to the difference of chemical potential $\delta\mu$, which is in first approximation linear with $T_c - T$, the temperature difference between the liquid–vapor transition temperature T_c and the actual temperature T. The strong dependence on $\delta\mu$ is due to the exponential[16] $\exp(-\epsilon(\ell^*)/T)$, so it is useful to write

$$I = I_0 e^{-\frac{\epsilon(\ell^*)}{T}} = I_0 e^{-\frac{4\gamma^3}{27(\delta\mu)^2 T}}, \tag{6.152}$$

with $I_0 \approx 10^{33}$. We can use the values $\gamma \simeq 0.17$ eV and $\delta\mu \simeq 10^{-3}(T_c - T)_K$ eV/K, where we have made explicit the linear dependence of $\delta\mu$ on $T_c - T$, and $(T_c - T)_K$ means the value in Kelvin of such a temperature difference. In conclusion, we obtain $I \approx 10^{\alpha_{BD}}$, where

$$\alpha_{BD} \simeq 33 - \frac{10^4}{(T_c - T)_K^2}. \tag{6.153}$$

Therefore, I is extremely sensitive to $\delta\mu$ through $(T_c - T)$. This fact also points out a weak point of the above treatment, where the quantity $\delta\mu$ is assumed to be constant during the nucleation process. In reality it depends on cluster density, due to the depletion effect.

Finally, we should consider that our treatment considers homogeneous nucleation, i.e., the formation of a nucleus of the new phase in a ideally perfect system. In a real system the escape from the metastable phase occurs via heterogeneous nucleation, induced by the walls of the container (which may reduce the surface tension) or by impurities (which may expedite the formation of critical nuclei).

6.7 Bibliographic Notes

The most classical reference on continuum theories of phase-ordering is the review paper by A. J. Bray, Theory of phase-ordering kinetics, *Advances in Physics*, **51** (2002) 481–587. Here one can find a lot on continuum models and different theoretical approaches to treat them, including the dynamical renormalization group. The main limitation is that nucleation and off-critical quenches are not considered.

Another classical reference is the review paper by P. C. Hohenberg and B. I. Halperin, Theory of dynamical critical phenomena, *Reviews of Modern Physics*, **49** (1977) 435–479.

The argument for finding the coarsening exponent $n = 1/3$ for the conserved kinetic Ising model has been derived from S. J. Cornell, K. Kaski, and R. B. Stinchcombe,

[15] With the notation used in Chapter 1, the thermal velocity is the square root of the average square velocity, $\langle v^2 \rangle^{1/2}$; see Eq. (1.6).

[16] For the gas–liquid transition, if we use Eq. (6.151) this is the only dependence on $\delta\mu$. In fact, we obtain $R(\ell^*) \sim (\delta\mu)^{-2}$, which cancels the $\delta\mu$-dependence of the Zel'dovich factor.

Domain scaling and glassy dynamics in a one-dimensional Kawasaki Ising model, *Physical Review B*, **44** (1991) 12263–12274.

A more recent collection of different contributions covering the topics of the present chapter is the book by S. Puri and V. Wadhawan, eds., *Kinetics of Phase Transitions* (Taylor & Francis, 2009).

An article with extensive simulations on the nonconserved Ising model and an analysis of the role of initial state and final quench temperature is F. Corberi and R. Villavicencio-Sanchez, Role of initial state and final quench temperature on aging properties in phase-ordering kinetics, *Physical Review E*, **93** (2016) 052105.

A book that is a useful bridge between this chapter and the following one is R. C. Desai and R. Kapral, *Dynamics of Self-Organized and Self-Assembled Structures* (Cambridge University Press, 2009). In particular, the authors discuss two topics that have been hardly touched on here: the droplet-like morphology appearing in a slightly off-critical quenching of a conserved model (see around Eq. (6.98)) and the self-similar distribution of clusters that results from Ostwald ripening.

For the nucleation process the reader is referred to J. S. Langer, An introduction to the kinetics of first-order phase transitions, in Claude Godrèche, ed., *Solids Far from Equilibrium* (Cambridge University Press, 1991), pp. 297–363.

A rapid overview of coarsening phenomena is given in L. F. Cugliandolo, Coarsening phenomena, *Comptes Rendus Physique*, **16** (2015) 257–266, where the author also discusses the relation between the morphology resulting from a quenching of a $d = 2$ Ising model and the percolation morphology.

More generally, the reader can browse all contributions to the special issue on coarsening dynamics, by Federico Corberi and Paolo Politi, eds., *Comptes Rendus Physique*, **16** (2015) 255–342.

7 Highlights on Pattern Formation

This last chapter deals with a topic, pattern formation, which, by itself, is covered by entire books. In the present and more general context of nonequilibrium phenomena, rather than providing an overview of the whole subject we have preferred to focus on a few relevant concepts. The first ones are the control parameter and the order parameter, which should be familiar to the reader after Chapters 3 and 4, dealing with equilibrium and nonequilibrium phase transitions. However, these concepts are introduced here again in reference to specific pattern-forming systems.

In brief, pattern formation appears when an out-of-equilibrium system passes from a homogeneous to a nonhomogeneous state while tuning a suitable external parameter. This transition (or, more precisely, this bifurcation) typically occurs because the homogeneous solution loses its stability, so that an infinitesimal perturbation is enough to drive the system far from there. We therefore need to perform a linear stability analysis of the homogeneous state. Usually, this is not a hard task, but in some cases, depending on the physics and the geometry of the problem, it certainly is. Two nontrivial examples are provided: the Turing patterns, Section 7.3, where difficulties arise from the vectorial character of the order parameter, and the Rayleigh–Bénard instability, Appendix P, whose description requires starting from the Navier–Stokes equations. The linear stability analysis not only determines the critical value of the control parameter beyond which a given solution loses stability; it also allows us to gain information on the new state and, from a general point of view, allows classification of the bifurcations.

But what is the fate of the instability? The answer concerns the nonlinear dynamics of the model and depends on the type of bifurcation, in particular, if the instability is stationary or oscillatory, if the unstable range of wave vectors includes $\boldsymbol{q} = 0$ or not, and if there are conservation laws. For the sake of simplicity and space, here we limit ourselves to stationary instabilities, and a special role is played by periodic steady states, which arise when the homogeneous state is unstable. Therefore, the first nonlinear step is to determine such periodic patterns and, afterward, to study their stability using suitable perturbative techniques. If periodic patterns are unstable, a secondary instability appears, the primary instability being the loss of stability of the homogeneous solution.

7.1 Pattern Formation in the Laboratory and the Real World

Previous chapters show that driven, nonequilibrium systems may display a rich and varied phenomenology, ranging from phase transitions with symmetry breaking to kinetic roughening. In the present, final chapter we consider pattern formation, i.e., the appearance of a nontrivial spatial pattern when tuning a suitable control parameter, related to the driving force.

Spatial patterns are widespread in nature. A well-known example in an extended system are sand dunes (including patterns underwater), whose appearance can be easily connected to wind or currents, which are the driving forces destabilizing the otherwise flat surface of the sand. Another example is the formation of snowflakes, with the growing ice nucleus that does not remain spherical as a water droplet but forms arms with the six-fold symmetry typical of the most common ice structure. In this case the driving force, supersaturation, is less evident to the unskilled observer. Patterns are also visible in living organisms; think of the formation of regular spots and stripes on animal skin or fur or of the regular structure of some leaves. Even the kitchen can be a place to observe patterns: if we cook a thin layer of oil in a pan over low heat we can notice the formation of hexagonal convection cells, which can be made more visible adding, e.g., some cinnamon to oil.

From now on, in order to be more specific and to introduce the most relevant concepts we make reference to a few experimental examples, shown in Fig. 7.1. The first example, Fig. 7.1(a), is present in any discussion of pattern formation: a liquid is confined in a closed container, whose height is much smaller than its horizontal sizes, and it is brought out of equilibrium by heating the bottom. The main difference with a pan on the fire is that the upper surface of the liquid is not free but confined. If $\Delta T = T_{\text{inf}} - T_{\text{sup}}$ is the temperature difference between the lower and upper surfaces, the system is out of equilibrium as soon as $\Delta T > 0$, but for small ΔT the fluid remains at rest and heat is transported by conduction. The experiment shows the existence of a critical value $(\Delta T)_c$ above which the fluid is set in motion and convection cells form.

Convection cells are known to exist in the atmosphere and in the Earth's mantle, but their size is orders of magnitude bigger than cells described here. In all cases a convection cell is characterized by a hot current of fluid going from a warm to a cold region and by a balancing cold current in the opposite direction. In the experiment discussed here, the process of cell formation is the result of an instability, called the Rayleigh–Bénard instability.[1] If we tune ΔT from $\Delta T < (\Delta T)_c$ to $\Delta T > (\Delta T)_c$, the resulting instability has two main features: at short times, the initially homogeneous (conductive) state is made unstable and convection cells appear; at longer times there is a readjustment of the size of convection cells through coalescence or splitting. Finally, the system settles in a nonequilibrium stationary state.

More formally, some key points can be stressed:

- We can define a control parameter, ΔT, and an order parameter, the field velocity $\mathbf{u}$ of the fluid.
- There exists a critical value of the control parameter, $(\Delta T)_c$, at which the order parameter passes from a homogeneous state (in this case, $\mathbf{u}(\mathbf{r}, t) \equiv 0$) to a nonhomogeneous state showing a spatial pattern (convection cells).
- The ensuing dynamics, if ΔT is not much higher than $(\Delta T)_c$, terminates at a spatially periodic, stationary state.

[1] After Lord Rayleigh and Henri Bénard.

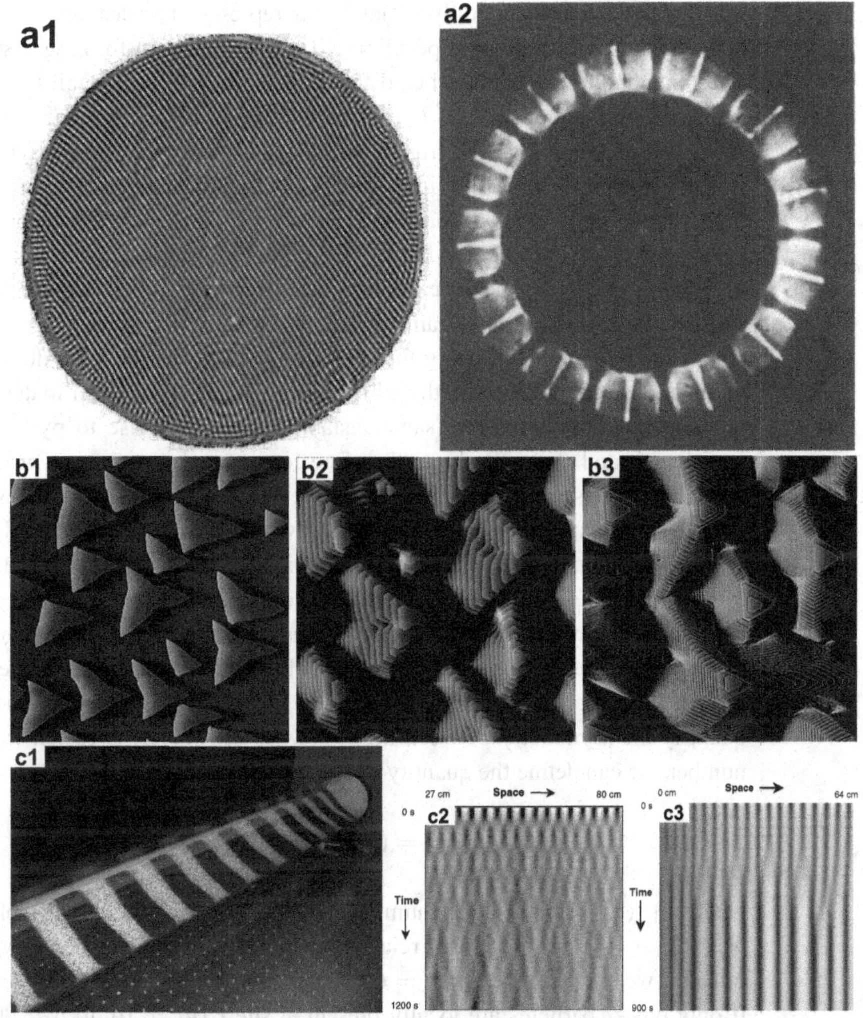

Fig. 7.1 (a1) Nearly straight rolls appear just above the threshold of the Rayleigh–Bénard instability ($r = 0.043$; see Eq. (7.3)) in a two-dimensional, circular cell. Reprinted from S. W. Morris et al., The spatio-temporal structure of spiral-defect chaos, *Physica D*, **97** (1996) 164–179. (a2) Convective rolls in a one-dimensional, annular geometry. Reprinted from S. Ciliberto, F. Bagnoli, and M. Caponeri, Pattern selection in thermal convection in an annulus, *Il Nuovo Cimento D*, **12** (1990) 781–792. (b1–b3) Scanning tunneling microscopy images of Pt/Pt(111) at (b1) 0.35 monoLayer (ML), (b2) 12 ML, and (b3) 90 ML. The area is 2900×2900 Å. Reprinted from M. Kalff et al., No coarsening in Pt(111) homoepitaxy, *Surface Science*, **426** (1999) L447–L453. (c1) Granular demixing. Segregation patterns of a mixture of different types of sands. In this system we may have (c2) traveling waves in the opposite direction so as to create standing waves or (c3) band merging. Panel (c1) courtesy of C. R. J. Charles and S. Morris. Panels (c2) and (c3) are reprinted from K. Choo et al., Dynamics of granular segregation patterns in a long drum mixer, *Physical Review E*, **58** (1998) 6115–6123.

The first two points are fully general and represent the distinctive features of pattern formation. The third point is peculiar and different pattern-forming systems may display different nonlinear dynamics, i.e., different outcomes of the instability.

The second example, Fig. 7.1(b), shows the result of a growth process of a crystal surface of platinum. Starting from a platinum substrate, whose high-symmetry orientation is characterized by the Miller indices (111), platinum atoms are deposited onto it in ultra-high vacuum conditions. Tuning the intensity Φ of the flux of deposited particles it is possible to make unstable the flat surface of the growing crystal, leading to the formation of mounds/pyramids. In this case Φ is the control parameter and the local height $z(\mathbf{r}, t)$ plays the role of the order parameter. The uniform state corresponds to $z(\mathbf{r}, t) = z_0 + \Phi t$. As for the dynamics following the instability, in the specific case discussed here we get mounds of constant wavelength and increasing height. With other materials or in different experimental conditions, the same instability may give rise to pyramids of increasing wavelength and constant slope, therefore leading to a coarsening process that strongly resembles a phase separation process for the order parameter $\mathbf{m}(\mathbf{r}, t) = \nabla z(\mathbf{r}, t)$.

In the third example, Fig. 7.1(c), we deal with a granular system, partly filling a rotating drum. It is a mixture of large (black) and small (transparent) glass spheres. In this case it is not possible to identify a general control parameter, but to stay at a simple level we can imagine that the role of the control parameter is played by the rotational speed Ω: if it is large enough, an initially homogeneous mixture will tend to segregate, producing a pattern of alternating black and white bands. Identifying the order parameter, which should quantify the asymmetry between the two types of spheres, is easier. If $N_{1,2}$ is their total number, we can define the quantity

$$u(\mathbf{r}, t) = (n_1 c_2 - n_2 c_1)/(n_1 c_2 + n_2 c_1), \tag{7.1}$$

where $n_{1,2}(\mathbf{r}, t)$ are the local number densities of large and small spheres and $c_{1,2} = N_{1,2}/(N_1 + N_2)$ are the global relative fractions, respectively. For a symmetric mixture, $c_1 = c_2$, we simply have $u(\mathbf{r}, t) = (n_1 - n_2)/(n_1 + n_2)$. In general u varies from $u = -1$, if only type-2 particles are locally present at site $\mathbf{r}$ ($n_1 = 0$), to $u = +1$, if only type-1 particles are at site $\mathbf{r}$ ($n_2 = 0$). If $u = 0$, the local concentrations are equal to the average concentrations, $n_i = c_i$. Once those bands have appeared (see Fig. 7.1(c1)), their size can remain constant or increase very slowly. It is also possible to have more complex dynamical behaviors, with the birth of oscillations or traveling waves.

A fourth and final example we want to mention here are the so-called Turing patterns, named after Alan Turing, who wrote a seminal paper in 1952 with the title "The Chemical Basis of Morphogenesis," proposing a basic model to account for the formation of natural patterns. According to Turing, such patterns can be the result of an instability of a homogeneous state, through a process of reaction–diffusion, where different ingredients may diffuse and react among them. A suitable combination of different diffusivity and different reaction properties can give rise to the emergence of a steady pattern of well-defined wavelength. Reaction–diffusion models will be studied with some detail in Section 7.3, but we can anticipate that they require a description based on more than one scalar order parameter.

In fact, throughout the rest of this chapter, the theoretical study of pattern formation in spatially extended systems will be based on a continuum approach, where the evolution of a scalar order parameter is described by a suitable partial differential equation, such as

$$\partial_t u(\mathbf{x}, t) = \mathcal{A}[u, r], \tag{7.2}$$

where $\mathcal{A}$ is a generic functional of the field u (which is assumed here to be a scalar, just for the sake of simplicity), and it also depends on the (reduced) control parameter r. The latter is usually defined so that its critical value is zero. For instance, in the case of the Rayleigh–Bénard instability we can define it as

$$r = \frac{\Delta T - (\Delta T)_c}{(\Delta T)_c}. \tag{7.3}$$

Making reference to the examples shown in Fig. 7.1 and discussed above, the homogeneous solution corresponds to heat conduction in (a), to a growing, flat surface in (b), and to a homogeneous mixture in (c). If $u(\mathbf{x}, t)$ represents, respectively, the velocity field in (a), the local height with respect to the average height ($u(\mathbf{x}, t) = z(\mathbf{x}, t) - \langle z(\mathbf{x}, t)\rangle$) in (b), and the quantity (7.1) in (c), then the homogeneous solution corresponds to $u(\mathbf{x}, t) \equiv 0$ in all cases.

We should stress that the homogeneous solution is expected to be a solution of the problem, $\mathcal{A}[0, r] \equiv 0$, for any r. What changes between negative and positive r (i.e., below and above the threshold) is the stability character of the homogeneous solution: conventionally, it should be stable for $r < 0$ and unstable for $r > 0$. Unstable means that some weak perturbation leads the system far from it. Since fluctuations are unavoidable in any real or simulated system, an unstable solution is physically unobservable; this is why the homogeneous solution disappears for $r > 0$ and a new, spatially modulated solution arises.

The above, simplified picture allows us to clarify why our study of pattern formation is basically made of two parts. The first part analyzes the stability character of the homogeneous solution with varying the control parameter and classifies the type of instability. This linear analysis also allows us to predict what the relevant space and time scales of the emerging pattern are. The second part goes beyond the linear stability analysis of the homogeneous solution and addresses the so-called nonlinear regime, trying to show the ensuing dynamics. The linear analysis is valid at short times, the nonlinear analysis at longer times.

7.2 Linear Stability Analysis and Bifurcation Scenarios

Two of the three experimental examples shown in Fig. 7.1 and discussed in the previous section display a phenomenology that resembles a phase separation process, discussed at length in Chapter 6. In phase-ordering, the homogeneous state becomes unstable after quenching the system from the disordered to the ordered phase. In a pattern-forming system, a homogeneous state is unstable when the system is driven out of equilibrium and a suitable control parameter exceeds a critical value. In phase separation, the instability produces ordered regions, whose size increases in time through a coarsening process. In cases shown in Fig. 7.1(b,c) we may have a spatial structure characterized by regions of constant order parameter, and it may happen that their size increases in time, therefore producing a dynamics similar to a phase separation process.

At first sight this statement is confusing, because phase-ordering dynamics in a system relaxing toward equilibrium is controlled by its free energy, but a driven, out-of-equilibrium system has no free energy. In spite of this, it appears that some out-of-equilibrium dynamics may be characterized by the decreasing of a suitable functional, called the Lyapunov functional or potential functional, and in some cases this functional looks like the Ginzburg–Landau free energy. In fact, this result is not so strange if we think of the GL free energy as a power expansion in the order parameter. However, there is an even simpler reason to reintroduce here the time-dependent Ginzburg–Landau (TDGL) and the Cahn–Hilliard (CH) models: these equations allow us to present the concepts of linear stability analysis and bifurcation scenarios.

Let us start by rewriting these two equations in the simple one-dimensional case, making explicit the dependence on the control parameter,

$$\frac{\partial u}{\partial t} = \frac{\partial^2 u}{\partial x^2} + ru - u^3 \equiv -\frac{\delta \mathcal{F}_{\rm GL}}{\delta u} \qquad \text{(TDGL)} \tag{7.4}$$

$$\frac{\partial u}{\partial t} = -\frac{\partial^2}{\partial x^2}\left(\frac{\partial^2 u}{\partial x^2} + ru - u^3\right) \equiv \frac{\partial^2}{\partial x^2}\frac{\delta \mathcal{F}_{\rm GL}}{\delta u} \qquad \text{(CH)} \tag{7.5}$$

$$\mathcal{F}_{\rm GL}[u] = \int dx \left[\frac{1}{2}\left(\frac{\partial u}{\partial x}\right)^2 - r\frac{u^2}{2} + \frac{u^4}{4}\right] \equiv \int dx \left[\frac{1}{2}\left(\frac{\partial u}{\partial x}\right)^2 + V(u)\right], \tag{7.6}$$

and recalling that dynamics minimizes the functional $\mathcal{F}_{\rm GL}[u]$, $d\mathcal{F}_{\rm GL}[u]/dt \leq 0$; see Section 6.4.2. The quantity r, the control parameter, allows passage from the single-well potential for $r < 0$, where $V(u)$ has a single minimum in $u = 0$, to a double-well potential for $r > 0$, where $V(u)$ has two equivalent minima in $u = \pm\sqrt{r}$; see Fig. 7.2.

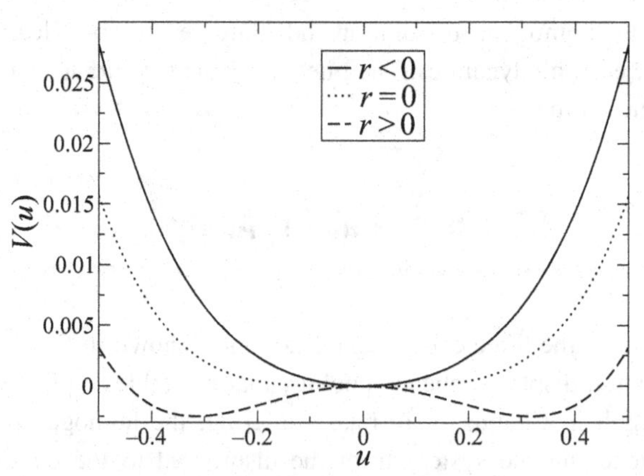

Fig. 7.2 The potential $V(u)$ for negative, vanishing, and positive r; see Eq. (7.6).

Let's now focus on the partial differential equations, observing that $u = 0$ is always solution of both equations, for any value of r. Its relevance in the dynamics depends on its stability character, i.e., on the evolution of a profile that is initially close to it. This analysis is called linear stability analysis, because it amounts to writing $u(x,t) = \epsilon\tilde{u}(x,t)$ and keeping only terms that are linear in the small parameter ϵ. The physical grounds of this approximation will be clear later. We can therefore write

$$\epsilon\partial_t\tilde{u}(x,t) = \epsilon\,(\tilde{u}_{xx} + r\tilde{u}) + O(\epsilon^3) \tag{7.7}$$

$$\epsilon\partial_t\tilde{u}(x,t) = \epsilon\,(-\tilde{u}_{xxxx} - r\tilde{u}_{xx}) + O(\epsilon^3), \tag{7.8}$$

where the terms we are going to disregard are of order ϵ^3. The resulting equations are, of course, linear and can be easily solved with the help of Fourier analysis. The evolution of a single Fourier component is determined by assuming $\tilde{u}(x,t) = u_0 e^{\sigma t}e^{iqx}$. It is straightforward to see that each spatial derivative corresponds to multiplying by (iq) and each time derivative corresponds to multiplying by σ,

$$\partial_t \leftrightarrow \sigma \tag{7.9}$$

$$\partial_x \leftrightarrow iq. \tag{7.10}$$

Therefore, for Eqs. (7.7) and (7.8) we obtain the stability spectra

$$\sigma = r - q^2 \qquad \text{(TDGL)} \tag{7.11}$$

$$\sigma = rq^2 - q^4. \qquad \text{(CH)} \tag{7.12}$$

If the TDGL/CH models are written in generic dimension d it is sufficient to replace ∂_{xx} with ∇^2 in Eqs. (7.4) and (7.5), and the linear stability analysis remains unchanged. For general d the perturbation has the form $\tilde{u}(\mathbf{x},t) = u_o e^{\sigma t}e^{i\mathbf{q}\cdot\mathbf{x}}$; Eq. (7.10) is replaced by $\nabla \leftrightarrow i\mathbf{q}$; and spectra (7.11) and (7.12) are still valid, with $q = |\mathbf{q}|$. We can now comment on such spectra, making reference to Fig. 7.3.

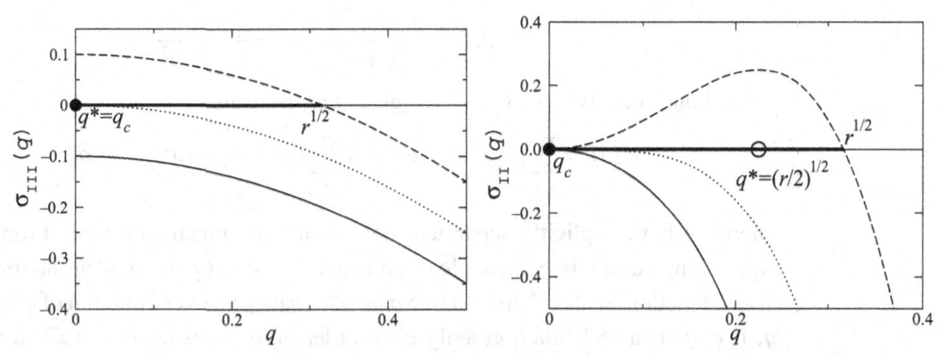

Fig. 7.3 The stability spectra for the TDGL equation (left) and the CH equation (right), with increasing the control parameter: $r < 0$ (solid line), $r = 0$ (dotted line), and $r > 0$ (dashed line). The thick line segments represent the instability region of the spectrum ($\sigma(q) > 0$), the solid circle indicates the critical wavevector, q_c, and the open circle (when different from the solid circle) indicates the most unstable wavevector, q^*. These two scenarios are representative of type-II (right panel) and type-III (left panel) bifurcations; see also Table 7.1.

The first, basic remark is that in both the nonconserved and conserved cases, $\sigma < 0$ for all $q \neq 0$ when $r \leq 0$ and $\sigma > 0$ for some q when $r > 0$. A positive (negative) $\sigma(q)$ means that the amplitude of the solution exponentially grows (decreases). A given stationary solution is stable if the amplitude of any perturbation goes to zero with time, otherwise it is unstable. The above results mean that the homogeneous solution $u = 0$ is stable for negative r and unstable for positive r, and this statement is valid for both the TDGL and CH equations. In other words, if we integrate one of such equations starting from a profile $u(x, 0)$ that is "small" everywhere, i.e., $|u(x, 0)| \ll 1\ \forall x$, we expect the evolution to converge to $u = 0$ if $r \leq 0$, while it increasingly moves away from $u = 0$ for $r > 0$. What the ensuing dynamics is for $r > 0$ depends on the nonlinear terms in the equation, which have been disregarded in the above analysis.

A second, basic remark is that $\sigma(q)$ is real. An imaginary part would correspond to an oscillating component of the amplitude. If $\sigma = \sigma_R + i\sigma_I$, we would have $\tilde{u}(x, t) = u_0 \cos(\sigma_I t) \exp(\sigma_R t)$. Therefore, in general terms, the real and imaginary parts of σ are different pieces of information about the stability of the solution. The above equations provide a purely real spectrum, because the linear part is just a sum of even spatial derivatives, therefore only terms of the form $(iq)^{2n} = (-1)^n q^{2n}$ appear in $\sigma(q)$. More precisely, in TDGL we have $n = 0, 1$ and in CH we have $n = 1, 2$.

We can generalize these considerations to any equation having the form

$$u_t = \sum_k c_k \partial_x^k u + \mathcal{N}[u] \equiv \mathcal{L}[u] + \mathcal{N}[u], \tag{7.13}$$

where $\mathcal{L}[u]$ is a generic, local, and linear operator and $\mathcal{N}[u]$ is the nonlinear part. We should stress that $\mathcal{L}$ and $\mathcal{N}$ might not be naturally separated, as in the following example:

$$u_t = u_{xx} + \frac{u}{1 + u^2}. \tag{7.14}$$

Here,

$$\mathcal{L}[u] = u + u_{xx} \tag{7.15}$$

$$\mathcal{N}[u] = \frac{u}{1 + u^2} - u = -\frac{u^3}{1 + u^2}. \tag{7.16}$$

The Fourier analysis of Eq. (7.13) gives the spectrum

$$\sigma(q) = \sum_{n \geq 0} c_{2n}(-1)^n q^{2n} + i \sum_{n \geq 0} (-1)^n c_{2n+1} q^{2n+1} \equiv \sigma_R + i\sigma_I, \tag{7.17}$$

where we have explicitly separated the real and the imaginary parts. From the previous expression, we can draw a couple of conclusions: (i) $\sigma_R(q)$ is an even function and $\sigma_I(q)$ is an odd function, so a real growth rate σ is necessarily an even function of q. (ii) Since $\partial_x \leftrightarrow iq$, if $\sigma(q)$ is a real function, only even-order derivatives appear in $\mathcal{L}[u]$, which implies (at least for the linear part) that the equation is invariant under the symmetry $x \to -x$. These conclusions are strictly correct for Eq. (7.13), i.e., a local equation for a scalar order parameter in $d = 1$. In the next section we consider the Turing instability, which is described by a pair of coupled order parameters. The linear analysis will show that an oscillatory behavior ($\sigma_I \neq 0$) may occur even if the model has the symmetry $x \to -x$.

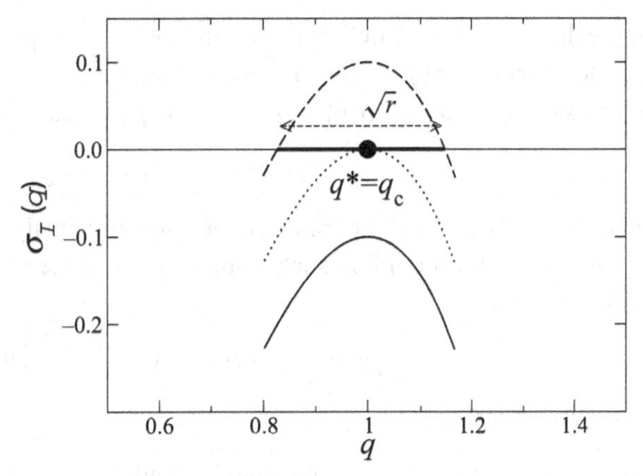

Fig. 7.4 Type I bifurcation scenario, valid for a static instability, with $q_c \neq 0$. The full, nonlinear equation we are referring to is the Swift–Hohenberg (SH) equation, introduced in Section 7.4. With increasing the control parameter, we pass from $r < 0$ (solid line), $r = 0$ (dotted line), and $r > 0$ (dashed line). The thick line segment represents the instability region of the spectrum ($\sigma(q) > 0$), which is of order $\sqrt{r}$, and the solid circle represents the critical wavevector, q_c, and the most unstable wavevector, q^*, which coincide in this model.

We have discussed the stability spectrum in some detail because, on the one hand, it plays a major role in determining the type of instability and, on the other hand, it determines the linear part of the partial differential equation. For this reason we proceed discussing further features of $\sigma(q)$ for TDGL and CH. Let us suppose that the control parameter is positive and very small, $0 < r \ll 1$.[2] The unstable part of the spectrum corresponds to $|q| < \sqrt{r}$, therefore a small interval around $q_c = 0$. In other words, Fourier modes of arbitrarily large wavelength are unstable. There is, however, a crucial difference between TDGL and CH, because the existence of a conservation law, $\partial_t \int dx u(x,t) = 0$, implies that a spatially constant profile cannot move, as explicitly seen from Eqs. (7.4) and (7.5) by assuming $u(x,t) \equiv u(t)$. TDGL gives $\dot{u} = ru - u^3$ and CH gives $\dot{u} = 0$. For $q = 0$, the linear stability analysis reduces to studying the stability of the solution $u = 0$ of these ordinary differential equations. For the CH equation, any constant value is a solution, which corresponds to saying that by perturbing $u = 0$, the amplitude neither grows nor decreases: $\sigma = 0$.

We can conclude that the form of $\sigma(q)$ depends on two main factors: (i) the value q_c of the most unstable mode at the instability threshold ($r \to 0^+$) and (ii) the presence or absence of the conservation law of the order parameter. TDGL and CH correspond to $q_c = 0$, the former without and the latter with the conservation law. It is now natural to wonder what an equation must look like if $q_c \neq 0$. Understanding this point is our next goal.

As an example, in Fig. 7.4 we schematically plot the expected behavior of $\sigma(q)$ when $q_c \neq 0$. We focus on the unstable region close to q_c, assuming that for q far from q_c, $\sigma(q) < 0$. More precisely, we assume that there is no conservation law ($\sigma(0) \neq 0$), that

[2] If r is not adimensional, we may always write $r = r_0\delta$, with $\delta \ll 1$.

we are dealing with a static instability, and that $\sigma = \sigma(q^2)$. These conditions are sufficient to write a function $\sigma(q)$ having a maximum in $q = q_c$ and such that $\sigma(q = q_c, r = 0) = 0$. In fact, since $q^2 - q_c^2 \simeq 2q_c(q - q_c)$ close to q_c, we can guess that

$$\sigma(q) = r - c_1(q^2 - q_c^2)^2, \tag{7.18}$$

with $c_1 > 0$. Now, we can perform a linear stability analysis in reverse order, asking what the linear operator providing such a spectrum is. Since we know that $\partial_x \leftrightarrow iq$, we easily obtain the equation

$$\partial_t u(x,t) = \left[r - c_1(\partial_{xx} + q_c^2)^2\right] u(x,t). \tag{7.19}$$

If we now rescale x and t,

$$x \to \alpha x \qquad \text{and} \qquad t \to \beta t, \tag{7.20}$$

we obtain

$$\partial_t u(x,t) = \left[\beta r - \frac{\beta c_1}{\alpha^4}(\partial_{xx} + \alpha^2 q_c^2)^2\right] u(x,t). \tag{7.21}$$

Therefore, we can choose α and β so that

$$\frac{\beta c_1}{\alpha^4} = 1 \qquad \text{and} \qquad \alpha^2 q_c^2 = 1. \tag{7.22}$$

However, for ease of presentation we prefer to keep q_c explicit. This corresponds to not rescaling x, $\alpha = 1$, while $\beta = 1/c_1$. There is no loss of generality to redefine $\beta r = r/c_1$ as r, so we finally obtain

$$\partial_t u(x,t) = \left[r - (\partial_{xx} + q_c^2)^2\right] u(x,t). \tag{7.23}$$

Our results are summarized in Table 7.1. They do not cover all possible scenarios of the rising of an instability, of course. They just represent the simplest cases of stationary instabilities, in one spatial dimension and for a single, scalar order parameter. The central idea is to classify them according to two criteria: (i) the value of the critical wavevector q_c, which can be zero (types II and III) or not zero (type I), and (ii) the conservation (type II)

Table 7.1 Different types of static bifurcations

Type[a]	Critical q [b]	σ	$\mathcal{L}$	q^* [c]
I	q_c	$r - (q^2 - q_c^2)^2$	$r - (\partial_{xx} + q_c^2)^2$	q_c
II	0	$rq^2 - q^4$	$-r\partial_{xx} - \partial_{xxxx}$	$\sqrt{\frac{r}{2}}$
III	0	$r - q^2$	$r + \partial_{xx}$	0

[a] We are using the classification introduced in M. C. Cross and P. C. Hohenberg, Pattern formation outside of equilibrium, *Reviews of Modern Physics*, **65** (1993) 851–1112.

[b] The critical wavevector is defined as the q value corresponding to the maximum of $\sigma(q)$ at the threshold, $r \to 0^+$.

[c] It is the possibly r-dependent wavevector maximizing $\sigma(q)$. $q^*(r = 0)$ is the critical wavevector.

or non-conservation (type I and III) of the order parameter. At a linear level, conservation implies that $\sigma(q=0)=0\,\forall r$. We do not consider here the conserved case with $q_c \neq 0$. The main message of the linear analysis is that it is possible to classify the linear spectrum $\sigma(q)$ according to the above criteria and that the knowledge of $\sigma(q)$ determines unambiguously the linear part of the partial differential equation. Therefore, the three classes of spectra, given in the third column of the table, determine the linear operators $\mathcal{L}$, explicitly written in the fourth column.

The linear analysis provides other relevant information about the instability. First of all, there is a clear difference between $q_c = 0$ and $q_c \neq 0$. In the former case, all modes with a wavelength larger than $2\pi/\sqrt{r}$ up to the size L of the physical system are unstable. That means very different unstable scales, which all contribute to the nonlinear behavior of the system. In the latter case, $q_c \neq 0$, unstable modes are in the approximate interval $(q_c-\delta, q_c+\delta)$, with $\delta \simeq \sqrt{r}/(2q_c)$. Therefore, even if the size of the unstable q-interval is approximately the same as for types II and III (i.e., of order $\sqrt{r}$), it is evident that for type I all length scales are approximately equal to $2\pi/q_c$. We will see that this fact has relevant consequences for the nonlinear dynamics.

If an unstable Fourier mode q has a time dependence $e^{\sigma t}$, the quantity $\tau_q = 1/\sigma(q)$ is the time scale to initiate the instability on that length scale. Therefore, we can expect that the mode q^*, corresponding to the shortest instability time, $\sigma'(q^*)=0$, will dominate the spatial pattern in the linear regime. Again, we should distinguish between $q^*=0$ (type III) and $q^* \neq 0$ (types I and II); see the column five of Table 7.1. If $q^*=0$, there is no well-defined length scale dominating over the others, so we expect that many different scales may appear, producing an initial "disordered" pattern. If $q^* \neq 0$, there is a clear length scale $\lambda^* = 2\pi/q^*$ that dominates. This scale is simply set by q_c and does not depend on r for type I instabilities, while it scales as $\lambda^* \simeq 1/\sqrt{r}$ for type II instabilities.

In all cases, regardless of q^*, the time $\tau^* = 1/\sigma(q^*)$ gives the typical time, because the instability develops. When $t \gtrsim \tau^*$, the amplitude of the modes is of order one and disregarding nonlinear terms is no longer justified. Therefore, our linear analysis is valid for times smaller than τ^*.

7.3 The Turing Instability

In this section we deal with an important class of pattern forming systems, the so-called *reaction–diffusion* models, which are primarily used to study the reaction dynamics of diffusing chemicals (hence their name). The main goal here is to investigate the conditions giving rise to a type I spectrum (see Table 7.1 and Fig. 7.4), because this is a necessary condition to get a pattern of defined size: in this section we use the wording "pattern formation" as a shorthand for systems displaying a type I spectrum.

Each chemical species is characterized by its concentration $u_i(\mathbf{x},t)$, whose time variation has a diffusive term plus a local term due to the interaction (reaction) among species,

$$\frac{\partial u_i}{\partial t} = f_i(u_1, u_2, \dots) + D_i \nabla^2 u_i. \tag{7.24}$$

In the limiting case of just one species we get

$$\frac{\partial u}{\partial t} = f(u) + \nabla^2 u, \tag{7.25}$$

which is the generalized TDGL equation, whose one-dimensional version is studied in Section 7.8; see Eq. (7.164). This equation cannot give rise to a type I spectrum in any dimension, as easily seen by its linearization around a constant solution $\bar{u}$, i.e., such that $f(\bar{u}) = 0$. If $u(\mathbf{x}, t) = \bar{u} + \epsilon(\mathbf{x}, t)$, we get

$$\partial_t \epsilon = f'(\bar{u})\epsilon + D\nabla^2 \epsilon, \tag{7.26}$$

whose solution $\epsilon(\mathbf{x}, t) = \epsilon_0 \exp(\sigma t + i\mathbf{q} \cdot \mathbf{x})$ gives the spectrum

$$\sigma(\mathbf{q}) = f'(\bar{u}) - D\mathbf{q}^2, \tag{7.27}$$

which is a type III spectrum, always centered in $\mathbf{q} = 0$.

This is the reason why we need to introduce more than one species. We also remark that if diffusion is isotropic, spatial dimension does not affect the linear stability analysis. So, in the following q will be the modulus of a d-dimensional wavevector, $q^2 = \mathbf{q} \cdot \mathbf{q}$.

The analysis of this section is focused on the linear stability of a general homogeneous solution, $u_i(\mathbf{x}, t) = \bar{u}_i$. We are searching the conditions for the homogeneous solution being unstable in a finite range of wavelengths, i.e., in a range (q_1, q_2) with $q_1 \neq 0$. Since it must be $\sigma(0) < 0$ and since diffusion certainly stabilizes the homogeneous solution for large enough q, we require some destabilizing mechanism induced by diffusion, but mediated by reactions. We are now going to study a two-species model with the goal of determining the physical ingredients to obtain pattern formation. The general conditions will be derived in Appendix Q.

7.3.1 Linear Stability Analysis

The most general reaction–diffusion equations for two species are

$$\begin{aligned} \frac{\partial u_1}{\partial t} &= f_1(u_1, u_2) + D_1 \nabla^2 u_1, \\ \frac{\partial u_2}{\partial t} &= f_2(u_1, u_2) + D_2 \nabla^2 u_2. \end{aligned} \tag{7.28}$$

We assume the existence of some homogeneous steady state $(\bar{u}_1, \bar{u}_2)$, which must therefore satisfy the conditions

$$\begin{aligned} f_1(\bar{u}_1, \bar{u}_2) &= 0 \\ f_2(\bar{u}_1, \bar{u}_2) &= 0. \end{aligned} \tag{7.29}$$

We are now going to perturb this state, writing

$$\begin{aligned} u_1(\mathbf{x}, t) &= \bar{u}_1 + \epsilon_1(\mathbf{x}, t) \\ u_2(\mathbf{x}, t) &= \bar{u}_2 + \epsilon_2(\mathbf{x}, t), \end{aligned} \tag{7.30}$$

then linearizing with respect to $\epsilon_{1,2}$,

$$\begin{aligned}\frac{\partial \epsilon_1}{\partial t} &= a_{11}\epsilon_1 + a_{12}\epsilon_2 + D_1\nabla^2\epsilon_1 \\ \frac{\partial \epsilon_2}{\partial t} &= a_{21}\epsilon_1 + a_{22}\epsilon_2 + D_2\nabla^2\epsilon_2,\end{aligned} \tag{7.31}$$

where

$$a_{ij} = \left.\frac{\partial f_i}{\partial u_j}\right|_{\bar{u}_1,\bar{u}_2}. \tag{7.32}$$

Eqs. (7.31) can be solved by assuming an exponential dependence on space and time,

$$\begin{aligned}\epsilon_1(\mathbf{x},t) &= \epsilon_1^0 e^{\sigma t} e^{i\mathbf{q}\cdot\mathbf{x}} \\ \epsilon_2(\mathbf{x},t) &= \epsilon_2^0 e^{\sigma t} e^{i\mathbf{q}\cdot\mathbf{x}},\end{aligned} \tag{7.33}$$

whose replacement in Eq. (7.31) gives

$$\begin{pmatrix} a_{11} - D_1 q^2 & a_{12} \\ a_{21} & a_{22} - D_2 q^2 \end{pmatrix} \begin{pmatrix} \epsilon_1^0 \\ \epsilon_2^0 \end{pmatrix} \equiv \mathbf{A_q} \begin{pmatrix} \epsilon_1^0 \\ \epsilon_2^0 \end{pmatrix} = \sigma \begin{pmatrix} \epsilon_1^0 \\ \epsilon_2^0 \end{pmatrix}. \tag{7.34}$$

The solvability condition is

$$\det(\mathbf{A_q} - \sigma\mathbf{I}) = 0, \tag{7.35}$$

which can be developed as

$$\sigma^2 - \mathrm{Tr}(\mathbf{A_q})\sigma + \det(\mathbf{A_q}) = 0. \tag{7.36}$$

This equation has two solutions, $\sigma_1(q)$ and $\sigma_2(q)$. The homogeneous steady state $(\bar{u}_1, \bar{u}_2)$ is linearly stable if both solutions have a negative real part for any q. Conversely, it is linearly unstable if at least one solution has a positive real part for some q. The next step is to discuss conditions under which pattern formation occurs.

Coupling between Chemicals

The first condition is somewhat, but not completely, trivial: we require that each species acts on the dynamics of the other species. In fact, if a_{12} (or a_{21}) vanishes, the product of the off-diagonal terms of $\mathbf{A_q}$ vanishes and Eq. (7.35) trivially gives ($i = 1, 2$)

$$\sigma_i(q) = a_{ii} - D_i q^2. \tag{7.37}$$

These are the type III spectra (see Table 7.1) of the single species problem, which are not what we are looking for. Therefore, both a_{12} and a_{21} must be nonvanishing in order to get pattern formation.

Different Diffusion Rates

The stability condition says that the real part of both solutions $\sigma_{1,2}(q)$ is negative. Since $\sigma_{1,2}$ are either real or complex conjugates, both of the following quantities,

$$\begin{aligned}\sigma_1 + \sigma_2 &= \mathrm{Tr}(\mathbf{A_q}) \equiv \mathcal{T}_q \\ \sigma_1\sigma_2 &= \det(\mathbf{A_q}) \equiv \mathcal{D}_q,\end{aligned} \tag{7.38}$$

are real. Therefore, the stability condition is equivalent to imposing

$$\mathcal{T}_q = a_{11} + a_{22} - (D_1 + D_2)q^2 < 0 \tag{7.39a}$$

$$\mathcal{D}_q = (a_{11} - D_1 q^2)(a_{22} - D_2 q^2) - a_{12}a_{21} > 0. \tag{7.39b}$$

We are interested in understanding the conditions for which these inequalities are satisfied for $q = 0$ and are *not* satisfied for finite q, having in mind that they are surely satisfied for large q, since in this limit diffusion is dominant, so that

$$\mathcal{T}_q \approx -(D_1 + D_2)q^2 \tag{7.40a}$$

$$\mathcal{D}_q \approx D_1 D_2 q^4. \tag{7.40b}$$

We are now going to show that for equal diffusion rates, stability at $q = 0$ implies stability at any q, which excludes pattern formation. This is trivial and even more general for the trace, because

$$\mathcal{T}_q = \mathcal{T}_0 - (D_1 + D_2)q^2, \tag{7.41}$$

so diffusion always has a stabilizing effect on the trace. This means that a change of stability with increasing q must be determined by a change of sign of $\mathcal{D}_q$.

As for the determinant, if $D_1 = D_2 = D$, we get

$$\mathcal{D}_q = \mathcal{D}_0 - \mathcal{T}_0(Dq^2) + (Dq^2)^2, \tag{7.42}$$

so that stability at $q = 0$ ($\mathcal{T}_0 < 0, \mathcal{D}_0 > 0$) also implies $\mathcal{D}_q > 0$ for any q.

In conclusion, we must consider different diffusion rates if we want to observe pattern formation. However, this is not enough, as seen in the next section.

Local Activation, Long-Range Inhibition

We have clarified that coupling and different diffusion coefficients are necessary conditions to have stability at vanishing q and instability for intermediate q. Let's take a further step. Since it must be $D_1 \neq D_2$, there is no loss of generality if we assume $D_1 > D_2$. While the general case will be considered in Appendix Q, here it is easier to focus on the limit $D_1 \gg D_2$. The stability condition (7.39b) on the determinant can be written

$$\mathcal{D}_q = \mathcal{D}_0 - (a_{22}D_1 + a_{11}D_2)q^2 + D_1 D_2 q^4 > 0. \tag{7.43}$$

If $\mathcal{D}_0 > 0$ and $D_1 \gg D_2$, we may have $\mathcal{D}_q < 0$, at increasing q if and only if[6] $a_{22} > 0$, which in turn imposes $a_{11} < 0$, because it must be that $\mathcal{T}_0 = a_{11} + a_{22} < 0$.

Since the sign of a_{ii} indicates whether the presence of the ith species reinforces ($a_{ii} > 0$, positive feedback) or suppresses ($a_{ii} < 0$, negative feedback) itself, we can summarize above results as follows: *We must have reaction between an activator that spreads slowly and an inhibitor that spreads quickly.* As for the cross coupling terms, a_{12} and a_{21}, since

$$\mathcal{D}_0 = a_{11}a_{22} - a_{12}a_{21} > 0, \tag{7.44}$$

[6] For fixed a_{11} and a_{22}, if D_1 is sufficiently large, we have $|a_{22}|D_1 \gg |a_{11}|D_2$.

the negative sign of the product $a_{11}a_{22}$ implies that two cross terms must have opposite signs. In conclusion, if $D_1 \gg D_2$, we have two possible cases for the emergence of the Turing instability,

$$a_{11} < 0, \quad a_{22} > 0, \quad \begin{cases} a_{12} > 0 & a_{21} < 0 \\ a_{12} < 0 & a_{21} > 0. \end{cases} \tag{7.45}$$

The two cases differ in the expression of the eigenvalues. In fact, at criticality, $\sigma_1(q_c) = 0$ and Eq. (7.34) is

$$(a_{11} - D_1 q_c^2)\epsilon_1^0 + a_{12}\epsilon_2^0 = 0. \tag{7.46}$$

For large D_1 it is manifest that

$$\frac{\epsilon_1^0}{\epsilon_2^0} = (\text{positive factor}) \times a_{12}, \tag{7.47}$$

so the ratio $\epsilon_1^0/\epsilon_2^0$ has the same sign as a_{12}. In Appendix Q we prove the validity of Eqs. (7.45) and (7.47) for the general case $D_1 > D_2$.

We can now describe in words the instability mechanism making reference to the upper case of (7.45): $a_{12} > 0$ and $a_{21} < 0$. A positive fluctuation of the activator (species 2) tends to grow and also to reinforce a positive fluctuation of the inhibitor (species 1), while the inhibitor tends to suppress both species, and the combined balance at $q = 0$ must result in a stable situation (because this is what we require). However, if the inhibitor has a large diffusion coefficient, it is no longer effective to suppress the growth of the activator, and an instability may appear. If the signs of the cross coupling terms, a_{12} and a_{21}, are reversed, it is enough to realize (see the linear Eqs. (7.31)) that it is equivalent to change the sign to one of the two perturbations, ϵ_1^0 or ϵ_2^0, which also explains why the ratio $\epsilon_1^0/\epsilon_2^0$ is related to the sign of a_{12}.

In Fig. 7.5 we give a pictorial summary of the bifurcation scenarios of the reaction–diffusion equations (7.28) for two species, plotting the space $(\mathcal{T}_q, \mathcal{D}_q)$, which define unambiguously $\sigma_{1,2}(q)$, i.e., the linear stability properties of Eqs. (7.28). For complex eigenvalues, $\sigma_{1,2} = R \pm iI$, we have $\mathcal{D}_q = \frac{1}{4}\mathcal{T}_q^2 + I^2$, while for real eigenvalues, $\sigma_{1,2} = R_{1,2}$, we have $\mathcal{D}_q = \frac{1}{4}\mathcal{T}_q^2 - \frac{1}{4}(R_1 - R_2)^2$. Therefore, the parabola $\mathcal{D}_q = \frac{1}{4}\mathcal{T}_q^2$ (dashed line) separates an upper region with complex eigenvalues and a lower region with real eigenvalues. Since the stable region corresponds to $\mathcal{T}_q < 0, \mathcal{D}_q > 0$ (upper left quadrant with dots), by tuning a suitable control parameter we may leave the stability region either (i) crossing the axis $\mathcal{D}_q = 0$ (vertical arrow) or (ii) crossing the axis $\mathcal{T}_q = 0$ (horizontal axis). In the former case the imaginary part of the eigenvalues is zero and we have a stationary bifurcation; in the latter case the imaginary part is nonzero and the bifurcation is oscillating. As shown above, a change of stability at finite q_c is determined by the change of sign of $\mathcal{D}_q$, while the change of sign of $\mathcal{T}_q = \mathcal{T}_0 - (D_1 + D_2)q^2$ implies $q_c = 0$. Therefore scenario (i) corresponds to a type I instability and scenario (ii) corresponds to a type III-o instability, i.e., an oscillating type III bifurcation. In order to obtain an oscillating Turing instability (type I-o), characterized by a finite q_c and a nonvanishing imaginary part $I(q_c)$, we should consider the dynamics of three chemical species.

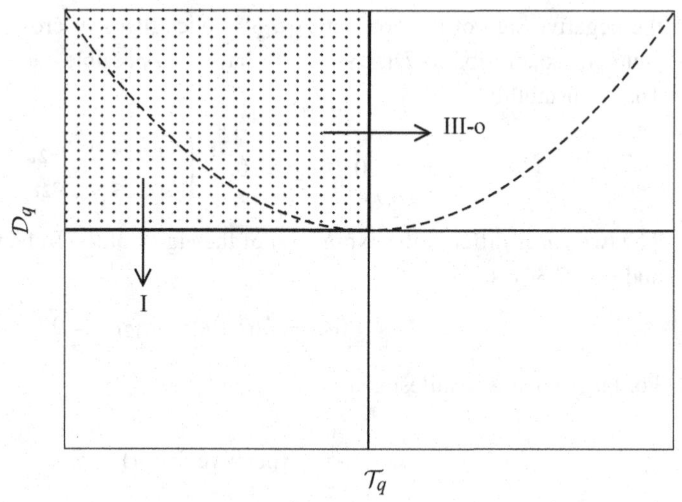

Fig. 7.5 The bifurcation scenarios in the plane $\mathcal{T}_q, \mathcal{D}_q$. The stability (dotted) region corresponds to $\mathcal{T}_q < 0, \mathcal{D}_q > 0$. We can pass from stability to instability following one of the two arrows, which correspond to a type I instability (vertical arrow crossing the axis $\mathcal{D}_q = 0$) or to a type III-o instability (horizontal arrow crossing the axis $\mathcal{T}_q = 0$).

7.3.2 The Brusselator Model

In this section we treat a specific example of the reaction–diffusion process of two species, the Brusselator model, presenting a detailed linear analysis of its unique homogeneous steady state.

The Brusselator is a minimal model for an autocatalytic reaction that involves two intermediates. Using the typical notation of chemical reactions, we write

$$A \to X \tag{7.48}$$

$$B + X \to Y + C \tag{7.49}$$

$$2X + Y \to 3X \tag{7.50}$$

$$X \to D. \tag{7.51}$$

Above, A and B are the initial reactants, C and D are the final products, and X and Y are the intermediates, which are the variables to be described by the coupled system (7.28). If u_1, u_2 represent the spatial densities of X, Y, respectively, and we include diffusion, we obtain the following pair of coupled equations,

$$\partial_t u_1 = \tilde{a} - (\tilde{b} + \tilde{d})u_1 + \tilde{c}u_1^2 u_2 + \tilde{D}_1 \nabla^2 u_1 \tag{7.52}$$

$$\partial_t u_2 = \tilde{b}u_1 - \tilde{c}u_1^2 u_2 + \tilde{D}_2 \nabla^2 u_2, \tag{7.53}$$

where the nonlinear terms originate from Eq. (7.50) and the (positive) quantities $\tilde{a}$, $\tilde{b}$, $\tilde{c}$, and $\tilde{d}$ denote the chemical reaction rates. It is possible to reduce the number of parameters by a suitable rescaling of u_1, u_2, t. More precisely, if

$$u_{1,2} \to \sqrt{\frac{\tilde{d}}{\tilde{c}}}\, u_{1,2} \quad \text{and} \quad t \to \frac{1}{\tilde{d}}\, t, \tag{7.54}$$

we can rid the system of two of the four reaction rates,

$$\partial_t u_1 = a - (b+1)u_1 + u_1^2 u_2 + D_1 \nabla^2 u_1 \tag{7.55}$$

$$\partial_t u_2 = b u_1 - u_1^2 u_2 + D_2 \nabla^2 u_2, \tag{7.56}$$

where $a = \tilde{a}\sqrt{\tilde{c}}/\tilde{d}^{3/2}$, $b = \tilde{b}/\tilde{d}$, $D_1 = \tilde{D}_1/\tilde{d}$, and $D_2 = \tilde{D}_2/\tilde{d}$. As for the diffusion constants, a rescaling of the space variable $\mathbf{x}$ would allow us to put either D_1 or D_2 equal to one. This is not common in the literature, so we maintain both coefficients, keeping in mind that only their ratio will be relevant to determining the stability of a given solution. In conclusion, the previous equations have the form (7.28), with

$$f_1 = a - (b+1)u_1 + u_1^2 u_2 \tag{7.57}$$

$$f_2 = b u_1 - u_1^2 u_2. \tag{7.58}$$

The only homogeneous fixed point $(\bar{u}_1, \bar{u}_2)$ is obtained when both f_1 and f_2 vanish, i.e.,

$$\bar{u}_1 = a, \qquad \bar{u}_2 = \frac{b}{a}, \tag{7.59}$$

and the elements $a_{ij} = (\partial f_i/\partial u_j)_{\bar{u}_1,\bar{u}_2}$ of the linearization matrix are

$$\mathbf{A_0} = \begin{pmatrix} b-1 & a^2 \\ -b & -a^2 \end{pmatrix}. \tag{7.60}$$

We observe that the second species (Y) is the inhibitor ($a_{22} < 0$), while the first species (X) is the activator ($a_{11} > 0$) if $b > 1$. We therefore expect that $D_2 > D_1$. As for the cross terms, a_{12} and a_{21}, they have opposite signs, as it should be.

We can now be more quantitative and impose the three conditions (see (Q.6) in Appendix Q) to determine the parameter region where a Turing pattern can appear. The condition $\mathcal{T}_0 = a_{11} + a_{22} < 0$ yields $b - 1 - a^2 < 0$, i.e.,

$$b < 1 + a^2, \tag{7.61}$$

while the condition $\mathcal{D}_0 = a_{11}a_{22} - a_{12}a_{21} > 0$ yields $-(b-1)a^2 + ba^2 = a^2 > 0$ and is therefore trivially satisfied. Finally, we should impose that the determinant $\mathcal{D}_q$, evaluated at its minimum $q = q_m$, is negative, i.e., $(b-1)D_2 - a^2 D_1 > 2a\sqrt{D_1 D_2}$. This condition can be rewritten as

$$b > \left(1 + a\sqrt{\frac{D_1}{D_2}}\right)^2. \tag{7.62}$$

The two conditions (7.61) and (7.62) are consistent if

$$1 + a^2 > \left(1 + a\sqrt{\frac{D_1}{D_2}}\right)^2, \tag{7.63}$$

and recalling that a is positive, we obtain

$$\sqrt{\frac{D_1}{D_2}} < \sqrt{\left(\frac{D_1}{D_2}\right)_c} \equiv \frac{1}{a}\left(\sqrt{1+a^2} - 1\right) - 1 < 1. \tag{7.64}$$

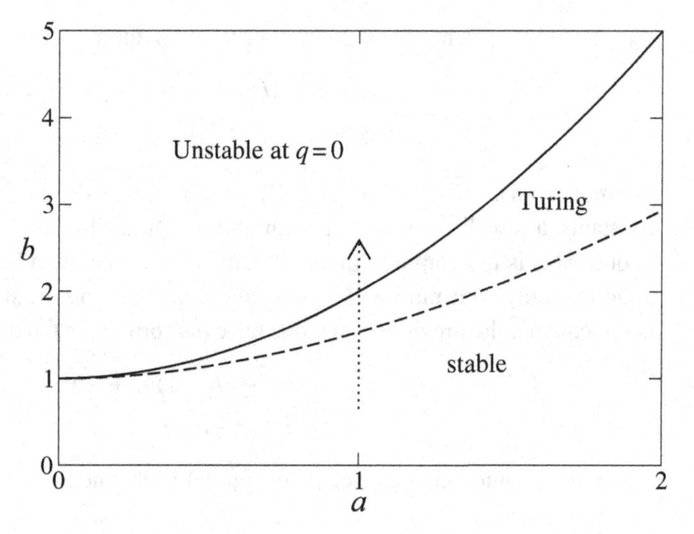

Fig. 7.6 The three regions of the parameter space (a, b) for fixed D_1/D_2. The solid line denotes the change of sign of $\mathcal{T}_0$, Eq. (7.61), which does not depend on the diffusion coefficients. The dashed line denotes the change of sign of $\mathcal{D}_{q_m}$, Eq. (7.62), for $D_1/D_2 = \frac{1}{3}(D_1/D_2)_c$. For $D_1/D_2 \to (D_1/D_2)_c$ the two lines coincide and the Turing region vanishes; for $D_1/D_2 \to 0$ the dashed line becomes the horizontal line $b = 1$.

This inequality confirms that the diffusion coefficient D_1 of the activator must be smaller than that of the inhibitor, D_2. How much smaller it is depends on the parameter a. If the ratio D_1/D_2 is just slightly lower than the critical value $(D_1/D_2)_c$, the Turing region is small, while for a vanishingly small ratio D_1/D_2 the Turing region is given by $1 < b < 1 + a^2$. In Fig. 7.6 we plot the different stability regions in the plane (a, b), for fixed D_1/D_2. Following the dotted arrow we first meet the stable region, where $\sigma_{1,2}(q) < 0$ for all q; then, we have a type I instability and enter the Turing region; finally, the unstable q-region extends up to $q = 0$.

7.4 Periodic Steady States

In Section 7.2 we have provided a sort of classification of stationary linear instabilities for a scalar order parameter $u(\mathbf{x}, t)$ and we explained that the validity of such linear analysis is limited to times shorter than some scale τ^*, because at longer times nonlinear terms are no longer negligible. Therefore, in this section we need to consider the full partial differential equation describing the dynamics of $u(\mathbf{x}, t)$, having in mind that in only a few cases can we derive the evolution equation from first principles. In most cases the evolution equation is the result of a phenomenological approach, where considerations based on symmetries and conservation laws play a major role.

We may wonder, for example, what nonlinear term we can add to the linear operator appearing in type I instabilities. The equation

$$\partial_t u = [r - (\partial_{xx} + q_c^2)^2]u \tag{7.65}$$

satisfies two symmetries, $x \to -x$ and $u \to -u$. The simplest possible nonlinear term still satisfying both symmetries is a cubic term, proportional to u^3, giving

$$\partial_t u = [r - (\partial_{xx} + q_c^2)^2]u + c_3 u^3. \tag{7.66}$$

If we require that the diverging amplitude of unstable Fourier modes, $u_0 e^{\sigma(q)t}$, is counterbalanced by the nonlinear term, it is necessary that its contribution to $\partial_t u$ is negative for $u > 0$ and positive for $u < 0$. This requires $c_3 < 0$, which can be set to $c_3 = -1$ by rescaling u. Therefore we get the equation

$$\partial_t u = [r - (\partial_{xx} + q_c^2)^2]u - u^3 , \qquad \text{(SH)} \tag{7.67}$$

which is well known in the literature as the Swift–Hohenberg (SH) equation.

We can observe that such a stabilizing term is necessary even if the symmetry $u \to -u$ is broken. In that case a quadratic term, $c_2 u^2$, should be added to the linear SH operator, but it cannot guarantee the saturation of the unstable amplitude, because it contributes to $\partial_t u$ with a constant sign, independent of u. Therefore, the SH equation (7.67) is the simplest nonlinear equation of type I satisfying the symmetry $u \to -u$. If this symmetry is broken, the term $c_2 u^2$ should be added to its right-hand side.

Once we have turned our attention to the nonlinear term u^3 for the type I instability, it should not be surprising that the very same nonlinearity appears in Eqs. (7.4) and (7.5). In conclusion, simple considerations of possible instability scenarios, accompanied by the effect of symmetries and conservation laws, allow us to focus on three partial differential equations: the Swift–Hohenberg (SH), the Cahn–Hilliard (CH), and the time-dependent Ginzburg–Landau (TDGL) equations, corresponding, respectively, to type I (SH), type II (CH), and type III (TDGL) instabilities.

The study of the dynamics close to the homogenous, steady solution $u = 0$ has revealed a rich source of information even for $r > 0$, i.e., when such a solution is linearly unstable: it has allowed us to classify instabilities and to extract relevant space and time scales. It is therefore not surprising that locating other steady states is often the first step beyond the linear study faced in Section 7.2, and it is also not surprising that spatially periodic steady states play a special role in pattern-forming systems. The rest of this section will be devoted to discussing the emergence of such periodic states.

Finding the steady states for TDGL is particularly simple, and in certain, physically relevant conditions they are also the steady states of the CH equation. In fact, TDGL and CH steady states are determined by the equations

$$u_{xx} + ru - u^3 = 0 \qquad \text{(TDGL)} \tag{7.68}$$

$$\partial_{xx}[u_{xx} + ru - u^3] = 0 . \qquad \text{(CH)} \tag{7.69}$$

Eq. (7.68) is easily acknowledged as Newton's equation for a (fictitious) particle of position u at "time" x, oscillating in the potential

$$\tilde{V}(u) = -V(u) = r\frac{u^2}{2} - \frac{u^4}{4}. \tag{7.70}$$

In this case periodic solutions are the only bounded steady states, and if we denote by A the amplitude of such oscillations, we can say that: (i) there is a periodic steady state for any

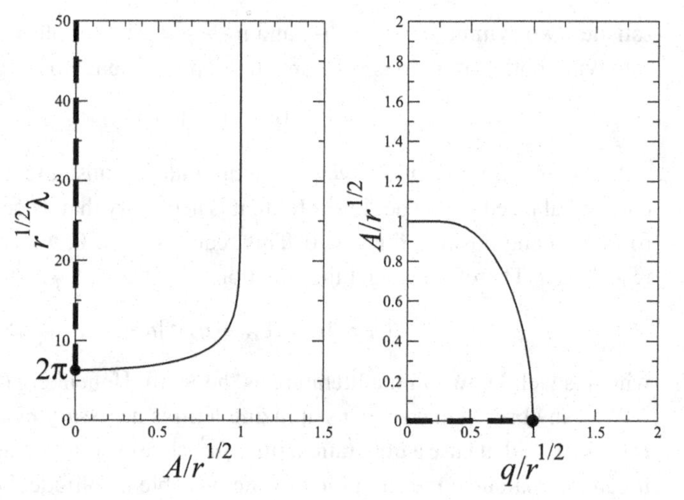

Fig. 7.7 The branch of steady states for TDGL, in the plane (λ, A) (left) and in the plane (A, q) (right). The dashed line intervals identify the unstable regions for the homogeneous solution, $u \equiv 0$.

amplitude $0 < A < \sqrt{r}$. (ii) The wavelength is an increasing function of A (because the potential is less and less steep). (iii) $\lambda(A \to 0) = 2\pi/\sqrt{r}$, which is the smallest unstable wavelength in the linear spectrum of the homogeneous solution, i.e., $\sigma_{\text{III}}(q) > 0 \; \forall q < 2\pi/\sqrt{r}$. (iv) $\lambda(A \to \sqrt{r}) = \infty$, because (in the mechanical analogy) the solution would correspond to the particle leaving the left maximum of the potential and reaching the right maximum of the potential after an infinite time (or the other way around). These limiting solutions correspond to the positive and negative kinks discussed in the previous chapter. For large, but finite λ, the periodic solution corresponds to a sequence of positive and negative kinks at distance $\lambda/2$.

The exact expression of $\lambda(A)$, derived in Appendix R, is

$$\lambda(A) = 4\sqrt{\frac{2}{2r - A^2}}\,K\left(\frac{A}{\sqrt{2r - A^2}}\right), \tag{7.71}$$

where $K(\cdots)$ is the complete elliptic integral of the first kind. In Fig. 7.7 we plot $\lambda(A)$ and $A(q)$ using rescaled variables so as to stress the intrinsic independence of these functions on r. In particular, the function $A(q)$ shows its similarity with the linear spectrum for the nonconserved model, $\omega(q) = r - q^2$, making evident that there is a periodic steady state for any q such that $\sigma_{\text{III}}(q) > 0$ and that the amplitude $A(q)$ vanishes when $\sigma(q)$ vanishes.

Let us now pass to the steady states of the CH model, Eq. (7.69), which satisfy the equation

$$u_{xx} + ru - u^3 = c_1 x + c_2. \tag{7.72}$$

The constant c_1 must vanish in order to get confined solutions. In this case the mechanical analogy is introduced by the asymmetric potential $\tilde{V}_{\text{CH}}(u) = \tilde{V}(u) - c_2 u$. Since CH dynamics conserves the order parameter and it is physically relevant to focus on symmetric steady states, i.e., such that their spatial average $\langle u(x) \rangle = 0$, we can choose $c_2 = 0$. Therefore, symmetric steady states for the CH model are the same steady states as for the TDGL model.

If we finally turn to the SH model, we realize that in this case it is not possible to find steady states analytically for general r. However, it is possible for vanishing r, because we can use a perturbative approach. Since the approach is fairly general, we discuss it in some detail using a formalism that is appropriate for a large class of equations having the form

$$\partial_t u = \mathcal{L}[u] - u^3. \tag{7.73}$$

In the absence of the nonlinear term the basic solution would be $u(x,t) = a(t)\cos(qx)$, with $\dot{a} = \sigma(q)a$. If q belongs to the unstable spectrum, i.e., $\sigma(q) > 0$, we need the nonlinear term to compensate for the linear, exponential growth. For a single harmonic we would have

$$u^3 = a^3(t)\cos^3(qx) = \frac{a^3(t)}{4}\left(3\cos(qx) + \cos(3qx)\right). \tag{7.74}$$

Therefore, the nonlinear term has a double effect: it contributes to the dynamics of the base harmonic through the term $\cos(qx)$ and it also induces the rising of higher order harmonics through the term $\cos(3qx)$. The first effect goes in the expected direction, because it counterbalances the unstable growth of $a(t)$. However, the appearance of the term $\cos(3qx)$ tells us we should modify the ansatz by writing

$$u(x,t) = a(t)\cos(qx) + a_3(t)\cos(3qx), \tag{7.75}$$

whose cube is now more involved,

$$\begin{aligned} u^3 = &\frac{3}{4}(a^3 + a^2 a_3 + 2aa_3^2)\cos(qx) + \frac{1}{4}(a^3 + 6a^2 a_3 + 3a_3^3)\cos(3qx) \\ &+ \frac{3}{4}(a^2 a_3 + aa_3^2)\cos(5qx) + \frac{3}{4}aa_3^2\cos(7qx) + \frac{1}{4}a_3^3\cos(9qx). \end{aligned} \tag{7.76}$$

As expected, even higher-order harmonics now appear, which implies we should write the whole series:

$$u(x,t) = a(t)\cos(qx) + \sum_{n \geq 1} a_{2n+1}(t)\cos((2n+1)qx). \tag{7.77}$$

The idea is that we can truncate the expansion (7.77) if $a \gg a_3 \gg a_5 \gg a_7 \cdots$. Let's now forget a_5 and higher-order harmonics and show that $a_3 \ll a$. So, let us approximate Eq. (7.76) with the first two terms on the right-hand side and replace them into Eq. (7.73). We get two coupled equations for $a(t)$ and $a_3(t)$,

$$\dot{a} = \sigma(q)a - \frac{3}{4}(a^3 + a^2 a_3 + 2aa_3^2), \tag{7.78}$$

$$\dot{a}_3 = \sigma(3q)a_3 - \frac{1}{4}(a^3 + 6a^2 a_3 + 3a_3^3). \tag{7.79}$$

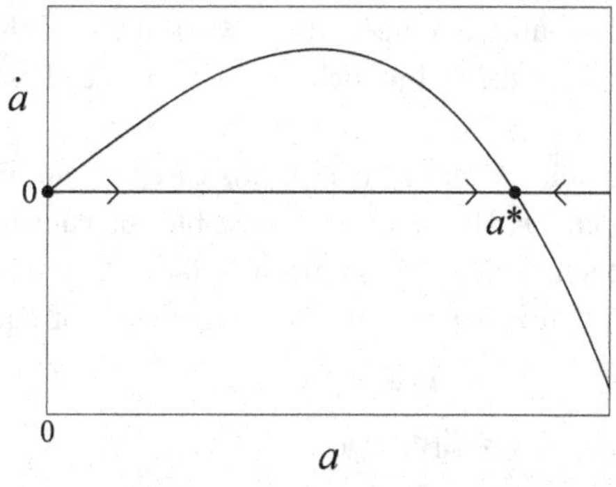

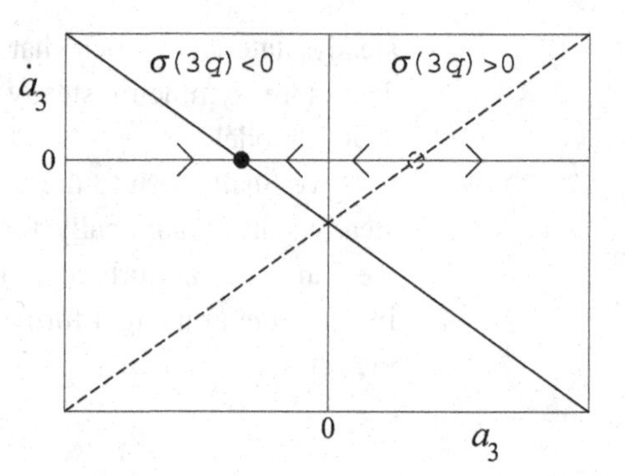

Fig. 7.8 Flows for $a(t)$ (see Eq. (7.80)) and $a_3(t)$ (see Eq. (7.81)).

As part of the hypothesis (to be confirmed) that $a_3 \ll a$, we also have $a_3^2 a \ll a^2 a_3 \ll a^3$, so that we can keep only nonlinear terms proportional to a^3,

$$\dot{a} = \sigma(q)a - \frac{3}{4}a^3, \tag{7.80}$$

$$\dot{a}_3 = \sigma(3q)a_3 - \frac{1}{4}a^3. \tag{7.81}$$

The equation for $a(t)$ has two fixed points (stationary solutions), $a = 0$ and $a = a^* = \sqrt{4\sigma(q)/3}$. The trivial one, $a = 0$, is unstable, because $\sigma(q) > 0$. The nontrivial solution, $a = a^*$ (which exists only if $\sigma(q) > 0$), is stable, as easily shown graphically in Fig. 7.8, left panel.

Let us now focus on the equation for the amplitude a_3. The key point is the sign of $\sigma(3q)$. In the SH equation close enough to the threshold, $3q$ is certainly outside the unstable interval, regardless of the value of q belonging to the unstable interval. Therefore, $\sigma(3q) < 0$ and the resulting (negative) fixed point $a_3 = a_3^* = (a^*)^3/(4\sigma(3q))$ is stable; see Fig. 7.8, right panel. If it was $\sigma(3q) > 0$, the now positive fixed point a_3^* would be unstable and we should take into account higher-order harmonics. The proof is completed once we remark that

$$\frac{a_3^*}{a^*} \simeq \frac{(a^*)^2}{\sigma(3q)} \simeq \frac{\sigma(q)}{\sigma(3q)}. \tag{7.82}$$

In conclusion, if $\sigma(q)$ is small and $\sigma(3q)$ is negative and not small, $|a_3^*/a^*| \ll 1$ and the above expansion is consistent.

We stress that the condition for the validity of the calculation (which goes beyond the specific SH equation) is twofold: $\sigma(q)$ must be small and $\sigma(3q)$ must be negative (and not small). These conditions are surely satisfied by the SH equation close to the threshold, i.e., for $r \to 0$, but they are also satisfied, e.g., by the TDGL equation for any r, if $q \to \sqrt{r}$; see Fig. 7.9.

A fairly similar perturbative approach can be performed for the CH equation, if $q \to \sqrt{r}$. In that case Eq. (7.73) is replaced by

$$\partial_t u = \mathcal{L}[u] + \partial_{xx}(u^3). \tag{7.83}$$

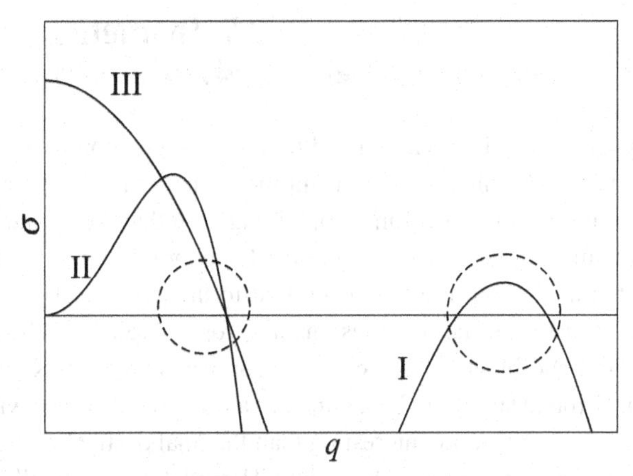

Fig. 7.9 Graphical sketch of the stability curves for the three models I, II, and III: the dashed areas highlight the q-regions, where the weakly nonlinear analysis can be applied.

If we start again from Eq. (7.75), Eq. (7.76) should be replaced by

$$-\partial_{xx}(u^3) = \frac{3q^2}{4}(a^3+a^2a_3+2aa_3^2)\cos(qx)+\frac{9q^2}{4}(a^3+6a^2a_3+3a_3^3)\cos(3qx), \tag{7.84}$$

where higher-order harmonics have been neglected.

We can now go directly to Eqs. (7.80) and (7.81) and rewrite them as

$$\dot{a} = \sigma(q)a - \frac{3q^2}{4}a^3 \tag{7.85}$$

$$\dot{a}_3 = \sigma(3q)a_3 - \frac{9q^2}{4}a^3. \tag{7.86}$$

Finally, we have

$$a^* = \sqrt{\frac{4\sigma(q)}{3q^2}} \tag{7.87}$$

$$a_3^* = \frac{9q^2(a^*)^3}{4\sigma(3q)} \tag{7.88}$$

$$\left|\frac{a_3^*}{a^*}\right| \simeq \frac{(a^*)^2}{|\sigma(3q)|} \simeq \frac{\sigma(q)}{|\sigma(3q)|} \ll 1. \tag{7.89}$$

We conclude that the same method can be consistently applied to the CH equation. In Fig. 7.9 we plot $\sigma(q)$ for the different models and highlight when the weakly nonlinear analysis can be applied: for model I, if $r \ll 1$, it can be applied to the whole unstable spectrum; for models II and III, it can be applied close to the upper limit of the unstable spectrum, for any r.

We should not be misled by this result and conclude that the above solutions are stable, because our analysis was limited to amplitude fluctuations. However, the above analysis is useful, because it proves the existence of periodic steady states in suitable limits and gives the explicit expression of these steady states.

7.5 Energetics

Dynamics can be accompanied by the minimization of a suitable Lyapunov functional. If so, we might think to circumvent the difficult task of determining the nonlinear evolution by minimizing such a functional, imagining that the "ground state" is the final outcome of dynamics. In fact, investigating the Lyapunov functional surely allows us to gain insights in the model, but it is not a shortcut to the dynamical problem for several reasons. First, as discussed in the previous chapter, for discrete models, metastable states may trap the system and block dynamics either forever (at $T = 0$ K) or for a long time. Second, the final ground state may be attained in a time that diverges with the system size; in this case, the way to it is more interesting than the final destination itself (think about the coarsening process occurring for TDGL and CH models). And finally, even transient dynamics may have its own interest.

For those who have not read Chapter 6 on phase-ordering, we start with the Lyapunov functional for the TDGL and CH equations; see Eqs. (7.4)–(7.6). It is straightforward to evaluate $\mathcal{F}_{\rm GL}$ for a periodic configuration of large wavelength, because both the "domain wall" term, $\frac{1}{2}(\partial_x u)^2$, and the excess potential energy with respect to the ground state, $V(u)-V(u \equiv \pm\sqrt{r})$, are concentrated in the finite region where the order parameter passes from a minimum of the potential to the other minimum. Therefore, we can write the average value of the free energy as

$$\langle \mathcal{F}_{\rm GL} \rangle = \frac{1}{\lambda}\int_0^\lambda dx \left[\frac{1}{2}u_x^2 + V(u) - V(\sqrt{r})\right] + V(\sqrt{r}) \tag{7.90}$$

$$= \frac{\rm const}{\lambda} + O\left(\frac{1}{\lambda^2}\right) + V(\sqrt{r}), \tag{7.91}$$

where we have used the fact that for large λ the integral $\int_0^\lambda[\cdots]$ of Eq. (7.90) is independent of λ. In conclusion, if we evaluate the average value of the Ginzburg–Landau free energy along the curve of the periodic steady state (see Fig. 7.7), energetic considerations based on the dynamical minimization of the Lyapunov functional would immediately lead us to establish a coarsening process where λ increases in time. Although this logical step is essentially true for TDGL/CH models, it is not true in general, as exemplified by the SH equation. So, we are now going to study the energetics for this equation, while the next sections are devoted to studying its dynamics.

The SH model has the variational structure

$$\begin{aligned}\frac{\partial u}{\partial t} &= \left[r - (\partial_{xx} + q_c^2)^2\right]u - u^3 \\ &\equiv -\frac{\delta \mathcal{F}_{\rm SH}}{\delta u},\end{aligned} \tag{7.92}$$

where

$$\mathcal{F}_{\rm SH} = \int dx \left[\left(q_c^4 - r\right)\frac{u^2}{2} + \frac{u^4}{4} - q_c^2 u_x^2 + \frac{1}{2}u_{xx}^2\right]. \tag{7.93}$$

The form of the pseudo free-energy $\mathcal{F}_{SH}$ is not easily understandable, in particular the negative sign of the "tension" term u_x^2. Another peculiar feature is related to the u-dependent, potential part, which has a positive quadratic term for small r; i.e., it is a single-well potential, not the classical double-well potential we are more used to. This is not surprising if we realize that the unstable q-interval of the homogeneous solution is centered in $q = q_c$ rather than in $q = 0$. In fact, for $r > q_c^4$ (strongly unstable regime, not considered here) the unstable interval extends up to $q = 0$ and the potential term gains the standard double-well form.

Let us now evaluate $\mathcal{F}_{SH}$ for a periodic steady state, in the approximation of small r, because in this limit we know the explicit expression of the steady state (see the previous section):

$$u(x) = a(q)\cos(qx) = \sqrt{\frac{4\sigma(q)}{3}}\cos(qx), \qquad \text{for } r \ll 1. \tag{7.94}$$

Since $\mathcal{F}_{SH}$ diverges in the thermodynamic limit, it is customary to evaluate its average value, as we did in Eq. (7.90),

$$\langle \mathcal{F}_{SH} \rangle = \frac{1}{L}\int_0^L dx[\cdots] = \frac{1}{\lambda}\int_0^\lambda dx[\cdots], \tag{7.95}$$

where the last equality is valid for periodic configurations of period λ. Using Eq. (7.93) we find

$$\langle \mathcal{F}_{SH} \rangle = \left(q_c^4 - r\right)\frac{\langle u^2 \rangle}{2} + \frac{\langle u^4 \rangle}{4} - q_c^2\langle u_x^2 \rangle + \frac{1}{2}\langle u_{xx}^2 \rangle, \tag{7.96}$$

where the average values can be evaluated using (7.94),

$$\begin{aligned}
\langle u^2 \rangle &= a^2\langle \cos^2(qx) \rangle = \frac{1}{2}a^2 \\
\langle u^4 \rangle &= a^4\langle \cos^4(qx) \rangle = \frac{3}{8}a^4 \\
\langle u_x^2 \rangle &= a^2q^2\langle \sin^2(qx) \rangle = \frac{q^2}{2}a^2 \\
\langle u_{xx}^2 \rangle &= a^2q^4\langle \cos^2(qx) \rangle = \frac{q^4}{2}a^2 .
\end{aligned} \tag{7.97}$$

Since $a^2(q) = \frac{4}{3}\sigma(q)$, we finally get

$$\begin{aligned}
\langle \mathcal{F}_{SH} \rangle &= -\frac{r}{4}a^2 + \frac{1}{4}a^2(q^2 - q_c^2)^2 + \frac{3}{32}a^4 \\
&= -\frac{1}{3}\sigma(q)\left[r - (q^2 - q_c^2)^2\right] + \frac{1}{6}\sigma^2(q) \\
&= -\frac{1}{6}\sigma^2(q),
\end{aligned} \tag{7.98}$$

where we have used the explicit expression for $\sigma(q)$, given in Table 7.1. Therefore, $\langle \mathcal{F}_{SH} \rangle$ is minimal where $\sigma(q)$ is maximal,[7] i.e., for $q = q_c$.

[7] Let us remember that steady states exist only where $\sigma(q) > 0$.

This result does not mean that the ground state, $u_{GS}(x) = a(q_c)\cos(q_c x)$ is attained, because the deterministic dynamics may stop at a stationary state, if it is locally stable. This is why it is important to perform a dynamical investigation, which is the goal of the next two sections. At the end we will resume this discussion on energetics versus dynamics.

7.6 Nonlinear Dynamics for Pattern-Forming Systems: The Envelope Equation

The instability scenario I, qualitatively depicted in Fig. 7.4, is the prototype of bifurcation giving rise to a pattern of well-defined wavelength, and the simplest equation characterized by such a scenario is the SH equation. However, what is most relevant here is the universality of the dynamics close to the threshold. We are going to derive the relevant equation, called the amplitude or envelope equation, starting from the SH model, and then we discuss it in terms of symmetries, which is the only way to bring universality out.

Type I bifurcation is characterized by an interval of unstable q-vectors of order $\sqrt{r}$ and growth rate of order $\sigma(q_c) = r$. In simple words, we expect to have a slow dynamics involving a weak modulation of an oscillation of wavelength $2\pi/q_c$. Dynamics is slow, because the time scale is set by the inverse of the growth rate, $\sigma(q_c)$, which is vanishingly small for $r \to 0$. The modulation is weak, because the active wave vectors, i.e., such that $\sigma(q) > 0$, have a small variability for $r \to 0$. The last point can be corroborated by a simple calculation. If we sum two oscillations,

$$u(x) = a_1 \cos(q_1 x) + a_2 \cos(q_2 x), \tag{7.99}$$

and define the mean wavevector $q_0 = (q_1 + q_2)/2$ and the half-difference $\delta = (q_1 - q_2)/2$, simple trigonometry gives

$$u(x) = (a_1 + a_2)\cos(\delta x)\cos(q_0 x) + (a_2 - a_1)\sin(\delta x)\sin(q_0 x). \tag{7.100}$$

In our problem $q_0 \simeq q_c$ and $\delta \simeq \sqrt{r} \ll q_c$, which means we have oscillations of wavevector q_0 with a weakly varying amplitude. With that in mind we can anticipate the form of the solution,

$$u(x,t) = A(x,t)e^{iq_c x} + \text{c.c.}, \tag{7.101}$$

where the amplitude $A(x,t)$ depends weakly on x and t. The correct formalism to treat this type of problem is multiscale analysis, which is briefly introduced in Appendix S.

The underlying idea (see Eq. (7.100)) is that the spatial variable x appears in the carrier wave as $q_c x$ and in the modulating amplitude as $\delta x \approx \sqrt{r}x$. It is therefore natural to introduce a fast scale $x_0 = x$ (because q_c is of order one) and a slow spatial scale $X = \sqrt{r}x$. As for time, the carrier wave is stationary, so no fast time scale appears, while the slow temporal scale is set by the product $\sigma(q_c)t \approx rt$, so we can define $T = rt$. We should finally remark that the amplitude A itself is small, close to the threshold: $A \approx u^* \approx \sqrt{\sigma(q_c)} \approx \sqrt{r}$.

To avoid handling square roots when performing the perturbative expansion, it is customary to put $r = \epsilon^2$. Under this notation, we can define the variables

$$\begin{aligned} &x_0 = x, && \text{fast scales} \\ &X = \epsilon x \quad T = \epsilon^2 t, && \text{slow scales} \end{aligned} \tag{7.102}$$

and perform a standard expansion of the solution u,

$$u = \epsilon u_1 + \epsilon^2 u_2 + \epsilon^3 u_3. \tag{7.103}$$

In fact, since the leading nonlinear term is of order ϵ^3 we expect the ϵ-expansion of u should arrive at such order.

We can now evaluate the time and space derivatives,

$$\partial_t = \epsilon^2 \partial_T \quad \text{and} \quad \partial_x = \partial_{x_0} + \epsilon \partial_X, \tag{7.104}$$

and apply them to the SH equation,

$$\partial_t u = [r - (\partial_{xx} + q_c^2)^2] u - u^3. \tag{7.105}$$

Finally, we can write

$$\epsilon^2 \partial_T u = \epsilon^2 u - \left((\partial_{x_0} + \epsilon \partial_X)^2 + q_c^2\right)^2 u - u^3 \tag{7.106}$$

$$\begin{aligned} &= \epsilon^2 u - \Big\{ \left(\partial^2_{x_0^2} + q_c^2\right)^2 + 4\epsilon \partial^2_{x_0 X}\left(\partial^2_{x_0^2} + q_c^2\right) \\ &\quad + \epsilon^2 \Big[2\partial^2_{XX}\left(\partial^2_{x_0^2} + q_c^2\right) + \left(2\partial^2_{x_0 X}\right)^2 \Big] \Big\} u - u^3 \end{aligned} \tag{7.107}$$

and expanding u we get

$$\begin{aligned} \epsilon^3 \partial_T u_1 &= \epsilon^3 u_1 - \left(\partial^2_{x_0^2} + q_c^2\right)^2 (\epsilon u_1 + \epsilon^2 u_2 + \epsilon^3 u_3) \\ &\quad - 4\epsilon \partial^2_{x_0 X}\left(\partial^2_{x_0^2} + q_c^2\right)(\epsilon u_1 + \epsilon^2 u_2) \\ &\quad - \epsilon^3 \left[2\partial^2_{XX}\left(\partial^2_{x_0^2} + q_c^2\right) + \left(2\partial^2_{x_0 X}\right)^2 \right] u_1 - \epsilon^3 u_1^3 + O(\epsilon^4) \end{aligned} \tag{7.108}$$

We should now proceed order by order.

Order ϵ

$$\left(\partial^2_{x_0^2} + q_c^2\right)^2 u_1 \equiv \mathcal{L}_0[u_1] = 0, \tag{7.109}$$

where we denote with $\mathcal{L}_0$ the unperturbed part of the linear evolution operator defined in Eq. (7.13). The general solution of (7.109) is

$$u_1 = A(X, T) e^{i q_c x_0} + \text{c.c.} \tag{7.110}$$

It is worth noting that this result, in terms of standard variables u, x, t means that

$$u(x, t) = \epsilon A(\epsilon x, \epsilon^2 t) e^{i q_c x} + \text{c.c.} + O(\epsilon^2). \tag{7.111}$$

This expression is clearly in agreement with the result for the steady state found in the previous section, as we are going to argue at the end of this analysis.

Order ϵ^2

$$-\mathcal{L}_0[u_2] - 4\partial^2_{x_0X}\left(\partial^2_{x_0^2} + q_c^2\right)u_1 = 0, \tag{7.112}$$

which implies

$$\mathcal{L}_0[u_2] = 0, \tag{7.113}$$

because the physically relevant solutions of the equation $(\partial^2_{x_0^2} + q_c^2)^2 u_1 = 0$ actually satisfy the stronger condition $(\partial^2_{x_0^2} + q_c^2)u = 0$. Finally, using that $\mathcal{L}_0[u_{1,2}] = 0$, at next order we obtain

Order ϵ^3

$$\partial_T u_1 = u_1 - \mathcal{L}_0[u_3] - \left(2\partial^2_{x_0X}\right)^2 u_1 - u_1^3 \tag{7.114}$$

or, equivalently

$$\mathcal{L}_0[u_3] = -\partial_T u_1 + u_1 - \left(2\partial^2_{x_0X}\right)^2 u_1 - u_1^3 \equiv f_3(x_0, X, T). \tag{7.115}$$

Since f_3 contains terms that are proportional to $e^{\pm iq_cx_0}$, which are solutions of the homogeneous equation $\mathcal{L}_0[u_3] = 0$, we must require that such terms vanish. Therefore, let us evaluate f_3 with $u_1 = Ae^{iq_cx_0} + A^*e^{-iq_cx_0}$,

$$\begin{aligned} f_3 = &- \left(\partial_T A e^{iq_cx_0} + \partial_T A^* e^{-iq_cx_0}\right) + \left(Ae^{iq_cx_0} + A^*e^{-iq_cx_0}\right) \\ &+ 4q_c^2\left(\partial^2_{XX}Ae^{iq_cx_0} + \partial^2_{XX}A^*e^{-iq_cx_0}\right) - \left(Ae^{iq_cx_0} + A^*e^{-iq_cx_0}\right)^3 \qquad (7.116) \\ = &\left(-\partial_T A + A + 4q_c^2\partial^2_{XX}A - 3|A|^2A\right)e^{iq_cx_0} - A^3e^{i3q_cx_0} + \text{c.c.} \qquad (7.117) \end{aligned}$$

The vanishing of the term between brackets in the last line implies the so-called envelope (or amplitude) equation,

$$\frac{\partial A}{\partial T} = A(1 - 3|A|^2) + 4q_c^2\frac{\partial^2 A}{\partial X^2}. \tag{7.118}$$

It is worth stressing the generality of this equation, which depends not on the explicit form of the SH equation (7.67) but on its symmetries and, of course, on the type I linear spectrum. Let us show how the amplitude equation may be determined using the two main symmetries of the SH equation: the spatial inversion symmetry, $x \to -x$, and the translational invariance, $x \to x + \bar{x}$. Eq. (7.118) has been derived from the starting ansatz

$$u(x, t) = A(X, T)e^{iq_cx_0} + \text{c.c.}, \tag{7.119}$$

where $x_0 = x$, $X = \epsilon x$, and $T = \epsilon^2 t$. The symmetry $x \to -x$ implies that the envelope equation for $A(X, T)$ must satisfy the same symmetry, $X \to -X$. Furthermore, since $x \to -x$ implies $x_0 \to -x_0$, A^* must satisfy the same equation. Finally, the translational invariance means that $A\exp(iq_c\bar{x})$ must be a solution, if A is. We can now write the simplest equation for $A(X, T)$ satisfying these constraints,

$$\partial_T A = c_1 A + c_2\partial_{XX}A + c_3|A|^2A, \tag{7.120}$$

where all coefficients c_i must be real. The signs of the coefficients of the linear terms are fixed by the condition that $A = 0$ is linearly unstable for large wavelength, so both c_1 and c_2

must be positive, while the coefficient of the nonlinear term must be negative in order to provide nonlinear saturation to the instability. The actual value of the coefficients is not relevant, because they can all be set equal to unity ($c_1 = c_2 = -c_3 = 1$) by rescaling X, T, and A. Finally, we remark that we have not considered the possible symmetry $u \to -u$. This means that Eq. (7.118) is still valid if we add a quadratic term to the SH equation (7.67).

Let us conclude this section by verifying that the amplitude equation (7.118) allows for simple steady state solutions, already found in the previous section. If we assume that $A(X) = A_0 e^{iQX}$ and replace it in Eq. (7.118), we find

$$|A_0|^2 = \frac{1 - 4q_c^2 Q^2}{3}, \tag{7.121}$$

so that the full solution, restoring old variables, is

$$u(x) = \epsilon A_0 e^{i(q_c + \epsilon Q)x} + \text{c.c.} \tag{7.122}$$

If we recognize that $q = q_c + \epsilon Q$, we can also write

$$\sigma(q) = r - (q^2 - q_c^2)^2 \tag{7.123}$$

$$\simeq \epsilon^2 - 4q_c^2 (q - q_c)^2 \tag{7.124}$$

$$= \epsilon^2 \left(1 - 4q_c^2 Q^2\right) \tag{7.125}$$

so that (apart from a constant, irrelevant phase factor)

$$\epsilon A_0 = \sqrt{\frac{\sigma(q)}{3}} \tag{7.126}$$

and

$$u(x) = \sqrt{\frac{\sigma(q)}{3}} e^{iqx} + \text{c.c.} = \sqrt{\frac{4\sigma(q)}{3}} \cos(qx), \tag{7.127}$$

which is the result found with a different approach in the previous section (see Eq. (7.94)). The key point here is that Eq. (7.118) gives more than the steady state; it gives the dynamics around the steady states, which allows us to determine their stability.

7.7 The Eckhaus Instability

We can now address the stability problem of the steady states of the amplitude equation. Let us rewrite the envelope equation,

$$\frac{\partial A}{\partial T} = A(1 - 3|A|^2) + 4q_c^2 \frac{\partial^2 A}{\partial X^2}, \tag{7.128}$$

whose time-independent solutions are

$$A_s(X) = \sqrt{\frac{1 - 4q_c^2 Q^2}{3}} e^{iQX} \equiv A_0 e^{iQX}. \tag{7.129}$$

We are going to perturb these solutions and study the resulting dynamics, retaining only the terms that are linear in the perturbations. However, rather than writing $A(X,T) = A_s(X) + \Delta(X,T)$, we perturb separately the amplitude and the phase of $A_s(X)$,

$$A(X,T) = A_0(1+\delta)e^{i(QX+\phi)} \tag{7.130}$$

where both $\delta = \delta(X,T)$ and $\phi = \phi(X,T)$ are real.

Keeping only terms that are linear in δ, ϕ, we can write

$$|A|^2 = A_0^2(1+2\delta) \tag{7.131}$$

$$\partial_T A = A_0 e^{i(qX+\phi)}\left[\frac{\partial\delta}{\partial T} + i\frac{\partial\phi}{\partial T}\right] \tag{7.132}$$

$$\partial_{XX} A = A_0 e^{i(qX+\phi)}\left[\frac{\partial^2\delta}{\partial X^2} - Q^2(1+\delta) - 2Q\frac{\partial\phi}{\partial X} + i\frac{\partial^2\phi}{\partial X^2} + 2iQ\frac{\partial\delta}{\partial X}\right]. \tag{7.133}$$

Assembling them in the envelope equation (7.130) and imposing the vanishing of the real and imaginary parts, we obtain two coupled, linear differential equations,

$$\frac{\partial\delta}{\partial T} = -6A_0^2\delta + 4q_c^2\frac{\partial^2\delta}{\partial X^2} - 8q_c^2 Q\frac{\partial\phi}{\partial X} \tag{7.134a}$$

$$\frac{\partial\phi}{\partial T} = 4q_c^2\frac{\partial^2\phi}{\partial X^2} + 8q_c^2 Q\frac{\partial\delta}{\partial X}. \tag{7.134b}$$

The stability analysis performed in Section 7.4 corresponds to perturb only the amplitude, making it dependent on time, which means that ϕ = constant and $\delta = \delta(T)$. In this case Eqs. (7.134) reduce to $\partial_T\delta = -6A_0^2\delta$; i.e., the solution $A_s(X)$ is stable for pure amplitude fluctuations. An equally simple analysis is not possible for the phase, because $\delta = 0$ implies that ϕ is a constant. In simple terms, the dynamics of the phase must take into account variations of the amplitude. It is possible to simplify the above equations to obtain an effective one for the phase only, but we will do that at the end of a more rigorous analysis.

The coupled linear equations (7.134) are studied by standard Fourier analysis, i.e., assuming

$$\delta = \bar{\delta}e^{\alpha T}e^{iKX} \tag{7.135}$$

$$\phi = \bar{\phi}e^{\alpha T}e^{iKX}, \tag{7.136}$$

which give a (2×2) linear system,

$$\begin{aligned}(\alpha + 6A_0^2 + 4q_c^2K^2)\bar{\delta} + 8iq_c^2QK\bar{\phi} &= 0\\ -8iq_c^2QK\bar{\delta} + (\alpha + 4q_c^2K^2)\bar{\phi} &= 0.\end{aligned} \tag{7.137}$$

For $K = 0$ we get a more formal version of the discussion of the previous paragraph, because the linear system simplifies to

$$(\alpha + 6A_0^2)\bar{\delta} = 0 \quad\text{and}\quad \alpha\bar{\phi} = 0, \tag{7.138}$$

whose two solutions are

$$\alpha = \alpha_1 = -6A_0^2 \qquad \text{with} \quad (\bar{\delta}, \bar{\phi}) = (1, 0) \tag{7.139}$$

$$\alpha = \alpha_2 = 0 \qquad \text{with} \quad (\bar{\delta}, \bar{\phi}) = (0, 1). \tag{7.140}$$

The first solution corresponds to stable amplitude fluctuations, the second solution to neutral phase fluctuations. When $K \neq 0$, we have two branches, $\alpha_{1,2}(K)$. Our goal is now to study $\alpha_2(K)$ to evaluate its sign at finite K.

The system (7.137) has solutions only if the determinant vanishes, i.e., if

$$(\alpha + 6A_0^2 + 4q_c^2K^2)(\alpha + 4q_c^2K^2) - (8q_c^2QK)^2 = 0 \tag{7.141}$$

or

$$\alpha^2 + 2\alpha(4q_c^2K^2 + 3A_0^2) + 4q_c^2K^2(6A_0^2 + 4q_c^2K^2 - 16q_c^2Q^2) = 0, \tag{7.142}$$

which gives the two branches

$$\alpha_{1,2} = -(4q_c^2K^2 + 3A_0^2) \mp \left[(4q_c^2K^2 + 3A_0^2)^2 - 8q_c^2K^2(1 + 2q_c^2K^2 - 12q_c^2Q^2)\right]^{1/2}. \tag{7.143}$$

The branch $\alpha_1(K)$, corresponding to the minus sign, is negative for any K, therefore giving stable dynamics. Instead, we are interested in the branch $\alpha_2(K)$, corresponding to the plus sign and such that $\alpha_2(0) = 0$. For large K, it is straightforward to check that Eq. (7.141) simplifies to $(\alpha + 4q_c^2K^2)^2 = 0$, giving $\alpha_{1,2} = -4q_c^2K^2$. Therefore, a possible instability may arise only from a positive α_2 at small K. In this limit we can neglect K^4 terms, so that the quantity in square brackets of Eq. (7.143), [. . .], and its square root are written as

$$[\cdots] = (1 - 4q_c^2Q^2)^2 + (8q_c^2QK)^2 \tag{7.144}$$

$$[\cdots]^{1/2} = (1 - 4q_c^2Q^2) + \frac{1}{2}\frac{(8q_c^2Q)^2}{1 - 4q_c^2Q^2}K^2. \tag{7.145}$$

Finally, we obtain

$$\alpha_2(K) = -4q_c^2\left(\frac{1 - 12q_c^2Q^2}{1 - 4q_c^2Q^2}\right)K^2 \equiv -4q_c^2\frac{\mathcal{N}(q)}{\mathcal{D}(q)}K^2. \tag{7.146}$$

Let us start by discussing the role of the different wavevectors appearing here: q_c, Q, and K.

- q_c appears in the SH equation and it is the critical wavevector: the dispersion relation of the linear stability analysis of the solution $u = 0$ gives $\sigma(q) = \epsilon^2 - (q^2 - q_c^2)^2$, therefore $\sigma > 0$ and the solution is unstable in the vicinity of $q = q_c$.
- Q labels the steady states appearing in the range where $\sigma(q) > 0$ and is related to q via the relation $q = q_c + \epsilon Q$. The explicit expression of the steady state is given in Eq. (7.129). The quantity at the denominator of Eq. (7.146) is proportional to $\sigma(q)$, therefore it is positive in the range of application of such equation.
- The stability of steady solutions labeled by Q has been studied in a linear approximation, where the perturbation has spatial and time dependences of the form e^{iKX} and $e^{\alpha T}$,

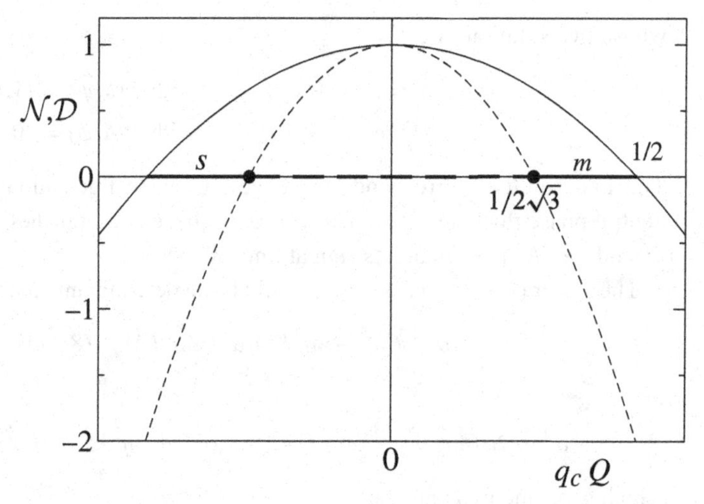

Fig. 7.10 $\mathcal{D}(q)$, solid line, and $\mathcal{N}(q)$, dashed line (see Eq. (7.146)). Thick dashed segment indicates Eckhaus stable steady states. Thick solid segments (regions *s* and *m*) indicate Eckhaus unstable steady states.

where $\alpha = \alpha(K)$ much in the same way that $\sigma = \sigma(q)$ for the stability analysis of the homogeneous solution $u = 0$. The main difference is that we have two branches, $\alpha_{1,2}(K)$, representing amplitude fluctuations (α_1) and phase fluctuations (α_2), with $\alpha_1(K) < 0$. Eq. (7.146) gives the spectrum of phase fluctuations at small but finite K.

We can now come back to $\alpha_2(K)$ (see Eq. (7.146)) and draw the relevant conclusions. Its sign depends on the numerator $\mathcal{N}(q)$. If Q is small enough, i.e., q is very close to the critical value q_c, $\alpha_2 < 0$ and the steady state is stable. On the contrary, if Q is "large" (but such that the denominator $\mathcal{D}(q)$ is positive, of course), then $\alpha_2(K)$ is positive and the steady state is unstable. In Fig. 7.10 we plot $\mathcal{D}(q)$ and $\mathcal{N}(q)$. $\mathcal{D}(q)$ is essentially equal to $\sigma(q)$ in the limit $\epsilon \to 0$; see Eq. (7.125). Steady state solutions exist if and only if $\mathcal{D}(q) > 0$ and they are stable if $\mathcal{N}(q) > 0$, i.e., for $q_c Q < 1/2\sqrt{3}$ (thick, dashed line segment), otherwise they are unstable (thick, solid line segments).

Since the stability depends on the branch $\alpha_2(K)$, which refers to phase fluctuations, the Eckhaus instability is a *phase* instability, leading to changes in the local wavevector of the structure. If the wavevector is too small (region *s*), it will tend to increase by the *splitting* of a cell in two (a cell is the spatial region corresponding to a single period of oscillation). Conversely, if the wavevector is too large (region *m*), it will tend to decrease by the *merging* of two neighboring cells. We should not think these processes (splitting and merging) involve all cells: if so, the wavevector of the structure would double (splitting) or it would halve (merging), leading the system to a region where the homogeneous solution is stable, which is not possible. We should rather think of local processes occurring, where cells, because of fluctuations, are smaller or larger than the optimal value, corresponding to q_c.

The above qualitative discussion of the Eckhaus instability can be accompanied by energetic considerations, according to which the minimum of the Lyapunov functional is attained for $q = q_c$, i.e., for $Q = 0$. Stability analysis tells us that a finite region around

$Q = 0$ is linearly stable, which may prevent the system from attaining the "ground state." In fact, we should distinguish between an adiabatic change of the control parameter and a sudden increase to a small but finite value of r. In the former case cells of the optimal value $q = q_c$ appear as soon as $r = 0^+$ and their size is maintained afterward. In the latter case, the system passes abruptly from a stable to an unstable state, characterized by a finite unstable interval. Because of fluctuations, the instability will lead the system to display cells of various size, and the splitting/merging process described above is visible. This readjustment of the local size is called Eckhaus instability and is a secondary instability, because it occurs after the Rayleigh–Bénard instability (the primary one) has led to passing from the homogeneous (conductive) state to a nonhomogeneous (convective) state. In the absence of thermal noise the system may be frozen in the Eckhaus stable region without attaining the ground state, i.e., $q = q_c$.

We can now go back to coupled Eqs. (7.134). As anticipated, we cannot study phase dynamics assuming a constant amplitude, $\delta = 0$. However, we know that for any given, constant value of the phase, $\phi = \phi_0$, the amplitude is stable, while phase dynamics has a diffusive character. This means that amplitude dynamics is *enslaved* to phase dynamics, and on the time scale relevant for phase dynamics, the amplitude can be considered as constant, so $\partial_T \delta = 0$ in Eq. (7.134a). Furthermore, we can focus on small K, so the term proportional to $\partial_{XX}\delta$ can be neglected with respect to the term proportional to δ. We are left with the simplified Eq. (7.134a),

$$-6A_0^2\delta - 8q_c^2 Q\frac{\partial\phi}{\partial X} = 0, \tag{7.147}$$

which gives the "slavery relationship" between δ and ϕ,

$$\delta = -\frac{4q_c^2 Q}{3A_0^2}\frac{\partial\phi}{\partial X}. \tag{7.148}$$

Finally, if we replace it in Eq. (7.134b), we get

$$\frac{\partial\phi}{\partial T} = 4q_c^2\frac{\partial^2\phi}{\partial X^2} + 8q_c^2 Q\frac{\partial\delta}{\partial X} \tag{7.149}$$

$$= \left(4q_c^2 - 8q_c^2 Q\frac{4q_c^2 Q}{3A_0^2}\right)\frac{\partial^2\phi}{\partial X^2} \tag{7.150}$$

$$= D(Q)\frac{\partial^2\phi}{\partial X^2}, \tag{7.151}$$

where the phase diffusion coefficient, $D(Q)$, is

$$D(Q) = 4q_c^2\frac{1 - 12q_c^2Q^2}{1 - 4q_c^2Q^2}. \tag{7.152}$$

This quantity is the same as that appearing in Eq. (7.146), where

$$\alpha_2(K) = -D(Q)K^2, \tag{7.153}$$

so the condition for phase instability, $\alpha_2(K) > 0$, is equivalent to having a negative phase diffusion coefficient, $D(Q) < 0$.

7.8 Phase Dynamics

All models we have considered so far, TDGL, CH, and SH, are characterized by a branch of periodic steady states, $u_s(x,q) = u_s(x+\frac{2\pi}{q}, q)$, for all q that make the homogeneous solution $u = 0$ unstable, i.e., $\sigma(q) > 0$. These solutions have been shown in suitable limits to be stable against amplitude fluctuations, and all relevant dynamics is phase dynamics, with local changes of the wavelength. This process may produce an endless coarsening – this is the case for TDGL and CH – or a transient rearrangement until the wavelength is in the stable domain – this is the case for SH. The stability of the amplitude implies its dynamics is adiabatically enslaved to phase dynamics. The key point is that phase dynamics is slow, which allows a multiscale perturbative treatment. Let us investigate why.

Translational invariance means that $u_s(x + c, q)$ is still a solution for any constant c. Therefore, if c acquires a weak spatial dependence such that c is almost constant on a scale $\lambda = 2\pi/q$, we expect the resulting dynamics to be slow. In a slightly more formal way, we can assume that dynamics occurs in the functional space

$$u(x,t) = u_s(x, q(X,T)), \tag{7.154}$$

where, as before, $X = \epsilon x$ and $T = \epsilon^2 t$. Two main features of this approach to phase dynamics should be stressed: first, it is applicable to patterns arbitrarily far from the threshold, and second, the meaning of ϵ is not obvious. In fact, we are used to perturbative treatments, where a small parameter appears in the differential equation or in the Hamiltonian, while here it does not appear explicitly. Its meaning is, rather, related to the weakness of the perturbation, or better still to the weakness of the variability of q. We are now going to be more precise.

The steady state $u_s(x,q)$ is defined by a constant q, and the phase of the pattern is simply $\phi = qx$. In the general case of a (slowly) varying q, we can define the phase

$$\phi = \int dx\, q \tag{7.155}$$

so that $q = \partial_x \phi$. We also define the slow phase

$$\psi = \epsilon\phi, \tag{7.156}$$

so $q = \partial_X \psi$ (actually, both ψ and X are of order ϵ and their ratio is of order one). The expansion of the order parameter is written as

$$u(x,X,T) = u_0(\phi, q(X,T)) + \epsilon u_1(\phi, q(X,T)) \tag{7.157}$$

and derivatives are written as

$$\partial_x = q\partial_\phi + \epsilon\partial_X \tag{7.158}$$

$$\partial_{xx} = q^2\partial_{\phi\phi} + \epsilon(q^2\partial_\phi\partial_X + \partial_X q\partial_\phi) + O(\epsilon^2) \tag{7.159}$$

$$= q^2\partial_\phi\partial_X + \epsilon(2q\partial_X + (\partial_X q))\partial_\phi \tag{7.160}$$

$$= q^2\partial_\phi\partial_X + \epsilon\psi_{XX}(1 + 2q\partial_q)\partial_\phi \tag{7.161}$$

$$\partial_t = \epsilon^2 \partial_{T\phi} \partial_\phi + \epsilon^2 \partial_T \quad (7.162)$$

$$= \epsilon \partial_T \psi + O(\epsilon^2). \quad (7.163)$$

We can now apply this method to any model displaying a branch of steady states. We could do that for the SH equation and we would get a result similar to, but more general than Eq. (7.151), because it would be valid for any value of the control parameter. However, we will illustrate the method and its power with a simple generalization of the TDGL equation,

$$\partial_t u = u_{xx} + F(u), \quad (7.164)$$

the standard model corresponding to $F(u) = u - u^3$.

If we now use the above expressions for derivatives and the u-expansion, we obtain

$$\epsilon \frac{\partial \phi}{\partial T} \partial_\phi u_0 = \left[q^2 \partial_{\phi\phi} + \epsilon \psi_{XX} (1 + 2q\partial_q) \partial_\phi \right] (u_0 + \epsilon u_1) + F(u_0 + \epsilon u_1) \quad (7.165)$$

$$= \left[q^2 \partial_{\phi\phi} u_0 + F(u_0) \right]$$

$$+ \epsilon \left\{ \left[q^2 \partial_{\phi\phi} + F'(u_0) \right] u_1 + \psi_{XX} (1 + 2q\partial_q) \partial_\phi u_0 \right\} + O(\epsilon^2) \quad (7.166)$$

The vanishing of the zero-order term,

$$\mathcal{N}[u_0] \equiv q^2 \partial_{\phi\phi} u_0 + F(u_0) = 0, \quad (7.167)$$

trivially corresponds to the condition defining steady states, so $u_0(\phi, q) = u_s(\phi = qx, q)$. The vanishing of the first-order term gives instead

$$\frac{\partial \phi}{\partial T} \partial_\phi u_0 = \left[q^2 \partial_{\phi\phi} + F'(u_0) \right] u_1 + \psi_{XX} (1 + 2q\partial_q) \partial_\phi u_0, \quad (7.168)$$

which is more transparent if rewritten as

$$\left[q^2 \partial_{\phi\phi} + F'(u_0) \right] u_1 = \frac{\partial \phi}{\partial T} \partial_\phi u_0 - \psi_{XX} (1 + 2q\partial_q) \partial_\phi u_0. \quad (7.169)$$

It therefore has the form

$$\mathcal{L}[u_1] = g(u_0, \psi_T, \psi_{XX}), \quad (7.170)$$

where the linear operator

$$\mathcal{L} \equiv q^2 \partial_{\phi\phi} + F'(u_0) \quad (7.171)$$

is the so-called Fréchet derivative of the operator $\mathcal{N}[u_0]$ defined in Eq. (7.167). In fact, if we take the derivation of such equation with respect to ϕ, we get

$$q^2 \partial_{\phi\phi\phi} u_0 + F'(u_0) \partial_\phi u_0 = \mathcal{L}[\partial_\phi u_0] = 0. \quad (7.172)$$

According to the Fredholm alternative theorem (see Appendix S), Eq. (7.170) has a solution only if g, defined as the right-hand side of Eq. (7.169), is orthogonal to $\partial_\phi u_0$,

$$\langle (\partial_\phi u_0) g \rangle \equiv \frac{1}{2\pi} \int_0^{2\pi} d\phi (\partial_\phi u_0) g = 0. \quad (7.173)$$

We must therefore require that

$$\left\langle (\partial_\phi u_0)^2 \right\rangle \partial_T \psi - \langle \partial_\phi u_0 (1 + 2q\partial_q) \partial_\phi u_0 \rangle \partial_{XX} \psi = 0, \tag{7.174}$$

which implies

$$\left\langle (\partial_\phi u_0)^2 \right\rangle \partial_T \psi = \partial_q \left\langle q (\partial_\phi u_0)^2 \right\rangle \partial_T \psi. \tag{7.175}$$

We finally get a diffusion equation for the slow phase ψ,

$$\frac{\partial \psi}{\partial T} = \frac{\partial_q \langle q(\partial_\phi u_0)^2 \rangle}{\langle (\partial_\phi u_0)^2 \rangle} \frac{\partial^2 \psi}{\partial X^2} \equiv D(q) \frac{\partial^2 \psi}{\partial X^2}, \tag{7.176}$$

where $D(q)$ is the phase diffusion coefficient and its sign is related to the stability ($D(q) > 0$) or instability ($D(q) < 0$) of the pattern of period $\lambda = 2\pi/q$ with respect to large wavelength modulations. Since it has the form $D(q) = (\partial_q D_1(q))/D_2(q)$, with $D_1(q) = \langle q(\partial_\phi u_0)^2 \rangle$ and $D_2(q) = \langle (\partial_\phi u_0)^2 \rangle > 0$, the sign of D depends on the increasing or decreasing behavior of $D_1(q)$. Before analyzing it, let us remark that the above procedure allows us to gain dynamical information from the analysis of the branch of steady states.

We have applied this method to Eq. (7.164), because its stationary solutions are particularly simple. In fact, the condition of time independence writes $u_{xx} + F(u) = 0$, which is Newton's equation of motion for a fictitious particle with position u at time x and subject to the force $-F(u)$, i.e., moving in the potential $V(u) = \int du F(u)$. For the standard TDGL equation, $F(u) = u - u^3$ and $V(u) = u^2/2 - u^4/4$, but here the only hypothesis is that the trivial solution $u \equiv 0$ is linearly unstable, which means that $F(u) \simeq c_1 u$ for small u, with $c_1 > 0$. In other words, $V(u)$ is a harmonic potential for small oscillations, but we make no assumptions about the nonlinear part of $F(u)$.

We are now going to prove that the phase diffusion coefficient $D(q)$ (see Eq. (7.176)) is related to the amplitude dependence of the wavelength of steady states.[8] If we revert to the old variable x, we find

$$D_1 = \langle q(\partial_\phi u_0)^2 \rangle = \frac{1}{2\pi} \int_0^{2\pi} d\phi q (\partial_\phi u_0)^2 = \frac{1}{2\pi} \int_0^{\lambda} dx (\partial_x u_0)^2 \equiv \frac{J}{2\pi}, \tag{7.177}$$

where J is the well-known action variable. In a similar way, we find $D_2 = (2\pi)^{-2} \lambda J$. Following classical mechanics results, the derivative of J with respect to the energy of the particle gives the period of oscillation, λ, so the following expression is finally established

$$D(q) = -\frac{\lambda^2 F(A)}{J(\partial_A \lambda)}, \tag{7.178}$$

where A is the amplitude of the oscillation, i.e., the (positive) maximal value of $u_0(x)$.

In conclusion, the stability is ruled by $\partial_A \lambda$. If the wavelength of the steady states is an increasing function of the amplitude, the steady states are unstable and a coarsening process takes place; otherwise, no phase instability occurs, the wavelength remains constant, and the amplitude diverges. This two scenarios are easily understood if $F(u) = c_1 u + c_3 u^3$, because when $c_3 < 0$ it is characteristic of the TDGL case, while for $c_3 > 0$ there

[8] In the mechanical analogy, this means the amplitude dependence of the period of oscillation of the particle.

is no nonlinear saturation of the linear stability. However, the approach is valid for *any* function $F(u)$, including the cases where $V(u)$ is such that $\partial_A \lambda$ changes sign. Furthermore, the above approach allows us to find the asymptotic coarsening law, $L \simeq t^n$, through a simple dimensional ansatz,

$$|D(q)| \sim \frac{L^2}{t}, \tag{7.179}$$

where $q = 2\pi/L$. If we apply this ansatz to the TDGL equation, $F(u) = u - u^3$, we recover the logarithmic coarsening, found in the previous chapter.

7.9 Back to Experiments

This chapter, more than the others, has referred to a few experimental examples in order to make explicit the meaning of phrases such as pattern formation and (in this context) control parameter and order parameter. We did not want to limit ourselves to a purely formal enunciation of such concepts. Although it is not our intention to provide a theory for specific pattern-forming systems, we want to close the chapter by going back to some classes of experiments.

Experiments on the convective instability in a fluid heated from below are fairly old and we would like to stress the difference between the setup where the fluid is confined and where its upper surface is free. In the confined geometry case we speak of Rayleigh–Bénard instability, which is due to the temperature dependence of the density of the fluid, which favors the exchange of the lower/warmer/lighter parcels with the upper/colder/heavier parcels. In the nonconfined geometry, the upper surface of the fluid is free and we speak of Bénard–Marangoni instability. Now the convection is triggered by the temperature dependence of the surface tension at the free surface, which is in contact with the air. According to the Marangoni effect, the gradient of surface tension induces a mass transfer at the interface, but mass conservation requires a simultaneous transfer of mass in the vertical direction, which causes the rising of convection cells.

Two main types of cells appear in convection experiments, rolls and hexagons, which differ in up/down symmetry (it is present in rolls and absent in hexagons). In fact, in the latter case the fluid rises (falls) in the center of the hexagons and falls (rises) at the edges. In standard conditions, Rayleigh–Bénard cells are roll-like, while hexagons appear when the up/down symmetry is broken by boundary conditions, either directly (the case of Bénard–Marangoni instability) or indirectly, because, e.g., the viscosity depends on temperature. Although convection cells have a constant wavelength, their orientation may vary from one area to another, and, for example, domains with rolls of different orientation may appear. In this case, the size of these domains increases in time through a coarsening process, similar to what we have discussed in the previous chapter on phase separation.

Finally, if we further increase the control parameter, i.e., the temperature difference ΔT between the lower and upper surfaces, new instabilities appear, until a turbulent state develops.

Experiments on patterns at the surface of a crystal surface brought out of equilibrium, either by a growth or by an erosion process, are more recent, because they need experimental conditions and probes that have been developed only since the 1980s. Furthermore, the popularity of such experiments has also been due to the possibility of testing the kinetic roughening models discussed in Chapter 5. In this chapter we have been interested in *deterministic* pattern-forming systems, whose phenomenology at a crystal surface is very broad. In fact, the mechanisms leading to an instability, then to pattern formation, range from kinetic to energetic mechanisms, passing through athermal ones. Unlike the hydrodynamic instabilities, which can be faced, in principle, by making use of the Navier–Stokes equations, there is no general, microscopic theory describing, for example, a growing surface far from equilibrium. We must therefore resort to using phenomenological equations or, in some cases, mesoscopic models based on a clear separation of time scales between the fast dynamics of surface atoms and the slow dynamics of steps. We may cite two different cases, the homoepitaxial growth[9] of a metal and the heteroepitaxial growth of a semiconductor. In the former case a kinetic instability may be at the origin of a growth dynamics with some resemblance to a phase separation process (see the previous chapter), where the local surface slope $\mathbf{m} = \nabla z(\mathbf{x}, t)$ plays the role of a (vectorial) order parameter, with the additional constraint to be an irrotational field, $\nabla \times \mathbf{m} \equiv 0$. The linear instability falls into the category II, but only if nonconservative processes, such as evaporation, are negligible. Nonlinear dynamics is a phase dynamics like that discussed in Section 7.8, with the additional complication that $\mathbf{m}$ is an irrotational vector field. In the latter case, heteroepitaxial growth of a semiconductor, we have an energetic instability driven by elasticity, whose final product is potentially of great interest, because it might be a self-organized way to produce quantum dots. Elasticity makes the problem fairly complicated even at a linear level, and the classes I–III introduced in Section 7.2 are not sufficient to cover the present problem. In fact, we need a type I scenario for a conserved order parameter.

The third example of pattern formation shown in panels (c1–c3) of Fig. 7.1 is a case of athermal dynamics in a granular system. Here the attempts to give a continuum description of dynamics, even at a linear level, are not satisfactory, and the system should rather be studied with simulations such as molecular dynamics. Despite the inherent difficulties due to the lack of common conceptual tools (free energy, temperature, etc.), in the dynamics of granular media it is possible to recognize phenomena that are familiar after reading Chapters 6 and 7.

We conclude this section with the celebrated Turing patterns. Together with those emerging from hydrodynamic Bénard instabilities, they are the most famous stationary patterns. Although the original article by Turing dates from 1952, it has been necessary to wait 40 years to have clear experimental evidence of Turing patterns, the main difficulty being the necessity to have species with significantly different diffusion coefficients. The experimental evidence of Turing patterns was obtained as a result of the chlorite–iodide–malonic acid (CIMA) reaction, where reagents are continuously fed on the top

[9] Homoepitaxy means an ordered growth of a material fitting to the lattice structure of a substrate of the same type. Heteroepitaxy means the substrate is of a different type.

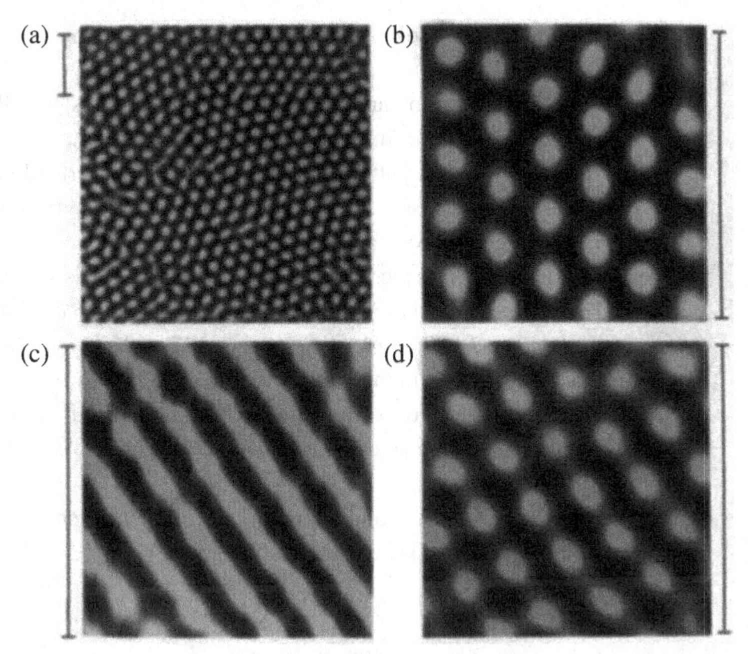

Fig. 7.11 Stationary chemical patterns. (a, b) Hexagons; (c) stripes (subject to a transverse instability); (d) mixed state (between spots and stripes). The bar beside each picture represents 1 mm. Reprinted from Q. Ouyang and H. L. Swinney, Transition from a uniform state to hexagonal and striped Turing patterns, *Nature*, **352** (1991) 610–612.

and the bottom of a thin hydrogel, where chemical species diffuse, while products are continuously removed. The continuous feeding and removal of chemicals is a necessary condition to have a stationary setup. In Fig. 7.11 we plot some stationary patterns in CIMA reactions, corresponding to different experimental conditions. The control parameter is the temperature, which allows us to pass from a homogeneous to a pattern configuration, when it is lowered below $T_c \simeq 18°\text{C}$.

7.10 Bibliographic Notes

The most important bibliographic resource to which we direct the reader is the book by M. Cross and H. Greenside, *Pattern Formation and Dynamics in Nonequilibrium Systems* (Cambridge University Press, 2009). Here you can find many physical considerations, discussion of experiments, and theoretical calculations. The historical review paper on the topic of this chapter is M. C. Cross and P. C. Hohenberg, Pattern formation outside of equilibrium, *Reviews of Modern Physics*, **65** (1993) 851–1112.

For a more mathematical introduction to methods, see R. Hoyle, *Pattern Formation: An Introduction to Methods* (Cambridge University Press, 2006).

The book by C. Misbah, *Complex Dynamics and Morphogenesis* (Springer, 2017), devotes much attention to bifurcation theory, which is the formally correct way to approach pattern formation. This book is a useful introduction to nonlinear physics and is suited to students.

Another book devoted to various topics of nonlinear physics is P. Manneville, *Instabilities, Chaos and Turbulence* (Imperial College Press, 2004).

The reader interested in pattern formation in the real world should consult Philip Ball, *Nature's Patterns: A Tapestry in Three Parts* (Oxford University Press, 2009), where the three parts are branches, flow, and shapes.

For more details on Turing patterns, the first reference is the original paper by A. M. Turing, The chemical basis of morphogenesis, *Philosophical Transactions of the Royal Society of London B: Biological Sciences*, **237** (1952), 37–72. Details about recent experiments can be found in the book by Cross and Greenside.

A detailed discussion of convective and other hydrodynamic instabilities is given in S. Chandrasekhar, *Hydrodynamic and Hydromagnetic Stability* (Dover Publications, 1981).

A formulation of the Fredhoml alternative theorem is given in D. Zwillinger, *Handbook of Differential Equations*, 3rd ed. (Academic Press, 1997).

Appendix A The Central Limit Theorem and Its Limitations

Let us consider the sequence of variables $x_1, \ldots, x_N$, all corresponding to the same random process whose probability distribution $p(x)$ is unknown, with the only condition that it has a finite standard deviation σ. Now define the stochastic variable X, equal to its arithmetic mean,

$$X = \frac{1}{N}\sum_{i=1}^{N} x_i. \tag{A.1}$$

If the x_i are not correlated, we have a simple expression for the variance:

$$\sigma_X^2 = \frac{1}{N^2}\sum_{i,j}\left(\langle x_i x_j\rangle - \langle x_i\rangle\langle x_j\rangle\right) = \frac{1}{N^2}\sum_i\left(\langle x_i^2\rangle - \langle x_i\rangle^2\right) = \frac{\sigma^2}{N}. \tag{A.2}$$

Therefore, the variance of the arithmetic mean vanishes for large N, and X tends to the average $\langle x\rangle$, evaluated according to $p(x)$. If the variables $x_i, x_{i+\Delta}$ are correlated but the correlation vanishes as $1/\Delta$ or faster, the variance σ_X^2 still vanishes with increasing N.

In conclusion, the above law of large numbers tells us that X has the average value $\langle x\rangle$ and variance σ^2/N. The central limit theorem we are going to discuss tells more: the distribution of the values of X is Gaussian regardless of $p(x)$ (with some caveats). In order to prove this statement we use the characteristic function of a probability distribution, $\Gamma(k)$, which is the average value of e^{ikx},

$$\Gamma(k) = \left\langle e^{ikx}\right\rangle = \int dx e^{ikx} p(x) = \sum_{n=0}^{\infty}\frac{(ik)^n}{n!}\langle x^n\rangle. \tag{A.3}$$

For a Gaussian distribution,

$$p_G(x) = \frac{1}{\sqrt{2\pi\sigma^2}} e^{-\frac{(x-x_0)^2}{2\sigma^2}}, \tag{A.4}$$

we obtain

$$\Gamma_G(k) = \frac{e^{ikx_0}}{\sqrt{2\pi\sigma^2}}\int dx e^{ikx} e^{-\frac{x^2}{2\sigma^2}} = e^{ikx_0 - \frac{k^2\sigma^2}{2}}. \tag{A.5}$$

Let us now evaluate the characteristic function for the distribution of the arithmetic mean, X,

$$\Gamma_X(k) = \left\langle e^{ik\frac{1}{N}\sum_i x_i}\right\rangle \tag{A.6}$$

$$= \left\langle e^{i\frac{k}{N}x}\right\rangle^N \tag{A.7}$$

$$= \left(1 + \frac{ik}{N}\langle x\rangle - \frac{k^2}{2N^2}\langle x^2\rangle + o\left(\frac{1}{N^2}\right)\right)^N \tag{A.8}$$

$$= \left[\exp\left(\frac{ik}{N}\langle x\rangle - \frac{k^2\sigma^2}{2N^2} + o\left(\frac{1}{N^2}\right)\right)\right]^N \tag{A.9}$$

$$= \exp\left(ik\langle x\rangle - \frac{k^2\sigma^2}{2N} + o\left(\frac{1}{N}\right)\right). \tag{A.10}$$

In conclusion, the characteristic function is equal to $\Gamma_G(k)$, with $\langle X\rangle = \langle x\rangle$ and $\sigma_X^2 = \sigma^2/N$. The main hypotheses are that both $\langle x\rangle$ and $\langle x^2\rangle$ must be finite. The fact that a binomial distribution can be approximated with a Gaussian function is a familiar example of the theorem.

The central limit theorem is based on the assumption that we are adding independent, randomly distributed variables. Therefore, the simplest reason why it may fail is the correlation between variables. This is the case when we follow the trajectory of a system at equilibrium performing a molecular dynamics or Monte Carlo simulation: if we sample the trajectory at time intervals Δt, the different values of a given observable are correlated unless Δt is much larger than its correlation time. However, the most interesting cases of the failing of the central limit theorem arise when the input of the problem is given by independent, randomly distributed variables and they "interact" to produce a non-Gaussian distribution.

The Tracy–Widom distributions are widespread, because they are related to the theory of random matrices, and many different problems in physics and combinatorics can be treated resorting to such a theory. A simple formulation of the problem is the following: we consider, e.g., the ensemble of unitary matrices, whose entries are independent, randomly distributed variables, and we wonder what the (suitably normalized) distribution of eigenvalues is. It turns out that this problem can be mapped to a gas of one-dimensional interacting particles, whose positions are the sought-after eigenvalues. Particles are subject to a potential well plus a repulsive, Coulomb interaction, which is what makes the distribution non-Gaussian. An interesting question to ask is about the distribution of the position of the outer particles, which correspond to the values of the maximal (and minimal) eigenvalue. Let's take the rightmost particle, which is repelled by the external potential from the right and by other particles from the left: their combination results in a distribution, which is not symmetric, and, therefore, it cannot be described by a Gaussian, which is instead symmetric.

Appendix B Spectral Properties of Stochastic Matrices

We consider a nonnegative matrix W, $W_{ij} \geq 0 \; \forall i,j$, such that

$$\sum_i W_{ij} = 1 \quad \forall j. \tag{B.1}$$

We say that $\mathbf{w}^{(\lambda)} = \left(w_1^{(\lambda)} w_2^{(\lambda)} \cdots w_j^{(\lambda)} \cdots\right)$ is the right-eigenvector of W with eigenvalue λ if the following relation holds:

$$\sum_j W_{ij} w_j^{(\lambda)} = \lambda w_i^{(\lambda)} \quad \forall i. \tag{B.2}$$

We want to prove the following properties:

(a) $|\lambda| \leq 1$.

Consider the (right) eigenvector $\mathbf{w}^{(\lambda)}$ of W; then, we can write

$$\sum_j W_{ij} |w_j^{(\lambda)}| \geq \left| \sum_j W_{ij} w_j^{(\lambda)} \right| = |\lambda| |w_i^{(\lambda)}|, \tag{B.3}$$

where the first inequality holds because $W_{ij} \geq 0$ and the last equality stems from the definition of eigenvector. By summing over the index i both sides of this inequality and considering (B.1), we finally obtain:

$$\sum_j |w_j^{(\lambda)}| \geq |\lambda| \sum_i |w_i^{(\lambda)}|, \tag{B.4}$$

i.e., $|\lambda| \leq 1$.

(b) There is at least one eigenvalue $\lambda = 1$.

Let us consider the N-component vector $\mathbf{v} = (1, 1, \ldots, 1, \ldots, 1)$. Taking into account (B.1), we can write

$$\sum_i v_i W_{ij} = \sum_i W_{ij} = 1 = v_j, \tag{B.5}$$

which implies that $\mathbf{v}$ is a left-eigenvector of W with eigenvalue 1. Since left and right eigenvalues coincide, there must be a (right) eigenvector with eigenvalue equal to 1.

(c) $\mathbf{w}^{(\lambda)}$ is either an eigenvector with eigenvalue 1, or it fulfills the condition $\sum_j w_j^{(\lambda)} = 0$.

The components of the eigenvector $\mathbf{w}^{(\lambda)}$ obey the relation

$$\sum_j W_{ij} w_j^{(\lambda)} = \lambda w_i^{(\lambda)}. \tag{B.6}$$

By summing both sides of this relation over i and considering (B.1), we obtain

$$\sum_j w_j^{(\lambda)} = \lambda \sum_i w_i^{(\lambda)}. \tag{B.7}$$

Accordingly, either $\lambda = 1$ or $\sum_j w_j^{(\lambda)} = 0$.

Appendix C Reversibility and Ergodicity in a Markov Chain

Let us consider a system with a discrete configuration space, where states are labeled by an index i. Let $p_i(t)$ be the probability of state i at time t and define the quantity

$$\mathcal{D}[t] = \sum_i \frac{1}{p_i^*}(p_i(t) - p_i^*)^2 = \sum_i \frac{p_i^2(t)}{p_i^*} - 1, \tag{C.1}$$

where the sum runs over all possible states of the system, p_i^* is the equilibrium distribution, and the last equality stems from the normalization condition

$$\sum_i p_i(t) = 1 \quad \forall t. \tag{C.2}$$

One can easily realize that $\mathcal{D} \geq 0$ and $\mathcal{D} = 0$ if and only if $p_i(t) = p_i^*\ \forall i$. Let us analyze how $\mathcal{D}$ changes in time,

$$\Delta\mathcal{D} \equiv \mathcal{D}[t+1] - \mathcal{D}[t] = \sum_i \frac{p_i^2(t+1)}{p_i^*} - \sum_i \frac{p_i^2(t)}{p_i^*}. \tag{C.3}$$

Since the evolution equation of a Markov chain reads (see Eq. (1.79))

$$p_i(t+1) = \sum_j W_{ij} p_j(t), \tag{C.4}$$

we can write (C.3) as

$$\begin{aligned}\Delta\mathcal{D} &= \sum_i \frac{1}{p_i^*}\left(\sum_j W_{ij} p_j(t)\right)\left(\sum_k W_{ik} p_k(t)\right) - \sum_i \frac{p_i^2(t)}{p_i^*} \\ &= \sum_{ijk} W_{ij} W_{ik} \frac{p_j(t) p_k(t)}{p_i^*} - \sum_i \frac{p_i^2(t)}{p_i^*}.\end{aligned} \tag{C.5}$$

Since p_i^* is the equilibrium distribution, the detailed balance condition (1.129), $p_a^* W_{ba} = p_b^* W_{ab}$, holds; a Markov chain whose stochastic matrix W obeys this condition is said to be reversible. We can use the detailed balance condition to write $W_{ij} W_{ik} = W_{ji} W_{ki} (p_i^*)^2 / (p_j^* p_k^*)$, while, in the second term, we insert the normalization condition $\sum_j W_{ji} = 1$ and we exchange the indices $i \leftrightarrow j$, thus obtaining

$$\Delta\mathcal{D} = \sum_{ijk} W_{ji} W_{ki} p_i^* \frac{p_j p_k}{p_j^* p_k^*} - \sum_{ij} W_{ij} \frac{p_j^2}{p_j^*}. \tag{C.6}$$

The shortened notation $p_i \equiv p_i(t)$ has also been used.

In the second addendum on the right-hand side we can use again the detailed balance condition $W_{ij} = (W_{ji} p_i^*)/p_j^*$ and introduce one more normalization condition $\sum_k W_{ki} = 1$, yielding

$$\Delta\mathcal{D} = \sum_{ijk} W_{ji} W_{ki} p_i^* \frac{p_j p_k}{p_j^* p_k^*} - \sum_{ijk} W_{ji} W_{ki} p_i^* \left(\frac{p_j}{p_j^*}\right)^2 . \tag{C.7}$$

Finally, the last addendum on the right-hand side of (C.7) can be rewritten as the sum of 1/2 of itself and 1/2 of the same expression with $j \leftrightarrow k$, and we finally obtain

$$\Delta\mathcal{D} = -\frac{1}{2} \sum_{ijk} W_{ji} W_{ki} p_i^* \left(\frac{p_j}{p_j^*} - \frac{p_k}{p_k^*}\right)^2 . \tag{C.8}$$

We can conclude that $\Delta\mathcal{D} \leq 0$; i.e., during the evolution the non negative quantity $\mathcal{D}$ either decreases or remains constant. In particular, if we want that $\Delta\mathcal{D} = 0$, each term must vanish separately. If we assume that the matrix W_{ij} is ergodic, for any initial condition after a time t_0 we have $(W^{t_0})_{ij} > 0\ \forall i, j$. Accordingly, from time t_0 the condition $\Delta\mathcal{D} = 0$ can be fulfilled if and only if $p_i = p_i^*\ \forall i$.

Appendix D The Diffusion Equation and Random Walk

D.1 The Diffusion Equation with Drift: General Solution

In this appendix we offer the general solution of the diffusion equation with drift (see Eq. (1.68)):

$$\frac{\partial \rho(\mathbf{x}, t)}{\partial t} = -\mathbf{v}_0 \cdot \nabla \rho(\mathbf{x}, t) + D\nabla^2 \rho(\mathbf{x}, t). \tag{D.1}$$

As we will see, the solution is the combination of a standard diffusion process in the plane orthogonal to $\mathbf{v}_0$ and a diffusion with drift along $\mathbf{v}_0$. Because of isotropy we can assume the drift $\mathbf{v}_0$ is oriented along the $\hat{x}$-axis and we limit ourselves to two spatial dimensions, the generalization to $d > 2$ being straightforward. The above equation therefore reads

$$\frac{\partial \rho(x, y, t)}{\partial t} = -v_0 \partial_x \rho(x, y, t) + D(\partial_{xx} + \partial_{yy})\rho(x, y, t), \tag{D.2}$$

whose solution can be found by separation of variables, $\rho(x, y, t) = p(x, t)q(y, t)$, obtaining

$$\partial_t p(x, t) = -v_0 \partial_x p(x, t) + D\partial_{xx} p(x, t) \tag{D.3}$$

$$\partial_t q(y, t) = D\partial_{yy} q(y, t), \tag{D.4}$$

which we are going to solve with initial conditions $p(x, 0) = \delta(x - x_0)$ and $q(y, 0) = \delta(y - y_0)$. In fact, because of the linear character of the equations, any different initial profile can be written as a superposition of the Dirac delta's evolving, each separately from the other.

The equation for $q(y, t)$ is called diffusion equation or heat equation,

$$\frac{\partial\, q(y, t)}{\partial\, t} = D\frac{\partial^2}{\partial\, y^2} q(y, t), \tag{D.5}$$

which can be easily solved by introducing the spatial Fourier transform of $q(y, t)$,

$$q(k, t) = \int_{-\infty}^{+\infty} dy\, e^{-iky} q(y, t), \tag{D.6}$$

where i is the standard notation for the imaginary constant, $i^2 = -1$. By applying the Fourier transform to both sides of (D.5), one obtains

$$\frac{\partial\, q(k, t)}{\partial\, t} = -Dk^2\, q(k, t), \tag{D.7}$$

which yields the solution

$$q(k, t) = q(k, 0)e^{-Dk^2 t}. \tag{D.8}$$

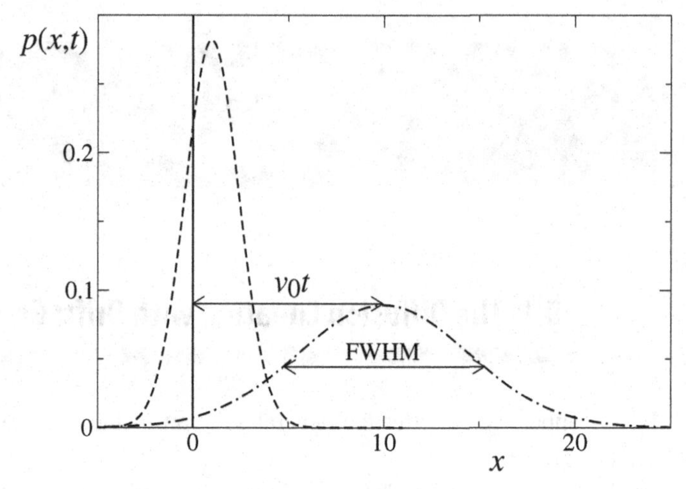

Fig. D.1 Plot of $p(x,t)$ (see Eq. (D.10)) at different times, for $v_0/D = 1$ and $x_0 = 0$: $Dt = 1$ (dashed curve) and $Dt = 10$ (dot-dashed curve). At $t = 0$ we have a Dirac delta, while at $t > 0$ we have a Gaussian centered in $x = v_0t$, with a full width at half maximum (FWHM) equal to $4\sqrt{\ln 2}\sqrt{Dt}$.

If we assume the initial condition $q(y,0) = \delta(y-y_0)$, we have $q(k,0) = \exp(-iky_0)$. The solution for $q(y,t)$ can be obtained by antitransforming expression (D.8):

$$q(y,t) = \frac{1}{2\pi}\int_{-\infty}^{+\infty} dk e^{ik(y-y_0)-Dk^2t} = \frac{1}{\sqrt{4\pi Dt}}\exp\left(-\frac{(y-y_0)^2}{4Dt}\right). \tag{D.9}$$

We can now turn to the equation for $p(x,t)$ and remark that for a constant drift v_0 it can be solved by the simple relation $p(x,t) = q(x - v_0t, t)$. In conclusion, we have

$$p(x,t) = \frac{1}{\sqrt{4\pi Dt}}\exp\left(-\frac{(x-v_0t-x_0)^2}{4Dt}\right) \tag{D.10}$$

$$q(y,t) = \frac{1}{\sqrt{4\pi Dt}}\exp\left(-\frac{(y-y_0)^2}{4Dt}\right). \tag{D.11}$$

Solution (D.10) (see Fig. D.1) corresponds to a Gaussian distribution, which broadens and flattens as time increases, while its center translates at constant speed v_0.

D.2 Gaussian Integral

In one spatial dimension, the most general, normalized Gaussian distribution has the following form:

$$\phi(x) = \frac{1}{\sqrt{2\pi\sigma^2}}\exp\left(-\frac{(x-\mu)^2}{2\sigma^2}\right). \tag{D.12}$$

This distribution appears several times throughout the book, with the first (μ) and the second (σ^2) momenta depending on time. For instance, $\phi(x)$ is the solution of the diffusion equation in the presence of a constant drift, Eq. (1.68), with $\mu = x_0 + v_0 t$ and $\sigma^2 = 2Dt$. It is also the solution of the Fokker–Planck equation of the Ornstein–Uhlenbech process, Eq. (1.195), with $\mu = x_0 e^{-kt}$ and $\sigma^2 = (D/k)(1 - e^{-2kt})$. Here we prove that $\phi(x)$ is normalized and that μ and σ^2 are the first two momenta:

$$\int_{-\infty}^{+\infty} \phi(x)dx = \frac{1}{\sqrt{2\pi\sigma^2}} \int_{-\infty}^{+\infty} \exp\left(-\frac{y^2}{2\sigma^2}\right) dy \tag{D.13}$$

$$= \frac{1}{\sqrt{\pi}} \int_{-\infty}^{+\infty} e^{-z^2} dz \tag{D.14}$$

$$= 1, \tag{D.15}$$

where we have performed the changes of variable $y = x - \mu$ and $z = y/\sqrt{2\sigma^2}$. The result,

$$\langle x \rangle = \int_{-\infty}^{+\infty} x\phi(x)dx = \mu, \tag{D.16}$$

is trivial, because $\phi(x)$ is symmetric with respect to $x = \mu$. As for the second momentum,

$$\langle x^2 \rangle = \int_{-\infty}^{+\infty} \frac{x^2}{\sqrt{2\pi\sigma^2}} \exp\left(-\frac{(x-\mu)^2}{2\sigma^2}\right) dx \tag{D.17}$$

$$= \int_{-\infty}^{+\infty} \frac{y^2 + \mu^2 + 2y\mu}{\sqrt{2\pi\sigma^2}} \exp\left(-\frac{y^2}{2\sigma^2}\right) dy \tag{D.18}$$

$$= \mu^2 + \int_{-\infty}^{+\infty} \frac{y^2}{\sqrt{2\pi\sigma^2}} \exp\left(-\frac{y^2}{2\sigma^2}\right) dy \tag{D.19}$$

$$= \mu^2 + \frac{2\sigma^2}{\sqrt{\pi}} \int_{-\infty}^{+\infty} z^2 e^{-z^2} dz \tag{D.20}$$

$$= \mu^2 + \sigma^2. \tag{D.21}$$

Above, in order, we have passed to $y = x - \mu$; we have used previous results for the normalization of the distribution and its symmetric form; and we have passed to $z = y/\sqrt{2\sigma^2}$. In conclusion, $\sigma^2 = \langle x^2 \rangle - \mu^2 = \langle x^2 \rangle - \langle x \rangle^2$.

D.3 Diffusion Equation: Fourier–Laplace Transform

The solution of the diffusion equation

$$\frac{\partial \rho(x,t)}{\partial t} = D \frac{\partial^2 \rho(x,t)}{\partial x^2} \tag{D.22}$$

has been found in Appendix D.1,

$$\rho(x,t) = \frac{1}{\sqrt{4\pi Dt}} \exp - \left[\frac{(x - x(0))^2}{4Dt}\right]. \tag{D.23}$$

Since $\rho(x,t)$ is actually a function of $(x - x(0), t)$, its Fourier–Laplace transform has the property

$$\rho(k,s) = e^{-ikx(0)} \rho_0(k,s), \tag{D.24}$$

where $\rho_0(x,t)$ corresponds to $x(0) = 0$. Here below we prove that the Laplace–Fourier transform of $\rho_0(x,t)$ is

$$\rho_0(k,s) = \frac{1}{s + Dk^2}, \tag{D.25}$$

without evaluating any integral, just using the fact that $\rho_0(x,t)$ satisfies Eq. (D.22). In fact, we can show that (D.23) is the antitransform of $\rho(k,s)$. As a first step we can rewrite Eq. (D.25) as

$$s\,\rho_0(k,s) = -D\,k^2\,\rho_0(k,s), \tag{D.26}$$

then we can antitransform with respect to Laplace both sides of this equation as

$$\frac{1}{2\pi\,i}\int_{a-i\infty}^{a+i\infty} ds\,e^{s\,t}\,s\,\rho_0(k,s) = -Dk^2\,\frac{1}{2\pi\,i}\int_{a-i\infty}^{a+i\infty} ds\,e^{s\,t}\,\rho_0(k,s), \tag{D.27}$$

where $a > a_c$, a_c is the abscissa of convergence of the inverse Laplace transform, i.e., the highest value of the real part among the singularities of $\rho_0(k,s)$ in the complex plane of the variable s. We want to point out that the result of the two integrals in (D.27) is independent of the actual value of a and that a_c can be set to zero if $\rho_0(k,s)$ is a regular function of s. We can now observe that the left-hand side is a time derivative and rewrite this equation as

$$\frac{\partial}{\partial t}\rho_0(k,t) = -Dk^2\,\rho_0(k,t), \tag{D.28}$$

where

$$\rho_0(k,t) = \frac{1}{2\pi\,i}\int_{a-i\infty}^{a+i\infty} e^{s\,t}\rho_0(k,s). \tag{D.29}$$

Now, we can antitransform with respect to Fourier both sides of (D.28),

$$\frac{\partial}{\partial t}\left[\frac{1}{2\pi}\int_{-\infty}^{+\infty} dk\,e^{i\,k\,x}\rho_0(k,t)\right] = -D\frac{1}{2\pi}\int_{-\infty}^{+\infty} dk\,k^2\,e^{i\,k\,x}\rho_0(k,t), \tag{D.30}$$

and observe that the right-hand side of this equation is a second derivative,

$$D\frac{\partial^2}{\partial x}\left[\frac{1}{2\pi}\int_{-\infty}^{+\infty} dk\,e^{i\,k\,x}\rho_0(k,t)\right], \tag{D.31}$$

so we finally obtain Eq. (D.22), where

$$\rho_0(x,t) = \frac{1}{2\pi}\int_{-\infty}^{+\infty} dk\,e^{i\,k\,x}\rho_0(k,t). \tag{D.32}$$

The reader who aims to prove (D.25) “more traditionally” should first Fourier-transform $\rho(x,t)$, which is a tabulated integral, so as to obtain $\rho_0(k,t) = e^{-k^2Dt}$, then Laplace-transform,

$$\rho_0(k,s) = \int_0^{\infty} e^{-st}e^{-k^2Dt}dt \tag{D.33}$$

$$= \frac{1}{s + Dk^2}. \tag{D.34}$$

D.4 Random Walk and Its Momenta

We describe here a generalization of the random walk problem discussed in Section 1.5.2, where the random walker at each unit time step walks a distance y, distributed according to a normalized distribution $\eta(y)$, i.e.,

$$\int_{-\infty}^{+\infty} dy\, \eta(y) = 1. \tag{D.35}$$

The probability that the walker is at position x at (discrete) time n is indicated by $p_n(x)$. Without prejudice of generality we can assume that the walk starts at the origin, and we can write $p_0(x) = \delta(x)$. At time $t = 1$ the distribution of the new position of the walker is provided by $\eta(x)$, i.e., $p_1(x) = \eta(x)$. More generally, we can conclude that $p_n(x)$ has to satisfy the following recursion relation:

$$p_n(x) = \int_{-\infty}^{+\infty} dy\, p_{n-1}(x-y)\, \eta(y) = \int_{-\infty}^{+\infty} dy\, p_{n-1}(y)\, \eta(x-y). \tag{D.36}$$

Fourier-transforming both sides of this equation, we can write

$$p_n(k) = p_{n-1}(k)\, \eta(k), \tag{D.37}$$

because the integral on the right-hand side is a convolution product that transforms into a standard product of the Fourier transformed functions, where

$$p_n(k) = \int_{-\infty}^{+\infty} dx\, e^{-ikx}\, p_n(x) \tag{D.38}$$

and

$$\eta(k) = \int_{-\infty}^{+\infty} dx\, e^{-ikx}\, \eta(x), \tag{D.39}$$

which is known in the literature as the characteristic function of the probability density distribution $\eta(x)$ and corresponds to $\langle \exp(ikx) \rangle$.[1]

The solution of (D.37) is

$$p_n(k) = \eta^n(k), \tag{D.40}$$

because $p_0(k) = 1$. It is useful to rewrite (D.39) by expanding the imaginary exponential inside the integral as

$$\begin{aligned}\eta(k) &= \int_{-\infty}^{+\infty} dx \left(1 - ikx - \frac{1}{2}k^2 x^2 + \frac{i}{6}k^3 x^3 + \cdots \right)\eta(x) \\ &= \sum_{n=0}^{+\infty} \frac{(-i)^n}{n!} \langle x^n \rangle\, k^n,\end{aligned} \tag{D.41}$$

where the expression

$$\langle x^m \rangle = \int_{-\infty}^{+\infty} dx\, x^m\, \eta(x) \tag{D.42}$$

[1] The sign in the exponential is irrelevant if the distribution $\eta(x)$ is an even function.

defines the mth momentum of the probability density function $\eta(x)$.[2] According to Eq. (D.41), we can also write

$$\langle x^m \rangle = i^m \left. \frac{d^m \eta(k)}{d k^m} \right|_{k=0}. \tag{D.43}$$

This is quite a useful formula, because it allows us to express the momenta of the probability density function $\eta(x)$ in terms of the derivatives of its Fourier transform $\eta(k)$.

D.5 Isotropic Random Walk with a Trap

The equation describing the distribution probability $p(x,t)$ of a random walker moving freely in one dimension, with diffusion coefficient D, is

$$\frac{\partial p}{\partial t} = D \frac{\partial^2 p}{\partial x^2}. \tag{D.44}$$

Its solution, if the particle is initially located in $x = x_0$, is given in Appendix D.1,

$$p_{x_0}(x,t) = \frac{1}{\sqrt{4\pi Dt}} \exp\left(-\frac{(x-x_0)^2}{4Dt}\right). \tag{D.45}$$

If there is a trap in $x = 0$, the probability $p(x,t)$ must vanish in $x = 0$ at any t. The correct solution is easily obtained by taking the linear combination of $p_{x_0}(x,t)$ and $p_{-x_0}(x,t)$, because it automatically satisfies the boundary condition $p(0,t) = 0$ at all times. This combination is simply

$$p(x,t) = \frac{1}{\sqrt{4\pi Dt}} \left[\exp\left(-\frac{(x-x_0)^2}{4Dt}\right) - \exp\left(-\frac{(x+x_0)^2}{4Dt}\right) \right]. \tag{D.46}$$

The probability that the particle is trapped in $x = 0$ in the time interval $(t, t+dt)$ is called first passage probability, $f(t)$, and it is equal to the current $|J|$ flowing in the origin, where $J = -Dp'(x=0)$, i.e.,

$$f(t) = |J(t)| = \frac{x_0}{\sqrt{4\pi Dt^3}} \exp\left(-\frac{x_0^2}{4Dt}\right). \tag{D.47}$$

This probability is normalized to 1,

$$\int_0^\infty f(t)dt = \int_0^\infty \frac{x_0}{\sqrt{4\pi Dt^3}} \exp\left(-\frac{x_0^2}{4Dt}\right) dt \tag{D.48}$$

$$= \frac{2}{\sqrt{\pi}} \int_0^\infty e^{-\tau^2} d\tau \tag{D.49}$$

$$= 1, \tag{D.50}$$

and the average trapping time diverges,

$$\langle t_{\text{tr}} \rangle = \int_0^\infty t f(t) dt = +\infty, \tag{D.51}$$

[2] We want to point out that (D.41) is a meaningful expression, provided all momenta $\langle x^m \rangle$ are finite.

because the integrand $tf(t) \sim 1/\sqrt{t}$ for large t. The median trapping time $t_{\rm med}$, defined through the relation

$$\int_0^{t_{\rm med}} f(t)dt = \int_{t_{\rm med}}^{\infty} f(t)dt, \tag{D.52}$$

scales as x_0^2/D.

Finally, the time dependence of the average position of the particle is easily evaluated from the definition:

$$\langle x(t)\rangle = \int_0^{\infty} dx \frac{x}{\sqrt{4\pi Dt}} \left[\exp\left(-\frac{(x-x_0)^2}{4Dt}\right) - \exp\left(-\frac{(x+x_0)^2}{4Dt}\right)\right] \tag{D.53}$$

and performing the change of variable $x \to -x$ in the integral of the right exponential,

$$\begin{aligned} \langle x(t)\rangle &= \int_0^{\infty} dx \frac{x}{\sqrt{4\pi Dt}} \exp\left(-\frac{(x-x_0)^2}{4Dt}\right) + \int_{-\infty}^{0} dx \frac{x}{\sqrt{4\pi Dt}} \exp\left(-\frac{(x-x_0)^2}{4Dt}\right) && \text{(D.54)} \\ &= \int_{-\infty}^{\infty} dx \frac{x}{\sqrt{4\pi Dt}} \exp\left(-\frac{(x-x_0)^2}{4Dt}\right) && \text{(D.55)} \\ &= x_0. && \text{(D.56)} \end{aligned}$$

In conclusion, the presence of a trap does not modify the average position of the walker.

D.6 Anisotropic Random Walk with a Trap

We have a random walker moving in one dimension, with diffusion coefficient D and a negative drift $-v$ (with $v > 0$). Its distribution probability $p(x,t)$ is described by the convection–diffusion equation,

$$\frac{\partial p}{\partial t} = v\frac{\partial p}{\partial x} + D\frac{\partial^2 p}{\partial x^2}. \tag{D.57}$$

If the initial condition is $p(x,0) = \delta(x - x_0)$, its solution, in the absence of traps, is given in Appendix D.1,

$$p_{x_0}(x,t) = \frac{1}{\sqrt{4\pi Dt}} \exp\left(-\frac{(x-x_0+vt)^2}{4Dt}\right). \tag{D.58}$$

If the particle has a trap in $x = 0$, this means the probability $p(x,t)$ must vanish in $x = 0$ at any t. This problem can be solved with the image method, which consists of finding a suitable linear combination of $p_{x_0}(x,t)$ and $p_{-x_0}(x,t)$ that satisfies the boundary condition $p(0,t) = 0$ at all times. This combination is

$$p(x,t) = \frac{1}{\sqrt{4\pi Dt}} \left[\exp\left(-\frac{(x-x_0+vt)^2}{4Dt}\right) - \exp\left(\frac{vx_0}{D}\right) \exp\left(-\frac{(x+x_0+vt)^2}{4Dt}\right)\right]. \tag{D.59}$$

The probability the particle is trapped in $x = 0$ in the time interval $(t, t + dt)$ is equal to the current $|J|$ flowing in the origin, where now $J = (-vp - Dp')_{x=0}$, i.e.,

$$|J| = D \left.\frac{\partial p}{\partial x}\right|_0 \tag{D.60}$$

$$= \frac{D}{\sqrt{4\pi Dt}} \left[\exp\left(-\frac{(x_0 - vt)^2}{4Dt}\right) \frac{(x_0 - vt)}{2Dt} + \exp\left(\frac{vx_0}{D}\right) \exp\left(-\frac{(x_0 + vt)^2}{4Dt}\right) \frac{(x_0 + vt)}{2Dt} \right] \tag{D.61}$$

$$= \frac{x_0}{\sqrt{4\pi Dt^3}} \exp\left(-\frac{(x_0 - vt)^2}{4Dt}\right). \tag{D.62}$$

The probability the particle is trapped within time t is

$$\int_0^t dt' |J(t')| = \frac{x_0}{\sqrt{4\pi D}} \exp\left(\frac{x_0 v}{2D}\right) \int_0^t \frac{dt'}{(t')^{3/2}} \exp\left(-\left(\frac{x_0^2}{4D}\frac{1}{t'} + \frac{v^2}{4D}t'\right)\right) \tag{D.63}$$

$$= \frac{2}{\sqrt{\pi}} e^c \int_{\bar{u}}^{\infty} du \exp\left(-\left(u^2 + \frac{c^2}{4u^2}\right)\right), \tag{D.64}$$

where $\bar{u} = x_0/\sqrt{4Dt}$, $c = vx_0/(2D)$, and the passage between (D.63) and (D.64) has been obtained with the change of variable $u^2 = x_0^2/(4Dt')$.

Since, for $c > 0$,

$$\int du \exp\left(-\left(u^2 + \frac{c^2}{4u^2}\right)\right) = \frac{\sqrt{\pi}}{4} \left[e^c \text{erf}\left(\frac{c}{2u} + u\right) - e^{-c} \text{erf}\left(\frac{c}{2u} - u\right) \right], \tag{D.65}$$

we can evaluate above integral, getting the result

$$\int_0^t dt' |J(t')| = \frac{e^c}{2} \left\{ e^c \left[1 - \text{erf}\left(\frac{c}{2\bar{u}} + \bar{u}\right) \right] + e^{-c} \left[1 + \text{erf}\left(\frac{c}{2\bar{u}} - \bar{u}\right) \right] \right\}. \tag{D.66}$$

Above, $\text{erf}(x) = (2/\sqrt{\pi}) \int_0^x ds e^{-s^2}$ is the error function.

The total probability to be trapped is one, regardless of v (i.e., regardless of c),

$$\int_0^{\infty} dt' |J(t')| = 1 \qquad \forall v, \tag{D.67}$$

because $\text{erf}(x \to +\infty) \to 1$. It is interesting to evaluate the average time to be trapped,

$$\langle t_{\text{tr}} \rangle = \int_0^{\infty} dt\, t |J(t)| = \frac{x_0}{\sqrt{4\pi D}} \int_0^{\infty} \frac{dt}{\sqrt{t}} \exp\left(-\frac{(x_0 - vt)^2}{4Dt}\right) \tag{D.68}$$

$$= \frac{x_0}{\sqrt{4\pi D}} \exp\left(\frac{x_0 v}{2D}\right) \int_0^{\infty} \frac{dt}{\sqrt{t}} \exp\left(-\left(\frac{x_0^2}{4D}\frac{1}{t} + \frac{v^2}{4D}t\right)\right) \tag{D.69}$$

$$= \sqrt{\frac{x_0^3}{\pi D v}} \exp\left(\frac{x_0 v}{2D}\right) K_{1/2}\left(\frac{x_0 v}{2D}\right), \tag{D.70}$$

where $K_{1/2}(x)$ is the modified Bessel function of the second kind. For small drift v, i.e., small asymmetry in the random walk, $K_{1/2}(x) \simeq \sqrt{\pi/(2x)}$ and

$$\langle t_{\text{tr}} \rangle \simeq \frac{x_0}{v}, \qquad \text{for } v \to 0. \tag{D.71}$$

We are now interested in evaluating the trapping probability (D.66) in the long t limit, i.e., in the limit of vanishing $\bar{u}$, in which case the arguments of the error functions diverge, so we can use the asymptotic expansion

$$\mathrm{erf}(x) \simeq 1 - \frac{e^{-x^2}}{\sqrt{\pi}x}, \qquad \text{for} \quad x \to +\infty \tag{D.72}$$

and get, for $\bar{u} \to 0$,

$$\int_0^t dt' |J(t')| \simeq 1 + \frac{e^c}{2\sqrt{\pi}} \exp\left(-\frac{c^2}{4\bar{u}^2}\right)\left[\frac{1}{\frac{c}{2\bar{u}} + \bar{u}} - \frac{1}{\frac{c}{2\bar{u}} - \bar{u}}\right] \tag{D.73}$$

$$\simeq 1 - \frac{4}{\sqrt{\pi}} \frac{e^c}{c^2} \bar{u}^3 \exp\left(-\frac{c^2}{4\bar{u}^2}\right). \tag{D.74}$$

We now pass to the survival probability $S(t) = 1 - \int_0^t dt' |J(t')|$, i.e., the probability that the particle has not been trapped within time t. Making explicit the expressions of c and $\bar{u}$, we finally obtain

$$S(t) \simeq \frac{2}{\sqrt{\pi}} \frac{x_0\sqrt{D}\, e^{\frac{vx_0}{2D}}}{v^2} \frac{1}{t^{3/2}} \exp\left(-\frac{v^2}{4D}t\right). \tag{D.75}$$

Appendix E The Kramers–Moyal Expansion

We illustrate the method known as Kramer–Moyal expansion to obtain the backward Kolmogorov equation (1.220). We start from the Chapman–Kolmogorov equation (1.86) in the form

$$W(X_0, t_0|X, t) = \int_{\mathbb{R}} dY\, W(X_0, t_0|Y, t_0 + \Delta t_0)\, W(Y, t_0 + \Delta t_0|X, t). \tag{E.1}$$

The Kramers–Moyal expansion makes use of the combination of the following two identities:

$$W(X_0, t_0|Y, t_0 + \Delta t_0) = \int_{\mathbb{R}} dZ\, \delta(Z - Y) W(X_0, t_0|Z, t_0 + \Delta t_0) \tag{E.2}$$

$$\delta(Z - Y) = \sum_{n=0}^{+\infty} \frac{(Z - X_0)^n}{n!} \left(\frac{\partial}{\partial X_0}\right)^n \delta(X_0 - Y), \tag{E.3}$$

where $\delta(x)$ denotes the Dirac-delta distribution of argument x and $\left(\frac{\partial}{\partial X_0}\right)^n$ is a shortened notation for the nth-order partial derivative with respect to X_0. While (E.2) is self-evident, (E.3) can be proved by considering a test function in the Schwartz space $\mathcal{S}$, $\phi(Y)$, for which the following relation holds:

$$\int dY\, \delta(Z - Y)\, \phi(Y) = \phi(Z). \tag{E.4}$$

In fact, the right-hand side of (E.3) yields

$$\begin{aligned}
&\int dY \sum_{n=0}^{+\infty} \frac{(Z - X_0)^n}{n!} \left(\frac{\partial}{\partial X_0}\right)^n \delta(X_0 - Y)\phi(Y) \\
&= \sum_{n=0}^{+\infty} \frac{(Z - X_0)^n}{n!} \left(\frac{\partial}{\partial X_0}\right)^n \int dY\, \delta(X_0 - Y)\phi(Y) \\
&= \sum_{n=0}^{+\infty} \frac{(Z - X_0)^n}{n!} \left(\frac{\partial}{\partial X_0}\right)^n \phi(X_0) \equiv \phi(Z),
\end{aligned}$$

where the left-hand side of last equality is the Taylor series expansion of $\phi(Z)$, thus proving the identity.

By substituting (E.3) into (E.2) we obtain

$$
\begin{aligned}
&W(X_0, t_0 | Y, t_0 + \Delta t_0) \\
&= \sum_{n=0}^{+\infty} \frac{1}{n!} \int_{\mathbb{R}} (Z - X_0)^n W(X_0, t_0 | Z, t_0 + \Delta t_0) dZ \left(\frac{\partial}{\partial X_0}\right)^n \delta(X_0 - Y) \\
&= \left[1 + \sum_{n=1}^{+\infty} \frac{1}{n!} \int_{\mathbb{R}} (Z - X_0)^n W(X_0, t_0 | Z, t_0 + \Delta t_0) dZ \left(\frac{\partial}{\partial X_0}\right)^n\right] \delta(X_0 - Y) \quad \text{(E.5)}
\end{aligned}
$$

where the last equality stems from the normalization condition

$$
\int_{\mathbb{R}} dZ\, W(X_0, t_0 | Z, t_0 + \Delta t_0) = 1, \tag{E.6}
$$

meaning that, starting from any initial condition (X_0, t_0), at any following time $t_0 + \Delta t_0$ the process will certainly be at some position Z among all the accessible ones.

We can insert (E.5) into the Chapman–Kolmogorov equation (E.1), thus obtaining

$$
\begin{aligned}
&W(X_0, t_0 | X, t) - W(X_0, t_0 + \Delta t_0 | X, t) \\
&= \sum_{n=1}^{+\infty} \frac{1}{n!} \int_{\mathbb{R}} dZ\, (Z - X_0)^n W(X_0, t_0 | Z, t_0 + \Delta t_0) \left(\frac{\partial}{\partial X_0}\right)^n \\
&\quad \times \int_{\mathbb{R}} dY\, \delta(X_0 - Y) W(Y, t_0 + \Delta t_0 | X, t) \\
&= \sum_{n=1}^{+\infty} \frac{1}{n!} \int_{\mathbb{R}} dZ\, (Z - X_0)^n\, W(X_0, t_0 | Z, t_0 + \Delta t_0) \left(\frac{\partial}{\partial X_0}\right)^n W(X_0, t_0 + \Delta t_0 | X, t),
\end{aligned} \tag{E.7}
$$

where we have assumed that $W(Y, t_0 + \Delta t_0 | X, t) \in \mathcal{S}$. We define the nth-order momentum

$$
\mu_n(X_0, t_0) = \lim_{\Delta t_0 \to 0} \frac{1}{\Delta t_0} \int_{\mathbb{R}} dZ (Z - X_0)^n\, W(X_0, t_0 | Z, t_0 + \Delta t_0), \tag{E.8}
$$

and using (E.7) we can write

$$
\begin{aligned}
\frac{\partial W(X_0, t_0 | X, t)}{\partial t_0} &= - \lim_{\Delta t_0 \to 0} \frac{W(X_0, t_0 | X, t) - W(X_0, t_0 + \Delta t_0 | X, t)}{\Delta t_0} \\
&= - \sum_{n=1}^{+\infty} \frac{\mu_n(X_0, t_0)}{n!} \left(\frac{\partial}{\partial X_0}\right)^n W(X_0, t_0 | X, t).
\end{aligned} \tag{E.9}
$$

If we assume that only the first two momenta of the Kramers–Moyal expansion, $\mu_1(X_0, t_0) \equiv a(X_0, t_0)$ and $\mu_2(X_0, t_0) \equiv b^2(X_0, t_0)$, are the leading contributions (i.e., $\mu_n(X_0, t_0) \approx 0$ for $n \geq 3$), we finally obtain the backward Kolmogorov equation (1.220).

Appendix F Mathematical Properties of Response Functions

Taking into account the property of time translation invariance of the response function $\Xi(t, t')$, the right-hand side of Eq. (2.44) has the form of a convolution product

$$\langle X(t) \rangle = \int dt' \, \Xi(t - t') \, h(t'), \quad t > t', \tag{F.1}$$

which becomes a standard product for the Fourier-transformed functions

$$\langle X(\omega) \rangle = \Xi(\omega) \cdot h(\omega), \tag{F.2}$$

where the frequency ω is the dual variable of time t. This relation indicates that, if we perturb a thermodynamic observable with a field at frequency ω, it responds linearly, at the same frequency ω.[1]

$\Xi(t)$ is a real quantity, while $\Xi(\omega)$ is a complex one. Let us introduce the following notation:

$$\begin{aligned} \Xi(\omega) &= \mathrm{Re}\, \Xi(\omega) + i \, \mathrm{Im}\, \Xi(\omega) \\ &\equiv \Xi^{\mathrm{R}}(\omega) + i \, \Xi^{\mathrm{I}}(\omega). \end{aligned} \tag{F.3}$$

Both $\Xi^{\mathrm{R}}(\omega)$ and $\Xi^{\mathrm{I}}(\omega)$ have a physical interpretation. In fact, we can write

$$\begin{aligned} \Xi^{\mathrm{I}}(\omega) &= -\frac{i}{2} \Big[\Xi(\omega) - \Xi^{*}(\omega) \Big] \\ &= -\frac{i}{2} \int_{-\infty}^{+\infty} dt \, \Xi(t) \Big[e^{-i\omega t} - e^{i\omega t} \Big] \\ &= -\frac{i}{2} \int_{-\infty}^{+\infty} dt \, e^{-i\omega t} \Big[\Xi(t) - \Xi(-t) \Big], \end{aligned} \tag{F.4}$$

where the superscript * is the complex conjugate symbol. We see that the imaginary part of the Fourier transform of the response function depends on the part of the response function that is antisymmetric under the *time-reversal*, therefore $\Xi^{\mathrm{I}}(\omega)$ is an odd function of its argument,

$$\Xi^{\mathrm{I}}(\omega) = -\, \Xi^{\mathrm{I}}(-\omega). \tag{F.5}$$

In the literature it is called the dissipative component of the response function, and it contains information about the dissipative processes associated with the perturbation field.

[1] This is not the case when nonlinear effects enter the game, making the overall treatment of the response problem much more complex.

Conversely, the real part of the response function is insensitive to the direction of time,

$$\Xi^{\mathrm{R}}(\omega) = \frac{1}{2}\int_{-\infty}^{+\infty} dt\, e^{-i\omega t}\Big[\,\Xi(t) + \Xi(-t)\Big], \tag{F.6}$$

namely, it is an even function of its argument:

$$\Xi^{\mathrm{R}}(\omega) = \Xi^{\mathrm{R}}(-\omega). \tag{F.7}$$

Causality imposes that

$$\Xi(t) = 0 \qquad \forall t < 0. \tag{F.8}$$

In the mathematical literature such a function is called backward Green's function. Condition (F.8) attributes specific properties to $\Xi(\omega)$. They can be inferred from the expression of the inverse Fourier transform:

$$\Xi(t) = \frac{1}{2\pi}\int_{-\infty}^{+\infty} d\omega\; e^{i\omega t}\; \Xi(\omega). \tag{F.9}$$

Without entering into mathematical detail, we can assume ω is a complex variable,[2] $\omega \to z = \omega - i\epsilon$ with $\epsilon \in \mathbb{R}^{+}$, by analytic extension of $\Xi(\omega)$ to the complex plane, i.e., $\Xi(\omega) \to \Xi(z)$. As a consequence of the Jordan lemma, for $t < 0$ the integral on the right-hand side of (F.9) can be computed by considering a closed integration path Γ as shown in Fig. F.1 (the half-circle of radius R lies in the lower-half complex plane):

$$\Xi(t) = \frac{1}{2\pi}\lim_{R\to+\infty}\int_{\Gamma(R)} dz\; e^{i\omega t}\, e^{\epsilon t}\; \tilde{\Xi}(\omega - i\epsilon). \tag{F.10}$$

The theorem of residues states that the right-hand side of (F.10) is equal to minus the sum of the residues of the poles of $\Xi(\omega - i\epsilon)$ in the lower-half complex plane. Accordingly, causality imposes that $\Xi(\omega - i\epsilon)$ is analytic in all of the lower-half plane of the complex variable $\omega - i\epsilon$ (i.e., no poles are present and the integral on the right-hand side of (F.10) is null for $\epsilon > 0$).

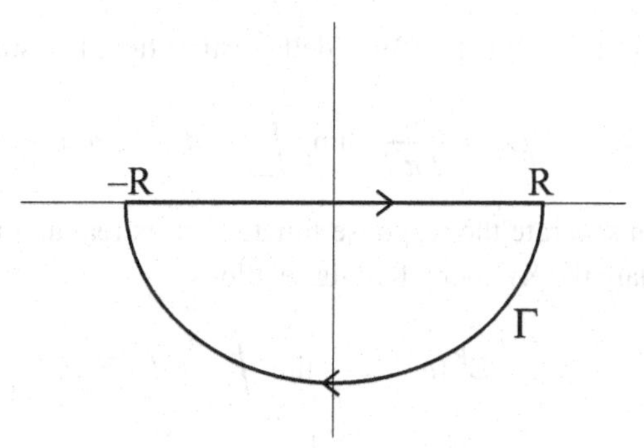

Fig. F.1 Integration path for Eq. (F.10).

[2] With our notation for the Fourier transform, the response function is analytic for negative imaginary argument, so we prefer to write $z = \omega - i\epsilon$, with positive ϵ.

Now, we are going to discuss how analyticity of $\Xi(\omega)$ in the lower-half complex plane engenders the so-called Kramers–Kronig relations between $\Xi^{R}(\omega)$ and $\Xi^{I}(\omega)$. The starting point is Cauchy's theorem, which provides us with the Cauchy integral representation of a complex function around any point z belonging to its analyticity domain,

$$\Xi(z) = \oint \frac{dz'}{2\pi i} \frac{\Xi(z')}{z' - z}, \tag{F.11}$$

where the contour integral is performed counter clockwise, i.e., in the opposite orientation of the path $\Gamma(R)$ in Fig. F.1.

Due to causality, the analyticity domain of the response function is the lower-half complex plane, so that Eq. (F.11) holds for any closed contour in the lower-half complex plane. On the other hand, we want to consider the case where $z = \omega - i\epsilon$ approaches the real axis, namely, the limit $\epsilon \to 0^{+}$. In such a case it is convenient to compute the right-hand side of (F.11) over a contour that is made by a segment of length $2R$, symmetric to the origin and parallel to the real axis at a distance $\delta < \epsilon$, closed by the semi-circle of radius R centered in the origin and lying in the lower-half of the complex plane (see Fig. F.1). If $|\Xi(z)| < 1/|z|$ for very large values of R, the integral on the semi-circle vanishes in the limit $R \to \infty$ and we can write

$$\Xi(\omega - i\epsilon) = -\frac{1}{2\pi i} \int_{-\infty}^{+\infty} d\omega' \frac{\Xi(\omega')}{\omega' - \omega + i\epsilon}, \tag{F.12}$$

where the minus sign in front of the integral appears, because the contour Γ is oriented clockwise. We can separate the real and imaginary part of the integrand by the relation

$$\frac{1}{\omega' - \omega + i\epsilon} = \frac{\omega' - \omega}{(\omega' - \omega)^2 + \epsilon^2} - i \frac{\epsilon}{(\omega' - \omega)^2 + \epsilon^2}. \tag{F.13}$$

Now we can make use of the following result,

$$\lim_{\epsilon \to 0^{+}} \frac{\epsilon}{(\omega' - \omega)^2 + \epsilon^2} = \pi\, \delta(\omega' - \omega), \tag{F.14}$$

where $\delta(\omega' - \omega)$ is the Dirac delta distribution. For small enough ϵ we can write

$$\Xi(\omega) = \frac{1}{2\pi i} \lim_{\epsilon \to 0^{+}} \int_{-\infty}^{+\infty} d\omega'\, \Xi(\omega') \frac{\omega - \omega'}{(\omega - \omega')^2 + \epsilon^2} + \frac{1}{2} \Xi(\omega). \tag{F.15}$$

We can separate the response function in its real and imaginary parts (see Eq. (F.3)), and we obtain the Kramers–Kronig relations

$$\Xi^{R}(\omega) = \frac{1}{\pi} \lim_{\epsilon \to 0} \int_{-\infty}^{+\infty} d\omega' \Xi^{I}(\omega') \frac{\omega - \omega'}{(\omega - \omega')^2 + \epsilon^2} \tag{F.16}$$

$$\Xi^{I}(\omega) = -\frac{1}{\pi} \lim_{\epsilon \to 0} \int_{-\infty}^{+\infty} d\omega'\, \Xi^{R}(\omega') \frac{\omega - \omega'}{(\omega - \omega')^2 + \epsilon^2}. \tag{F.17}$$

They are a direct consequence of causality and show that the dissipative (imaginary) and the reactive (real) components of the response function are related to each other by a nonlocal dependence in frequency space.

There is an alternative formulation of the Kramers–Kronig relations, namely,

$$\Xi(\omega) = \lim_{\epsilon \to 0^+} \int_{-\infty}^{+\infty} \frac{d\omega'}{\pi} \frac{\Xi^{\mathrm{I}}(\omega')}{\omega - \omega' - i\epsilon}, \tag{F.18}$$

where the response function is expressed just in terms of its dissipative (imaginary) component. Eq. (F.18) can be obtained by considering the relation

$$\lim_{\epsilon \to 0^+} \frac{1}{\omega - i\epsilon} = \lim_{\epsilon \to 0^+} \left[i \frac{\epsilon}{\omega^2 + \epsilon^2} + \frac{\omega}{\omega^2 + \epsilon^2} \right] = i\pi\,\delta(\omega) + \mathrm{PV}\left(\frac{1}{\omega}\right), \tag{F.19}$$

which holds in the sense of distributions. PV (principal value) is defined as

$$\mathrm{PV} \int_{-\infty}^{+\infty} \frac{d\omega}{\omega} \phi(\omega) = \lim_{\epsilon \to 0^+} \int_{-\infty}^{-\epsilon} \frac{d\omega}{\omega} \phi(\omega) + \int_{\epsilon}^{+\infty} \frac{d\omega}{\omega} \phi(\omega), \tag{F.20}$$

where $\phi(\omega)$ is a test function in Schwarz space, i.e., a regular rapidly decreasing function of its argument.

In fact, if we specialize (F.19) to the case $\omega \to \omega - \omega'$ and assume that $\Xi^{\mathrm{I}}(\omega')$ has the properties of a test function in Schwarz space, we can write

$$\begin{aligned} \lim_{\epsilon \to 0^+} \int_{-\infty}^{+\infty} \frac{d\omega'}{\pi} \frac{\Xi^{\mathrm{I}}(\omega')}{\omega - \omega' - i\epsilon} &= i\,\Xi^{\mathrm{I}}(\omega) + \mathrm{PV} \int_{-\infty}^{+\infty} \frac{d\omega'}{\pi} \frac{\Xi^{\mathrm{I}}(\omega')}{\omega - \omega'} \\ &= i\,\Xi^{\mathrm{I}}(\omega) + \lim_{\epsilon \to 0^+} \int_{-\infty}^{+\infty} \frac{d\omega'}{\pi} \Xi^{\mathrm{I}} \frac{\omega - \omega'}{(\omega - \omega')^2 + \epsilon^2}. \end{aligned} \tag{F.21}$$

Making use of Eq. (F.16), we finally find

$$\lim_{\epsilon \to 0^+} \int_{-\infty}^{+\infty} \frac{d\omega'}{\pi} \frac{\Xi^{\mathrm{I}}(\omega')}{\omega - \omega' - i\epsilon} = i\,\Xi^{\mathrm{I}}(\omega) + \Xi^{\mathrm{R}}(\omega') \equiv \Xi(\omega)\,. \tag{F.22}$$

Appendix G The Van der Waals Equation

The van der Waals equation for real gases can be obtained by considering that particles are subject to a reciprocal attracting interaction, which can be approximated by a square-well potential $U(r)$, as shown in Fig. G.1: E_0 is the well depth and d is the range of the interaction. At equilibrium, all particles have the same average energy, so the kinetic energy of a particle outside the interaction range (E_{out}) is related to the same quantity for a particle inside the interaction range (E_{in}) by the relation

$$E_{out} = E_{in} - E_0. \tag{G.1}$$

Accordingly, the average kinetic energy of a particle $\langle E \rangle$ is given by

$$\langle E \rangle = (1-p)E_{out} + pE_{in} = E_{out} + pE_0, \tag{G.2}$$

where p is the fraction of particles inside the interaction range. This quantity can be estimated by considering that the volume of the interaction sphere associated with the square-well potential is given by $V_I = 4/3\pi d^3$. By assuming space isotropy and homogeneity one can write

$$p = \frac{NV_I}{V}, \tag{G.3}$$

where N is the number of molecules in the gas and V is the volume of the container.

For an ideal gas, the pressure exerted by the particles on the walls of the container is given by

$$P = \frac{2}{3}\langle E \rangle \frac{N}{V}, \tag{G.4}$$

and it can be derived in kinetic theory, assuming $N/6$ particles travel toward a given wall and the force exerted on it is determined by the variation of momentum of each particle bouncing on it. This formula can now be modified for a real gas, assuming that when particles collide with the walls of the container they are typically far from the other particles. This means that $\langle E \rangle \to E_{out}$ and V should be replaced by the effective volume available to particles, $V - Nb$, where $b = \frac{4}{3}\pi d^3$ is the volume occupied by a particle, to which is associated a finite, physical radius d. In conclusion,

$$P = \frac{2}{3}E_{out}\frac{N}{V - Nb} \tag{G.5}$$

$$= \frac{2}{3}\left(\langle E \rangle \frac{N}{V - Nb} - E_0 \frac{N^2 V_I}{V(V - Nb)}\right). \tag{G.6}$$

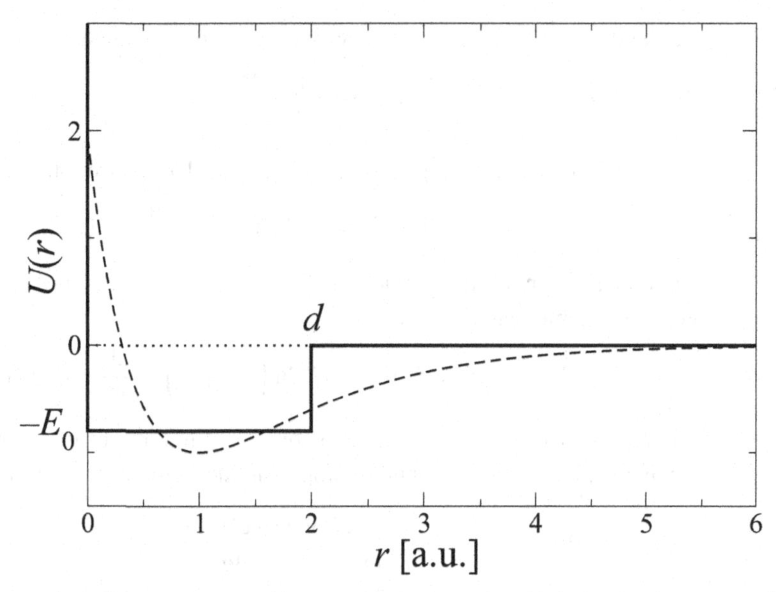

Fig. G.1 The phenomenological interaction potential between particles in a real gas (dashed line) and its square-well approximation (solid line).

According to the equipartition theorem,

$$\frac{2}{3}\langle E\rangle = T, \tag{G.7}$$

where T is the temperature of the gas. We can finally write the van der Waals equation for a real gas as

$$P = \frac{T}{v-b} - \frac{a}{v^2}, \tag{G.8}$$

where $a = (2/3)E_0 V_I$, $v = V/N$ is the volume per particle, and in the second term of the right-hand side we have neglected b (the "physical" volume of a particle) with respect to v, because, considering a and b as small parameters, it would be a sort of second-order effect.

The critical point is determined by the relations

$$\left.\frac{\partial P}{\partial v}\right|_{v_c,T_c} = 0, \qquad \left.\frac{\partial^2 P}{\partial v^2}\right|_{v_c,T_c} = 0, \qquad P_c = P(v_c, T_c), \tag{G.9}$$

which give

$$v_c = 3b, \qquad T_c = \frac{8a}{27b}, \qquad P_c = \frac{a}{27b^2}. \tag{G.10}$$

Normalizing thermodynamical quantities with respect to their values at the critical point,

$$v^* = \frac{v}{v_c}, \qquad P^* = \frac{P}{P_c}, \qquad T^* = \frac{T}{T_c}, \tag{G.11}$$

the van der Waals equation can be written in a free-parameter form,

$$P^* = \frac{8}{3}\frac{T^*}{v^* - \frac{1}{3}} - \frac{3}{(v^*)^2}. \tag{G.12}$$

The Helmholtz free energy per particle can be derived from the relation $P = -\partial f/\partial v|_T$,

$$f(v, T) = -T\ln(v - b) - \frac{a}{v} + f_0(T), \tag{G.13}$$

where $f_0(T)$ is an arbitrary function. The Gibbs free energy per particle, $g = f + Pv$, in reduced variables can be written as

$$g^* = P^* v^* - \frac{8}{3}T^* \ln\left(v^* - \frac{1}{3}\right) - \frac{3}{v^*} + f_0^*(T^*). \tag{G.14}$$

For $T^* < 1$ (i.e., $T < T_c$) the two equivalent minima of g^*, corresponding to the two coexisting phases, can be found by imposing the following conditions,

$$g^*(v_1^*) = g^*(v_2^*) \tag{G.15}$$

$$\left.\frac{\partial g^*}{\partial v^*}\right|_{v^*=v_1^*} = \left.\frac{\partial g^*}{\partial v^*}\right|_{v^*=v_2^*} = 0, \tag{G.16}$$

which are equivalent to the condition

$$P^*(v_1^* - v_2^*) = \int_{v_1^*}^{v_2^*} P^* dv^*, \tag{G.17}$$

corresponding to the so-called Maxwell construction, graphically depicted by the horizontal segment in Fig. 6.18, whose extrema determine the points on the coexistence curve.

The critical exponents can be estimated by the equation of state (G.12) close to the critical point $P^* = v^* = T^* = 1$. This can be obtained by expanding up to leading order Eq. (G.12) around the critical point, namely,

$$\delta p = \left.\frac{\partial P^*}{\partial T^*}\right|_c \delta t + \frac{1}{6}\left.\frac{\partial^3 P^*}{\partial (v^*)^3}\right|_c (\delta v)^3 = 4\delta t - \frac{3}{2}(\delta v)^3, \tag{G.18}$$

where $\delta p = P^* - 1$, $\delta t = T^* - 1$, and $\delta v = v^* - 1$. The terms of order δv and $(\delta v)^2$ in the expansion are absent, because $\left.\frac{\partial P^*}{\partial v^*}\right|_c = \left.\frac{\partial^2 P^*}{\partial (v^*)^2}\right|_c = 0$. For $T = T_c$, i.e., $\delta t = 0$, Eq. (G.18) indicates that the pressure is a cubic function of the volume. Accordingly, we can conclude that close to the critical point the volumes v_1^* and v_2^* are symmetric with respect to the critical value, namely, $v_1^* = 1 + \delta v$ and $v_2^* = 1 - \delta v$. By substituting into (G.16), that is equivalent to

$$P^*(v_1^*) = P^*(v_2^*) \tag{G.19}$$

and expanding up to leading order in δv (that is, $(\delta v)^2$) we obtain

$$\delta v \sim |\delta t|^{1/2} \sim \left(\frac{T_c - T}{T_c}\right)^{1/2}. \tag{G.20}$$

We can conclude that the scaling law of the order parameter close to the critical temperature is characterized by a critical exponent $\beta = \frac{1}{2}$, which is the classical exponent predicted by the Landau mean-field theory (see Section 3.2.3). Notice that close to the critical point the heat capacity at constant volume $C_V = T^2 \frac{\partial^2 F}{\partial T^2}$, where $F = Nf$, cannot be determined by the Helmholtz free energy (G.13), because this is defined up to an arbitrary function $F_0(T) = Nf_0(T)$. The discontinuity of C_V at the critical point predicted by the Landau mean-field theory, corresponding to a classical critical exponent $\alpha = 0$, can be recovered by assuming that the critical isothermal line ($T^* = 1$) for large values of v^* recovers the behavior of an ideal gas; i.e., the van der Waals equation of state simplifies to

$$P^* = \frac{8}{3}\frac{T^*}{v^*}. \tag{G.21}$$

Appendix H The Ising Model

The Ising model on a regular lattice is defined by the Hamiltonian

$$\mathcal{H}_I = -J\sum_{\langle ij\rangle}\sigma_i\sigma_j - H\sum_i \sigma_i, \tag{H.1}$$

where the variable $\sigma_i = \pm 1$ is a binary variable and the first sum runs over all unordered pairs of nearest-neighbor sites of the lattice. For establishing the equivalence with different models it is useful to write

$$\mathcal{H}_I = \sum_{\langle ij\rangle}\epsilon_{ij}, \tag{H.2}$$

with

$$\epsilon_{ij} = -J\sigma_i\sigma_j - \frac{H}{\gamma}(\sigma_i + \sigma_j) + K. \tag{H.3}$$

Here above γ is the coordination number of the lattice (i.e., the number of nearest neighbors per lattice site) and K is a constant for future use, which vanishes for the Ising model.

The Ising model is usually discussed using a magnetic language where $\sigma_i \pm 1$ is a spin variable and the ground state is a completely ferromagnetic state with all spins parallel. If $H = 0$ the two ferromagnetic states are degenerate, if $H \neq 0$ there is one single ground state with all spins parallel to H, i.e., $\sigma_i = \text{sign}\,(H)$. The model with $H = 0$, in dimension $d > 1$, undergoes a second-order phase transition between a disordered, paramagnetic phase at high T and an ordered, ferromagnetic phase at low T.

If the order parameter is conserved, the total magnetization is a constant of motion and the energy term proportional to H is constant as well. In this case the ground state corresponds to a complete phase separation between two oppositely magnetized, ferromagnetic regions. There are two models, the binary alloy and the lattice gas, where the order parameter is naturally conserved, because magnetization is replaced by a matter density. Let us define these two models, which can be reformulated using the Ising language.

In the model of binary mixture each lattice site may be occupied by a type 1 atom (e.g., Zn) or by a type 2 atom (e.g., Cu). Each pair of nearest-neighbor atoms has a different energy, according to the type: e_1, e_2, e_{12} for pairs (11), (22), and (12), respectively. If we associate a type 1 atom with an up spin and a type 2 atom with a down spin, we can make the identification

$$e_1 = \epsilon_{++} = -J - \frac{2H}{\gamma} + K \tag{H.4}$$

$$e_2 = \epsilon_{--} = -J + \frac{2H}{\gamma} + K \tag{H.5}$$

$$e_{12} = \epsilon_{+-} = J + K \tag{H.6}$$

and obtain $J = -\frac{1}{4}(e_1+e_2)+\frac{1}{2}e_{12}$, $H = \frac{\gamma}{4}(e_2-e_1)$, and $K = \frac{1}{4}(e_1+e_2)+\frac{1}{2}e_{12}$. Therefore, we have a ferromagnetic interaction ($J > 0$) if the mean energy coupling between two particles of the same type is lower than the energy coupling between two different particles. The phase transition corresponds to a passage from a disordered mixture at high T to a separated phase at low T.

The lattice gas model describes a fluid, and the spin variable $s_i = \pm 1$ is replaced by the variable $n_i = 0, 1$, indicating whether the site is occupied by a particle or is empty. The energy is

$$\mathcal{H}_{\rm LG} = -\epsilon_0 \sum_{\langle ij \rangle} n_i n_j. \tag{H.7}$$

If we identify the particles with type 1 particles (or spin up) and the holes with type 2 particles (or down spins), i.e., $n_i = (s_i + 1)/2$, the model can be mapped to the binary mixture case with $e_1 = -\epsilon_0$ and $e_2 = e_{12} = 0$, so that the corresponding Ising model has the parameters $J = \epsilon_0/4$, $H = (\gamma/4)\epsilon_0$, and $K = -\epsilon_0/4$. The phase transition corresponds to the passage from a gas phase at high T to a condensed phase at low T.

H.1 The Ising Model in One Dimension

As mentioned in Section 3.2.2 the Ising model (3.10) in $d = 1$ for $H = 0$ exhibits a phase transition at $T_c = 0$. This can be easily shown by computing explicitly the partition function (3.12), while assuming periodic boundary conditions ($\sigma_{N+1} = \sigma_1$)

$$Z(N, T, J, H = 0) \equiv \sum_{\{\sigma_i\}} \exp(-\beta \mathcal{H}_{\rm I}) = \sum_{\{\sigma_i\}} \exp\left(\beta J \sum_{i=1}^{N} \sigma_i \sigma_{i+1}\right), \tag{H.8}$$

where $\beta = 1/T$ and the volume V has been replaced by the number of spins in the one-dimensional lattice, whose spacing a can be set equal to 1 without prejudice of generality, so that $V \equiv N$. We can rewrite (H.8) as

$$\begin{aligned} Z(N, T, J, H = 0) &= \sum_{\sigma_1} \sum_{\sigma_2} \cdots \sum_{\sigma_N} \prod_{i=1}^{N} \exp\left(\beta J \sigma_i \sigma_{i+1}\right) \\ &= \sum_{\sigma_1} \sum_{\sigma_2} \cdots \sum_{\sigma_N} c^N \prod_{i=1}^{N} (1 + t\sigma_i \sigma_{i+1}), \end{aligned} \tag{H.9}$$

where $c = \cosh(\beta J)$ and $t = \tanh(\beta J)$. Once we have noticed that $\sigma_i \sigma_{i+1} = \pm 1$, the last expression stems from the identity $\exp(\pm K) = \cosh(K)[1 \pm \tanh(K)]$ and from pulling c out of the product.

There are only two factors from the product of the N binomials that survive: those independent of any σ_i. In fact, any factor depending (linearly) on σ_i cancels out when performing the sums over the $\sigma_i = \pm 1$. So, the two surviving terms are the product of all "1"s and the product $\prod_{i=1}^{N} \sigma_i \sigma_{i+1} = \prod_{i=1}^{N} \sigma_i^2 = 1$, due to periodic boundary conditions. Then, we can write

$$Z(N, T, J, H = 0) = 2^N c^N (1 + t^N), \tag{H.10}$$

where the factor 2^N derives from the summations over σ_i. According to (3.13) and (3.23) the free energy density has the expression

$$f(T, J, H = 0) = T \lim_{N \to \infty} \frac{1}{N} \ln Z = T \ln[2 \cosh(\beta J)], \tag{H.11}$$

where we have used the result, valid for $|t| < 1$,

$$\lim_{N \to \infty} \frac{1}{N} \ln(1 + t^N) = 0. \tag{H.12}$$

The free energy density does not exhibit any singularity at finite temperature, because it is an analytic function of its variables for $0 < T \leq \infty$, so the Ising model in $d = 1$ has no phase transition for $T > 0$. On the other hand, for $T = 0$ the free energy reduces just to the internal energy and the equilibrium state coincides with the minimum of $\mathcal{H}_I$ for $H = 0$. In fact, there are two minima of $\mathcal{H}_I$ for $T = 0$, obtained for $\sigma_i = +1$ and $\sigma_i = -1$ for all i, that correspond to a magnetization density $m(T = 0, H = 0) = \pm 1$. We can conclude that the Ising model in $d = 1$ exhibits a discontinuous phase transition at $T = 0$, because it preserves $m(T, H = 0) = 0$ for any $T > 0$, while at $T = 0$ it exhibits a completely positive or negative magnetization.

This singular behaviour is a peculiar feature of thermodynamic fluctuations in $d = 1$, which can be illustrated by a simple argument. Let us suppose a fully magnetized state with $m = +1$ on a chain of length N with periodic boundary conditions and we want to evaluate the change of the free energy due to a fluctuation that makes n consecutive spins flip to $\sigma_i = -1$. The internal energy increases by an amount $4J$, due to the opposite orientation of spins at the boundaries of the flipped domain, while the entropy of the new state, i.e., the logarithm of the number of equivalent states, is of the order $\ln N$, because the equivalent states can be obtained by moving the reversed domain along the chain. Accordingly, for any finite T and in the thermodynamic limit the flipped state has a lower free energy than the fully magnetized one, because, despite the internal energy increasing by a finite amount, the entropic factor (i.e., the one that estimates fluctuations) increases by a macroscopic amount, proportional to $\ln N$. By iterating this argument, i.e., by further flipping new domains, we can conclude that for any finite T the equilibrium state corresponds to $m(T, H = 0) = 0$, because the entropic contribution to minimizing the free energy is always favored with respect to the increase of the internal energy. This argument does not apply to the Ising model in $d > 1$, because in order to reverse a macroscopic fraction of spins we need a domain wall, whose energy cost increases with the number of spins. Therefore, for $d > 1$ the internal energy and the entropy compete on an equal ground.

Since we know the exact solution of the Ising model in $d = 1$, studying its renormalization by the procedure discussed in Section 3.2.6 has only a pedagogical interest. Let us consider the partition function of the Ising model in $d = 1$ with an external magnetic field H,

$$Z(N, J, H) = \sum_{\{\sigma_i\}} \exp\left[\sum_{i=1}^{N} \left(J\sigma_i\sigma_{i+i} + \frac{H}{2}(\sigma_i + \sigma_{i+i})\right)\right], \tag{H.13}$$

where we have adopted the simplified notation $\beta J \to J$ and $\beta H \to H$. We can now construct a block transformation, in the spirit of the renormalization procedure described in Section 3.2.6. In particular, we can find an explicit expression of the renormalization transformation,

$$\mathcal{R}(J,H) = (J',H') \tag{H.14}$$

for a known value of the scale factor ℓ. This task can be accomplished by integrating the partition function (H.13) over all spins with even index. For this purpose, it is useful to rewrite (H.13) as

$$Z(N,J,H) = \sum_{\{\sigma_{2i+1}\}} \prod_{i=1}^{N/2} \sum_{\sigma_{2i}} \exp\Big(J\sigma_{2i}(\sigma_{2i-1}+\sigma_{2i+1}) + \frac{H}{2}(\sigma_{2i-1}+2\sigma_{2i}+\sigma_{2i+1})\Big). \tag{H.15}$$

Each factor of the product has the form

$$\sum_{\sigma_{2i}=\pm 1} e^{A\sigma_{2i}+B} = e^B \sum_{\sigma_{2i}=\pm 1} e^{A\sigma_{2i}} = 2e^B \cosh(A) \tag{H.16}$$

with

$$A = J(\sigma_{2i-1}+\sigma_{2i+1}) + H, \qquad B = \frac{H}{2}(\sigma_{2i-1}+\sigma_{2i+1}). \tag{H.17}$$

We can also use the equation

$$2\cosh(A) = 2\cosh\left(J(\sigma_{2i-1}+\sigma_{2i+1}) + H\right) \tag{H.18}$$

$$= A_0 \exp\Big(J'\sigma_{2i-1}\sigma_{2i+1} + \frac{\eta}{2}(\sigma_{2i-1}+\sigma_{2i+1})\Big), \tag{H.19}$$

where

$$J' = \frac{1}{4}\ln\left(\frac{\cosh(2J+H)\cosh(2J-H)}{\cosh^2(H)}\right) \tag{H.20}$$

$$\eta = \frac{1}{2}\ln\left(\frac{\cosh(2J+H)}{\cosh(2J-H)}\right) \tag{H.21}$$

$$A_0 = 2\cosh(H)\exp(J'), \tag{H.22}$$

to recover the same form of the Ising Hamiltonian after having traced over all the spins of even index $2i$, with J' given by (H.20) and $H' = H + \eta$.

We want to point out that all of these equations provide an explicit definition of the renormalization transformation $\mathcal{R}$ with $\ell = 2$ (see Section 3.2.6). By this decimation procedure the original Ising Hamiltonian defined on a lattice of N sites has been transformed into the new Ising Hamiltonian defined on a lattice of $N/2$ sites, whose spacing has been increased by a factor $\ell = 2$ and that has acquired an additional constant, thus,

$$\mathcal{H}(J,H) \to \mathcal{H}(J',H') + \ln A_0. \tag{H.23}$$

As a consequence, the free-energy densities of the original Hamiltonian and of the *decimated* Hamiltonian satisfy the relation

$$f(J,H) = \frac{1}{2}\ln A_0 + \frac{1}{2}f(J',H'), \tag{H.24}$$

where the factor 2 comes from $\ell = 2$ (see Eq. (3.96)), while the first term on the right-hand side is the additional constant mentioned in note 9 of Chapter 3.

Let us analyze the result of the renormalization procedure in the absence of an external field, i.e., $H = 0$. The above equations simplify to

$$J' = \frac{1}{2} \ln \cosh(2J) \tag{H.25}$$

$$H' = 0 \tag{H.26}$$

$$A_0 = 2\sqrt{\cosh(2J)}. \tag{H.27}$$

The first equation can be interpreted as a mapping $J' = F(J)$ that has only two fixed points, i.e., two solutions of the equation $J^* = F(J^*)$, $J_1^* = 0$ (which corresponds to $T = \infty$), and $J_2^* = +\infty$ (which corresponds to $T = 0^+$). On the other hand, starting from any finite value of J the iteration of the renormalization transformation maps J to the stable fixed point $J_1^* = 0$, because, for any finite J, $J' < J$. The physical interpretation is that for any finite initial coupling J the Ising Hamiltonian in $d = 1$ is renormalized to an infinite temperature state, where $m = 0$, with the exception of the case $T = 0$, where the model is fully magnetized, because its ground state energy is equivalent to the minimum of the free-energy density. This result, as expected, is consistent with the exact solution of the model.

H.2 The Renormalization of the Two-Dimensional Ising Model

The decimation procedure sketched in the previous section amounts to computing explicitly the contribution to the partition function of the spins located at even lattice sites in a spin chain with periodic boundary conditions. Apart from the additional constant, this procedure exactly transforms the original Hamiltonian into a renormalized one, which has the same form with new couplings that are functions of the original ones. This is a peculiar feature of the case $d = 1$ for a model with nearest-neighbor and local interactions, as in the Ising case with an external magnetic field. Here, we proceed in a pedagogical illustration of the technical problems inherent in the construction of an explicit renormalization procedure in $d > 1$ by considering the Ising model in $d = 2$ in the absence of an external magnetic field. In particular, we want to provide an example of how a decimation procedure requires us to include additional interactions and coupling constants in the renormalized Hamiltonian of the Ising model, thus yielding an effective description of the renormalization procedure that has to operate on an abstract space of renormalized Hamiltonians.

We consider a square lattice made of $N \times N$ sites and we impose periodic boundary conditions; i.e., the lattice has the topology of a toroidal surface. In analogy with the decimation procedure adopted for the case $d = 1$ (see Appendix H.1) we can decimate the partition function by summing over all of the spins whose coordinates (i,j) are such that $|i + j|$ is an even integer. These sites form the sublattice $\mathcal{S}_e$ of solid circles in the left panel of Fig. H.1, and the scale factor associated with the renormalization transformation is $\ell = \sqrt{2}$; see the right panel of the figure.

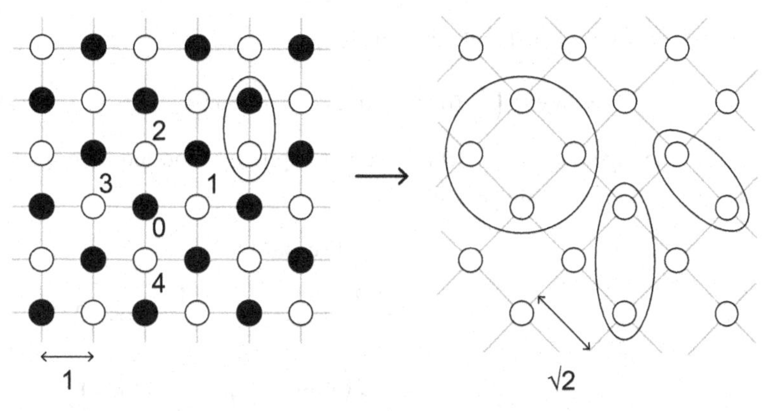

Fig. H.1 Illustration of one step of the renormalization group procedure for the Ising model on a square lattice, which allows passage from the left panel (lattice spacing equal to 1) to the right panel (lattice spacing equal to $\sqrt{2}$). The decimation procedure is obtained by summing over the configurations of the spins located at the solid circles. Starting from a Hamiltonian involving two-spin interactions between nearest-neighbor spins, we obtain two-spin interaction between nearest and next-to-nearest neighbors and four-spin interactions.

For the sake of simplicity we can fix our attention on one of this sites, identified by the symbolic index 0, and its four nearest-neighbor sites, identified by the symbolic indices from 1 to 4, which, by construction, do not belong to $\mathcal{S}_e$. We can write the partition function in the following form:

$$Z(N^2, J) = \sum_{\sigma_1,\dots,\sigma_4} \prod_{0\in\mathcal{S}_e} \sum_{\sigma_0=\pm 1} \exp\Big(J\sigma_0(\sigma_1+\sigma_2+\sigma_3+\sigma_4)\Big), \tag{H.28}$$

where the index "0" means a generic site belonging to $\mathcal{S}_e$ and "1, 2, 3, 4" are the indices of the neighboring sites of "0." Performing the sum over σ_0, we obtain

$$\sum_{\sigma_0=\pm 1} \exp\Big(J\sigma_0(\sigma_1+\sigma_2+\sigma_3+\sigma_4)\Big) = 2\cosh\Big(J(\sigma_1+\sigma_2+\sigma_3+\sigma_4)\Big). \tag{H.29}$$

We could recover the original Hamiltonian if it was possible to rewrite the right-hand side in the form

$$A_0 \exp\Big(\frac{J'}{2}(\sigma_1\sigma_2+\sigma_2\sigma_3+\sigma_3\sigma_4+\sigma_4\sigma_1)\Big), \tag{H.30}$$

where the factor $1/2$ is due to the fact that decimated squares provide two contributions to each one of the new links between nearest-neighbor sites.

We can easily realize that this is not the case, because the function on the right-hand side of (H.29) can take three different values for $\sigma_i = \pm 1$ and $i = 1, \dots, 4$,[1] so that we need at least three free parameters to be adjusted, not just two. Moreover, (H.29) is invariant under permutations of the coordinate indices $1, \dots, 4$, and if we want to rewrite it in exponential form we have to include all interaction terms among the undecimated spins that keep this property, while maintaining the Z_2 symmetry ($\sigma \to -\sigma$) of the original Hamiltonian. Therefore, we should rather think to rewrite the right-hand side of Eq. (H.29) as

[1] In fact, $\sum_{i=1}^4 \sigma_i = \pm 4, \pm 2, 0$ and $\cosh(x)$ is an even function of its argument.

$$2\cosh\Big(J(\sigma_1+\sigma_2+\sigma_3+\sigma_4)\Big) \tag{H.31}$$
$$= A_0 \exp\Big(\frac{J'}{2}(\sigma_1\sigma_2+\sigma_2\sigma_3+\sigma_3\sigma_4+\sigma_4\sigma_1+\sigma_1\sigma_3+\sigma_2\sigma_4)+K\sigma_1\sigma_2\sigma_3\sigma_4\Big).$$

Some algebra, whose details are not explicitly reported here, yields the equations of the renormalization transformation:

$$J' = \frac{1}{4}\ln\cosh(4J) \tag{H.32}$$

$$K = \frac{1}{8}\ln\cosh(4J) - \frac{1}{2}\ln\cosh(2J) \tag{H.33}$$

$$A_0 = 2\Big(\cosh(2J)\Big)^{\frac{1}{2}}\Big(\cosh(4J)\Big)^{\frac{1}{8}}. \tag{H.34}$$

We want to point out that the new Hamiltonian of the renormalized partition function contains not only interactions between nearest-neighbor spins on the rescaled square lattice with spacing $\ell = \sqrt{2}$, but also those between next-to-nearest neighbors (i.e., spins on the diagonals of the square cell) and the interaction among the four spins together (see Fig. H.1). If we iterate the decimation procedure, the new renormalized Hamiltonian is expected to contain further and more involved interaction terms, thus yielding a larger and larger set of equations necessary to define the renormalization transformation. It is clear that we have to devise a consistent strategy to reduce this avalanche of equations to an effective equation for the basic coupling J in the form[2]

$$J' = R(J), \tag{H.35}$$

where R should epitomize the contribution of the new coupling constants generated by the renormalization procedure. The simplest way of projecting the renormalization transformation onto the subspace of the parameter J is to consider just Eq. (H.32). As in the case $d = 1$, this relation yields only two trivial fixed points, the stable one $J_1^* = 0$ (i.e., $T = +\infty$) and the unstable one $J_2^* = \infty$ (i.e., $T = 0$), because $J' < J$ for any finite J. Once again, we obtain no phase transition at finite temperature, thus failing our expectations.

A heuristic recipe to obtain a nontrivial result amounts to defining an effective renormalized nearest-neighbor coupling on the square lattice, $\bar{J}'$, such that its contribution to the ground-state energy (all aligned spins) is the same as the one given, in the decimated Hamiltonian, by the nearest- and next-to-nearest neighbor terms,[3]

$$\bar{J}'(\sigma_1\sigma_2+\sigma_2\sigma_3+\sigma_3\sigma_4+\sigma_4\sigma_1) = J'(\sigma_1\sigma_2+\sigma_2\sigma_3+\sigma_3\sigma_4+\sigma_4\sigma_1+\sigma_1\sigma_3+\sigma_2\sigma_4), \tag{H.36}$$

thus yielding the relation

$$\bar{J}' = \frac{3}{2}J' = \frac{3}{8}\ln\cosh(4J). \tag{H.37}$$

[2] In principle J should be a vector of different coupling constants, including the original nearest-neighbor coupling. We make things simple just considering one single coupling, the basic one.

[3] This might appear to be a reasonable physical assumption, because it is based on the idea of keeping constant the minimum value of the internal energy of the original and of the decimated Hamiltonian, while ignoring the four-body interaction term, usually called plaquette. On the other hand, the degree of arbitrariness of this heuristic recipe becomes evident as soon as we observe that also including the plaquette contribution to the ground state energy, we recover only the two trivial fixed points $J_1^* = 0$ (attractive) and $J_1^* = +\infty$ (repulsive).

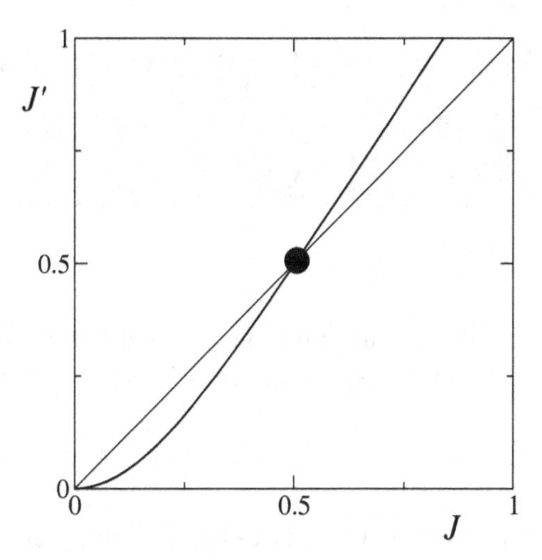

Fig. H.2 Plot of Eq. (H.38) (thick line), showing the existence of a nontrivial fixed point $J' = J$, located at the intersection with the diagonal (thin line).

Now, we identify $\bar{J}'$ with J' and finally obtain the effective renormalization transformation,

$$J' = \frac{3}{8} \ln \cosh(4J). \tag{H.38}$$

As sketched in Fig. H.2, in addition to the usual fixed points of infinite temperature ($J_1^* = 0$) and zero temperature ($J_2^* = \infty$), this map admits a nontrivial fixed point, located at $J_c^* = 0.50689$: it corresponds to the expected finite critical temperature $T_c = (J_c^*)^{-1}$. It is an unstable fixed point, because $J' < J$ for $J < J_c^*$ and $J' > J$ for $J > J_c^*$. Despite the adopted heuristic recipe necessarily introducing some level of approximation, we want to point out that the estimated value of J_c^* is not far from the exact one of $J_c^* = 0.44068$ obtained by the Onsager solution of the Ising model in two dimensions.

Making use of Eqs. (3.99) and (3.101) we can obtain an estimate of the critical exponent ν. We can first estimate τ by expressing (3.99) in the form

$$J' - J_c^* = \left.\frac{dJ'}{dJ}\right|_{J_c^*} \left(J - J_c^*\right) + o\left((J - J_c^*)^2\right), \tag{H.39}$$

where we have assumed $|J - J_c^*| \ll 1$, so that

$$\tau = \left.\frac{dJ'}{dJ}\right|_{J_c^*} = 1.44892. \tag{H.40}$$

It follows that

$$\frac{1}{\nu} = \frac{\ln \tau}{\ln \ell} = \frac{\ln 1.44892}{\ln \sqrt{2}} \approx 1.06996, \tag{H.41}$$

to be compared with the exact result $\nu = 1$ and with the classical exponent of the Landau mean-field theory, $\nu = 1/2$.

This result allows us to appreciate the effectiveness of the renormalization group approach to the study of critical phenomena, even if affected by some tricky approximation.

Appendix I Derivation of the Ginzburg–Landau Free Energy

Let us consider an Ising model in general dimension d, with a ferromagnetic interaction, $J_{ij} > 0$, which depends on the distance between the lattice sites i and j,

$$\mathcal{H} = -\frac{1}{2}\sum_{i,j} J_{ij}\sigma_i\sigma_j, \tag{I.1}$$

In a continuum spirit, we replace σ_i by $m(\mathbf{x})$, the average magnetization in a cube V of side ℓ, centered in $\mathbf{x}$,

$$m(\mathbf{x}) = \frac{M(\mathbf{x})}{\ell^d} \equiv \frac{\sum_{i\in V}\sigma_i}{\ell^d} \tag{I.2}$$

and

$$\begin{aligned}
\mathcal{H} &= -\frac{1}{2}\int d\mathbf{x}d\mathbf{x}' J(|\mathbf{x}-\mathbf{x}'|)m(\mathbf{x})m(\mathbf{x}') & \text{(I.3)}\\
&= \frac{1}{4}\int d\mathbf{x}d\mathbf{x}' J(|\mathbf{x}-\mathbf{x}'|)\left[\left(m(\mathbf{x})-m(\mathbf{x}')\right)^2 - \left(m^2(\mathbf{x})+m^2(\mathbf{x}')\right)\right] & \text{(I.4)}\\
&\simeq \frac{1}{4}\int d\mathbf{x}d\mathbf{r}J(r)\,(\nabla m\cdot\mathbf{r})^2 - \frac{1}{2}\int d\mathbf{x}d\mathbf{r}J(r)m^2(\mathbf{x}) & \text{(I.5)}\\
&= \frac{1}{4}\sum_i^d \int d\mathbf{r}J(r)r_i^2 \int d\mathbf{x}\left(\frac{\partial m}{\partial x_i}\right)^2 - \frac{1}{2}\int d\mathbf{r}J(r)\int d\mathbf{x}m^2(\mathbf{x}). & \text{(I.6)}
\end{aligned}$$

The integral $\int d\mathbf{r}J(r)r_i^2$ does not depend on the direction i, so we can define the quantities

$$K = \frac{1}{2}\int d\mathbf{r}J(r)r_i^2 = \frac{1}{2d}\int d\mathbf{r}J(r)r^2 \tag{I.7}$$

$$\tilde{J} = \int d\mathbf{r}J(r) \tag{I.8}$$

and finally write

$$\mathcal{H} = \frac{K}{2}\int d\mathbf{x}(\nabla m)^2 - \frac{\tilde{J}}{2}\int d\mathbf{x}\, m^2(\mathbf{x}). \tag{I.9}$$

Let us now evaluate the entropy term counting the number of microscopic configurations corresponding to the same value of $m(\mathbf{x})$. This task is easily accomplished if we observe that $M(\mathbf{x}) = N_+ - N_-$, where $N_\pm$ are the number of up and down spins in the volume V and $N = N_+ + N_- = (\ell/a)^d$ is the total number of spins in V, a being the lattice constant. In fact, the entropy of volume V is given by $S(\mathbf{x}) = \ln W$, where

$$W = \frac{N!}{N_+! N_-!}. \tag{I.10}$$

Using the Stirling approximation $N! \simeq (N/e)^N$, we find

$$S(\mathbf{x}) = N \ln 2 - \frac{1}{2}(N+M) \ln\left(1 + \frac{M}{N}\right) - \frac{1}{2}(N-M) \ln\left(1 - \frac{M}{N}\right), \tag{I.11}$$

where the spatial dependence of M on $\mathbf{x}$ has not been made explicit to simplify the notation. Finally, it is easier to normalize the magnetization m so that its maximum value is ± 1,

$$u(\mathbf{x}) \equiv \frac{M}{N} = a^d m(\mathbf{x}). \tag{I.12}$$

If $\mathcal{S} = \int d\mathbf{x} S(\mathbf{x})$ is the total entropy, the Helmholtz free energy $\mathcal{F} = \mathcal{H} - T\mathcal{S}$ is written as

$$\begin{aligned} \mathcal{F} &= \int \frac{d\mathbf{x}}{a^{2d}} \left[\frac{K}{2}(\nabla u)^2 - \frac{\tilde{J}}{2} u^2(\mathbf{x}) \right] \\ &\quad - \frac{T}{V} \int d\mathbf{x} N \left[\ln 2 - \frac{1}{2}(1+u)\ln(1+u) - \frac{1}{2}(1-u)\ln(1-u) \right] \\ &= \int d\mathbf{x} \left[\frac{K}{2a^{2d}} (\nabla u)^2 + \mathcal{V}(u) \right], \end{aligned} \tag{I.13}$$

where

$$\mathcal{V}(u) = \frac{\tilde{J}}{a^{2d}} \left\{ -\frac{u^2}{2} + \frac{Ta^d}{\tilde{J}} \left[-\ln 2 + \frac{1}{2}(1+u)\ln(1+u) + \frac{1}{2}(1-u)\ln(1-u) \right] \right\}. \tag{I.14}$$

The function $\mathcal{V}(u)$ has a single minimum in $u = 0$ if $T > T_c = \tilde{J}/a^d$, and two symmetric minima if $T < T_c$. In Fig. I.1, left panel, we plot $\mathcal{V}^*(u) = (a^{2d}/\tilde{J})\mathcal{V}(u)$, for different values of $T^* = T/T_c$,

$$\mathcal{V}^*(u) = -\frac{u^2}{2} + T^* \left[-\ln 2 + \frac{1}{2}(1+u)\ln(1+u) + \frac{1}{2}(1-u)\ln(1-u) \right]. \tag{I.15}$$

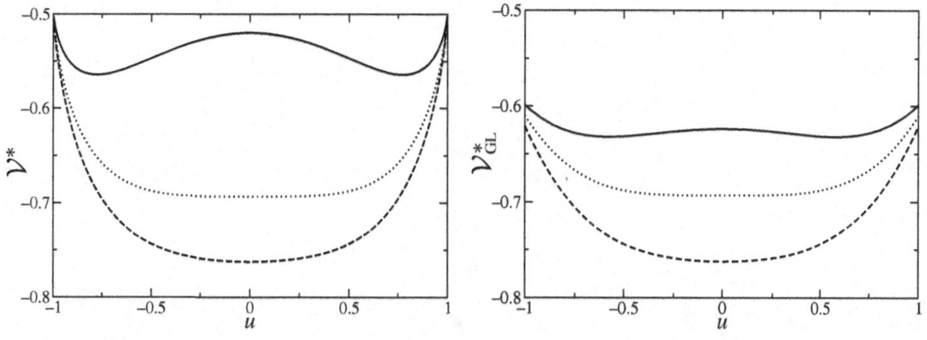

Fig. I.1 The potentials $\mathcal{V}^*(u)$ (left) and $\mathcal{V}^*_{GL}(u)$ (right) for different values of T^*: $T^* = 1.1$ (dashed line), $T^* = T^*_c = 1$ (dotted line), $T^* = 0.9$ (solid line).

It is worth noting that physically it must be $|u| \leq 1$. In the limit $u \to 1$

$$\mathcal{V}^*(u) = -\frac{1}{2} + \frac{T^*}{2}(1-u)\ln(1-u) + (1-u)\left(1 - \frac{T^*}{2}(1+\ln 2)\right) + O((1-u)^2). \quad (I.16)$$

Therefore $\mathcal{V}^*(1)$ is finite, but $(\mathcal{V}^*)'(1) \to +\infty$, which implements the above constraint $|u| \leq 1$. Furthermore, minimizing the above expression we find that for small T^* the minima of the potential occur for $\bar{u} = \pm(1 - \exp(-2/T^*))$.

In the literature, the potential $\mathcal{V}^*(u)$ is rarely used. Rather, it is common to use the classical quartic potential, which comes out analytically from $\mathcal{V}^*(u)$ in the limit of small u, i.e., in the vicinity of T_c. Using the Taylor expansion $\ln(1+u) = \sum_{n\geq 1}(-1)^{n+1}u^n/n$ up to fourth order, we obtain

$$\mathcal{V}^*_{\rm GL}(u) = -T^* \ln 2 + \frac{(T^*-1)}{2}u^2 + \frac{T^*}{12}u^4, \quad (I.17)$$

which is also plotted in Fig. I.1, right panel.

If we enter this potential within Eq. (I.13), we get

$$\mathcal{F}_{\rm GL} = \frac{\tilde{J}}{a^{2d}} \int d\mathbf{x} \left[\frac{K}{\tilde{J}}(\nabla u)^2 + \frac{T^*-1}{2}u^2 + \frac{T^*}{12}u^4 \right], \quad (I.18)$$

where a constant in the integrand has been removed.

It is worth noting that by rescaling the space, $\mathbf{x} \to b\mathbf{x}$, and the order parameter, $u \to cu$, it is possible to rewrite $\mathcal{F}_{\rm GL}$ so as to display one parameter only, the scale of energy:

$$\mathcal{F}_{\rm GL} = f_0 \int d\mathbf{x} \left[\frac{1}{2}(\nabla u)^2 \pm \frac{1}{2}u^2 + \frac{1}{4}u^4 \right], \quad (I.19)$$

where the sign in front of the quadratic term is the same sign of $(T - T_c)$ and the numerical prefactors $\frac{1}{2}$ and $\frac{1}{4}$ can be set to any other positive values.

It is useful to also rewrite the expression (I.13) in rescaled variables,

$$\mathcal{F}_{\rm GL} = f_0 \int d\mathbf{x} \left\{ \frac{1}{2}(\nabla u)^2 - \frac{u^2}{2} + \frac{T}{T_c}\left[-\ln 2 + \frac{1}{2}(1+u)\ln(1+u) + \frac{1}{2}(1-u)\ln(1-u) \right] \right\}. \quad (I.20)$$

Appendix J Kinetic Monte Carlo

We describe here the basics of a kinetic Monte Carlo simulation, using the original formulation by A. B. Bortz, M. H. Kalos, and J. L. Lebowitz and applying it to the bridge model; see Fig. J.1.

A discrete and stochastic model is first of all defined by specifying the configuration space, then by assigning the transition rates $\nu_{i\to j}$ between state i and state j. For a given microscopic state i at time t, we can think to list *all* allowed transitions, whose rates are described for simplicity by $\nu_1, \nu_2, \ldots, \nu_M$. Making explicit reference to Fig. J.1 for the initial configuration, we have $M = 6$, with $\nu_1 = \nu_2 = \nu_3 = 1$ (hops toward empty sites), $\nu_4 = q$ (exchange between a positive and a negative particle), $\nu_5 = \alpha$ (input of a negative particle), and $\nu_6 = \beta$ (removal of a negative particle). We can now define $\nu_{\text{tot}} = \sum_{k=1}^{M} \nu_k$ and represent a segment of length ν_{tot} as in Fig. J.1.

We now extract a random number, uniformly distributed in the interval $[0, \nu_{\text{tot}}]$, select the corresponding transition, and perform it. It is clear that a given transition k has the probability $p_k = \nu_k/\nu_{\text{tot}}$ to occur. This is a so-called rejection-free algorithm, because it provides an effective evolution at any time step.

The last step is to increment time by a quantity Δt that is exponentially distributed, according to

$$p(\Delta t) = \nu_{\text{tot}} \exp(-\nu_{\text{tot}} \Delta t). \tag{J.1}$$

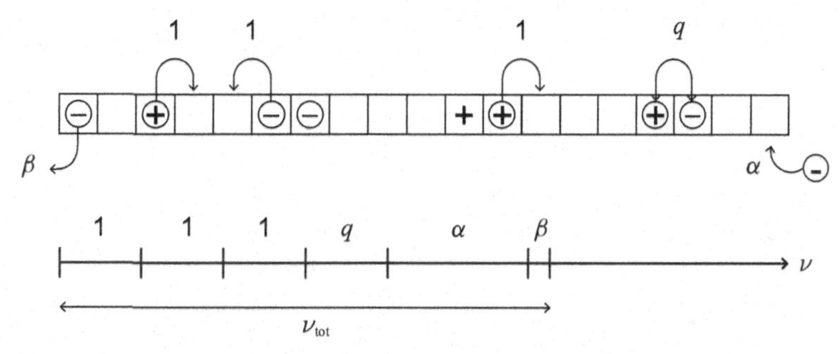

Fig. J.1 Top: Sketch of a specific microscopic configuration for the bridge model, with three positive particles and four negative particles and the indication of allowed transitions and of their rates: three hops toward empty sites (rate 1), one exchange between a positive and a negative particle (rate q), the input of a negative particle from the right (rate α), and the output of a negative particle from the left (rate β). Bottom: Each transition rate $\nu \neq 0$ is indicated by a segment of length ν. The choice of a random number in the interval $[0, \nu_{\text{tot}}]$ allows us to select (and implement) a specific transition.

The use of this distribution is only useful if we need to study correlations on time scales of the order or shorter than $1/\nu_{tot}$; otherwise, we can take Δt equal to its average value, $\langle \Delta t \rangle = 1/\nu_{tot}$, at any step.

The above dynamics is particularly effective if there are processes with transition rates that vary by orders of magnitude. This occurs, for example, in the conserved, Kawasaki dynamics of an Ising model (see the discussion in Section 6.2.2) or in the bridge model in the regime $\beta \ll \alpha$ (see Section 4.5). If the nonvanishing transition rates are not so different, simplified algorithms are available. For example, we may list the allowed processes, choose one with its probability, and increment the time by $1/K$, where K is the total number of allowed processes. Finally, we may even choose disallowed processes, therefore getting an algorithm that is not rejection free. For example, think of the evolution rules of the TASEP model with open boundary conditions (see Section 4.4.2): we choose at random a site i and check if hopping/insertion/extraction of the particle in site i is allowed. Then, we increment the time by $1/L$, the inverse of the system size.

Appendix K The Mean-field Phase Diagram of the Bridge Model

In the main text we found that for $q = 1$ the bridge model is equivalent to two independent TASEP models (each model for a class of particles), with input/output parameters equal to (α_+, β) and (α_-, β) for positive and negative particles, respectively, where

$$\alpha_+ = \frac{J_+}{\dfrac{J_+}{\alpha} + \dfrac{J_-}{\beta}} \qquad \text{and} \qquad \alpha_- = \frac{J_-}{\dfrac{J_-}{\alpha} + \dfrac{J_+}{\beta}}. \tag{K.1}$$

We now assume that the bridge model is in a given phase and use the above equations to find if such solution exists for some values of α and β. Since symmetric solutions are easier to study, we start with them.

K.1 Symmetric Solutions

If $J_+ = J_- = \tilde{J}$, we also have

$$\alpha_+ = \alpha_- = \tilde{\alpha} = \frac{\alpha\beta}{\alpha + \beta}. \tag{K.2}$$

The symmetric low-density (LD) phase exists if $\tilde{\alpha} < \beta$ and $\tilde{\alpha} < \frac{1}{2}$. The former condition (see Eq. (K.2)) is always satisfied, while the latter gives

$$\boxed{\text{LD} \quad \frac{1}{\alpha} + \frac{1}{\beta} > 2} \tag{K.3}$$

We can also check formally that the symmetric high-density (HD) phase cannot exist. In fact, it would require $\tilde{\alpha} > \beta$ and $\beta < \frac{1}{2}$, but the first condition cannot be satisfied. Finally, the symmetric maximal current (MC) phase exists if $\tilde{\alpha} \geq \frac{1}{2}$ and $\beta \geq \frac{1}{2}$. The former condition gives

$$\boxed{\text{MC} \quad \frac{1}{\alpha} + \frac{1}{\beta} \leq 2} \tag{K.4}$$

so the latter is automatically satisfied.

K.2 The Asymmetric HD–LD Phase

We now consider the interesting possibility that one class of particles (e.g., the positive ones) is in the HD phase and the other class (the negative ones) is in the LD phase. This choice implies the following relations and conditions:

$$\begin{array}{llll} \oplus \text{ HD} & J_+ = \beta(1-\beta), & \alpha_+ > \beta, & \beta < \frac{1}{2} \\ \ominus \text{ LD} & J_- = \alpha_-(1-\alpha_-), & \alpha_- < \beta, & \alpha_- < \frac{1}{2}. \end{array} \tag{K.5}$$

We immediately remark that the second condition on α_- is redundant. Then, we can replace the explicit expressions for $J_\pm$ in Eq. (K.1) and get

$$\alpha_+ = \frac{\beta(1-\beta)}{\frac{\beta(1-\beta)}{\alpha} + \frac{\alpha_-(1-\alpha_-)}{\beta}}, \qquad \alpha_- = \frac{\alpha_-(1-\alpha_-)}{\frac{\alpha_-(1-\alpha_-)}{\alpha} + \frac{\beta(1-\beta)}{\beta}}. \tag{K.6}$$

The second equation is a simple quadratic equation for α_-, whose solutions is

$$\alpha_- = \frac{(1+\alpha) \pm \sqrt{(1+\alpha)^2 - 4\alpha\beta}}{2} \tag{K.7}$$

and whose discriminant is certainly positive if $\beta < \frac{1}{2}$. The solution with the positive sign must be rejected, because α_- would be larger than $\frac{1}{2}$. Summarizing, we have

$$\alpha_- = \frac{1+\alpha - \sqrt{(1+\alpha)^2 - 4\alpha\beta}}{2}, \qquad \alpha_+ = \frac{\beta(1-\beta)}{\frac{\beta(1-\beta)}{\alpha} + \frac{\alpha_-(1-\alpha_-)}{\beta}}. \tag{K.8}$$

Let us first test the condition $\alpha_- < \beta$, i.e.,

$$\frac{1+\alpha - \sqrt{(1+\alpha)^2 - 4\alpha\beta}}{2} < \beta. \tag{K.9}$$

Isolating the square root, we find

$$1 + \alpha - 2\beta < \sqrt{(1+\alpha)^2 - 4\alpha\beta}. \tag{K.10}$$

Since $\beta < \frac{1}{2}$ and $\alpha > 0$, the left-hand side is positive and we can take the square of both terms, obtaining the condition $4\beta^2 < 4\beta$, which is manifestly satisfied for $\beta < \frac{1}{2}$. Therefore, we are left with the conditions $\beta < \frac{1}{2}$ and $\alpha_+ > \beta$. The latter inequality (see Eq. (K.8), right) is rewritten as

$$\frac{\beta(1-\beta)}{\alpha} + \frac{\alpha_-(1-\alpha_-)}{\beta} < 1 - \beta. \tag{K.11}$$

Replacing the expression of $\alpha_-(\alpha, \beta)$ given in (K.8), left, we get

$$\boxed{\text{HD–LD} \quad \beta^2(1+\alpha-\beta) + \frac{\alpha^2}{2}\left(2\beta + \sqrt{(1+\alpha)^2 - 4\alpha\beta} - \alpha - 1\right) - \alpha\beta < 0} \tag{K.12}$$

Let us focus analytically on the case $\beta \ll 1$. In this limit, we get

$$-\left(\frac{\alpha}{1+\alpha}\right)\beta + O(\beta^2) < 0, \tag{K.13}$$

which is surely satisfied if α is not vanishing. If α also is vanishing, higher-order terms should be accounted for. The limit $\alpha \to \infty$ requires expansion of the square root in (K.12) up to α^{-2} terms and gives the condition $\beta < \frac{1}{3}$. The numerical solution of Eq. (K.12) gives the lower line of Fig. 4.11, $\beta_1(\alpha)$. Since $\beta_1(\alpha) < \frac{1}{3}$ for any α, the condition $\beta < \frac{1}{2}$ is redundant and the asymmetric LD–HD phase exists for $\beta < \beta_1(\alpha)$.

K.3 The Asymmetric MC–LD Phase

We assume, without loss of generality, that positive particles are in the MC phase and negative particles in the LD phase, so that

$$\begin{aligned} &\oplus\,\text{MC} \quad J_+ = \tfrac{1}{4}, \qquad && \alpha_+ \geq \tfrac{1}{2}, \qquad && \beta \geq \tfrac{1}{2} \\ &\ominus\,\text{LD} \quad J_- = \alpha_-(1-\alpha_-), \qquad && \alpha_- < \beta, \qquad && \alpha_- < \tfrac{1}{2}\,. \end{aligned} \tag{K.14}$$

We easily remark that the first condition on α_- is redundant. We can replace the explicit expressions for $J_\pm$ in Eq. (K.1) and get

$$\alpha_+ = \frac{\frac{1}{4}}{\dfrac{1}{4\alpha} + \dfrac{\alpha_-(1-\alpha_-)}{\beta}}, \qquad \alpha_- = \frac{\alpha_-(1-\alpha_-)}{\dfrac{\alpha_-(1-\alpha_-)}{\alpha} + \dfrac{1}{4\beta}}. \tag{K.15}$$

If we define the quantity $\gamma = \alpha_-(1-\alpha_-)$, the two previous equations (K.15) can be rewritten in compact form as

$$\alpha_+^{-1} = \alpha^{-1} + 4\gamma\beta^{-1}, \qquad \alpha_-^{-1} = \alpha^{-1} + \frac{\beta^{-1}}{4\gamma}, \tag{K.16}$$

whence

$$\alpha_+^{-1} - \alpha_-^{-1} = \beta^{-1}\left(4\gamma + \frac{1}{4\gamma}\right) \geq 0, \tag{K.17}$$

where we have used that both γ and β are positive. However, from the conditions $\alpha_+ \geq \frac{1}{2}$ and $\alpha_- < \frac{1}{2}$, we deduce that $\alpha_+^{-1} - \alpha_-^{-1} < 0$, which is in contradiction with Eq. (K.17). Therefore, there are no solutions satisfying the condition for the existence of the MC–LD phase, which is forbidden.

K.4 The Asymmetric LD–LD Phase

As strange as it may seem, this asymmetric phase is legitimate: both types of particles are in the LD phase, but they have different densities (which also hints at the impossibility of having an asymmetric MC–MC phase). In short, we must have

$$\begin{array}{llll} \oplus \text{ LD} & J_+ = \alpha_+(1-\alpha_+), & \alpha_+ < \beta, & \alpha_+ < \frac{1}{2} \\ \ominus \text{ LD} & J_- = \alpha_-(1-\alpha_-), & \alpha_- < \beta, & \alpha_- < \frac{1}{2} \end{array} \tag{K.18}$$

with

$$\alpha_+ = \frac{\alpha_+(1-\alpha_+)}{\dfrac{\alpha_+(1-\alpha_+)}{\alpha} + \dfrac{\alpha_-(1-\alpha_-)}{\beta}}, \qquad \alpha_- = \frac{\alpha_-(1-\alpha_-)}{\dfrac{\alpha_-(1-\alpha_-)}{\alpha} + \dfrac{\alpha_+(1-\alpha_+)}{\beta}}, \tag{K.19}$$

which can be simplified to the expressions

$$\alpha_+ = 1 - \frac{\alpha_+(1-\alpha_+)}{\alpha} - \frac{\alpha_-(1-\alpha_-)}{\beta}, \qquad \alpha_- = 1 - \frac{\alpha_-(1-\alpha_-)}{\alpha} - \frac{\alpha_+(1-\alpha_+)}{\beta}. \tag{K.20}$$

Defining the quantities $S = \alpha_+ + \alpha_-$ and $D = \alpha_+ - \alpha_-$ and taking the sum and the difference of $\alpha_\pm$ as given above we obtain

$$S = 1 - \left(\frac{\alpha\beta}{\alpha - \beta}\right), \qquad D^2 = (2 - S)\left(S - \frac{2\alpha\beta}{\alpha + \beta}\right). \tag{K.21}$$

Since $S = \alpha_+ + \alpha_- < 1$, the condition $D^2 > 0$ implies $S > 2\alpha\beta/(\alpha + \beta)$, i.e.,

$$\boxed{\text{LD–LD} \quad \frac{\alpha\beta}{\alpha + \beta} < \frac{1}{2}\left(1 - \frac{\alpha\beta}{\alpha - \beta}\right)} \tag{K.22}$$

The line $\beta = \beta_2(\alpha)$, where the two sides are equal, corresponds to the vanishing of D^2, i.e., $\alpha_+ = \alpha_-$. Therefore, it means we are passing toward the symmetric LD phase. As for the line $\beta = \beta_1(\alpha)$, corresponding to the vanishing of the left-hand side of Eq. (K.12), it means the transition between the HD–LD phase toward the LD–LD phase. In conclusion, the LD-LD phase is expected to exist in the (α, β) region defined by the conditions $\beta_1(\alpha) < \beta < \beta_2(\alpha)$. We avoid giving a more rigorous proof of this statement.

Appendix L The Deterministic KPZ Equation and the Burgers Equation

In this appendix we provide some insights into the one-dimensional deterministic KPZ (dKPZ) equation,

$$\partial_t h(x,t) = \nu h_{xx} + \frac{\lambda}{2} h_x^2, \tag{L.1}$$

and into the Burgers equation, obtained from it by taking the spatial derivative of both sides,

$$\partial_t m(x,t) = \nu m_{xx} + \lambda m m_x. \tag{L.2}$$

We are going to argue that dynamics leads to the formation of negative curvature parabolas, which are connected by small regions of high positive curvature, corresponding to moving fronts in the Burgers equation. Below we prove that a parabola is a self-similar solution of the dKPZ equation and we derive the expression of a front solution for the Burgers equation. Finally, we use the Cole–Hopf transformation to provide a formally exact solution of the dKPZ equation for any initial condition and in any dimension d.

Qualitative considerations of the evolution of a parabola can be easily drawn by considering the different contributions of the linear and nonlinear terms. The linear term is constant, either positive or negative depending on the sign of the curvature. The nonlinear term is always positive and increases with the distance from the vertex of the parabola. The combination of the two terms is graphically depicted in Fig. L.1: a parabola of negative curvature moves in the negative direction and flattens, while a parabola of positive curvature moves in the positive direction and steepens. Let us be more quantitative, assuming the functional form

$$h(x,t) = A(t) + C(t)(x - x_0)^2. \tag{L.3}$$

Evaluating its temporal and spatial derivatives, we find

$$\dot{A} + \dot{C}(x - x_0)^2 = 2\nu C + 2\lambda C^2 (x - x_0)^2, \tag{L.4}$$

whose solution is

$$A(t) = A(0) - \frac{\nu}{\lambda} \ln \left| \frac{1}{C(0)} - 2\lambda t \right| \tag{L.5}$$

and

$$C(t) = \frac{1}{\dfrac{1}{C(0)} - 2\lambda t}. \tag{L.6}$$

Therefore, as anticipated, a parabola with negative curvature moves downward and it flattens with time, $|C(t)| \approx 1/t$, while a parabola with positive curvature moves upward and

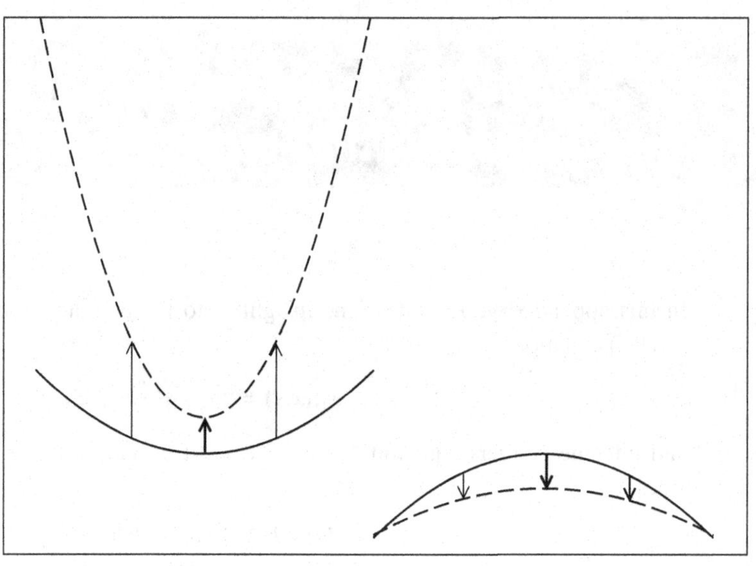

Fig. L.1 Evolution of two parabolas, with negative and positive curvature. The thick arrow at the vertex of the paraboloid is the constant contribution of the linear term, the nonlinear term vanishing there. Moving away from the center we must add the increasingly important contribution of the nonlinear term, which is always positive (for $\lambda > 0$).

its curvature increases in time, diverging at a finite time $\tau_c = 1/(2\lambda C(0))$. In a continuum profile, where regions with positive and negative curvature alternate, this divergence disappears and is replaced by a finite curvature dip, easily derivable in the context of the Burgers equation where it corresponds to a kink or a front. So, let us turn to Eq. (L.2) and look for a front-type solution, i.e., a function $m(x,t) = u(x - vt)$ with $u(x) \to \pm u_\pm$ for $x \to \pm\infty$. Since $\partial_t u(x - vt) = -v\partial_x u$, the sought-after function $u(x)$ satisfies the equation

$$- vu_x = \nu u_{xx} + \lambda u u_x. \tag{L.7}$$

Integrating between $-\infty$ and $+\infty$ we find

$$v = -\frac{\lambda}{2}(u_+ - u_-). \tag{L.8}$$

So, perfectly antisymmetric boundary conditions ($u_+ = u_- = u_0$) induce a static kink, while an asymmetry sets the front in motion.

The form of the kink can be found by integrating Eq. (L.7). We do it explicitly for the static kink only, i.e., $v = 0$. In this case, after integration we obtain $\nu u_x = \frac{\lambda}{2}(u_0^2 - u^2)$, whose solution is

$$u(x) = u_0 \tanh\left(\frac{\lambda u_o}{2\nu}x\right), \tag{L.9}$$

which can also be integrated to provide the dip-like profile for the dKPZ equation,

$$h(x,t) = \frac{2\nu}{\lambda} \ln\cosh\left(\frac{\lambda u_o}{2\nu}x\right) + \frac{\lambda u_0^2}{2}t. \tag{L.10}$$

The first term on the right-hand side represents a function that interpolates between the straight lines $\mp u_0 x$, and the second term is a drift, quite similar to the drifts felt by the self-similar parabolas. In a "real" profile, the continuity between the pieces of parabola and the pieces of dip-like profile generally imposes asymmetric boundary conditions on the latter, therefore determining their drift and a sort of process where bigger parabolas "eat" smaller ones.

We conclude this appendix using the Cole–Hopf transformation to solve the dKPZ equation exactly in any dimension d,

$$\partial_t h(\mathbf{x}, t) = \nu \nabla^2 h + \frac{\lambda}{2}(\nabla h)^2. \tag{L.11}$$

The Cole–Hopf transformation is

$$H(\mathbf{x}, t) = e^{\frac{\lambda}{2\nu} h(\mathbf{x}, t)}, \tag{L.12}$$

which transforms the nonlinear Eq. (L.11) into a much simpler diffusion equation,

$$\partial_t H = \nu \nabla^2 H. \tag{L.13}$$

If H evolves according to the diffusion equation, it is obvious to conclude that a paraboloid is a self-similar solution of the dKPZ: in fact a Gaussian is a self-similar solution of the diffusion equation, and the logarithm of a Gaussian, $h(\mathbf{x}, t) = (2\nu/\lambda) \ln H(\mathbf{x}, t)$, is a paraboloid.

The general solution of H is straightforward if we recall that a Gaussian is the solution of the diffusion equation corresponding to a Dirac delta as the initial condition. Since any initial condition $H(\mathbf{x}, 0)$ is a linear combination of Dirac delta functions centered at all points $\mathbf{x}'$, $H(\mathbf{x}, 0) = \int d\mathbf{x}' \delta(\mathbf{x} - \mathbf{x}') H(\mathbf{x}', 0)$, the general solution of H is the same linear combination of the evolution of such a Dirac delta, i.e., Gaussian functions:

$$H(\mathbf{x}, t) = \int \frac{d\mathbf{x}'}{(4\pi \nu t)^{d/2}} \exp\left[-\frac{(\mathbf{x} - \mathbf{x}')^2}{4\nu t}\right] H(\mathbf{x}', 0). \tag{L.14}$$

If we now invert the Cole–Hopf transformation, Eq. (L.12), we get the general solution of the dKPZ equation:

$$h(\mathbf{x}, t) = \frac{2\nu}{\lambda} \ln \left\{ \int \frac{d\mathbf{x}'}{(4\pi \nu t)^{d/2}} \exp\left[-\frac{(\mathbf{x} - \mathbf{x}')^2}{4\nu t} + \frac{\lambda}{2\nu} h(\mathbf{x}', 0)\right] \right\}. \tag{L.15}$$

This result might suggest a strong simplification of the stochastic equation (5.41) as well. This does not happen, because of noise transformation. If we start from

$$\partial_t h = \nu \nabla^2 h + \frac{\lambda}{2}(\nabla h)^2 + \eta, \tag{L.16}$$

we obtain

$$\partial_t H = \frac{\lambda}{2\nu} H \partial_t h \tag{L.17}$$

$$= \frac{\lambda}{2\nu} H\left(\nu \nabla^2 h + \frac{\lambda}{2}(\nabla h)^2 + \eta\right) \tag{L.18}$$

$$= \nu \nabla^2 H + \frac{\lambda}{2\nu} H \eta(\mathbf{x}, t), \tag{L.19}$$

which is known as the stochastic heat equation. Therefore, the stochastic term now depends on H itself: it is called multiplicative noise and can also be found in the mean-field description of directed percolation; see Section 3.6.3. Multiplicative noise, which is more difficult to treat than standard additive noise, also poses the problem of choosing the Itô or the Stratonovich representation of stochastic noise.

L.1 The Functional Derivative

Throughout the book we meet quantities such as the following,

$$\mathcal{F}[h] = \int d\mathbf{x} f(\mathbf{x}, h(\mathbf{x}), \nabla h(\mathbf{x}), \dots), \tag{L.20}$$

and we may need to evaluate the variation of $\mathcal{F}[h]$ induced by an infinitesimal variation of $h(\mathbf{x})$,

$$\delta\mathcal{F}[h] = \mathcal{F}[h+\delta h] - \mathcal{F}[h] \equiv \int d\mathbf{x} \frac{\delta\mathcal{F}}{\delta h(\mathbf{x})} \delta h(\mathbf{x}). \tag{L.21}$$

The equality above is the definition of the functional derivative $\delta\mathcal{F}/\delta h(\mathbf{x})$.

In order to evaluate it we introduce a small parameter ϵ that allows us to tune the strength of the variation of h, $\delta h(\mathbf{x}) = \epsilon\varphi(\mathbf{x})$, where $\varphi(\mathbf{x})$ is any smooth function (to be further commented on below). Therefore, we obtain

$$\int d\mathbf{x} \frac{\delta\mathcal{F}}{\delta h(\mathbf{x})} \varphi(\mathbf{x}) = \lim_{\epsilon\to 0} \frac{\mathcal{F}[h+\epsilon\varphi] - \mathcal{F}[h]}{\epsilon} = \frac{d}{d\epsilon} \mathcal{F}[h+\epsilon\varphi]\bigg|_{\epsilon=0}. \tag{L.22}$$

In order to be more specific, let us assume that f depends at most on the first spatial derivatives of h, so

$$\frac{d}{d\epsilon} \mathcal{F}[h+\epsilon\varphi]\bigg|_{\epsilon=0} = \frac{d}{d\epsilon} \int d\mathbf{x} f(\mathbf{x}, h+\epsilon\varphi, \nabla h + \epsilon\nabla\varphi)\bigg|_{\epsilon=0} \tag{L.23}$$

$$= \int d\mathbf{x} \left(\frac{\partial f}{\partial h}\varphi + \frac{\partial f}{\partial(\partial_i h)} \partial_i \varphi \right) \tag{L.24}$$

$$= \int d\mathbf{x} \left(\frac{\partial f}{\partial h}\varphi + \partial_i \left(\frac{\partial f}{\partial(\partial_i h)}\varphi \right) - \left(\partial_i \frac{\partial f}{\partial(\partial_i h)} \right) \varphi \right) \tag{L.25}$$

$$= \int d\mathbf{x} \left(\frac{\partial f}{\partial h} - \partial_i \frac{\partial f}{\partial(\partial_i h)} \right) \varphi. \tag{L.26}$$

Above we use the standard notation to sum over repeated indices, and from the third to the fourth line we remove the total derivative, $\int d\mathbf{x}\partial_i(\cdots)$, that vanishes either for periodic boundary conditions or for perturbations $\varphi(\mathbf{x})$ which vanish at infinity.

Since above result must hold for any function $\varphi(\mathbf{x})$, we can finally write

$$\frac{\delta\mathcal{F}}{\delta h(\mathbf{x})} = \frac{\partial f}{\partial h} - \partial_i \frac{\partial f}{\partial(\partial_i h)} \equiv \frac{\partial f}{\partial h} - \nabla \cdot \frac{\partial f}{\partial(\nabla h)}. \tag{L.27}$$

To avoid cumbersome notation we extend this result to a dependence of f on derivatives of generic order for one spatial dimension only. If

$$\mathcal{F}[h] = \int dx\, f(x, h, \partial_x h, \partial_x^2 h, \partial_x^3 h, \dots), \tag{L.28}$$

we have

$$\frac{\delta \mathcal{F}}{\delta h(x)} = \frac{\partial f}{\partial h} + \sum_{n=1}^{\infty} (-)^n \frac{\partial^n}{\partial x^n} \frac{\partial f}{\partial(\partial_x^n h)}. \tag{L.29}$$

Should the above derivation be too formal, we provide a simpler one that uses the discrete representation, the drawback being we focus on a specific function $\mathcal{F}$ in one spatial dimension,

$$\mathcal{F}[h] = \int dx \left[U(h) + \frac{K}{2}(\partial_x h)^2 \right], \tag{L.30}$$

which is rewritten as a discrete sum,

$$\mathcal{F}[h] = \sum_i \Delta \left[U(h_i) + \frac{K}{2} \left(\frac{h_{i+1} - h_{i-1}}{2\Delta} \right)^2 \right], \tag{L.31}$$

where Δ is the discrete lattice constant.

Let us now evaluate the quantity $\delta\mathcal{F}[h]$, Eq. (L.21), using δ_i as the discrete version of $\delta h(x)$,

$$\begin{aligned}
\delta\mathcal{F}[h] =& \sum_i \Delta \Bigg[U(h_i + \delta_i) - U(h_i) \\
&+ \frac{K}{2} \left(\left(\frac{h_{i+1} + \delta i + 1 - h_{i-1} - \delta_{i-1}}{2\Delta} \right)^2 - \left(\frac{h_{i+1} - h_{i-1}}{2\Delta} \right)^2 \right) \Bigg] \\
=& \sum_i \Delta \left[\frac{\partial U}{\partial h_i} \delta_i + \frac{K}{2} \frac{(h_{i+1} - h_{i-1})(\delta_{i+1} - \delta_{i-1})}{2\Delta^2} \right] \qquad (L.32) \\
=& \sum_i \Delta \left[\frac{\partial U}{\partial h_i} - K \frac{(h_{i+2} + h_{i-2} - 2h_i)}{(2\Delta)^2} \right] \delta_i. \qquad (L.33)
\end{aligned}$$

We can finally go back to continuum and write

$$\delta\mathcal{F}[h] = \int dx \left[\frac{\partial U}{\partial h} - K \frac{\partial^2 h}{\partial x^2} \right] \delta h(x), \tag{L.34}$$

that is to say,

$$\frac{\delta \mathcal{F}}{\delta h(x)} = \frac{\partial U}{\partial h} - K \frac{\partial^2 h}{\partial x^2}, \tag{L.35}$$

in agreement with the general formula (L.27).

In reference to Section 5.7.2, it is useful to define the functional derivative of f (see Eq. (L.20)) through the relation

$$\frac{\delta \mathcal{F}}{\delta h(\mathbf{x})} \equiv \int d\mathbf{x}' \frac{\delta f(h(\mathbf{x}'))}{\delta h(\mathbf{x})}. \tag{L.36}$$

For example, if $f = U(h(\mathbf{x}')) + \frac{K}{2}(\nabla h(\mathbf{x}'))^2$, we have

$$\frac{\delta f(h(\mathbf{x}'))}{\delta h(\mathbf{x})} = [U'(h(\mathbf{x}') + K\nabla^2 h(\mathbf{x}')]\delta(\mathbf{x} - \mathbf{x}'). \tag{L.37}$$

Appendix M The Perturbative Renormalization Group for KPZ: A Few Details

We limit ourselves to giving a few details here about the method of the dynamical renormalization group for the KPZ equation. The starting point is the linear relation between noise and interface profile in the **k**-space, for the EW equation; then, using the EW propagator, i.e., the ratio between profile and noise, we construct a perturbative theory for the KPZ equation. The RG is a method to get rid of the divergences appearing in the naive perturbation theory.

The linear character of the EW equation allows for a simple and direct relationship between noise and interface profile. Such a relation is better shown in the (space and time) Fourier space, by writing

$$h(\mathbf{x}, t) = \int_{-\infty}^{\infty} \frac{d\omega}{2\pi} \int_{V_\mathbf{q}} \frac{d\mathbf{q}}{(2\pi)^d} e^{i(\mathbf{q}\cdot\mathbf{x}+\omega t)} h(\mathbf{q}, \omega) \tag{M.1}$$

and replacing it in Eq. (5.40). We obtain

$$h(\mathbf{q}, \omega) = G_0(\mathbf{q}, \omega)\eta(\mathbf{q}, \omega), \tag{M.2}$$

where

$$G_0(\mathbf{q}, \omega) = \frac{1}{\nu q^2 + i\omega} \tag{M.3}$$

is the propagator of the EW equation. If we do the same thing for KPZ, we obtain

$$\begin{aligned} i\omega h(\mathbf{k}, \omega) &= -\nu k^2 h(\mathbf{k}, \omega) \\ &\quad -\frac{\lambda}{2} \int \frac{d\omega}{2\pi} \int_{V_\mathbf{q}} \frac{d\mathbf{q}}{(2\pi)^d} \mathbf{q}\cdot(\mathbf{k}-\mathbf{q}) h(\mathbf{q}, \Omega) h(\mathbf{k}-\mathbf{q}, \omega - \Omega) \\ &\quad +\eta(\mathbf{k}, \omega)\,. \end{aligned} \tag{M.4}$$

The quantities appearing in the first and third lines are simple, because they are the Fourier transform of linear terms (time derivative, Laplacian, and noise, respectively). The term in the second line derives from the KPZ nonlinearity. Using the definition of G_0, we can rewrite as

$$\begin{aligned} h(\mathbf{k}, \omega) &= G_0(\mathbf{k}, \omega)\eta(\mathbf{k}, \omega) \\ &\quad -\frac{\lambda}{2} G_0(\mathbf{k}, \omega) \int \frac{d\omega}{2\pi} \int_{V_\mathbf{q}} \frac{d\mathbf{q}}{(2\pi)^d} \mathbf{q}\cdot(\mathbf{k}-\mathbf{q}) h(\mathbf{q}, \Omega) h(\mathbf{k}-\mathbf{q}, \omega - \Omega). \end{aligned} \tag{M.5}$$

This expression can be easily iterated perturbatively, getting a sequence of terms in increasing powers of λ, the perturbation parameter: it is sufficient to replace the above expression for $h(\mathbf{k}, \omega)$ in the λ term and repeat the same operation. The key point is that this expansion produces terms that are divergent for $d < 2$, making useless such naive approach.

The renormalization group is a method of circumventing these divergences. The idea is to change the spatial scale b of the system, integrating out small scales. As a result of this coarse-graining procedure, the system is described by new coupling parameters ν, λ and by a new strength of noise Γ, much in the same way a coarse-graining procedure of an equilibrium Ising system leads to new couplings $J_1, J_2, \ldots$ and to a new temperature. If we define, $\ell = \ln b$, the result of the dynamical RG machinery is

$$\frac{d\nu}{d\ell} = \nu\left[z - 2 + \frac{K_d(2-d)}{4d}g^2\right] \equiv \nu f_\nu(g) \tag{M.6a}$$

$$\frac{d\lambda}{d\ell} = \lambda(\alpha + z - 2) \tag{M.6b}$$

$$\frac{d\Gamma}{d\ell} = \Gamma\left[z - d - 2\alpha + \frac{K_d}{4}g^2\right] \equiv \Gamma f_\Gamma(g), \tag{M.6c}$$

where $K_d = S_d/(2\pi)^d$, S_d is the surface area of a d-dimensional sphere of radius equal to one, and

$$g^2 = \frac{\Gamma\lambda^2}{\nu^3}. \tag{M.7}$$

These equations define the so-called RG flow. On large scales, the system can be described by stable fixed points, which also fix the exponents α, z. Let us first consider the EW case, where $\lambda = 0 = g^2$. Eqs. (M.6) simplify to

$$\frac{d\nu}{d\ell} = \nu(z-2) \tag{M.8a}$$

$$\frac{d\Gamma}{d\ell} = \Gamma(z - d - 2\alpha), \tag{M.8b}$$

and imposing the condition for a fixed point, $d\nu/d\ell = 0 = d\Gamma/d\ell$, we find the well-known values $z = 2$ and $\alpha = (2-d)/2$. Furthermore, any pair ν, Γ turns out to be a fixed point, because parameters of the EW equation do *not* renormalize, because of the linear character of the equation.

When we consider the KPZ equation, we can see immediately from Eq. (M.6b) that λ does not renormalize, because the fixed-point condition, $d\lambda/d\ell = 0$, does not depend on λ and gives the general condition[1]

$$\alpha + z = 2. \tag{M.9}$$

This relation is not limited to our perturbative approach; it is an exact relation depending on a physical invariance of the KPZ equation, as discussed in Section 5.7.1.

Let us now come back to the RG flow. Eqs. (M.6a) and (M.6c) can be combined, in order to find the RG flow of the coupling g,

[1] More precisely, the condition $d\lambda/d\ell = 0$ also has the solution $\lambda = 0$, which corresponds to the EW limit we have just discussed.

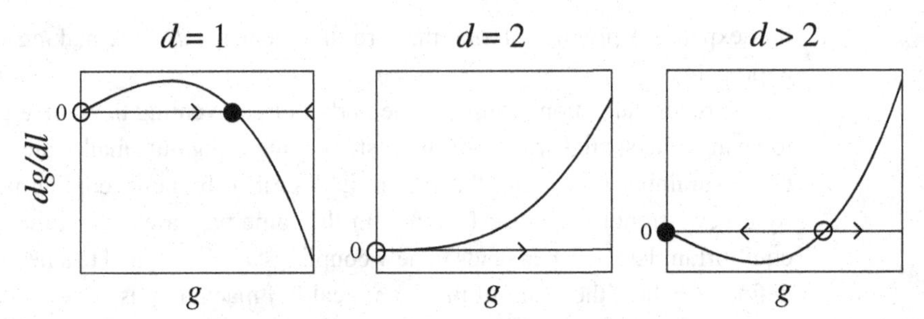

Fig. M.1 Plots of $dg/d\ell$ for different dimensions. Solid circles mean stable fixed points, open circles mean unstable fixed points.

$$\begin{aligned}\frac{dg^2}{d\ell} &= \frac{\lambda^2}{\nu^3}\frac{d\Gamma}{d\ell} - 3\frac{\lambda^2\Gamma}{\nu^4}\frac{d\nu}{d\ell} \\ &= g^2(f_\Gamma(g) - 3f_\nu(g))\,. \end{aligned} \tag{M.10}$$

Since $dg^2/d\ell = 2g(dg/d\ell)$, using the relation $(\alpha + z) = 2$ we finally obtain

$$\frac{dg}{d\ell} = \frac{2-d}{2}g + \frac{(2d-3)K_d}{4d}g^3. \tag{M.11}$$

The function $dg/d\ell$ versus g is plotted in Fig. M.1 for $d = 1$, $d = 2$, and $d > 2$. It depicts the RG flow, the resulting fixed points, and their stability. The point $g = 0$ is always a fixed point, i.e., a constant solution of the RG flow, and it corresponds to the EW fixed point. For $d \leq 2$ it is unstable, as signaled by the positive coefficient of the linear term in the right-hand side of (M.11): however small is $g(0)$ (i.e., however small the coefficient λ of the KPZ nonlinearity is), a coarse graining of the system leads to increasing values of g, because the nonlinearity is relevant and dominant for $d < 2$. Instead, for $d > 2$ the coefficient of the linear term is negative, meaning that a small g is renormalized to zero under coarse graining: the KPZ nonlinearity is irrelevant and exponents α, z are the same as for the EW equation. Therefore, the RG analysis confirms the expectations based on the simple scale transformation discussed at the beginning of this section.

If we want to go beyond the small g limit, i.e., make a perturbative analysis of the EW fixed point, we need to take into account the cubic term in (M.11). We focus on $d = 1$, the only relevant (integer) dimension with a stable fixed point for $g = g^* \neq 0$. Since $K_1 = \frac{1}{2\pi}$, Eq. (M.11) is written $\frac{dg}{d\ell} = \frac{1}{2}g - \frac{1}{8\pi}g^3$, which gives the stable fixed point,

$$g^* = 2\sqrt{\pi}, \qquad d = 1. \tag{M.12}$$

If we impose $f_\nu(g^*) = 0$ (see Eq. (M.6a)), because we must have $d\nu/d\ell = 0$ at the fixed point, we find

$$z_{\text{KPZ}} = \frac{3}{2}, \qquad d = 1, \tag{M.13}$$

and using the relation $(\alpha + z) = 2$,

$$\alpha_{\text{KPZ}} = \frac{1}{2}, \qquad d = 1. \tag{M.14}$$

Therefore, in $d = 1$, $\alpha_{\text{KPZ}} = \alpha_{\text{EW}}$. This equality has a physical motivation, as explained in Section 5.7.2, but it is not true in $d > 1$. The value of $\beta = \alpha/z$ is larger, instead: $\beta_{\text{KPZ}} = \frac{1}{3} > \beta_{\text{EW}} = \frac{1}{4}$. For $d = 2$ we can say nothing, while for $d > 2$ the point $g = 0$ is stable, but its basin of attraction is confined to $g < g^*$. If $g(0) > g^*$, the nonlinearity is not renormalized to zero. Rather, it increases forever. A perturbative RG approach, such as the one discussed here, is unable to say more than that.

Appendix N The Gibbs–Thomson Relation

In this appendix we derive how the density of a gas in equilibrium with a droplet of the condensed phase depends on the radius R of the droplet. This result is used to write boundary conditions (6.36) in Section 6.3.2.

Let us imagine that particles are condensed in a droplet of radius R, surface S, volume V, mass M_s, and density ρ_s. The droplet is in equilibrium with a gas of mass M_g and density ρ. If g_s and g are the mass densities of the Gibbs free energy of the two phases and σ is the interface free energy, we have

$$G_{\text{TOT}} = g_s M_s + g M_g + \sigma S. \tag{N.1}$$

If we vary the mass of the droplet by a quantity δM_s, we have

$$\delta G_{\text{TOT}} = \delta M_s \left(g_s - g + \sigma \frac{1}{\rho_s} \frac{\partial S}{\partial V} \right) \tag{N.2}$$

$$= \delta M_s \left(g_s - g + \sigma \frac{d-1}{\rho_s} \frac{1}{R} \right). \tag{N.3}$$

At equilibrium we must have $\delta G_{\text{TOT}} = 0$, i.e.,

$$g - g_s = \frac{d-1}{\rho_s} \frac{\sigma}{R}. \tag{N.4}$$

Taking the derivative of both sides with respect to the pressure P at constant temperature T and using the thermodynamic relations

$$\frac{1}{\rho_s} = \left. \frac{\partial g_s}{\partial P} \right|_T, \qquad \frac{1}{\rho} = \left. \frac{\partial g}{\partial P} \right|_T, \tag{N.5}$$

we find

$$\frac{1}{\rho} - \frac{1}{\rho_s} = -\frac{(d-1)\sigma}{\rho_s R} \left[\frac{1}{R} \left. \frac{\partial R}{\partial P} \right|_T + \frac{1}{\rho_s} \left. \frac{\partial \rho_s}{\partial P} \right|_T \right]. \tag{N.6}$$

Assuming the condensed phase is incompressible, $\partial \rho_s / \partial P = 0$, and $\rho_s \gg \rho$, we finally get

$$\frac{1}{\rho} = -\frac{(d-1)\sigma}{\rho_s R^2} \left(\left. \frac{\partial P}{\partial R} \right|_T \right)^{-1}. \tag{N.7}$$

If we use the equation of state for the ideal gas, $\rho = mP/T$, we find

$$\frac{1}{\rho} = -\frac{(d-1)\sigma m}{\rho_s T R^2} \frac{1}{\rho'(R)}, \tag{N.8}$$

whose solution is

$$\rho(R) = \rho_\infty(T) \exp\left(\frac{(d-1)\sigma m}{\rho_s T}\frac{1}{R}\right). \tag{N.9}$$

For large R we can approximate,

$$\rho(R) \simeq \rho_\infty(T)\left[1 + \frac{(d-1)\sigma m}{\rho_s T}\frac{1}{R}\right], \tag{N.10}$$

which is equivalent to the first boundary condition (6.36). The second boundary condition corresponds to exchanging the droplet and the gas. Within this geometry, a positive δM_s implies a negative surface term in Eq. (N.2), equivalent to changing the sign of σ in Eq. (N.10).

Appendix O The Allen–Cahn Equation

Eq. (6.18) is a special case of a more general equation describing the dynamics of a domain wall in the nonconserved case, whose derivation is the goal of this appendix. Let us start with Eq. (6.9),

$$\frac{\partial m}{\partial t} = \nabla^2 m - U'(m).$$

In proximity to a domain wall, m varies only in the direction perpendicular to the wall. For this reason it is useful to introduce a local unit vector $\hat{\mathbf{g}}$, perpendicular to the wall and oriented in the direction of increasing m; see Fig. O.1. Along such direction the local coordinate is denoted g, much in the same way we have the coordinate x and the unit vector $\hat{\mathbf{x}}$ in a Cartesian coordinate system. Under the new coordinate system,

$$\nabla m = \left(\frac{\partial m}{\partial g}\right)_t \hat{\mathbf{g}}, \tag{O.1}$$

$$\nabla^2 m = \left(\frac{\partial^2 m}{\partial g^2}\right)_t + \left(\frac{\partial m}{\partial g}\right)_t \nabla \cdot \hat{\mathbf{g}}, \tag{O.2}$$

$$\left.\frac{\partial m}{\partial t}\right|_g = -\left(\frac{\partial m}{\partial g}\right)_t \left(\frac{\partial g}{\partial t}\right)_m. \tag{O.3}$$

The first equation is trivial, because m varies in space only in the $\hat{\mathbf{g}}$ direction and the second is just its reapplication plus the fact that $\hat{\mathbf{g}}$ itself varies along the wall. Finally, the third equation means that m depends on time, because the wall moves, i.e., g depends on time;

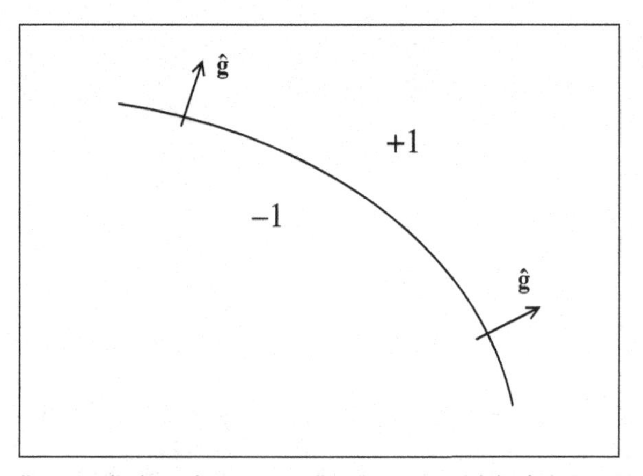

Fig. O.1 The unit vector $\hat{\mathbf{g}}$ is locally perpendicular to the domain wall and points toward the domain $m = +1$.

the minus sign is there, because a wall moving so that $(\partial_t g)_m > 0$ determines a local decrease of m, since $(\partial_g m)_t > 0$ by definition of $\hat{\mathbf{g}}$. In fact, the minus sign has the same physical motivation as the minus sign in the left-hand side of Eq. (6.14), where R plays the role of g. If we assemble previous formulas, we obtain the exact equation,

$$-\left(\frac{\partial m}{\partial g}\right)_t \left(\frac{\partial g}{\partial t}\right)_m = \left(\frac{\partial^2 m}{\partial g^2}\right)_t + \left(\frac{\partial m}{\partial g}\right)_t \nabla \cdot \hat{\mathbf{g}} - U'(m). \tag{O.4}$$

The steady profile $m(g)$ of a straight wall is determined by imposing the stationarity condition, $\partial_t g = 0$, and the condition of straight wall, $\nabla \cdot \hat{\mathbf{g}} = 0$:

$$\left(\frac{\partial^2 m}{\partial g^2}\right)_t - U'(m) = 0. \tag{O.5}$$

We assume that dynamics is slow enough not to perturb this profile, so the above relation is still valid in out-of-equilibrium condition, which implies

$$\left(\frac{\partial g}{\partial t}\right)_m = -\nabla \cdot \hat{\mathbf{g}}. \tag{O.6}$$

The quantity $(\partial_t g)_m$ is the domain wall velocity v and $\nabla \cdot \hat{\mathbf{g}}$ is the curvature K of the domain wall. Finally, we obtain the so-called Allen–Cahn equation,

$$v = -K. \tag{O.7}$$

For a spherical domain wall of radius R, $K = (d-1)/R$ and we recover Eq. (6.18). However, Eq. (O.7) is more general and can be used to obtain valuable information on the statistics of domains; see Section 6.4.4.

Appendix P The Rayleigh–Bénard Instability

In Fig. P.1 we sketch the geometry of the Rayleigh–Bénard instability: a fluid is confined between two horizontal plates at a distance d and kept at different temperatures, the lower plate being warmer than the upper one. Heat is transported by conduction for a small temperature difference ΔT and by convection when ΔT exceeds a critical value $(\Delta T)_c$.

The starting point for our analysis is the Navier–Stokes equations for an incompressible fluid,

$$\rho\left(\frac{\partial}{\partial t}+\mathbf{u}\cdot\nabla\right)\mathbf{u}=-\nabla p+\rho\nu\nabla^2\mathbf{u}-\rho g\hat{\mathbf{x}}_3, \tag{P.1}$$

where ρ is the density, $\mathbf{u}(\mathbf{x},t)$ is the velocity field, g is the acceleration of gravity (which is oriented downward in the vertical direction $\hat{\mathbf{x}}_3$), and ν is the kinematic viscosity. We do not derive the Navier–Stokes equations here and we limit ourselves to their interpretation. The left-hand side has the form of a material derivative,

$$\frac{d}{dt}\equiv\frac{\partial}{\partial t}+\mathbf{u}\cdot\nabla, \tag{P.2}$$

which corresponds to evaluating the time variation of a quantity, which is transported by the fluid. On the right-hand side we have the forces *per unit volume* acting on a fluid parcel: the force induced by a pressure gradient, the viscous force, and the gravity force. It is worth stressing that the first two terms both derive from the stress tensor $\mathbb{P}$ whose form, for an incompressible fluid, is

$$\mathbb{P}_{ij}=-p\delta_{ij}+\rho\nu\left(\frac{\partial u_i}{\partial x_j}+\frac{\partial u_j}{\partial x_i}\right). \tag{P.3}$$

Applying the divergence to the stress tensor, we obtain the first two terms on the right-hand side.

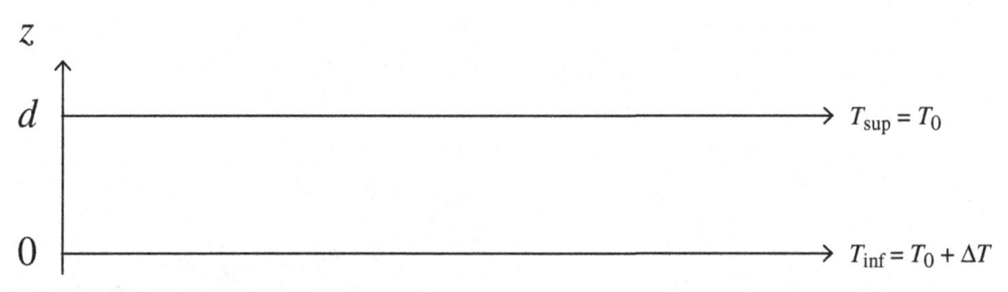

Fig. P.1 Sketch of the Rayleigh–Bénard instability.

The key physical process triggering convection is the dependence of the fluid density on temperature. Because of that, heating the system from below determines an unstable situation, where lighter slices of fluid are located below heavier slices of fluid. Therefore, we need to introduce the temperature field $T(\mathbf{x}, t)$, whose evolution is ruled by the heat equation,

$$\frac{dT}{dt} \equiv \frac{\partial T}{\partial t} + (\mathbf{u} \cdot \nabla)T = \kappa \nabla^2 T, \qquad \text{(P.4)}$$

where κ is the thermal diffusivity. In conclusion, we have four scalar equations, Eqs. (P.1) and (P.4), with four unknowns, the three components of $\mathbf{u}$ and T.

Matter conservation implies

$$\frac{\partial \rho}{\partial t} + \nabla \cdot (\rho \mathbf{u}) = 0 \qquad \text{(P.5)}$$

or

$$\frac{\partial \rho}{\partial t} + (\mathbf{u} \cdot \nabla)\rho + \rho \nabla \cdot \mathbf{u} \equiv \frac{d\rho}{dt} + \rho \nabla \cdot \mathbf{u} = 0. \qquad \text{(P.6)}$$

Incompressibility means $d\rho/dt = 0$, so it must be that

$$\nabla \cdot \mathbf{u} = 0. \qquad \text{(P.7)}$$

In the Navier–Stokes equations (P.1) the density ρ appears in the inertial term, in the gravity term, and in the viscous force. It varies very weakly with T, because the coefficient of thermal expansion, $\alpha = -\frac{1}{\rho}\frac{\partial \rho}{\partial T}$, is of order 10^{-4}K^{-1} for a fluid. As explained above, the temperature dependence of ρ cannot be neglected, otherwise no convective instability would appear. However, we can assume a simple linear dependence,

$$\rho = \rho_0(1 - \alpha(T - T_0)) \qquad \text{(P.8)}$$

and we can limit ourselves to considering such dependence in the gravity term, which is the primary cause of the instability. This simplification is known as Boussinesq approximation (after Joseph Valentin Boussinesq). In conclusion, we have to consider the following set of equations:

$$\nabla \cdot \mathbf{u} = 0 \qquad \text{(P.9)}$$

$$\rho_0 \left(\frac{\partial}{\partial t} + \mathbf{u} \cdot \nabla \right) \mathbf{u} = -\nabla p + \rho_0 \nu \nabla^2 \mathbf{u} - \rho g \hat{\mathbf{x}}_3 \qquad \text{(P.10)}$$

$$\frac{\partial T}{\partial t} + (\mathbf{u} \cdot \nabla)T = \kappa \nabla^2 T \qquad \text{(P.11)}$$

$$\rho = \rho_0(1 - \alpha(T - T_0)). \qquad \text{(P.12)}$$

The first step is to determine the temperature and pressure profiles of the conductive state, $\mathbf{u}_c \equiv 0$. The second step will be to perturb it and analyze its linear stability spectrum. Since the conductive state is time independent and translational invariant in the (x, y)-plane, Eq. (P.11) reduces to $d^2T(z)/dz^2 = 0$, whose solution is

$$T = T_c(z) = T_0 + \Delta T \left(1 - \frac{z}{d} \right), \qquad \text{(P.13)}$$

where we have used the following boundary conditions for the temperature, $T(z=0) = T_0 + \Delta T$ and $T(z=d) = T_0$.

The pressure profile is determined by Eq. (P.10), which reduces in the conductive state to $dp/dz = -\rho g$. Its solution is

$$p = p_c(z) = p_0 - g\int_0^z \rho(T_c(z))dz = p_0 - \rho_0 g z\left[1 - \alpha\Delta T\left(1 - \frac{z}{2d}\right)\right]. \qquad \text{(P.14)}$$

Therefore, in the conductive state the temperature profile has a linear behavior that interpolates between $T_{\mathrm{inf}} = T_0 + \Delta T$ and $T_{\mathrm{sup}} = T_0$. The pressure profile is a downward parabolic profile.

We should now perturb the conductive state, writing

$$\mathbf{u} = \mathbf{u}_c + \mathbf{v}^*(\mathbf{x}, t) \qquad \text{(P.15)}$$

$$T = T_c(z) + \theta^*(\mathbf{x}, t) \qquad \text{(P.16)}$$

$$p = p_c(z) + p^*(\mathbf{x}, t) \qquad \text{(P.17)}$$

and linearizing the resulting equations, keeping only terms linear in starred quantities. This procedure gives

$$\rho_0\frac{\partial \mathbf{u}^*}{\partial t} = -\nabla p^* + \alpha\rho_0 g\theta^*\hat{\mathbf{z}} + \nu\rho_0\nabla^2\mathbf{u}^* \qquad \text{(P.18)}$$

$$\frac{\partial\theta^*}{\partial t} = \frac{\Delta T}{d}u_z^* + \kappa\nabla^2\theta^*. \qquad \text{(P.19)}$$

It is now useful to rescale all variables so as to make them dimensionless,

$$\tilde{\mathbf{x}} = \frac{\mathbf{x}}{d}, \quad \tilde{t} = \frac{\kappa t}{d^2}, \quad \tilde{\mathbf{u}} = \frac{d\mathbf{u}^*}{\kappa}, \quad \tilde{\theta} = \frac{\theta^*}{\Delta T}, \quad \tilde{p} = \frac{d^2 p^*}{\rho_0\kappa^2}. \qquad \text{(P.20)}$$

While using d for rescaling space and ΔT for rescaling temperature is straightforward, other rescaling operations deserve some comment. Thermal diffusivity κ defines a time scale $\tau_c = d^2/\kappa$ that is used to rescale t; $d/\tau_c = \kappa/d$ is used to rescale the velocity field. Finally, since p has the dimension of a density multiplied by a square velocity, it can be rescaled with $\rho_0(d/\tau_c)^2 = \rho_0\kappa^2/d^2$.

In the following we omit the tilde and the linear equations are written as[1]

$$\frac{\partial\mathbf{u}}{\partial t} = -\nabla p + RP\theta\hat{\mathbf{z}} + P\nabla^2\mathbf{u} \qquad \text{(P.21)}$$

$$\frac{\partial\theta}{\partial t} = u_z + \nabla^2\theta, \qquad \text{(P.22)}$$

where

$$R = \frac{\alpha\Delta T g d^3}{\nu\kappa} \quad \text{and} \quad P = \frac{\nu}{\kappa} \qquad \text{(P.23)}$$

are, respectively, the (dimensionless) Rayleigh and the Prandtl number.

It is now possible to get rid of the pressure field by applying the curl operator to both sides of Eq. (P.21). More precisely, we apply it twice, because the exact relation

[1] With the new variables, the height z varies in the interval [0, 1].

$$\nabla \times (\nabla \times \mathbf{a}) = \nabla(\nabla \cdot \mathbf{a}) - \nabla^2 \mathbf{a} \tag{P.24}$$

simplifies if $\mathbf{a}$ is a solenoidal vector field ($\nabla \cdot \mathbf{a} = 0$), which is the case for $\mathbf{u}$ and $\nabla^2 \mathbf{u}$. In conclusion, we get

$$-\frac{\partial \nabla^2 \mathbf{u}}{\partial t} = RP\left(\nabla \cdot \frac{\partial \theta}{\partial z} - \nabla^2 \theta \hat{\mathbf{z}}\right) - P\nabla^4 \mathbf{u}. \tag{P.25}$$

The $\hat{\mathbf{z}}$ component of previous equation, together with Eq. (P.22), gives a coupled system for u_z, θ,

$$\frac{\partial \nabla^2 u_z}{\partial t} = RP\nabla_{\parallel}^2 \theta + P\nabla^4 u_z \tag{P.26}$$

$$\frac{\partial \theta}{\partial t} = u_z + \nabla^2 \theta. \tag{P.27}$$

We should now supplement above equations with six boundary conditions, because the highest order derivative for u_z (four) sums to the highest order derivative for θ (two), giving six. There are three boundary conditions at each plate ($z = 0$ and $z = 1$), one for θ and two for u_z. Since θ is the perturbation of a conductive solution satisfying boundary conditions for T, we must require $\theta(z = 0) = \theta(z = 1) = 0$. As for the velocity field, which vanishes in the conductive solution, the physical boundary conditions are $\mathbf{u}(x, y, 0) = \mathbf{u}(x, y, 1) = 0$. Since $\nabla \cdot \mathbf{u} = 0$, they imply the vanishing of $\partial_z u_z(z = 0, 1)$. In conclusion, θ, u_z, and $\partial_z u_z$ must vanish in $z = 0, 1$ (i.e., $z = 0, d$ in dimensional variables).

These conditions make the problem too complicated to be treated here in full detail, so we make the common choice of considering different boundary conditions for u_z: instead of vanishing its first derivative (no-slip conditions, valid at a rigid boundary), we impose the vanishing of its second derivative (stress-free conditions, valid at a free boundary). Therefore, we are going to impose that $\theta, u_z, \partial_{zz} u_z$ all vanish for $z = 0, 1$.

The standard procedure for a linear stability analysis is to expand the perturbations in Fourier components. Translational invariance along x, y implies that any component $e^{i(k_x x + k_y y)}$ is permitted, while the z-dependence must satisfy boundary conditions, which implies a discretization of k_z because the wave must have nodes. We can finally write

$$u_z(\mathbf{x}, t) = u_n \sin(n\pi z) e^{i(k_x x + k_y y)} e^{\sigma t} + \text{c.c.} \tag{P.28}$$

$$\theta(\mathbf{x}, t) = \theta_n \sin(n\pi z) e^{i(k_x x + k_y y)} e^{\sigma t} + \text{c.c.} \tag{P.29}$$

The wavevector in the (x, y)-plane determines the (horizontal) size and orientation of emerging convection cells, while n determines the number of convection cells in the z-direction. It is reasonable to expect the first mode to destabilize the convective state, while increasing ΔT corresponds to $n = 1$, so we limit ourselves to this case. Replacing the above expressions for $n = 1$ in Eqs. (P.21) and (P.22), we get the linear system

$$\left(\sigma(k^2 + \pi^2) + P(k^2 + \pi^2)^2\right) u_1 - RPk^2 \theta_1 = 0 \tag{P.30}$$

$$u_1 - (\sigma + k^2 + \pi^2)\theta_1 = 0. \tag{P.31}$$

Imposing the vanishing of the determinant to assure the existence of nontrivial solutions, we can find the relation between the growth rate σ of the perturbation and its wavevector k,

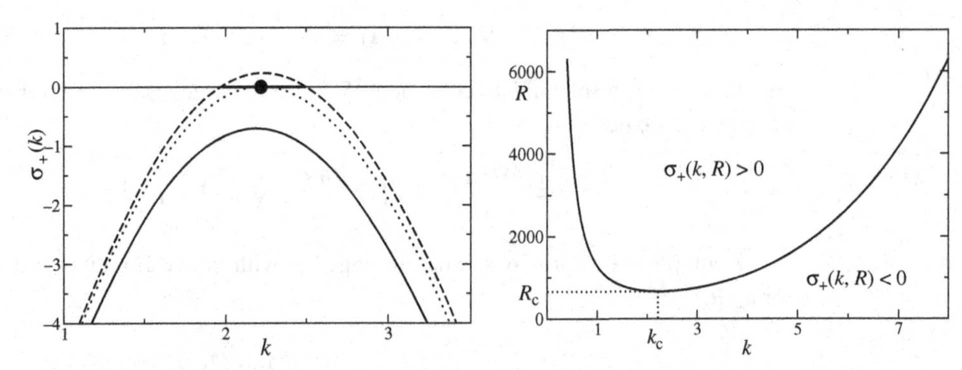

Fig. P.2 Left: The curve $\sigma_+(k)$, Eq. (P.32), for $R < R_c$ (solid line), $R = R_c$ (dotted line), and $R > R_c$ (dashed line). The thick, horizontal segment indicates the unstable interval of wavevectors, and the solid circle indicates the critical wavevector k_c. Compare this figure with Fig. (7.4). Right: The neutral curve, Eq. (P.35), separating the stable region (below the curve) from the unstable region (above the curve).

$$\sigma_\pm(k) = -\frac{1}{2}(1+P)(k^2+\pi^2) \pm \sqrt{\frac{1}{4}(P-1)^2(k^2+\pi^2)^2 + \frac{RPk^2}{k^2+\pi^2}}. \quad \text{(P.32)}$$

The solution $\sigma_-(k)$ is certainly negative for all k, therefore stable. In Fig. P.2 we plot $\sigma_+(k)$ for $P = 5$ (the Prandtl number for water) and different values of R. We can observe how the growth rate starts to become positive when R is larger than a critical value of the Rayleigh number, $R_c = 27\pi^4/4$ (see below). Just above R_c, $\sigma_+(k)$ is positive in a small interval around a critical wavevector $k_c = \pi/\sqrt{2}$, therefore labeling the Rayleigh–Bénard instability as a type I instability.

Both R_c and k_c can be determined analytically in the stress-free case. However, recall that the physical picture is represented by the no-slip case, which gives

$$R_c \simeq 1708, \qquad k_c \simeq 3.117, \qquad \lambda_c = \frac{2\pi}{k_c} \simeq 2.016 \qquad \text{(no-slip case)} \quad \text{(P.33)}$$

Since λ_c is twice the size of a convection cell, the above result means that at the threshold, convection cells have an aspect ratio almost equal to one.

Let us now complete the stress-free case, by observing that each mode k becomes unstable at large enough R, $R > R^*$, where $\sigma_+(k, R^*) = 0$. It is interesting to check that R^* does not depend on the Prandtl number. In fact,

$$R^*(k) = \frac{(k^2+\pi^2)^3}{k^2}. \quad \text{(P.34)}$$

The sought-after values k_c, R_c correspond to the values minimizing $R^*(k)$,

$$\left.\frac{dR^*}{dk}\right|_{k_c} = 0 \qquad \text{and} \qquad R_c = R^*(k_c), \quad \text{(P.35)}$$

which give

$$R_c = \frac{27\pi^4}{4} \simeq 658, \qquad k_c = \frac{\pi}{\sqrt{2}}, \qquad \lambda_c = \frac{2\pi}{k_c} = 2\sqrt{2}, \qquad \text{(stress-free case)} \quad \text{(P.36)}$$

Let us conclude this appendix by providing a simple physical explanation of the existence of a minimal Rayleigh number, because convective cells appear. Heat conduction and convection are two distinct channels to transfer heat between the two plates at different temperature. The evidence that convection appears only above a certain threshold $(\Delta T)_c$ means that below this value conduction is more efficient and above this value convection is more efficient. We can quantify their efficiency by evaluating the typical times τ_c, τ_v of the conductive and convective channel, respectively. The smaller τ is, the more efficient the channel.

The typical conductive time is easily found, because conduction is a diffusive process, so the time τ_c relevant for diffusion on a length scale d via thermal diffusivity κ is simply $\tau_c \simeq d^2/\kappa$. Evaluating the typical convection term requires instead determining the speed v_0 of a parcel of liquid. Once it is known, we can write the ballistic expression $\tau_v = d/v_0$. The velocity v_0 can be inferred by the Navier–Stokes equation, by balancing the force due to Archimedes' principle with the viscous force,

$$(\alpha \Delta T) g \rho_0 = \nu \frac{v_0}{d^2} \rho_0, \tag{P.37}$$

which gives $v_0 = (\alpha \Delta T g d^2)/\nu$. Finally, we get $\tau_v = \nu/(\alpha \Delta T d g)$. The ratio between the conductive and the convective time is nothing but the Rayleigh number,

$$\frac{\tau_c}{\tau_v} = \frac{\alpha \Delta T g d^3}{\kappa \nu} \equiv R. \tag{P.38}$$

Therefore, the condition $R > R_c$ for triggering convection cells means that τ_v is sufficiently smaller than τ_c.

Appendix Q General Conditions for the Turing Instability

The Turing instability appears when $\mathcal{D}_q$ changes sign from positive to negative (see Section 7.3.1), assuming that no instability occurs for $q = 0$. The expression of $\mathcal{D}_q(q^2)$ (see Eq. (7.39b)),

$$\mathcal{D}_q = (a_{11} - D_1 q^2)(a_{22} - D_2 q^2) - a_{12}a_{21} \tag{Q.1}$$

$$= D_1 D_2 q^4 - (D_1 a_{22} + D_2 a_{11}) q^2 + (a_{11}a_{22} - a_{12}a_{21}), \tag{Q.2}$$

is an upward-facing parabola, whose minimum $q_{\rm m}^2$ satisfies the condition

$$\left.\frac{\partial \mathcal{D}_q}{\partial q^2}\right|_{q_{\rm m}} = 2D_1 D_2 q_{\rm m}^2 - (D_1 a_{22} + D_2 a_{11}) = 0, \tag{Q.3}$$

so that

$$q_{\rm m}^2 = \frac{D_1 a_{22} + D_2 a_{11}}{2D_1 D_2}. \tag{Q.4}$$

It is worth noting that $q_{\rm m}$ is an extremum of $\mathcal{D}_q$ (the other, trivial extremum being $q = 0$), and for generic values of the coefficients a_{ij} it differs from the critical wavevector q_c, which is the maximum of the eigenvalue $\sigma_1(q)$ at the threshold, defined by the conditions $\sigma_1(q_c) = 0$ and $\sigma_1'(q_c) = 0$. However, at the threshold $q_{\rm m} = q_c$, as is clear from the expression $\mathcal{D}_q = \sigma_1(q)\sigma_2(q)$. In fact, at the threshold, $\sigma_1(q_c) = \mathcal{D}_{q_c} = 0$ and $\sigma_2(q_c) < 0$. Therefore, we also have $\mathcal{D}'_{q_c} = \sigma_1'(q_c) = 0$.

The condition $\mathcal{D}_{q_{\rm m}} = 0$ defines the criticality, and the condition

$$\mathcal{D}_{q_{\rm m}} = -\frac{(D_1 a_{22} + D_2 a_{11})^2}{4D_1 D_2} + (a_{11}a_{22} - a_{12}a_{21}) < 0 \tag{Q.5}$$

defines the parameter region, where the Turing instability may appear. We can finally summarize the conditions causing a Turing pattern to appear:

$$\begin{aligned} \mathcal{T}_0 < 0 \quad & \text{(i)} \quad a_{11} + a_{22} < 0 \\ \mathcal{D}_0 > 0 \quad & \text{(ii)} \quad a_{11}a_{22} - a_{12}a_{21} > 0 \\ \mathcal{D}_{q_{\rm m}} < 0 \quad & \text{(iii)} \quad D_1 a_{22} + D_2 a_{11} > 2\sqrt{D_1 D_2 (a_{11}a_{22} - a_{12}a_{21})}, \end{aligned} \tag{Q.6}$$

with the last condition derived from Eq. (Q.5), taking into account that

$$D_1 a_{22} + D_2 a_{11} = 2D_1 D_2 q_{\rm m}^2 > 0. \tag{Q.7}$$

This same equation, if rewritten as

$$(D_1 - D_2) a_{22} + D_2 (a_{22} + a_{11}) > 0, \tag{Q.8}$$

provides two results: first, if $D_1 = D_2$, this condition cannot agree with (i), as already shown in the main text; second, if we assume $D_1 > D_2$, we must have $a_{22} > 0$ and consequently (i) gives $a_{11} < 0$. As a consequence of that, (ii) gives that crossing couplings a_{12}, a_{21} must have opposite signs.

Finally, the equation for eigenvalues at criticality gives Eq. (7.46),

$$(a_{11} - D_1 q_c^2)\epsilon_1^0 + a_{12}\epsilon_2^0 = 0.$$

Using the expression of q_c^2 (see Eq. (Q.4)), we can evaluate the sign of the coefficient multiplying ϵ_1^0

$$a_{11} - D_1 q_c^2 = \frac{D_2 a_{11} - D_1 a_{22}}{2D_2} < 0, \qquad \text{(Q.9)}$$

because $a_{22} > 0$ and $a_{11} < 0$. Therefore, the fluctuations $\epsilon_1^0, \epsilon_2^0$ have the same sign if $a_{12} > 0$ and opposite signs if $a_{12} < 0$.

Appendix R Steady States of the One-Dimensional TDGL Equation

The TDGL equation is written as

$$u_{xx} + ru - u^3 = 0, \tag{R.1}$$

where the explicit dependence on the control parameter r can be removed by rescaling x and u. If

$$x = \frac{y}{\sqrt{r}}, \qquad u(x) = \sqrt{r}g(y) = \sqrt{r}g(\sqrt{r}x), \tag{R.2}$$

we obtain

$$g_{yy} = -g - g^3, \tag{R.3}$$

which can be integrated by multiplying both sides by g_y. We obtain

$$\frac{1}{2}g_y^2 = -\frac{g^2}{2} + \frac{g^4}{4} + E, \tag{R.4}$$

where $E = \frac{1}{2}\bar{A}^2 - \frac{1}{4}\bar{A}^4$, $\bar{A}$ being the maximal amplitude of the oscillation (in reduced variables). For $g_y > 0$,

$$dy = \frac{dg}{\sqrt{2E - g^2 + \frac{g^4}{2}}}. \tag{R.5}$$

The reduced wavelength of the oscillation is therefore defined as

$$\bar{\lambda} = 2\int_{-\bar{A}}^{\bar{A}} \frac{dg}{\sqrt{2E - g^2 + \frac{g^4}{2}}} \tag{R.6}$$

$$= \frac{4}{\sqrt{1 - \frac{\bar{A}^2}{2}}} \int_0^1 \frac{ds}{\sqrt{(1-s^2)(1-k^2s^2)}} \tag{R.7}$$

$$= 4\sqrt{\frac{2}{2-\bar{A}^2}}K(k), \tag{R.8}$$

where $k = \bar{A}/\sqrt{2-\bar{A}^2}$ and $K(k)$ is the complete elliptic integral of the first kind.

Coming back to old variables, i.e., to Eq. (R.1), we find

$$\lambda(A) = 4\sqrt{\frac{2}{2r - A^2}}K\left(\frac{A}{\sqrt{2r - A^2}}\right). \tag{R.9}$$

Appendix S Multiscale Analysis

The equation for a one-dimensional anharmonic oscillator is

$$\ddot{x} + x = -\epsilon f(x, \dot{x}), \tag{S.1}$$

where $f(x, \dot{x})$ is a generic function of position and velocity and ϵ is the parameter for the perturbation expansion. A naive perturbation theory simply assumes $x(t) = \sum_{n \geq 0} \epsilon^n x_n(t)$ and replaces it in Eq. (S.1). Let's do it by limiting ourselves to zero and first-order terms,

$$\ddot{x}_0 + \epsilon \ddot{x}_1 + x_0 + \epsilon x_1 = -\epsilon f(x_0, \dot{x}_0). \tag{S.2}$$

At zero (unperturbed) order, we have the harmonic oscillator equation, $\ddot{x}_0 + x_0 = 0$, whose solution is $x_0 = A_0 \cos(t + \phi_0)$. At first order the nonlinear term f acts as a forcing term. To be specific we suppose having $f(x, \dot{x}) = x^3$ (Duffing oscillator), in which case

$$\ddot{x}_1 + x_1 = -A_0^3 \cos^3(t + \phi_0) \tag{S.3}$$

$$= -\frac{A_0^3}{4} \cos(3t + \phi_0) - \frac{3A_0^3}{4} \cos(t + \phi_0). \tag{S.4}$$

The second term on the right-hand side is resonant with the natural frequency of the harmonic oscillator, and this determines a secular term in the solution, which grows linearly in time,

$$x_1(t) = -\frac{3A_0^3}{8} t \sin(t + \phi_0) + \text{nonsecular terms}. \tag{S.5}$$

A linear term in the expansion means that the perturbation theory breaks down when $t \sim 1/\epsilon$, because $\epsilon x_1(1/\epsilon)$ is of the same order as $x_0(1/\epsilon)$. The multiple-scale analysis aims at a perturbation expansion without secular terms, which is obtained by removing the resonant terms. This is done by introducing multiple temporal scales, whose physical meaning is clarified by the following example.

Let us consider the damped harmonic oscillator

$$\ddot{x} + x = -\epsilon \dot{x}, \tag{S.6}$$

whose exact solution (in complex notation) is

$$x(t) = A \exp\left(i\sqrt{1 - \frac{\epsilon^2}{4}} t - \frac{\epsilon}{2} t \right) = A \exp\left(i\left(t - \frac{\epsilon^2}{8} t\right) - \frac{\epsilon}{2} t + O(\epsilon^3) \right). \tag{S.7}$$

This expression makes explicit that the solution depends on time through $t_0 = t$, $t_1 = \epsilon t$, $t_2 = \epsilon^2 t$, and so on. The "times" t_n represent different time scales over which different phenomena may occur. So, the oscillation period is of order $t_0 \approx 1$, the amplitude is dumped on a time of order $t_1 \approx 1$, the oscillation period changes on a time of order $t_2 \approx 1, \ldots$ From a formal point of view, the multiple scale analysis introduces an expansion of times in addition to the expansion of the solution, and these additional ϵ terms allow us to impose the vanishing of the resonant terms. Let us see the method at work for the Duffing oscillator.

Introducing additional times means that

$$\frac{d}{dt} = \partial_{t_0} + \epsilon \partial_{t_1} + \cdots \tag{S.8}$$

$$\frac{d^2}{dt^2} = \partial_{t_0 t_0} + 2\epsilon \partial_{t_1 t_0} + \cdots . \tag{S.9}$$

The zero-order perturbation does not change and the solution is the same as before, with the caveat that the integration constants A_0 and ϕ_0 now can depend on the slower time scale, $x_0(t_0, t_1) = A_0(t_1)\cos(t_0 + \phi_0(t_1))$. Instead, the first-order perturbation is written as

$$\frac{\partial^2 x_1}{\partial t_0^2} + x_1 = -2\frac{\partial^2 x_0}{\partial t_1 \partial t_0} - A_0^3(t_1)\cos^3(t_0 + \phi_0(t_1)) \tag{S.10}$$

$$= 2\frac{\partial}{\partial t_1}\Big(A_0(t_1)\sin(t_0 + \phi_0(t_1))\Big) - A_0^3(t_1)\cos^3(t_0 + \phi_0(t_1)) \tag{S.11}$$

$$= 2\dot{A}_0 \sin(t_0 + \phi_0) + \left[2A_0\dot{\phi}_0 - \frac{3}{4}A_0^3\right]\cos(t_0 + \phi_0) - \frac{A_0^3}{4}\cos(3(t_0 + \phi_0)),$$

where in the last expression, for ease of notation, we have not made explicit the t_1-dependence of A_0, ϕ_0, and the dot over such quantities means the derivative with respect to t_1.

We can first remark that if no slow time scales are introduced, A_0 and ϕ_0 are constant, and for $\dot{A}_0 = 0 = \dot{\phi}_0$ the above expression reduces to Eq. (S.4), which contains resonant terms. Now we have the freedom to impose the vanishing of such terms, writing

$$\dot{A}_0 = 0, \tag{S.12}$$

$$2A_0\dot{\phi}_0 - \frac{3}{4}A_0^3 = 0. \tag{S.13}$$

The former condition implies that A_0 is a constant (or, better, it depends on t_2 and slower time scales, which are neglected in this first-order calculation). The latter condition implies $\phi_0(t_1) = \frac{3}{8}A_0^2 t_1 + \bar{\phi}_0(t_2)$.

We can now turn to the equation for x_1,

$$\frac{\partial^2 x_1}{\partial t_0^2} + x_1 = -\frac{A_0^3}{4}\cos(3(t_0 + \phi_0)), \tag{S.14}$$

whose solution (if $x_1(0) = \dot{x}_1(0) = 0$) is $x_1(t_0) = \frac{A_0^3}{32}[\cos(3t_0) - \cos(t_0)]$.

Reassembling the different pieces and getting back to the variable t, we obtain, at order ϵ,

$$x(t) = A_0 \cos\left[\left(1 + \frac{3}{8}A_0^2\epsilon\right)t\right] + \epsilon\frac{A_0^3}{32}\left[\cos(3t) - \cos(t)\right]. \tag{S.15}$$

In this appendix we have made use of the simple anharmonic oscillator to show step by step how the method works, but in the main text we use the multiple-scale method in the more complicated context of partial differential equations. In that case, two main difficulties arise: first, we have temporal and spatial multiple scales, because the order parameter now depends on both x and t; second, the condition of absence of resonant terms might not be that easy to solve, because the unperturbed equation is itself nonlinear. This is the reason why we must conclude this section by giving a nonrigorous formulation of the Fredholm alternative theorem, which is used in the main text. In brief, let us suppose having the linear problem

$$\mathcal{L}[u] = g(x), \tag{S.16}$$

where $\mathcal{L}$ is a linear differential operator. If the homogeneous problem, $\mathcal{L}[u_0] = 0$, has a nontrivial solution $u_0 \neq 0$, then also the adjoint problem $\mathcal{L}^\dagger[v_0] = 0$ has a nontrivial solution and Eq. (S.16) has a solution if and only if g is orthogonal to v_0,

$$\langle g|v_0\rangle \equiv \int dx g^*(x) v_0(x) = 0. \tag{S.17}$$

Index

affinity, 100
Allen–Cahn equation, 293, 404–405
Arrhenius formula, 49, 54
  Kramers approximation, 53
asymmetric exclusion process (ASEP), 226–228
Avogadro number, 2

Bénard–Marangoni instability, 347
Bak–Sneppen (BS) model, 192
Bak–Tang-Wiesenfeld (BTW) model, 191
ballistic deposition (BD) model, 240, 241
Becker–Döring theory, 304–309
binary alloy, *see* binary mixture
binary mixture, 141, 261, 262, 292, 376
Boltzmann distribution, 47, 159
bridge model, 203
  exact solution, 207–212
  mean-field (MF) phase diagram, 389–392
  mean-field (MF) solution, 205–207
Brownian motion, 13
  Einstein theory, 14
  Fokker–Planck equation, 19
  generalized, 75
  Langevin equation, 15
  sedimentation equilibrium, 16
Brownian particle, *see* Brownian motion
Brusselator model, 326–328
Burgers equation, 227, 393

Cahn–Hilliard (CH) equation, 288, 295, 317, 332
cellular automaton (CA), 168
central limit theorem, 351–352
Chapman–Kolmogorov equation
  continuous time, 47, 366
  discrete time, 24
coarsening, 269, 313, 334, 344
  conserved scalar case, 275–280
  nonconserved scalar case, 270–275
  nonscalar systems, 296–303
constitutive relation, 89
contact processes, 172, 179, 187
critical exponents
  equilibrium
    Fisher scaling law, 151, 154
    Ising, 149
    Josephson scaling law, 150, 153
    mean-field (MF), 140, 149
    Rushbrooke scaling law, 152, 153
    Widom scaling law, 151, 153
  nonequilibrium
    directed percolation, 173–176
    mean-field (MF), 178

detailed balance, 26, 32, 35, 54, 159, 160, 306, 355
diffusion equation, 20, 278, *see also* random walk
  Fourier–Laplace transform, 359–360
  general solution, 357–358
diffusion limited aggregation (DLA) model, 256
directed percolation (DP), 164–168
  compact (CDP), 185
  mean-field (MF) theory, 177–181
Domany–Kinzel (DK) model, 168–172
dynamical percolation (DyP), 187

Eckhaus instability, 339–343
Edwards–Wilkinson (EW) equation, 226, 229–238
Einstein formula, 15
electric conductivity, 115, 122
  quantum field theory (QFT), 132
entropy production rate, 101, 105
  variational principle, 105–106
envelope equation, 336, 338, 339
equipartition theorem, 147
Ettinghausen effect, 124

Family–Vicsek scaling, 218, 232, 236, 251
first exit time, 49
  formula, 51
fluctuation–dissipation relation, 17, 76, 80–82, 95, 127, 286, 287
fluctuation–dissipation theorem, 83, 96, 128
Fokker–Planck equation, 19–21, 47–49, 76
  conservation law (as), 42
  general, 40–42
  growth models, 244
  particle with a mechanical force, 46
  stationary solution
    absorbing barriers, 43
    growth models, 245
    mechanical force, 47

Fredholm alternative theorem, 417
free energy, 142
functional derivative, 396–397

Galilean transformation, 242
galvanomagnetic effect, 121
Gibbs–Thomson relation, 402–403
Ginzburg criterion, 148, 180
Ginzburg–Landau (GL) free energy, 145, 384–386
Glauber dynamics, 265–266, 290
grass model, 253, 254
Green's function, 85
Green–Kubo relation, 89–95
Gutenberg–Richter law, 189

Hall effect, 123
harmonic oscillator
  damped, 85–89, 415
heat conductivity, 115, 123
Herring–Mullins equation, 235, 237

ideal gas, 4
Ising model, 141–144, 376–383

Jordan's Lemma, 87

Kardar–Parisi–Zhang (KPZ) equation, 226, 238–250
  deterministic (dKPZ), 393–396
  experimental results $d = 1$, 250–251
  experimental results $d = 2$, 251–252
  renormalization group (RG), 398
Kawasaki dynamics, 266–269, 290
kinetic theory, 4–9
  mean free path, 6
Kolmogorov equation
  backward, 49, 366
  forward, 49
Kramers–Kronig relations, 83, 96, 370
Kramers–Moyal expansion, 49, 366–367
Kubo relation
  Brownian particle, 73

Lévy flights, 62
Lévy walks (LW), 64–68
Landau free energy, *see* Ginzburg–Landau free energy
Landau theory, 145–149
Langevin equation, 15–19
  bridge model, 211
  conserved phase separation, 288
  contact processes, 179
  detailed balance, 54
  generalized, 75
  growth models, 223
  nonconserved phase separation, 274, 288
  phase separation, 286
lattice gas (LG), 141, 161, 262, 377
  driven (DLG), 162

linear response theory, 70–134
Lorentzian function, 88, 92
Lyapunov functional, 245, 316, 334

Markov chains, 22–32
  ergodicity, 30, 355
master equation
  Brownian particle, 19
  continuous time, 37, 159
  Markov chains, 32
Maxwell construction, 140, 283, 374
Maxwell distribution, 6
mechanothermal effect, 109
Metropolis algorithm, 35, 160, 162
molecular dynamics, 93, 159
Monte Carlo, 33–35
  kinetic (KMC), 387
multiscale analysis, 415–417

Nernst effect, 123
nonequilibrium steady state (NESS), 159
  bridge model, 206
  Domany–Kinzel (DK) model, 170
  nucleation, 306
  TASEP, 203
  vs. equilibrium states, 159
nucleation
  classical theory, 303–304
  Zel'dovich factor, 308

Onsager matrix, 103, 110
Onsager reciprocity relation, 103
Onsager regression relation, 97
Onsager theorem, 113
Ornstein–Uhlenbeck process, 45, 46
Ornstein–Zernike formula, 148
Ostwald ripening, 280

parity-conserving (PC) model, 188
pattern formation, 311–349
  linear stability analysis, 315–321
  periodic steady states, 328–333
  weakly nonlinear analysis, 331–333
Peltier effect, 118
phase diffusion equation, 346
Prandtl number, 408

quenching, 280–285
  off-critical, 294–296

random deposition (RD) model, 228–229
random deposition with relaxation (RDR) model, 217, 219, 230, 239, 240
random walk, 2, 7, 222, 229, 254, 266, 268, 275, 357
  absorbing barriers, 28
  anisotropic with a trap, 363

random walk (cont.)
continuous time (CTRW), 36, 56
Domany–Kinzel (DK) model, 185
generalized, 55
isotropic with a trap, 362
momenta, 361
ring geometry, 26
Rayleigh number, 408, 411
Rayleigh–Bénard instability, 312, 406–411
reaction–diffusion equations, 322
renormalization group (RG)
block transformation, 155, 379
equilibrium, 154
Ising model, 377–383
nonequilibrium
KPZ, 249
response function, 81, 83
generalized, 95, 98
mathematical properties, 368–371
quantum, 125, 126, 128, 131, 133, 134
restricted solid-on-solid (RSOS) model, 249, 250
Righi–Leduc effect, 124
roughness, 216, 219
critical dimension, 220, 221
exponents, 218, 219
Edwards–Wilkinson model, 232
KPZ model, 245, 246, 251
RSOS model, 251

scale invariance, 154
scaling theory
equilibrium, 152
nonequilibrium, 176
Seebeck effect, 116
self-affinity, 221
self-organized criticality (SOC), 189–195
self-similarity, 221, 262
single step model, 227, 228
specific heat, 140, 146, 150
standard model, 161–164
stochastic Burgers equation, 228
stochastic matrices, 23
spectral properties, 353–354
stochastic processes
continuous time, 36–68
discrete time, 21–35
subdiffusive, 56, 61, 62
superdiffusive, 56, 64, 67
Stokes's law, 14
susceptibility
harmonic oscillator, 86
linear response, 79, 85, 126
magnetic, 143, 144, 146, 151
thermodynamic, 77
Swift–Hohenberg (SH) equation, 329, 334, 337
symplectic algorithms, 93

TASEP model, 195–203
mean-field (MF), 200
thermodynamic limit, 144
thermodynamic sum rule, 85
thermoelectric effect, 113
thermomagnetic effect, 121
thermomechanical effect, 109
Thomson–Joule effect, 119
time-dependent Ginzburg–Landau (TDGL) equation, 271, 288, 294, 316, 317, 322, 329, 345
steady states $d = 1$, 414
Tracy–Widom (TW) distribution, 248, 352
transport
charged particles, 113
coefficients, 11, 12, 93
kinetic energy, 12
mass, 11
momentum, 12
coupled, 111, 112
mass, 14
phenomena, 10
quantum, 131
Turing instability, 321–328
general conditions, 412–413
linear stability analysis, 322

van der Waals gas, 138, 372
equation of state, 139, 373
Van t'Hoff law, 14
viscosity, quantum field theory (QFT), 133

weak ergodicity, 94
Wiener process, 38
Wiener–Khinchin theorem, 82, 88

Printed in the United States
By Bookmasters